MATHEMATICS OF THE BOND MARKET

Mathematical models of bond markets are of interest to researchers working in applied mathematics, especially in mathematical finance. This book concerns bond market models in which random elements are represented by Lévy processes. These are more flexible than classical models and are well suited to describing prices quoted in a discontinuous fashion.

The book's key aims are to characterize bond markets that are free of arbitrage and to analyze their completeness. Nonlinear stochastic partial differential equations (SPDEs) are an important tool in the analysis. The authors begin with a relatively elementary analysis in discrete time, suitable for readers who are not familiar with finance or continuous time stochastic analysis. The book should be of interest to mathematicians, in particular to probabilists, who wish to learn the theory of the bond market and to be exposed to attractive open mathematical problems.

Encyclopedia of Mathematics and Its Applications

This series is devoted to significant topics or themes that have wide application in mathematics or mathematical science and for which a detailed development of the abstract theory is less important than a thorough and concrete exploration of the implications and applications.

Books in the **Encyclopedia of Mathematics and Its Applications** cover their subjects comprehensively. Less important results may be summarized as exercises at the ends of chapters. For technicalities, readers can be referred to the bibliography, which is expected to be comprehensive. As a result, volumes are encyclopedic references or manageable guides to major subjects.

All the titles listed below can be obtained from good booksellers or from Cambridge University Press. For a complete series listing visit www.cambridge.org/mathematics.

124 F. W. King *Hilbert Transforms I*
125 F. W. King *Hilbert Transforms II*
126 O. Calin and D.-C. Chang *Sub-Riemannian Geometry*
127 M. Grabisch *et al. Aggregation Functions*
128 L. W. Beineke and R. J. Wilson (eds.) with J. L. Gross and T. W. Tucker *Topics in Topological Graph Theory*
129 J. Berstel, D. Perrin and C. Reutenauer *Codes and Automata*
130 T. G. Faticoni *Modules over Endomorphism Rings*
131 H. Morimoto *Stochastic Control and Mathematical Modeling*
132 G. Schmidt *Relational Mathematics*
133 P. Kornerup and D. W. Matula *Finite Precision Number Systems and Arithmetic*
134 Y. Crama and P. L. Hammer (eds.) *Boolean Models and Methods in Mathematics, Computer Science, and Engineering*
135 V. Berthé and M. Rigo (eds.) *Combinatorics, Automata and Number Theory*
136 A. Kristály, V. D. Rădulescu and C. Varga *Variational Principles in Mathematical Physics, Geometry, and Economics*
137 J. Berstel and C. Reutenauer *Noncommutative Rational Series with Applications*
138 B. Courcelle and J. Engelfriet *Graph Structure and Monadic Second-Order Logic*
139 M. Fiedler *Matrices and Graphs in Geometry*
140 N. Vakil *Real Analysis through Modern Infinitesimals*
141 R. B. Paris *Hadamard Expansions and Hyperasymptotic Evaluation*
142 Y. Crama and P. L. Hammer *Boolean Functions*
143 A. Arapostathis, V. S. Borkar and M. K. Ghosh *Ergodic Control of Diffusion Processes*
144 N. Caspard, B. Leclerc and B. Monjardet *Finite Ordered Sets*
145 D. Z. Arov and H. Dym *Bitangential Direct and Inverse Problems for Systems of Integral and Differential Equations*
146 G. Dassios *Ellipsoidal Harmonics*
147 L. W. Beineke and R. J. Wilson (eds.) with O. R. Oellermann *Topics in Structural Graph Theory*
148 L. Berlyand, A. G. Kolpakov and A. Novikov *Introduction to the Network Approximation Method for Materials Modeling*
149 M. Baake and U. Grimm *Aperiodic Order I: A Mathematical Invitation*
150 J. Borwein *et al. Lattice Sums Then and Now*
151 R. Schneider *Convex Bodies: The Brunn-Minkowski Theory (Second Edition)*
152 G. Da Prato and J. Zabczyk *Stochastic Equations in Infinite Dimensions (Second Edition)*
153 D. Hofmann, G. J. Seal and W. Tholen (eds.) *Monoidal Topology*
154 M. Cabrera García and Á. Rodríguez Palacios *Non-Associative Normed Algebras I: The Vidav–Palmer and Gelfand–Naimark Theorems*
155 C. F. Dunkl and Y. Xu *Orthogonal Polynomials of Several Variables (Second Edition)*
156 L. W. Beineke and R. J. Wilson (eds.) with B. Toft *Topics in Chromatic Graph Theory*
157 T. Mora *Solving Polynomial Equation Systems III: Algebraic Solving*
158 T. Mora *Solving Polynomial Equation Systems IV: Buchberger Theory and Beyond*
159 V. Berthé and M. Rigo (eds.) *Combinatorics, Words and Symbolic Dynamics*
160 B. Rubin *Introduction to Radon Transforms: With Elements of Fractional Calculus and Harmonic Analysis*
161 M. Ghergu and S. D. Taliaferro *Isolated Singularities in Partial Differential Inequalities*
162 G. Molica Bisci, V. D. Radulescu and R. Servadei *Variational Methods for Nonlocal Fractional Problems*
163 S. Wagon *The Banach–Tarski Paradox (Second Edition)*
164 K. Broughan *Equivalents of the Riemann Hypothesis I: Arithmetic Equivalents*
165 K. Broughan *Equivalents of the Riemann Hypothesis II: Analytic Equivalents*
166 M. Baake and U. Grimm (eds.) *Aperiodic Order II: Crystallography and Almost Periodicity*
167 M. Cabrera García and Á. Rodríguez Palacios *Non-Associative Normed Algebras II: Representation Theory and the Zel'manov Approach*
168 A. Yu. Khrennikov, S. V. Kozyrev and W. A. Zúñiga-Galindo *Ultrametric Pseudodifferential Equations and Applications*
169 S. R. Finch *Mathematical Constants II*
170 J. Krajíček *Proof Complexity*
171 D. Bulacu, S. Caenepeel, F. Panaite and F. Van Oystaeyen *Quasi-Hopf Algebras*
172 P. McMullen *Geometric Regular Polytopes*
173 M. Aguiar and S. Mahajan *Bimonoids for Hyperplane Arrangements*

Encyclopedia of Mathematics and Its Applications

Mathematics of the Bond Market

A Lévy Processes Approach

MICHAŁ BARSKI
Faculty of Mathematics, Informatics and Mechanics,
University of Warsaw

JERZY ZABCZYK
Institute of Mathematics,
Polish Academy of Sciences

CAMBRIDGE
UNIVERSITY PRESS

CAMBRIDGE
UNIVERSITY PRESS

University Printing House, Cambridge CB2 8BS, United Kingdom

One Liberty Plaza, 20th Floor, New York, NY 10006, USA

477 Williamstown Road, Port Melbourne, VIC 3207, Australia

314–321, 3rd Floor, Plot 3, Splendor Forum, Jasola District Centre, New Delhi – 110025, India

79 Anson Road, #06–04/06, Singapore 079906

Cambridge University Press is part of the University of Cambridge.

It furthers the University's mission by disseminating knowledge in the pursuit of education, learning, and research at the highest international levels of excellence.

www.cambridge.org
Information on this title: www.cambridge.org/9781107101296
DOI: 10.1017/9781316181836

First published 2020

Printed in the United Kingdom by TJ International Ltd, Padstow Cornwall

A catalogue record for this publication is available from the British Library.

ISBN 978-1-107-10129-6 Hardback

To our wives Anna and Barbara

Contents

Preface

The Field

The book is devoted to the mathematical theory of the bond market, which is a part of mathematical finance. It is addressed to mathematicians, especially to probabilists who are not necessarily familiar with mathematical finance. In fact, Part I – out of the four parts of this book – treats the subject in discrete time and the knowledge of classical probability, as presented in Feller [51], is sufficient for its understanding.

Mathematical finance is today a part of stochastic analysis. Such concepts as stochastic integral and martingales play a fundamental role in finance. For instance, the mathematical theory of stochastic integration is well developed for large classes of integrators and integrands, and general concepts are ideally suited to financial modelling. Integrators are *price processes* of financial commodities, integrands describe *trading strategies* and the integrals represent *accumulated wealth.*

Basic objects of the theory are two random fields $P(t,T), f(t,T)$, $0 \leq t \leq T$, and a stochastic process $R(t), t \geq 0$, defined on a filtered probability space $(\Omega, \mathcal{F}, (\mathcal{F}_t), \mathbb{P})$. They are related to each other by the formulas

$$P(t,T) = e^{-\int_t^T f(t,s)ds}, \quad 0 \leq t \leq T, \qquad R(t) = f(t,t), \ t \geq 0,$$

and interpreted as, respectively, *bond prices*, *forward rates* and *short rate*. In particular, $P(t,T)$ is the price of a bond at time t that matures at time T, that is, the owner of the bond will receive cash $P(T,T)$ at time T.

The theory is relatively young, approximately 40 years old, and poses new mathematical questions. An important one is about the *absence of arbitrage.* Intuitively, the market should not allow agents to accumulate wealth, by clever investments, without the possibility of facing losses. This property of bond models is mathematically expressed in the concept of non-arbitrage. A related question concerns conditions under which there exists a *martingale measure* for the bond prices, that is, a probability measure $\mathbb{Q}$ equivalent to $\mathbb{P}$ such that for each $T \geq 0$, the process of discounted bond prices

$$\hat{P}(t,T) = e^{-\int_0^t R(s)ds} P(t,T), \quad t \in [0,T]$$

is a local martingale under $\mathbb{Q}$. Problems of this type have never been asked earlier. Another question is that of *completeness* of the market. Mathematically it is equivalent to the condition that each, say, bounded $\mathcal{F}_{T^*}$-measurable random variable, with $T^* > 0$, can be represented as a sum of a constant and a stochastic integral, over the interval $[0, T^*]$, with integrator $\hat{P}(t,\cdot)$, $t \in [0, T^*]$.

The time evolution of bond prices, short rates and forward rates is studied using the theory of Lévy processes and stochastic differential equations. In fact, applications of the theory of stochastic partial differential equations with Lévy noise – a relatively young branch of stochastic processes – are discussed in the book in great detail.

For the reader's convenience the book starts with an extensive treatment of discrete time models. Here the role of Lévy processes is played by random walks.

Lévy Modelling

A good model of bond prices should satisfy several conditions and allow easy confrontation with reality. Stochastic processes used in applications are numerically "tractable" if they are of Markov type or, more specifically, if they are solutions of stochastic equations. For them, at least theoretically, one can find finite dimensional distributions by solving parabolic equations of Kolmogorov type.

As already mentioned, the book is concerned with models in which random elements are represented through Lévy processes that are natural generalizations of the Wiener process. There are several reasons to go outside the classical paradigm. Models based on Lévy processes allow one to treat situations leading to heavy-tailed distributions. Moreover, they allow exploiting the full strength of Markovian modelling because the most general Markov processes are solutions of stochastic differential equations driven by Lévy processes. Since Lévy processes admit jumps, they are well suited to describing prices quoted on exchanges in a discontinuous fashion.

The mathematical theory of the bond market sets a specific area in financial mathematics. Its analysis involves an infinite dimensional setting because basic objects of the theory, bond prices and forward rates, are function-valued processes. Such a framework can hardly be found in classical stock market models.

The research literature on the Lévy bond market is very extensive and growing with an increasing speed. The starting point was the seminal 1997 papers by Björk, Kabanov and Runggaldier [20] and Björk, Di Masi, Kabanov and Runggaldier [19] that laid down the foundations for the analysis of the bond market in a stochastic model with a general discontinuous noise and prompted further research in that direction. Important contributions describing basic properties of the bond market with Lévy noise are due to Eberlein, Jacod and Raible [48], [47]. Interesting results were published in particular by Filipović, Tappe and Teichmann [52], [54], [56], [57]. Several issues were treated by the authors of the present book [5], [3], [7], [6], [8] and [9] and together with Jakubowski [76], [4]. As the results are mathematically rather involved, it seemed that a book on the subject giving solid foundations for future research would be a welcome contribution.

There are rather few books containing material on Lévy modelling of the financial market and there is none devoted to the bond market. The well-known book by Cont and Tankov [29] deals with stock markets. Only in the final comments does it indicate Lévy bond markets as a possible direction of research. Similarly Applebaum [2] considers some problems of Lévy stock markets limiting his discussion of the bond market to some far-reaching suggestions. In the book [100] by Peszat and Zabczyk a more extensive treatment is available, but many questions were left for further study. The well-known books of Carmona and Tehranchi [25] and Filipović [52] as well as part of the classical monograph of Björk [16] are devoted to the bond market, but all deal with models based on the Wiener process.

Aims of the Book

Our first aim is to mathematically characterize those Lévy bond markets that are free of arbitrage. Intuitively, a market is arbitrage free if a trader is not able to generate profit without taking risk. A sufficient condition for that is the existence of the so-called martingale probability measure equivalent to the basic one.

The second main concept we analyze is completeness of the market. Again, intuitively, a market is complete if a trader can construct a strategy that reproduces any prespecified financial contract.

It turns out that a useful tool to construct arbitrage-free bond market models is provided by stochastic equations. The stochastic equations that appear here are nonlinear and sometimes with partial derivatives. Their analysis is one of the main novelties of the book.

The analysis of the mentioned issues is mathematically rather involved. To make the material more accessible we begin by considering a discrete time setting. It is of independent interest, and almost all results from the continuous time framework are proven here in a more direct way.

Structure of the Book

The book consists of four parts preceded by an Introduction that, in particular, contains some financial background. Part I deals with discrete time models and it is aimed at those readers who have had no previous contact with mathematical finance. The randomness is generated by a sequence of independent identically distributed random variables, a counterpart of the increments of Lévy processes. The results described in this part suggest what can be obtained in the much more challenging continuous time setting. Part II is an overview of results from stochastic analysis required for the continuous time framework. In Part III we treat in detail bond markets driven by Lévy processes, covering such topics as non-arbitrage conditions including the derivation of the general Heath–Jarrow–Morton conditions as well as the existence of martingale measures and completeness of the models. Special attention is paid to the important class of models with affine term structure and general models with Markovian factors. In Part IV we construct arbitrage-free models with the use of stochastic partial differential equations with Lévy noise. The equations that appear there are of unusual type as their coefficients, both linear and nonlinear, are of nonlocal character.

Acknowledgements

It is a pleasure to thank our colleagues Tomas Björk, Nikos Frangos, Jacek Jakubowski, Szymon Peszat, Anna Rusinek and Thorsten Schmidt for discussions on the topics of this book. We also thank Jerzy Trzeciak for language consultations.

The first author thanks Leipzig University and Warsaw University for good working conditions and the Institute of Mathematics Polish Academy of Sciences for constant support. The second author is grateful to his home institution, the Institute of Mathematics Polish Academy of Sciences for providing a stimulating research environment. Financial support from the Warsaw Center of Mathematical and Computer Sciences is gratefully acknowledged.

Any comments and remarks from the readers are welcome and can be sent to mbarski@mimuw.edu.pl.

Introduction

I.1 Bonds

Bonds are financial assets issued by governments, central banks or companies. Their holders receive some fixed payments at future dates. The lifetime of a bond is specified by its *maturity* – the date when the *nominal value* of the bond is paid. All previous payments are called *coupons* and they are usually fixed as fractions of the nominal value of the bond. The payments received by the holder, although fixed, can, however, be influenced by the *credit rating* of the issuer. This means that in case of the issuer's bankruptcy the promised payments can be reduced or even cancelled.

There are many kinds of bonds depending on the length of maturity, the frequency of coupon dates and the credit rating of the issuer. Bonds with maturities between 2 and 5 years are called *short-term bonds* or *bills*, those with maturities between 6 and 12 years are *medium-term bonds* or *notes*. Maturities of the *long-term bonds* exceed 12 years but usually are not longer than 30 years. *Perpetual bonds* called also *consols* have infinite maturities, so they pay a stream of coupons forever. The credit rating of the issuer, which describes his/her *default probability*, is assigned by rating agencies and usually denoted by a combination of letters A, B, C, D corrected by $+$ or $-$. The highest rank AAA is followed by $AA+$, AA and so on till D. Coupons of a bond with a high credit rating offer lower payments than those with a low credit rating but the probability that they will be paid without reduction is higher. Real gamblers who have no risk aversion may invest money in *junk bonds* offering profitable coupons that are, however, biased by a critical rating value. Bonds and related financial contracts constitute an enormous market with trading volume exceeding that of the shares. Instead of going deeper into classifying the variety of bonds, we will now focus on their mathematical description.

In this book we consider *zero coupon risk-free bonds* with nominal value 1, which means that 1 unit of cash is paid to the holder at maturity. There are no coupons and the default probability of the issuer disappears. A bond with maturity $T > 0$ will also be called a *T-bond* as it is uniquely characterized by its maturity, and its price at time

$t \in [0, T]$ will be denoted by $P(t, T)$. So, $P(0, T)$ stands for the initial price of the T-bond and $P(T, T) = 1$ is its nominal value. The set of all maturities will be assumed to be $[0, +\infty)$, and by a *bond market* we mean the family of T-bonds with $T \geq 0$. Our model framework with an infinite number of bonds is a kind of mathematical idealization of the real bond market where only a finite number of bonds are traded, but it can be justified by a huge variety of available bonds. Consideration of bonds without coupons is not really restrictive. In fact, every non-zero coupon bond can be represented as a combination of zero coupon bonds no matter what its coupon scheme. A property that does not feature in our study is the default possibility of the issuer. So, the standing assumption in the whole book is that the nominal value of each bond will be paid with probability one.

The family of prices

$$P(t, T), \quad t \in [0, T]; \qquad T \geq 0$$

is called the *term structure of zero coupon bond prices*. The number $P(t, T)$ can be identified with a *risk-free investment* with two dates of payment given by the pair (t, T), where $0 \leq t < T$. Indeed, buying the T-bond for $P(t, T)$ units of cash at time t provides the payoff $P(T, T) = 1$ at T. Since $P(t, T)$ and the nominal value are known at time t, the deal is free of risk. Although $P(0, T)$ and $P(T, T)$ are known at $t = 0$, the price evolution $t \mapsto P(t, T)$ on $(0, T)$ is random and is affected by the state of the economy. One should realize that bonds and stocks represent two competitive parts of the security market that combine investment gain and risk in a different way. In a good economical situation the stock market is developing well and its low investment risk attracts investors. In this situation the bond market, to be competitive, must offer high gains, i.e. the difference between current prices and nominal values of bonds should be high. This means that bond prices are low. Conversely, high bond prices correspond to high uncertainty on the stock market related to economical perturbations.

I.2 Models

It is of prime importance to develop stochastic models that describe the evolution of bond price processes in a way that reflects their real behaviour. Now we briefly introduce models investigated in the book.

Heath–Jarrow–Morton Models

A forward rate is a random function of two variables

$$f(t, T) = f(\omega, t, T), \qquad t \in [0, T], \quad T \geq 0,$$

such that for each $t \geq 0$ the trajectory $T \mapsto f(t, T)$ is known at time t. The bond prices are then given by

$$P(t, T) = e^{-\int_t^T f(t,u)du}, \quad t \in [0, T], \quad T \geq 0.$$

The previous bond price formula reflects two important properties observed on the real market. The bond price $P(t,T)$ behaves in a regular way in T and is chaotic in t providing that time fluctuations of the forward rate are sufficiently rough. In the seminal paper [67] of Heath, Jarrow and Morton the forward rate dynamics has the form

$$\begin{aligned} df(t,T) &= \alpha(t,T)dt + \sigma(t,T)dW(t), \qquad t \in [0,T], \quad T \geq 0, \\ f(0,T) &= f_0(T), \quad T \geq 0, \end{aligned} \tag{I.2.1}$$

where W is a Wiener process. In this approach the forward rate is viewed as a family of stochastic processes $t \mapsto f(t,T)$ parametrized by $T \geq 0$. Then (I.2.1) is a system of separate differential equations with coefficients $\alpha(\cdot,T), \sigma(\cdot,T)$ and initial condition $f_0(T)$ for each T. Our aim is to extend (I.2.1) by replacing W by an $\mathbb{R}^d$-valued Lévy process $Z = (Z_1, \dots, Z_d)$. Then (I.2.1) boils down to

$$\begin{aligned} df(t,T) &= \alpha(t,T)dt + \sum_{i=1}^{d} \sigma_i(t,T)dZ_i(t), \qquad t \in [0,T], \quad T \geq 0, \\ f(0,T) &= f_0(T), \quad T \geq 0. \end{aligned} \tag{I.2.2}$$

The equation (I.2.2) extends the previous model framework significantly by admitting a large class of noise distributions and incorporating new path properties of forward rates, like jumps.

Factor Models

Bond prices and forward rates can also be treated as functions of *time to maturity*. For a fixed date t we focus now on the functions

$$x \mapsto P(t,t+x), \quad x \mapsto f(t,t+x), \quad x \geq 0,$$

where $x := T - t$ with $T \geq t$. Modelling the shapes of the preceding functions and their stochastic evolution in time is encompassed by the *factor models*

$$P(t,T) = F(T-t, X(t)), \quad f(t,T) = G(T-t, X(t)), \quad 0 \leq t \leq T, \tag{I.2.3}$$

where F, G are deterministic functions and X is some stochastic process bringing randomness to the model. The process X is called a *factor* and should be interpreted as consisting of observed economical parameters. In particular, it can be given by the short-rate process $R(t)$.

We study models (I.2.3) where X is a Markov process and characterize admissible functions F, G in (I.2.3), in terms of the transition semigroup of X.

Of prime interest are factors specified by stochastic equations, like the well-known Cox–Ingersol–Ross short-rate model

$$dR(t) = (aR(t) + b)dt + \sqrt{c}\sqrt{R(t)}dW(t), \quad R(0) = R_0, \quad t > 0,$$

or Vasiček, Ho–Lee and Hull–White models (see Björk [16], Filipović [52] for details). We go, however, beyond the continuous paths framework and deal also with multiplicative factors of the form

$$dX(t) = aX(t)dt + bX(t)dZ(t),\ X(0) = x,$$

as well as with the Ornstein–Uhlenbeck short-rate process

$$dR(t) = (a + bR(t))dt + dZ(t), \quad R(0) = R_0, \quad t \geq 0,$$

where Z is a Lévy process and a, b some constants.

Affine Term Structure Models

In the affine term structure model the bond prices have the form

$$P(t, T) = e^{-C(T-t)-D(T-t)R(t)}, \quad 0 \leq t \leq T, \tag{I.2.4}$$

where C, D are deterministic regular functions and R stands for the short-rate process. In fact, (I.2.4) is a particular case of (I.2.3) with

$$G(u, x) = C'(u) + D'(u)x \tag{I.2.5}$$

and the random factor given by the short-rate process R. The linear dependence over x in (I.2.5) implied by (I.2.4) allows us to characterize Lévy processes Z, which generate short rates of the form

$$dR(t) = F(R(t))dt + \sum_{i=1}^{d} G_i(R(t-))dZ_i(t), \quad t \geq 0,\ R(0) = x \tag{I.2.6}$$

that are admissible for affine models. Among real valued Lévy martingales, the only ones turn out to be the Wiener process and the α-stable martingale with Lévy measure

$$\nu(dy) = \frac{1}{y^{1+\alpha}} \mathbf{1}_{[0,+\infty)}(y)dy, \quad \alpha \in (1, 2).$$

In the multidimensional case the coordinates of Z can be given by the Wiener process, the α-stable martingales with $\alpha \in (1, 2)$, α-stable subordinators with $\alpha \in (0, 1)$ and an arbitrary subordinator that enters (I.2.6) in the additive way.

We also present a general characterization of admissible Markov short rates that generate affine models in terms of their generators. This part of the material is based on the paper [53] of Filipović and also provides a characterization of the functions C, D in (I.2.4).

Constructing Models

An efficient way to construct arbitrage-free models is by using the theory of partial differential equations for forward rate processes. The no-arbitrage requirement leads to equations with nonlocal and nonlinear coefficients. A typical example is the following equation for the forward rate

$$r(t,x) = \left(\frac{\partial}{\partial x} r(t,x) - \left(\int_0^x r(t,v)dv\right)^{\alpha} r(t,x)\right) dt + r(t,x)dZ(t), \quad x \geq 0, t \geq 0,$$

where

$$r(t,x) := f(t,t+x), \quad x \geq 0, t \geq 0,$$

and Z is an α-stable martingale. Equations of this type with the Wiener process Z were introduced by Musiela [97]. The equations prompt interesting mathematical questions about existence and uniqueness of solutions and their positivity, discussed in Part IV.

I.3 Content of the Book

The book consists of four parts: (I) "Bond Market in Discrete Time"; (II) "Fundamentals of Stochastic Analysis"; (III) "Bond Market in Continuous Time"; and (IV) "Stochastic Equations in the Bond Market". The first part has a more elementary character than the remaining three. It uses classical probability concepts and results rather than more advanced stochastic analysis, as in the rest of the book. The book ends with Appendices containing the proof of the martingale representation theorem in the pure jump case, material on generators of equations with Lévy processes and on evolution equations. Special care is devoted to the following models of the bond market: the HJM model in which forward rates are defined by stochastic equations; factor models in which price curves are moved by stochastic processes of economic factors; and affine models in which bond prices are exponential functions of the short rate.

Part I starts from preliminaries on the discrete time financial market in Chapter 1. Arbitrage-free models are studied in Chapter 2. We derive, in particular, a discrete time version of the CIR equation of the continuous time theory. Practically, all Markovian short-rate processes of the affine term structure are determined. Completeness of the bond market is studied in Chapter 3. Bond curves are vectors with an infinite number of coordinates and only those models with curves evolving in finite dimensional spaces might be complete. Specific conditions for approximate completeness of the main models are deduced.

Part II is divided into three substantial chapters. Chapter 4 recalls concepts and results from stochastic analysis like semimartingales, square integrable martingales and Doob–Meyer decomposition. Stochastic integration with respect to semimartingales and random measures as well as Ito's formula are treated. They will be of constant use later. Chapter 5 concerns Lévy processes, our basic tool. We first apply the general stochastic analysis theory to this class of processes and describe specific subclasses. Then in Chapter 6 we formulate the integral representation theorem for local martingales with respect to the Lévy filtration due to Kunita. Essential, although rather classical elements of the proof, like chaos expansion and multiple Itô–Wiener integrals are presented in Appendix A. The second part of this chapter is concerned with Girsanov's formula for densities of equivalent measures.

Part III concerning the continuous time bond market starts with a mathematical description of the models and their elementary properties in Chapter 7. Arbitrage-free Heath–Jarrow–Morton models of the bond market are analyzed in Chapter 8. The main results here are general non-arbitrage conditions of the HJM type for an arbitrary physical probability measure. Chapter 9 investigates the non-arbitrage problem when the models are given in the form of forward curves moved by Markovian factor processes. The main result is the *term structure equation*. Some applications to special factor processes, like multiplicative or Ornstein–Uhlenbeck processes are presented as well. Chapter 10 is devoted to non-arbitrage conditions for affine models of bond prices. It consists of two major sections concerned, respectively, with short rates given as solutions of general stochastic equations and short rates that are general Markov processes. We present results due to Filipović. We also give a generalization to discontinuous short-rate processes of the Cox–Ingersol–Ross theorem. The final Chapter 11 is on completeness of the bond market. The hedging problem is formulated in terms of the solvability of the so-called *hedging equation*. It is discussed in various settings related to special forms of the Lévy process. Approximate completeness is discussed as well.

Part IV focuses on building arbitrage-free markets through stochastic equations. In Chapter 12 the equations are introduced. General equations for the forward curve under the martingale measure are analyzed by the methods of stochastic evolution equations in Chapter 13. Conditions for local and global existence of solutions are established. Some applications to the so-called Morton–Musiela equation are presented as well. Chapter 14 treats the case when volatility in the HJM model is a linear function of the forward curve. Then the forward rate satisfies the so-called Morton's equation. The equation has a unique solution for a large class of Lévy processes characterized in terms of the logarithmic growth conditions of the Lévy exponent. We develop the method introduced by Morton, who treated the Wiener case and obtained a negative result. The Morton–Musiela equation is treated in Chapter 15.

Part I

Bond Market in Discrete Time

1

Elements of the Bond Market

Here we introduce basic concepts of the bond market model in discrete time, like bond prices, rates, portfolios and strategies. Contingent claims and non-arbitrage conditions are discussed as well.

1.1 Prices and Rates

A *bond market* consists of four stochastic processes

$$P(t,T), \quad T \in \mathbb{N}_0, \quad t = 0, 1, \ldots, T,$$

$$B(t), \quad t \in \mathbb{N}_0,$$

$$f(t,T), \quad T \in \mathbb{N}_0, \quad t = 0, 1, \ldots, T,$$

$$R(t), \quad t \in \mathbb{N}_0,$$

called, respectively, *bond prices*, *bank account*, *forward rates* and *short rate*, defined on a probability space $(\Omega, \mathcal{F}, \mathbb{P})$, equipped with a filtration $\{\mathcal{F}_t,\ t \in \mathbb{N}_0\}$, $\mathbb{N}_0 := \{0, 1, \ldots\}$, and adapted to the filtration. This means that for each $t \in \mathbb{N}_0$ the random variables

$$P(t,T),\ T \geq t; \quad f(t,T),\ T \geq t; \quad B(t); \quad R(t);$$

are $\mathcal{F}_t$-measurable. Often $\mathcal{F}_t$ is assumed to be generated by $P(t,T)$, $T \geq t$:

$$\mathcal{F}_t = \sigma\{P(s,T);\ s = 0, 1, \ldots, t,\ T \geq s\}. \tag{1.1.1}$$

In this case the filtration is the *minimal* one.

The value $P(t,T)$ is interpreted as the price, at time t, of the bond that *matures* at time T, called *maturity* of the bond. That means that the owner of this T*-bond* will receive at time T the so-called *nominal value*, specified on the bond, which we assume to be 1. Hence it is natural to assume that

$$P(t,T) \geq 0 \quad \text{and} \quad P(T,T) = 1. \tag{1.1.2}$$

Other elements of the bond market are related to the bond prices in some specific ways and satisfy some natural requirements.

The value $B(t)$ is interpreted as the amount of money in the bank account at time t resulting from depositing 1 at time 0. It is convenient to assume that the deposit grows as a result of converting it at each moment t into bonds that mature at $t+1$. Thus, if the deposit at time t was D, then one buys

$$D \cdot \frac{1}{P(t, t+1)},$$

$t+1$-bonds and the deposit at time $t+1$ is

$$D \cdot \frac{1}{P(t, t+1)} \cdot P(t+1, t+1) = D \cdot \frac{1}{P(t, t+1)}.$$

This leads to the recurrent identity

$$B(t+1) = B(t) \frac{1}{P(t, t+1)}, \tag{1.1.3}$$

and to the definition of the short rate $R(t)$

$$e^{R(t)} := \frac{1}{P(t, t+1)}, \tag{1.1.4}$$

which clearly shows that R is an adapted process. It follows also that

$$B(0) = 1, \quad B(t) = \frac{1}{P(0,1)P(1,2) \cdot \ldots \cdot P(t-1, t)}, \quad t = 1, 2, \ldots, \tag{1.1.5}$$

which implies that $B(t)$ is $\mathcal{F}_{t-1}$-measurable for each $t \geq 1$, hence B is a *predictable* process. As a consequence of (1.1.5) and (1.1.4) one obtains

$$B(t) = e^{\sum_{s=0}^{t-1} R(s)}, \quad t = 1, 2, \ldots, \quad B(0) = 1. \tag{1.1.6}$$

Of great importance are the so-called *discounted bond prices* $\hat{P}(t, T)$ defined by

$$\hat{P}(t, T) := \frac{P(t, T)}{B(t)}, \quad t \leq T. \tag{1.1.7}$$

The *forward rate* process $f(t, T)$ is determined by the identity

$$P(t, T) = e^{-\sum_{s=t}^{T-1} f(t,s)}, \quad t \leq T. \tag{1.1.8}$$

Thus, for $t \leq T$,

$$e^{-f(t,T)} = \frac{P(t, T+1)}{P(t, T)}, \quad \text{and} \quad f(t, T) = \ln \frac{P(t, T)}{P(t, T+1)}.$$

In particular,

$$f(t, t) = R(t), \ \ t \in \mathbb{N}_0. \tag{1.1.9}$$

The relations (1.1.2), (1.1.4), (1.1.6), (1.1.8) and (1.1.9) are assumed to be always true.

In fact, the definitions of bond prices $P(t,T)$, discounted bond prices $\hat{P}(t,T)$ and forward rates $f(t,T)$ can be extended to the case $t > T$ by assuming that nominal values of bonds are transferred into the bank account at their maturity times. Thus, for $t > T$ one defines

$$P(t,T) = P(T,T) \cdot e^{\sum_{s=T}^{t-1} R(s)} = e^{\sum_{s=0}^{t-1} R(s) - \sum_{s=0}^{T-1} R(s)} = B(t)/B(T).$$

Consequently,

$$\hat{P}(t,T) = P(t,T)/B(t) = 1/B(T), \quad t > T.$$

Defining forward rates $f(t,T)$ for $t > T$ by the same formula as for $t \leq T$, i.e. (1.1.8), we obtain

$$f(t,T) = \ln \frac{P(t,T)}{P(t,T+1)} = \ln \frac{B(t)/B(T)}{B(t)/B(T+1)} = \ln e^{R(T)} = R(T), \quad t > T. \quad (1.1.10)$$

Notice that (1.1.10) leads to the following formula for the discounted bond prices

$$\hat{P}(t,T) = e^{-\sum_{s=0}^{t-1} R(s)} P(t,T) = e^{-\sum_{s=0}^{t-1} f(t,s)} \cdot e^{-\sum_{s=t}^{T-1} f(t,s)} \quad (1.1.11)$$

$$= e^{-\sum_{s=0}^{T-1} f(t,s)}, \quad t,T \in \mathbb{N}_0. \quad (1.1.12)$$

Summarizing, the extended processes are given by

$$P(t,T) = B(t)/B(T), \quad \hat{P}(t,T) = 1/B(T), \quad f(t,T) = f(T,T) = R(T), \quad t > T. \quad (1.1.13)$$

Typically bond prices are bounded, in the sense

$$P(t,T) \leq 1, \qquad t \leq T, \quad (1.1.14)$$

and monotone with respect to T, i.e.

$$P(t,T) \geq P(t,T+1). \quad (1.1.15)$$

This means that bonds with longer maturities should be cheaper. If (1.1.14) and (1.1.15) hold, then the market will be called *regular*. On the real bond market, both properties, (1.1.14) and (1.1.15), can, however, break down when the economy slumps. It is clear that (1.1.15) is equivalent to the *positivity* of forward rate, i.e.

$$f(t,s) \geq 0, \qquad t \leq s. \quad (1.1.16)$$

If the forward rate is positive, then (1.1.14) holds as well, so positive forward rates generate regular markets.

1.2 Models of the Bond Market

There are four models of the bond market discussed in the sequel:

(a) The *discrete time HJM model* is a version of the famous Heath, Jarrow and Morton (HJM) model in continuous time introduced in [67]. One stipulates that for each T, the forward rates $f(t,T)$ change stochastically in t and

$$f(t+1,T) = f(t,T) + \alpha(t,T) + \langle \sigma(t,T), \xi_{t+1} \rangle, \quad 0 \le t < T. \tag{1.2.1}$$

The previously mentioned $\xi_1, \xi_2, \ldots$ are independent identically distributed random variables taking values in $U = \mathbb{R}^d$. Its partial sum

$$Z(t) = \xi_1 + \xi_2 + \cdots + \xi_t, \quad t = 1, 2, \ldots, \tag{1.2.2}$$

can be viewed as a discrete time counterpart of a Lévy process in continuous time. For each T the processes $\alpha(\cdot,T)$ and $\sigma(\cdot,T)$ are adapted to the filtration $\{\mathcal{F}_t\}$, which is given by

$$\mathcal{F}_0 = \{\Omega, \emptyset\}, \quad \mathcal{F}_t := \sigma\{\xi_1, \xi_2, \ldots, \xi_t\}, \quad t = 1, 2, \ldots,$$

and $\langle \cdot, \cdot \rangle$ stands for the scalar product in U. (1.2.1) means that the future value of the forward rate $f(t+1,T)$ arises from the current one $f(t,T)$ by shifting it by $\alpha(t,T)$, which is known, and perturbing it by $\langle \sigma(t,T), \xi_{t+1} \rangle$, which is random. This interpretation justifies $\alpha(\cdot,\cdot), \sigma(\cdot,\cdot)$ to be called *drift*, or *volatility* of the forward rate. The initial forward curve $f(0,T)$, $T = 0, 1, \ldots$, is regarded as known at time zero.

It is often convenient to study the bond market in the *moving frame* based on the so-called *Musiela parametrization* and write $f(t,\cdot)$ in terms of *time to maturity* $T - t$. Thus, one defines

$$r(t,j) := f(t, t+j), \quad j = 0, 1, \ldots. \tag{1.2.3}$$

In particular, the short rate is given by

$$R(t) = r(t,0), \quad t \ge 0.$$

(b) In the *forward rate model with Markovian trace* one defines the *γ-trace* of the forward rate r by

$$r^\gamma(t) := (r(t,0), \ldots, r(t,\gamma)), \quad t = 0, 1, \ldots, \tag{1.2.4}$$

where $\gamma \in \mathbb{N}_0 := \{0, 1, 2, \ldots\}$, which is simply a random vector consisting of the first γ coordinates of r given by (1.2.3). The model is based on the assumption that r^γ is a Markov chain in the space $\mathbb{R}_+^{\gamma+1}$ and examined with the use of its transition operator $\mathcal{P}(\cdot,\cdot)$. The particular case with $\gamma = 0$ corresponds to the Markovian short-rate process $R(t) = r^0(t)$, $t \ge 0$.

(c) The *affine model*, called also *affine term structure*, is a particular case of the model with Markovian trace in which the forward rate is of the form

$$r(t,k) = C(k+1) - C(k) + \langle (D(k+1) - D(k)), r^{\mathcal{Y}}(t)\rangle, \quad t \geq 0 \qquad (1.2.5)$$

for some deterministic functions C, D such that $C(0) = 0$, $D(0) = 0$. Thus, the shape of the *forward curve* $k \mapsto r(t,k)$ is given and controlled in a linear way by the trace only. Condition (1.2.5) can be equivalently written with the use of the bond prices, i.e.

$$P(t,T) = e^{-C(T-t)-\langle D(T-t), r^{\mathcal{Y}}(t)\rangle}, \quad t \leq T.$$

(d) In the *Markovian factor model*, which is a generalization of the affine model, one requires that

$$P(t,T) = F(T-t, X(t)), \qquad t, T = 0, 1, \ldots, t \leq T,$$

or, equivalently,

$$f(t,T) = G(T-t, X(t)), \quad t, T = 0, 1, \ldots, t \leq T,$$

where $X(t)$, $t \geq 0$ is a Markov chain on $(E, \mathcal{E})$, called *factor*, and $F(\cdot, x)$, $G(\cdot, x)$ are deterministic functions describing possible shapes of the *bond curves*, or *forward curves* for a fixed value x of the factor. It is clear that

$$F(0,x) = 1, \quad F(k,x) = e^{-\sum_{j=0}^{k-1} G(j,x)}, \quad k = 1, \ldots, \; x \in E.$$

In particular, the short-rate process may serve as the factor. The basic issue in the factor model is the interplay between the transition operator of the factor process and the shapes of forward or bond curves.

1.3 Portfolios and Strategies

A *portfolio* at time $t \geq 0$ is a family of $\mathcal{F}_t$-measurable random variables

$$\left(b(t), \; \varphi_t(j)_{j=t+2,t+3,\ldots,}\right),$$

where $b(t)$ is the amount of money deposited in the bank account and $\varphi_t(j)$ stands for the number of bonds that will mature at time j, bought by the investor. Note that at time t one can trade with bonds maturing at $t+2$ or later. The reason for that is that the evolution at time t of bonds that have matured or will mature at $t+1$ is governed by the short-rate process $R(t)$. Indeed, by (1.1.4),

$$P(t,t+1) = e^{-R(t)} = \frac{1}{B(t)},$$

and for bonds that have matured earlier than $t+1$ we can assume that their nominal values are automatically transferred to the bank. Hence investing in those bonds is equivalent to putting money in the savings account. *Finite portfolios* are based on a finite number of bonds, that is, $\varphi_t(j) = 0$ for all $j > j(t)$ and some number $j(t)$. The proper framework for the bond market relies, however, on portfolios involving an infinite number of bonds, i.e. $\varphi_t = \{\varphi_t(j)\}$ may have infinitely many non-vanishing coordinates. Then the corresponding *portfolio wealth* at time t is given by

$$X(t) = b(t) + \sum_{j \geq t+2} \varphi_t(j) P(t,j), \quad t = 0, 1, \ldots, \tag{1.3.1}$$

providing that the above sum is finite. To guarantee that, let us assume that forward rates are positive. It follows then from the formula

$$P(t,T) = e^{-\sum_{s=t}^{T-1} f(t,s)} \leq 1, \quad t \leq T$$

that the price process $P(t) := (P(t, t+2), (t, t+3), \ldots)$ takes values in the set of *bounded sequences* m, i.e. in

$$m := \{x = (x_1, x_2, \ldots) : \sup_i \mid x_i \mid < +\infty\}.$$

Consequently, portfolios may take values in l^1, where

$$l^1 := \{x = (x_1, x_2, \ldots) : \sum_{n=1}^{+\infty} \mid x_n \mid < +\infty\},$$

that is, $\varphi_t = (\varphi_t(t+2), \varphi_t(t+3), \ldots) \in l^1$, because then

$$\left| \sum_{j \geq t+2} \varphi_t(j) P(t,j) \right| \leq \sum_{j \geq t+2} \mid \varphi_t(j) \mid = \mid \varphi_t \mid_{l^1} < +\infty,$$

and the portfolio wealth is also well defined.

By (1.3.1), portfolio wealth at time $t+1$ equals

$$X(t+1) = b(t+1) + \sum_{j \geq t+3} \varphi_{t+1}(j) P(t+1, j).$$

If $X(t+1)$ arises from $X(t)$ only through the fluctuations of bond prices and bank account, then we say that the *self-financing condition* at time $t+1$ is preserved. Specifically, this means that

$$X(t+1) = b(t) e^{R(t)} + \sum_{j \geq t+2} \varphi_t(j) P(t+1, j). \tag{1.3.2}$$

A *self-financing strategy* is a sequence of portfolios $(b(t), \varphi_t), t = 0, 1, 2, \ldots$ for which the self-financing condition holds at any time. If this is the case, then from the identity

$$X(t) = X(0) + \sum_{s=0}^{t-1} (X(s+1) - X(s)),$$

and by (1.3.1), (1.3.2), we obtain

$$X(t+1) - X(t) = (e^{R(t)} - 1)b(t) + \sum_{j \geq t+2} \varphi_t(j)(P(t+1,j) - P(t,j)).$$

Using more compact notation, we have thus

$$\triangle X(t+1) = (e^{R(t)} - 1)b(t) + \langle \varphi_t, \triangle P(t+1) \rangle,$$

where

$$\triangle P(t+1) := \triangle P(t+1,j) = P(t+1,j) - P(t,j), \quad j \geq t+2.$$

So, we see that the portfolio wealth of a self-financing strategy $(b(t), \varphi_t)$ evolves according to the formula

$$X(t) = X(0) + \sum_{s=0}^{t-1} (e^{R(s)} - 1)b(s) + \sum_{s=0}^{t-1} \langle \varphi_s, \triangle P(s+1) \rangle. \tag{1.3.3}$$

Since $\triangle P(s)$, for each $s = 0, \ldots, t$, takes values in the space of bounded sequences m, the last sum is clearly well defined. A self-financing strategy is thus defined by (1.3.1) and (1.3.3). In the sequel we will use the following alternative characterization of the self-financing condition in terms of the *discounted portfolio wealth* and the discounted bond prices

$$\hat{X}(t) := \frac{X(t)}{B(t)}, \quad \hat{P}(t,j) := \frac{P(t,j)}{B(t)}, \quad t = 0, 1, \ldots, j = 0, 1, \ldots.$$

It is useful to note and easy to check with the use of (1.1.13), that for any $s = 0, 1, \ldots$,

$$\triangle \hat{P}(s+1,j) = 0, \quad j = 0, 1, \ldots, s+1,$$

where

$$\triangle \hat{P}(s+1,j) := \hat{P}(s+1,j) - \hat{P}(s,j), \; s = 0, 1, \ldots, \; j = 0, 1, \ldots.$$

Proposition 1.3.1 *(a) If $(b(t), \varphi_t)$ is a self-financing strategy, then*

$$\hat{X}(t) = X(0) + \sum_{s=0}^{t-1} \langle \varphi_s, \triangle \hat{P}(s+1) \rangle, \quad t = 1, 2, \ldots. \tag{1.3.4}$$

(b) For any initial capital x and a strategy φ_t there exists a unique $b(t)$ such that the strategy $(b(t), \varphi_t)$ is self-financing and $X(0) = x$.

Let us notice that the sum in (1.3.4) is well defined, because φ_s, for each s, takes values in l^1 and $\triangle \hat{P}(s+1)$ in m.

Proof (*a*) From (1.3.1) and (1.3.2) we have

$$\hat{X}(t) = \frac{b(t)}{B(t)} + \sum_{j\geq t+2} \varphi_t(j)\frac{P(t,j)}{B(t)} = \frac{b(t)}{B(t)} + \sum_{j\geq t+2} \varphi_t(j)\hat{P}(t,j),$$

$$\hat{X}(t+1) = e^{R(t)}\frac{b(t)}{B(t+1)} + \sum_{j\geq t+2} \varphi_t(j)\frac{P(t+1,j)}{B(t+1)} = \frac{b(t)}{B(t)} + \sum_{j\geq t+2} \varphi_t(j)\hat{P}(t+1,j),$$

which implies

$$\Delta\hat{X}(t+1) = \sum_{j\geq t+2} \varphi_t(j)\Delta\hat{P}(t+1,j) = \langle\varphi_t, \Delta\hat{P}(t+1)\rangle.$$

Formula (1.3.4) can be obtained by adding the increments of $\hat{X}(t)$.
(b) The sequence $b(\cdot)$ is defined inductively by setting first $b(0)$:

$$b(0) = x - \sum_{j\geq 2} \varphi_0(j)P(0,j),$$

and then using the equation:

$$b(t+1) + \sum_{j\geq t+3} \varphi_{t+1}(j)P(t+1,j) = b(t)e^{R(t)} + \sum_{j\geq t+2} \varphi_t(j)P(t,j).$$

The strategy $(b(t), \varphi_t)$ is self-financing and it is the only self-financing strategy that satisfies $X(0) = x$. □

Proposition 1.3.1 is of great importance because it simplifies the problem of construction of self-financing strategies $(b(t), \varphi_t)$ to a free choice of the initial capital $X(0) = x$ and investment in bonds, i.e. (φ_t) only. In the sequel we will often identify any self-financing strategy $(b(t), \varphi_t)$ with the pair $(X(0), \varphi_t)$.

1.4 Contingent Claims

Let us consider a financial contract that obliges its *seller* to pay at a fixed future date $t > 0$ some random amount of money X to the buyer of the contract. The contract is formulated at time 0 and X is assumed to be an $\mathcal{F}_t$-measurable random variable. It is called *contingent claim* at time t for short, *claim* at t. Typically, claims are some functions of the prices of bonds that are available on the market up to time t. In this case, the filtration is assumed to be the minimal one, i.e.

$$\mathcal{F}_s = \sigma\{\ P(u,T);\ u = 0, 1, \ldots, s,\ T = 0, 1, 2, \ldots\}.$$

The goal of the seller might be to find a self-financing strategy $(b(u), \varphi_u)$ such that for the corresponding wealth process,

$$X(s) = X(0) + \sum_{s=0}^{s-1}(e^{R(u)} - 1)b(u) + \sum_{u=0}^{s-1}\langle\varphi_u, \Delta P(u+1)\rangle, \quad s = 0, 1, \ldots, t,$$

one has at time t

$$X(t) = X. \tag{1.4.1}$$

Recall that $(b(u), \varphi_u)$ are adapted to $\{\mathcal{F}_u\}$ and φ_u lives in l^1. Condition (1.4.1) means that the seller could eliminate the risk arising from paying X at time t by starting from $X(0)$ and following the strategy $(b(u), \varphi_u)$. Therefore, $X(0)$ is called a *fair price* of the contract. A strategy $(b(u), \varphi_u)$ for which (1.4.1) holds is called a *replicating strategy*, or *hedging strategy*, for X and the claim X is called *attainable*.

It is convenient to reformulate the problem of looking for replicating strategies in terms of discounted values, i.e. $\hat{X} = X/B(t), \hat{P}(s,j) = P(s,j)/B(s)$.

Proposition 1.4.1 *Let X be a contingent claim.*

(a) *If $(b(s), \varphi_s)$ is a replicating strategy for X, then the discounted claim admits the representation*

$$\hat{X} = X(0) + \sum_{s=0}^{t-1} \langle \varphi_s, \Delta\hat{P}(s+1)\rangle. \tag{1.4.2}$$

(b) *If there exists a pair (x, φ_s) such that*

$$\hat{X} = x + \sum_{s=0}^{t-1} \langle \varphi_s, \Delta\hat{P}(s+1)\rangle \tag{1.4.3}$$

then there exists a replicating strategy $(b(s), \varphi_s)$ for X with initial wealth $X(0) = x$.

Proof (*a*) If $(b(s), \varphi_s)$ is a replicating strategy for X, then the related wealth process clearly satisfies

$$X(t)/B(t) = \hat{X}.$$

By Proposition 1.3.1 we obtain (1.4.2).
(*b*) For the pair (x, φ_s) satisfying (1.4.3) we can find, by Proposition 1.3.1, $b(s)$ such that $(b(s), \varphi_s)$ is self-financing and $X(0) = x$. Again, by Proposition 1.3.1, the final discounted wealth process of this strategy $\hat{X}(t)$ equals the right side of (1.4.3). So, (1.4.3) means that $\hat{X}(t) = \hat{X}$, which implies that $X(t) = X$. Thus, $(b(s), \varphi_s)$ is a replicating strategy for X. □

Proposition 1.4.1 is a key result for the problem of determining replicating strategies. It allows us to forget the self-financing condition and first find any pair (x, φ_t) satisfying (1.4.3). If we do this, then it is always possible to construct a replicating strategy for X.

1.5 Arbitrage

A self-financing strategy $(X(0), \varphi_t)$ is an *arbitrage* (or *arbitrage strategy*) if, for some $t_0 > 0$, the corresponding portfolio wealth process satisfies

$$X(0) = 0, \quad \mathbb{P}(X(t_0) \geq 0) = 1, \quad \mathbb{P}(X(t_0) > 0) > 0. \tag{1.5.1}$$

The preceding conditions can be interpreted as a risk-free possibility of realizing positive gain starting from zero initial endowment. The possibility of constructing such strategies in the model should be excluded. A bond market that does not allow arbitrage strategies is called *arbitrage free*.

Let us first specify two classes of self-financing strategies that will be used for examining the absence of arbitrage. A self-financing strategy (φ_t) belongs to $\mathcal{A}_1$ if there exist constants M and K such that, for each t,

$$| \varphi_t(j) | \leq M, \quad j \geq t+2, \quad \text{and} \quad \varphi_t(j) = 0 \quad \text{for} \quad j \geq K.$$

Clearly, any element of $\mathcal{A}_1$ is a finite portfolio. The class $\mathcal{A}_2$, by definition, consists of self-financing strategies that are bounded in l^1, i.e. for some M and each $t \geq 0$

$$| \varphi_t |_{l^1} = \sum_{j \geq t+2} | \varphi_t(j) | \leq M.$$

One of the main problems of the bond market theory is to find sufficient and necessary conditions under which the market is arbitrage free. We present two basic and rather simple results on arbitrage-free markets.

Theorem 1.5.1 *Let us assume that, for any $T \in \mathbb{N}_0$, the discounted price process of the T-bond is a martingale. Then there are no arbitrage strategies in the class $\mathcal{A}_1$. If, additionally, forward rates are positive, then there are no arbitrage strategies in the class $\mathcal{A}_2$.*

Proof Let (φ_t) be a self-financing strategy. The corresponding discounted portfolio wealth $\hat{X}(t)$ is a martingale if and only if

$$\mathbb{E}(\hat{X}(t+1) - \hat{X}(t) \mid \mathcal{F}_t) = 0, \quad t = 0, 1, 2, \ldots,$$

or, by (1.3.4), equivalently,

$$\mathbb{E}(\langle \varphi_t, \Delta \hat{P}(t+1) \rangle \mid \mathcal{F}_t) = \mathbb{E}\Big(\sum_{j \geq t+2} \varphi_t(j) \left(\hat{P}(t+1, j) - \hat{P}(t, j) \right) \mid \mathcal{F}_t \Big)$$

$$= \sum_{j \geq t+2} \varphi_t(j) \mathbb{E} \left(\hat{P}(t+1, j) - \hat{P}(t, j) \mid \mathcal{F}_t \right) = 0, \quad t = 0, 1, 2, \ldots,$$

providing that exchange of summation and conditional expectation is allowed. This is clearly the case when φ_t belongs to $\mathcal{A}_1$. The same is true if forward rates are positive and φ_t belongs to $\mathcal{A}_2$ because then $\hat{P}(t,T) \leq 1$ for all $t \leq T$. Since $\hat{P}(t,j)$ is a martingale for each j,

$$\mathbb{E}\left(\hat{P}(t+1,j) - \hat{P}(t,j) \mid \mathcal{F}_t\right) = 0,$$

and it follows that for both classes $\mathcal{A}_1$ and $\mathcal{A}_2$ the process $\hat{X}(t)$ is a martingale. Let us assume that φ is an arbitrage strategy. Then, by passing to discounted values in (1.5.1), for some $t_0 > 0$,

$$X(0) = 0, \quad \mathbb{P}(\hat{X}(t_0) \geq 0) = 1, \quad \mathbb{P}(\hat{X}(t_0) > 0) > 0.$$

Since $\hat{X}(t)$ is a martingale,

$$0 = \mathbb{E}\hat{X}(t_0) = \mathbb{E}[\hat{X}(t_0)\mathbf{1}_{\{\hat{X}(t_0)>0\}}] > 0,$$

which is a contradiction. □

An important sufficient condition for the absence of arbitrage can be formulated in terms of the existence of the so-called *martingale measure*. A measure $\mathbb{Q}$ on $(\Omega, \mathcal{F})$ is a *martingale measure* for a bond market defined on $(\Omega, \mathcal{F}, \{\mathcal{F}_t\}, \mathbb{P})$ if

$$\mathbb{Q}(A) = 0 \quad \Longleftrightarrow \quad \mathbb{P}(A) = 0, \quad A \in \mathcal{F},$$

and, for each $T > 0$ the process

$$\hat{P}(t,T), \ 0 \leq t \leq T, \quad \text{is a martingale under } \mathbb{Q}.$$

In fact, in Theorem 1.5.1 we assumed that $\mathbb{P}$ is a martingale measure. However, each step in the proof remains true if we replace $\mathbb{P}$ with an arbitrary martingale measure. Hence, the following generalization of Theorem 1.5.1 is true.

Theorem 1.5.2 *Let us assume that there exists a martingale measure $\mathbb{Q}$. Then there are no arbitrage strategies in the class $\mathcal{A}_1$. If, additionally, forward rates are positive, then there are no arbitrage strategies in the class $\mathcal{A}_2$.*

A Counterexample

In the classical stock market, where the number of trading assets is finite, the absence of arbitrage implies the existence of a martingale measure. This fact constitutes the well-known First Fundamental Theorem of Asset Pricing. In the bond market setting, where the number of tradeable bonds is infinite, this implication turns out not to be true. We will construct an example of an arbitrage-free regular bond market that admits no martingale measures.

Proposition 1.5.3 *There exists a one-period regular bond market that does not admit a martingale measure and is arbitrage free in the class of strategies that are only assumed to generate finite initial wealth.*

Proof We adapt here the idea from the example presented in Schachermayer [115] to the bond market setting. The prices at time $t = 0$ are given by a deterministic sequence $P(0,T), T = 0, 1, 2, ..$ satisfying

$$P(0,0) = 1, \quad 0 < P(0,T) \leq 1, \quad P(0,T) \geq P(0,T+1), \quad T = 0, 1, \ldots. \tag{1.5.2}$$

For regularity we need the random sequence $P(1,T), T = 1, 2, \ldots$, to satisfy

$$P(1,1) = 1, \quad 0 < P(1,T) \leq 1, \quad P(1,T) \geq P(1,T+1), \quad T = 1, 2, \ldots. \tag{1.5.3}$$

In fact, (1.5.3) can be expressed in terms of increments of the discounted prices

$$\eta(T) := \hat{P}(1,T) - \hat{P}(0,T) = P(1,T)P(0,1) - P(0,T), \quad T = 1, 2, \ldots.$$

Since

$$P(1,T) = \frac{\eta(T) + P(0,T)}{P(0,1)}, \quad T = 1, 2, \ldots,$$

(1.5.3) is equivalent to

$$\begin{aligned} &\eta(1) = 0, \quad \eta(T) + P(0,T) > 0, \\ &\eta(T+1) - \eta(T) \leq P(0,T) - P(0,T+1), \quad T = 1, 2, \ldots. \end{aligned} \tag{1.5.4}$$

The model is defined on the probability space $\Omega = \{\omega_1, \omega_2, \ldots\}$ of natural numbers, i.e. $\omega_k = k, k = 1, 2, \ldots$, with the measure $\mathbb{P}(\{\omega_k\}) = \frac{1}{2^k}, k = 1, \ldots$. The initial prices are given by

$$P(0,T) = \frac{1}{2^T}, \quad T = 0, 1, \ldots,$$

and $\eta(T)$ by

$$\eta(1) \equiv 0, \quad \eta(T)(\omega_k) := \begin{cases} \frac{1}{n^T} & \text{for } k = T, \\ -\frac{1}{n^T} & \text{for } k = T+1, \\ 0 & \text{elsewhere,} \end{cases} \quad T = 2, 3, \ldots,$$

where $n > 4$. It is clear that (1.5.2) is satisfied. Since

$$\eta(1) = 1 > -P(0,1), \quad \eta(T) \geq -\frac{1}{n^T} > -\frac{1}{2^T} = -P(0,T), \quad T = 2, 3, \ldots,$$

and

$$\eta(2) - \eta(1) \leq \frac{1}{n^2} < \frac{1}{2^2} = P(0,1) - P(0,2),$$

$$\eta(T+1) - \eta(T) \leq \frac{1}{n^{T+1}} + \frac{1}{n^T} = \frac{n+1}{n^{T+1}} \leq \frac{1}{2^{T+1}} = P(0,T) - P(0,T+1),$$
$$T = 2, 3, \ldots,$$

so (1.5.4) is satisfied as well. Let us assume that $\mathbb{Q}$ is a martingale measure. Then

$$0 = \mathbb{E}^{\mathbb{Q}}[\eta(T)] = q_T \frac{1}{n^T} - q_{T+1} \frac{1}{n^T}, \quad T = 2, 3, \ldots,$$

where $q_k := \mathbb{Q}(\{\omega_k\}), k = 1, 2, \ldots$. But this implies that $q_T = q_{T+1}$ for $T = 2, 3, \ldots$, which is impossible because $\sum_{k=1}^{+\infty} q_k = 1$.

Now let us assume that $\varphi(T), T = 1, 2, \ldots$ is an arbitrage strategy. The discounted portfolio wealth at time $t = 1$ satisfies

$$\hat{X}(1) = \sum_{T=1}^{+\infty} \varphi(T)\eta(T) \geq 0.$$

In particular,

$$\hat{X}(1)(\omega_2) = \varphi(2)\frac{1}{n^2} \geq 0$$

and

$$\hat{X}(1)(\omega_3) = -\varphi(2)\frac{1}{n^2} + \varphi(3)\frac{1}{n^3} \geq 0.$$

Consequently, $\varphi(2) \geq 0$ and $\varphi(3) \geq n\varphi(2)$. By induction, we obtain $\varphi(T) \geq n^{T-2}\varphi(2), T = 2, 3, \ldots$. Then the initial wealth equals

$$\sum_{T=1}^{+\infty} \varphi(T)P(0, T) = \varphi(1)\frac{1}{2} + \sum_{T=2}^{+\infty} \varphi(T)\frac{1}{2^T} \geq \frac{\varphi(1)}{2} + \varphi(2)\sum_{T=2}^{+\infty} \frac{n^{T-2}}{2^T} = +\infty.$$

This shows that there are no arbitrage strategies with final initial capital. □

Remark 1.5.4 In the literature on financial modelling, a more general definition of a martingale measure $\mathbb{Q}$ can be found. Very often one only requires that $\hat{P}(t, T)$ are *local martingales* under $\mathbb{Q}$. In fact, in the present setting both definitions are equivalent. This follows from the fact that $\hat{P}(t, T)$ is positive and Proposition 1.5.5.

Proposition 1.5.5 *If $X(t), t = 0, 1, 2, \ldots$, is a positive local martingale, then it is a martingale.*

Proof Let $\{\tau_n\}$ be a localizing sequence of stopping times, that is, $\tau_n \leq \tau_{n+1}$, $\tau_n \uparrow +\infty$ and $X(\tau_n \wedge t)$ t $= 0, 1, \ldots$, is a martingale for each n. In particular, X_0 is integrable. Since X is positive, application of Fatou's lemma in the formula

$$\mathbb{E}[X(\tau_n \wedge t) \mid \mathcal{F}_{t-1}] = X(\tau_n \wedge (t-1)), \quad t = 1, 2, \ldots, n = 1, 2, \ldots, \tag{1.5.5}$$

yields

$$\mathbb{E}[X(t) \mid \mathcal{F}_{t-1}] \leq X(t-1), \quad t = 1, 2, \ldots ,$$

and

$$X(0) \geq \mathbb{E}[X(1)] \geq \ldots.$$

Since

$$| X(\tau_n \wedge t) | \leq | X(1) | + \cdots + | X(t) |$$

and

$$\mathbb{E}\big[\, | X(1) | + \cdots + | X(t) | \, \big] \leq tX(0) < +\infty,$$

we let again $n \uparrow +\infty$ in (1.5.5) and, by dominated convergence, obtain

$$\mathbb{E}[X(t) \mid \mathcal{F}_{t-1}] = X(t-1), \quad t = 1, 2, \ldots. \qquad \square$$

2

Arbitrage-Free Bond Markets

In this chapter we characterize bond markets that are arbitrage free. Conditions for the existence of martingale measures are studied. Special attention is paid to Markovian models of forward rates, in particular to factor models and affine models.

2.1 Martingale Modelling

Economists believe that the market economy does not allow arbitrage and therefore realistic models of bond prices should be arbitrage free (see (1.5.1)). Sometimes models satisfy a stronger condition than the existence of a martingale measure (see Theorem 1.5.2), which we denote (MP) as abbreviation for *martingale prices*:

For arbitrary $T = 1, 2, \ldots$ the discounted bond price process

$$\hat{P}(t,T) = \frac{P(t,T)}{B(t)}, \qquad t = 0, 1, \ldots, T \tag{MP}$$

is a martingale.

So, (MP) tells us that under the original measure $\mathbb{P}$ the discounted bond prices are martingales on the underlying probability space $(\Omega, \mathcal{F}, \mathcal{F}_t; t = 0, 1, \ldots)$. Although (MP) is a strong requirement, theoretical models satisfying (MP) are important for determining prices of *contingent claims*. Let X be an $\mathcal{F}_{T^*}$-measurable, with some $T^* > 0$, contingent claim that is attainable. Then, by Proposition 1.4.1, its discounted value can be represented in the form

$$x + \sum_{s=0}^{T^*-1} \langle \varphi(s), \hat{P}(s+1) - \hat{P}(s) \rangle = \hat{X} \tag{2.1.1}$$

for some adapted process $\varphi(s)$, and x in (2.1.1) defines the *price* of X (see Section 1.4). Taking expectations in (2.1.1) yields

$$x = \mathbb{E}[\hat{X}] = \mathbb{E}[Xe^{-\sum_{s=0}^{T^*-1} R(s)}], \tag{2.1.2}$$

which is a simple formula for the fair price of the claim X.

Additional Markovian type assumptions under (MP) make the model easier to handle and, in particular, lead to more explicit formulas for financial quantities like x in (2.1.2). If the bond market does not satisfy (MP) but one can modify it by measure change such that (MP) holds under a new measure, then one can still use (2.1.2) for calculating x. Hence we will consider also a weaker condition than (MP), which also precludes arbitrage. We denote it by (MM) as abbreviation for *martingale measure* property.

There exists a measure $\mathbb{Q} \sim \mathbb{P}$ such that for each $T = 1, 2, \ldots,$

$$\hat{P}(t,T), \quad t = 0, 1, \ldots, T \tag{MM}$$

is a $\mathbb{Q}$-martingale.

In this chapter we examine conditions (MM) and (MP) in models introduced in Section 1.2. All of them are formulated in terms of forward rates and incorporate some kind of Markovianity concepts.

2.2 Martingale Measures for HJM Models

In describing martingale measures in the HJM model it is convenient to extend the definition of the forward rate $f(t,T)$ also for $t > T$. As described in Section 1.1 this can be done by assuming that nominal values of bonds are kept on the bank account after their maturity times. This boils down to writing the HJM model (see Section 1.2 (a)) in the form

$$f(t+1,T) = f(t,T) + \alpha(t,T) + \langle \sigma(t,T), \xi_{t+1} \rangle, \quad t, T = 0, 1, \ldots, \tag{2.2.1}$$

where we put

$$\alpha(t,T) = 0, \quad \sigma(t,T) = 0, \qquad t \geq T.$$

Consequently,

$$f(s,t) = f(t,t) = R(t), \quad t < s$$

and

$$\hat{P}(t,T) = e^{-\sum_{s=0}^{T-1} f(t,s)}, \quad t \leq T \tag{2.2.2}$$

(see (1.1.11)).

2.2.1 Existence of Martingale Measures

We analyze the (MM) condition in the HJM model with running time bounded by T^*, where $T^* > 0$. So, we are looking for a measure $\mathbb{Q}$ such that each discounted T-bond price,

$$\hat{P}(t,T), \quad t = 0, 1, \ldots, T \wedge T^*,$$

is a $\mathbb{Q}$-martingale for $T = 0, 1, 2, \ldots$. Since $\mathbb{Q}$ is equivalent to $\mathbb{P}$, its density $\rho = \rho_{T^*}$ is $\mathcal{F}_{T^*}$-measurable and there exists a function ψ such that

$$\rho = \psi(\xi_1, \xi_2, \ldots, \xi_{T^*}). \tag{2.2.3}$$

The corresponding *density process*

$$\rho_t := \left.\frac{d\mathbb{Q}}{d\mathbb{P}}\right|_{\mathcal{F}_t}, \quad t = 0, 1, \ldots, T^*$$

has the form

$$\rho_t = \psi_t(\xi_1, \xi_2, \ldots, \xi_t), \quad t = 0, 1, \ldots, T^*, \tag{2.2.4}$$

where $\{\psi_t\}$ is given by

$$\psi_t(x_1, x_2, \ldots, x_t) = \int \psi(x_1, x_2, \ldots, x_t, y_1, y_2, \ldots, y_{T^*-t})\mu(dy_1) \ldots \mu(dy_{T^*-t}),$$
$$t = 0, 1, \ldots, T^* - 1$$

and $\psi_{T^*} = \psi$. Above μ stands for the distribution of ξ_1 under $\mathbb{P}$. We use also the Laplace exponent φ_ξ and the Laplace transform L_ξ of ξ_1 which are given by

$$L_\xi(u) = \mathbb{E}[e^{-\langle u, \xi \rangle}] = e^{\varphi_\xi(u)}, \qquad u \in U = \mathbb{R}^d, \tag{2.2.5}$$

providing that the expectation above is finite.

The following result shows how ψ, α and σ should be related to each other to guarantee that ψ corresponds to a martingale measure. Specifically, it provides conditions for the (MM) property to be satisfied.

Theorem 2.2.1 *Let $\mathbb{Q}$ be a measure such that $\mathbb{Q} \sim \mathbb{P}$ and ψ be the function describing its density via* (2.2.3). *Let us define the functions*

$$\psi_{t,t+1}(x_1, x_2, \ldots, x_t, \lambda) := \int_U e^{-\langle \lambda, y \rangle} \cdot \frac{\psi_{t+1}(x_1, x_2, \ldots, x_t, y)}{\psi_t(x_1, x_2, \ldots, x_t)} \mu(dy),$$
$$t = 0, 1, \ldots, T^* - 1.$$

Then the processes of discounted bond prices $\hat{P}(t, T)$, $T = 0, 1, 2, \ldots$ are $\mathbb{Q}$-martingales if and only if

$$\alpha(t, T) = \ln \left(\frac{\psi_{t,t+1}(\xi_1, \ldots, \xi_t, \sum_{s=t}^{T} \sigma(t, s))}{\psi_{t,t+1}(\xi_1, \ldots, \xi_t, \sum_{s=t}^{T-1} \sigma(t, s))} \right), \quad t = 0, 1, \ldots, T-1, \quad \mathbb{P} - a.s. \tag{2.2.6}$$

In particular, $\hat{P}(t, T)$, $T = 0, 1, 2, \ldots$ are $\mathbb{P}$-martingales if and only if

$$\alpha(t, T) = \varphi_\xi \left(\sum_{s=t}^{T} \sigma(t, s) \right) - \varphi_\xi \left(\sum_{s=t}^{T-1} \sigma(t, s) \right), \qquad t = 0, 1, \ldots, T-1. \tag{2.2.7}$$

Proof We use the fact that $\hat{P}(t,T)$ is a $\mathbb{Q}$-martingale if and only if $\hat{P}(t,T)\rho_t$ is a $\mathbb{P}$-martingale. By (2.2.2), (2.2.1) and (2.2.4) we have

$$\mathbb{E}[\rho_{t+1}\hat{P}(t+1,T) \mid \mathcal{F}_t] = \mathbb{E}[e^{-\sum_{s=0}^{T-1} f(t+1,s)}\psi_{t+1}(\xi_1,\xi_2,\ldots,\xi_{t+1}) \mid \mathcal{F}_t]$$

$$= e^{-\sum_{s=0}^{T-1} f(t,s)}\psi_t(\xi_1,\xi_2,\ldots,\xi_t)\cdot \mathbb{E}\left[e^{-\sum_{s=0}^{T-1}\{\alpha(t,s)+\langle\sigma(t,s),\xi_{t+1}\rangle\}}\frac{\psi_{t+1}(\xi_1,\xi_2,\ldots,\xi_{t+1})}{\psi_t(\xi_1,\xi_2,\ldots,\xi_t)}\middle|\mathcal{F}_t\right]$$

$$= \rho_t\hat{P}(t,T)e^{-\sum_{s=0}^{T-1}\alpha(t,s)}\mathbb{E}\left[e^{-\langle\sum_{s=0}^{T-1}\sigma(t,s),\xi_{t+1}\rangle}\frac{\psi_{t+1}(\xi_1,\xi_2,\ldots,\xi_{t+1})}{\psi_t(\xi_1,\xi_2,\ldots,\xi_t)}\;\middle|\;\mathcal{F}_t\right].$$

Since the conditional expectation above equals $\psi_{t,t+1}(\xi_1,\xi_2,\ldots,\xi_t,\sum_{s=0}^{T-1}\sigma(t,s))$, we see that $\hat{P}(t,T)\rho_t$ is a martingale if and only if

$$e^{-\sum_{s=t}^{T-1}\alpha(t,s)}\cdot\psi_{t,t+1}\left(\xi_1,\xi_2,\ldots,\xi_t,\sum_{s=t}^{T-1}\sigma(t,s)\right)=1,\quad t=0,1,\ldots,T-1,\quad \mathbb{P}-a.s.$$

It follows from the formula above that

$$\sum_{s=t}^{T-1}\alpha(t,s) = \ln\left(\psi_{t,t+1}(\xi_1,\ldots,\xi_t,\sum_{s=t}^{T-1}\sigma(t,s))\right),\quad t=0,1,\ldots,T-1,\tag{2.2.8}$$

and repeating the same arguments for the $(T+1)$-bond we obtain

$$\sum_{s=t}^{T}\alpha(t,s) = \ln\left(\psi_{t,t+1}(\xi_1,\ldots,\xi_t,\sum_{s=t}^{T}\sigma(t,s))\right),\quad t=0,1,\ldots,T.\tag{2.2.9}$$

Subtracting (2.2.8) from (2.2.9) yields (2.2.6).

To treat the case when $\hat{P}(t,T)$ are $\mathbb{P}$-martingales we put $\psi \equiv 1$. Then $\psi_t \equiv 1$ for each $t = 0,1,\ldots,T^*$ and consequently $\psi_{t,t+1}(x_1,x_2,\ldots,x_t,\lambda) = e^{\varphi_\xi(\lambda)}$. The assertion follows directly from (2.2.6). □

It is of interest to describe models for which the drift $\alpha(t,T)$ is a deterministic function of volatilities and does not depend on the past noise $\xi_1,\xi_2,\ldots,\xi_t$. With the use of Theorem 2.2.1 we get the following result.

Proposition 2.2.2 *Let $\{h_t\}$ be a sequence of measurable functions satisfying*

$$h_t(y)>0,\quad \mu-a.s.,\quad \int_U h_t(y)\mu(dy)=1,\quad t=1,2,\ldots,T^*.$$

Let us define

$$g_t(\lambda) := \int_U e^{-\langle\lambda,y\rangle}h_{t+1}(y)\mu(dy),\quad \lambda\in U,\quad t=0,1,\ldots,T^*-1.$$

Then the model with drift given by

$$\alpha(t,T) = \ln\left(\frac{g_t(\sum_{s=1}^{T}\sigma(t,s))}{g_t(\sum_{s=1}^{T-1}\sigma(t,s))}\right), \quad t = 0,1,\ldots,T-1, \quad T = 0,1,\ldots, \tag{2.2.10}$$

satisfies (MM). Moreover, there exists a martingale measure $\mathbb{Q}$ such that $\xi_1, \xi_2, \ldots, \xi_{T^*}$ are independent under $\mathbb{Q}$.

Proof Let us notice that for

$$\psi(x_1, x_2, \ldots, x_{T^*}) := h_1(x_1)h_2(x_2)\ldots h_{T^*}(x_{T^*}), \tag{2.2.11}$$

one obtains

$$\psi_0 = 1, \quad \psi_t(x_1, x_2, \ldots, x_t) = h_1(x_1)h_2(x_2)\ldots h_t(x_t), \quad t = 1,2,\ldots,T^*,$$

and consequently

$$\begin{aligned}\psi_{t,t+1}(x_1,\ldots,x_t,\lambda) &= \int_U e^{-\langle\lambda,y\rangle}\cdot\frac{\psi_{t+1}(x_1,x_2,\ldots,x_t,y)}{\psi_t(x_1,x_2,\ldots,x_t)}\mu(dy)\\ &= \int_U e^{-\langle\lambda,y\rangle}\cdot h_{t+1}(y)\mu(dy)\\ &= g_t(\lambda), \quad t = 0,1,\ldots,T^*-1.\end{aligned}$$

It follows from (2.2.6) that for the model with drift given by (2.2.10) the measure $\mathbb{Q}$ with density (2.2.11) is a martingale measure. By (2.2.11) we see that $\xi_1, \ldots, \xi_{T^*}$ are independent also under $\mathbb{Q}$. □

2.2.2 Uniqueness of the Martingale Measure

With the use of Theorem 2.2.1 we deduce now conditions for the uniqueness of a martingale measure. They are formulated in Theorems 2.2.3 and 2.2.4 below.

Theorem 2.2.3 *Let the forward rate be given by* (2.2.1) *with real valued factors* $\{\xi_i\}$. *If*

$$\sum_{s=t}^{+\infty} |\,\sigma(t,s)\,| < +\infty, \quad t = 0,1,\ldots,T^*-1,$$

then the arising bond market admits exactly one martingale measure or there are no martingale measures.

Theorem 2.2.4 *If ξ_1 in* (2.2.1) *takes $K < +\infty$ different values in $\mathbb{R}$ and, for any $t = 0,1,\ldots,T^*-1$, the function*

$$T \longrightarrow \sum_{s=t}^{T}\sigma(t,s), \quad T = t, t+1, \ldots,$$

takes at least K different values, then the arising bond market admits exactly one martingale measure or there are no martingale measures.

Proof of Theorem 2.2.3 We use the notation from Theorem 2.2.1. Recall that a measure with density $\psi(\xi_1, \xi_2, \ldots, \xi_{T^*})$ is a martingale measure if and only if the drift α is determined by (2.2.6). So, the model admits a martingale measure if and only if (2.2.6) holds with some function ψ. Let us assume that there exists another martingale measure with density $\tilde{\psi}(\xi_1, \xi_2, \ldots, \xi_{T^*})$. Then by (2.2.6), or, equivalently, by (2.2.9), for each $t = 0, 1, \ldots, T^* - 1$, we obtain

$$\psi_{t,t+1}(\xi_1, \ldots, \xi_t, \sum_{s=t}^{T} \sigma(t,s)) = \tilde{\psi}_{t,t+1}(\xi_1, \ldots, \xi_t, \sum_{s=t}^{T} \sigma(t,s)), \quad T = t, t+1, \ldots. \tag{2.2.12}$$

For the case $t = 0$ this yields

$$\psi_{0,1}\left(\sum_{s=0}^{T} \sigma(0,s)\right) = \tilde{\psi}_{0,1}\left(\sum_{s=0}^{T} \sigma(0,s)\right), \quad T = 0, 1, \ldots.$$

But

$$\psi_{0,1}(\lambda) = \int_{\mathbb{R}} e^{-\lambda y} \psi_1(y) \mu(dy), \quad \tilde{\psi}_{0,1}(\lambda) = \int_{\mathbb{R}} e^{-\lambda y} \tilde{\psi}_1(y) \mu(dy)$$

are analytic functions which are equal on the convergent sequence $\lambda(T) := \sum_{s=0}^{T} \sigma(0,s)$. Hence they are equal in the whole domain. Since they are Laplace transforms of the measures $\psi_1(y)\mu(dy)$, $\tilde{\psi}_1(y)\mu(dy)$, also the measures are identical. Hence

$$\psi_1(\xi_1) = \tilde{\psi}_1(\xi_1). \tag{2.2.13}$$

Now we use (2.2.12) with $t = 1$, which yields

$$\psi_{1,2}(\xi_1, \sum_{s=1}^{T} \sigma(1,s)) = \tilde{\psi}_{1,2}(\xi_1, \sum_{s=1}^{T} \sigma(1,s)), \quad T = 1, 2, \ldots.$$

Since

$$\psi_{1,2}(x_1, \lambda) = \int_{\mathbb{R}} e^{-\lambda y} \frac{\psi_2(x_1, y)}{\psi_1(x_1)} \mu(dy), \quad \tilde{\psi}_{1,2}(x_1, \lambda) = \int_{\mathbb{R}} e^{-\lambda y} \frac{\tilde{\psi}_2(x_1, y)}{\tilde{\psi}_1(x_1)} \mu(dy)$$

and (2.2.13) holds, we obtain

$$\int_{\mathbb{R}} e^{-\lambda y} \psi_2(\xi_1, y) \mu(dy) = \int_{\mathbb{R}} e^{-\lambda y} \tilde{\psi}_2(\xi_1, y) \mu(dy)$$

for each $\lambda = \lambda(T) = \sum_{s=1}^{T} \sigma(1,s)$. Since the last sum converges, again, by the analyticity of the Laplace transform, we obtain that

$$\psi_2(\xi_1,\xi_2) = \tilde{\psi}_2(\xi_1,\xi_2).$$

It is clear that iterative application of the preceding arguments yields

$$\psi(\xi_1,\xi_2,\ldots,\xi_{T^*}) = \tilde{\psi}(\xi_1,\xi_2,\ldots,\xi_{T^*}),$$

which means that the martingale measure is unique. □

For the proof of Theorem 2.2.4 we use a corollary from the following auxiliary result.

Lemma 2.2.5 *The function*

$$f(x) := c_1 x^{a_1} + c_2 x^{a_2} + \cdots + c_K x^{a_K}, \quad x > 0,$$

where $c_k, a_k \in \mathbb{R}$, for $k = 1,2,\ldots,K$ and $a_1 < a_2 < \cdots < a_K$, not all c_k are zero, has at most $K-1$ positive roots.

Proof We follow Gantmacher and Krein [61, p.76] and prove the assertion by induction. The result is clearly true for $K = 1$. Assuming that it is true for $K-1$ we show it for K. Equivalently, we have to show that

$$x^{-a_1} f(x) = c_1 + c_2 x^{a_2 - a_1} + \cdots + c_K x^{a_K - a_1}$$

has at most $K-1$ positive roots. But we know from the previous inductive step that

$$\frac{d}{dx}\left(x^{-a_1} f(x)\right) = (a_2 - a_1)c_2 x^{a_2 - a_1 - 1} + \cdots + (a_K - a_1)c_K x^{a_K - a_1 - 1}$$

has at most $K-2$ positive roots. However, between two consecutive roots of $x^{-a_1} f(x)$ there is at least one root of its derivative. So, if $x^{-a_1} f(x)$ had at least K roots then its derivative would have at least $K-1$ roots, which is a contradiction. □

Corollary 2.2.6 *Let $a_1 < a_2 < \cdots < a_K$ be real numbers. Then the vectors*

$$v_1 := \begin{pmatrix} x_1^{a_1} \\ \vdots \\ x_K^{a_1} \end{pmatrix}, \quad v_2 := \begin{pmatrix} x_1^{a_2} \\ \vdots \\ x_K^{a_2} \end{pmatrix}, \ldots, \quad v_K := \begin{pmatrix} x_1^{a_K} \\ \vdots \\ x_K^{a_K} \end{pmatrix} \tag{2.2.14}$$

are linearly independent for any positive reals $x_1, x_2, \ldots, x_K$ such that $x_i \neq x_j, i \neq j$. Indeed, let us assume that $v_1, v_2, \ldots, v_K$ are not linearly independent for some $x_1, x_2, \ldots, x_K$. Then there exist constants $c_1, c_2, \ldots, c_K$, not all equal 0, such that

$$c_1 v_1 + c_2 v_2 + \cdots + c_K v_K = 0.$$

This, however, means that each $x_k, k = 1, 2, \ldots, K$ is a root of the function

$$f(x) := c_1 x^{a_1} + c_2 x^{a_2} + \cdots + c_K x^{a_K}, \quad x \geq 0,$$

which is impossible by Lemma 2.2.5.

Proof of Theorem 2.2.4 Let us assume that $\psi(\xi_1, \xi_2, \ldots, \xi_{T^*})$ and $\tilde{\psi}(\xi_1, \xi_2, \ldots, \xi_{T^*})$ are densities of two different martingale measures. As in the proof of Theorem 2.2.3 we show that the condition

$$\psi_{t,t+1}(\xi_1, \ldots, \xi_t, \sum_{s=t}^{T} \sigma(t, s)) = \tilde{\psi}_{t,t+1}(\xi_1, \ldots, \xi_t, \sum_{s=t}^{T} \sigma(t, s)), \quad T = t, t+1, \ldots, \tag{2.2.15}$$

for each $t = 0, 1, \ldots, T^* - 1$, implies that

$$\psi(\xi_1, \xi_2, \ldots, \xi_{T^*}) = \tilde{\psi}(\xi_1, \xi_2, \ldots, \xi_{T^*}). \tag{2.2.16}$$

Let us denote the values of ξ_1 by $a_1 < a_2 < \cdots < a_K$ and

$$p_k := \mathbb{P}(\xi_1 = a_k) = \mu(\{a_k\}), \quad k = 1, 2, \ldots, K.$$

Condition (2.2.15) with $t = 0$ yields

$$\psi_{0,1}\left(\sum_{s=0}^{T} \sigma(0, s)\right) = \tilde{\psi}_{0,1}\left(\sum_{s=0}^{T} \sigma(0, s)\right), \quad T = 0, 1, \ldots. \tag{2.2.17}$$

But

$$\psi_{0,1}(\lambda) = \int_{\mathbb{R}} e^{-\lambda y} \psi_1(y) \mu(dy) = \sum_{k=1}^{K} e^{-\lambda a_k} \psi_1(a_k) p_k$$

and

$$\tilde{\psi}_{0,1}(\lambda) = \int_{\mathbb{R}} e^{-\lambda y} \tilde{\psi}_1(y) \mu(dy) = \sum_{k=1}^{K} e^{-\lambda a_k} \tilde{\psi}_1(a_k) p_k,$$

so (2.2.17) means that

$$\sum_{k=1}^{K} e^{-\lambda(T) a_k} \psi_1(a_k) p_k = \sum_{k=1}^{K} e^{-\lambda(T) a_k} \tilde{\psi}_1(a_k) p_k$$

for each $\lambda(T) := \sum_{s=0}^{T} \sigma(0, s)$. Thus, for any K different values $\lambda(T_1), \ldots, \lambda(T_K)$, we have

$$\sum_{k=1}^{K} x_i^{a_k} \psi_1(a_k) p_k = \sum_{k=1}^{K} x_i^{a_k} \tilde{\psi}_1(a_k) p_k, \quad i = 1, 2, \ldots, K,$$

where $x_k := e^{-\lambda(T_k)} > 0, k = 1, 2, \ldots, K$. By Corollary 2.2.6, $\psi_1(a_k) = \tilde{\psi}_1(a_k)$, so

$$\psi_1(\xi_1) = \tilde{\psi}_1(\xi_1). \tag{2.2.18}$$

The use of (2.2.15) with $t = 1$ yields

$$\psi_{1,2}(\xi_1, \sum_{s=1}^{T} \sigma(1,s)) = \tilde{\psi}_{1,2}(\xi_1, \sum_{s=1}^{T} \sigma(1,s)), \quad T = 1, 2, \ldots. \tag{2.2.19}$$

Since

$$\psi_{1,2}(x_1, \lambda) = \sum_{k=1}^{K} e^{-\lambda a_k} \frac{\psi_2(x_1, a_k)}{\psi_1(x_1)} p_k, \quad \tilde{\psi}_{1,2}(x_1, \lambda) = \sum_{k=1}^{K} e^{-\lambda a_k} \frac{\tilde{\psi}_2(x_1, a_k)}{\tilde{\psi}_1(x_1)} p_k$$

and (2.2.18) holds, we obtain from (2.2.19) that

$$\sum_{k=1}^{K} e^{-\lambda(T) a_k} \left(\psi_2(x_1, a_k) - \tilde{\psi}_2(x_1, a_k) \right) p_k = 0$$

for each $\lambda(T) = \sum_{s=1}^{T} \sigma(1, s)$. Taking K different values $\lambda(T_1), \lambda(T_2), \ldots, \lambda(T_K)$ we conclude again from Corollary 2.2.6 that

$$\psi_2(\xi_1, \xi_2) = \tilde{\psi}_2(\xi_1, \xi_2).$$

Further application of the preceding arguments for $t = 2, 3, \ldots, T^* - 1$ yields finally (2.2.16). □

2.3 Martingale Measures and Martingale Representation Property

In this section we characterize equivalent measures in a way that is specific for continuous time models. It differs from the framework used in Section 2.2, where the density process of an equivalent to $\mathbb{P}$ measure $\mathbb{Q}$ was represented simply in the form

$$\rho_t = \psi_t(\xi_1, \xi_2, \ldots, \xi_t), \quad t = 0, 1, \ldots, T^*,$$

where $\{\xi_t\}$ is a sequence of $\mathbb{R}^d$-valued independent and identically distributed random variables on $(\Omega, \mathcal{F}, \mathbb{P})$ and ψ_t is some function. Our current framework will exploit *martingale representation properties* related to the martingale Z defined by

$$Z_0 := 0, \qquad Z_t := \sum_{s=1}^{t} \xi_s, \qquad \mathbb{E}[\xi_t] = 0, \qquad t = 1, 2, \ldots, T^*.$$

Since the increments of Z are independent and stationary, Z can be viewed as a discrete time counterpart of a Lévy process in continuous time. We say that Z has the *martingale representation property* if any martingale X adapted to $\{\mathcal{F}_t\}$, where $\mathcal{F}_0$ is trivial and $\mathcal{F}_t := \sigma\{\xi_1, \xi_2, \ldots, \xi_t\}; t = 1, 2, \ldots, T^*$, can be written as a discrete time stochastic integral over Z, that is, in the form

$$X_t = X_0 + \sum_{s=0}^{t-1} \langle Y_s, Z_{s+1} - Z_s \rangle, \quad t = 1, \ldots, T^*, \tag{2.3.1}$$

where Y is some adapted process. The martingale representation property is examined in Section 2.3.1, where we show that it requires very restrictive distributional conditions for Z. Therefore, also a *generalized martingale representation property* of Z will be used to cover the general situation. It turns out that any martingale X can be represented in the following way

$$X_t = X_0 + \sum_{s=0}^{t-1} \int_U \psi(s, y) \tilde{\pi}(\{s+1\}, dy), \quad t = 1, \ldots, T^*, \tag{2.3.2}$$

where $\psi(s, y)$ is an adapted process and $\tilde{\pi}$ is the so-called *compensated jump measure* of the process Z. The precise definition of the preceding integral and the proof of (2.3.2) is presented in Section 2.3.2. The concept of jump measure comes from the continuous time setting and is used to prove martingale representation property for Lévy processes.

We use both martingale representations, i.e. (2.3.1) and (2.3.2), to describe the density process ρ_t of an equivalent measure. This leads to the classical and generalized Girsanov's theorems described in Section 2.3.3, which are used in Section 2.3.4 to characterize martingale measures in the HJM model.

2.3.1 Martingale Representation Property

The martingale representation property will be examined in a more general setting than announced previously. Let $\{M_t\}_{t=0,1,\ldots,T^*}, M_0 = 0$ be an $U = \mathbb{R}^d$-valued martingale on a probability space $(\Omega, \mathcal{F}, \mathbb{P})$ equipped with the filtration $\mathcal{F}_t := \sigma(M_0, M_1, \ldots, M_t), t = 0, 1, \ldots, T^*$. The increments of M are not necessarily assumed neither to be independent nor identically distributed. Our goal is to formulate conditions for M such that any $\mathcal{F}_t$-adapted martingale X starting from zero admits the representation

$$X_t = \sum_{s=0}^{t-1} \langle Y_s, M_{s+1} - M_s \rangle, \quad t = 1, \ldots, T^*, \tag{2.3.3}$$

where Y is some adapted process.

Theorem 2.3.1 *The martingale M has the martingale representation property if and only if for any $t = 0, 1, \ldots, T^* - 1$ the conditional distribution of $M_{t+1} - M_t$ given $M_1, M_2, \ldots, M_t$ is concentrated on a finite set A_t in U such that*

$$\sharp A_t = n(t) < +\infty, \quad \textit{and} \quad \textit{dim}(\textit{span}\, A_t) = n(t) - 1. \tag{2.3.4}$$

If, additionally, $n(t) - 1 = d$ *for each* $t = 0, 1, \dots, T^* - 1$, *then for any martingale* X *the representation* (2.3.3) *is unique. Above* $\sharp A_t$ *stands for the number of elements of* A_t *and span* A_t *for the linear space spanned by the set* A_t.

Example 2.3.2 A real valued martingale M of the form

$$M_t = \sum_{s=1}^{t} \xi_s, \quad t = 1, 2, \dots,$$

where $\{\xi_s\}$ is a sequence of independent random variables with zero mean, has a martingale representation property if and only if the distribution of each ξ_s is concentrated on two points only. If this is the case, then the representation (2.3.3) is unique.

Corollary 2.3.3 *The martingale representation property of* M *implies that any adapted process* $\{X_t\}$ *is finite valued. In particular, all moments of* X_t *are finite. Indeed, it follows from Theorem 2.3.1 that the number of paths of* M *on a finite time interval is finite. Since* X_t *is of the form* $X_t = x_t(M_0, M_1, \dots, M_t)$, $t = 0, 1, \dots, T^* - 1$, *where* $x_t(\cdot)$ *is some function, its values form a finite set.*

The proof of Theorem 2.3.1 is based on the following auxiliary result.

Lemma 2.3.4 *Let* M *be a zero mean* $U - \mathbb{R}^d$*-valued random variable. Then for any function* $\psi : U \longrightarrow \mathbb{R}$ *such that* $\mathbb{E}[\psi(M)] = 0$ *there exists* $y \in U$ *such that*

$$\psi(M) = \langle y, M \rangle \quad \mathbb{P} - a.s. \tag{2.3.5}$$

if and only if the distribution of M *is concentrated on a finite set of vectors* $m_1, m_2, \dots, m_n$ *such that dim(span*$\{m_1, m_2, \dots, m_n\}$*)* $= n - 1$. *The representation* (2.3.5) *is unique if* $n - 1 = d$.

Recall that the *support* of a measure μ defined on the Borel subsets of a set I is the smallest closed set A such that $\mu(A) = \mu(I)$.

Proof *(Sufficiency)* Let us assume that $m_1, m_2, \dots, m_{n-1}$ are linearly independent. Since (2.3.5) has the form

$$\psi(m_i) = \langle y, m_i \rangle, \quad i = 1, 2, \dots, n, \tag{2.3.6}$$

we can find $y \in U$ such that the first $n - 1$ equations are satisfied. The zero mean conditions

$$\sum_{i=1}^{n} \mu_i m_i = 0, \qquad \sum_{i=1}^{n} \mu_i \psi(m_i) = 0,$$

with $\mu_i = \mathbb{P}(M = m_i), i = 1, 2, \ldots, n$, imply that

$$m_n = -\frac{1}{\mu_n}\sum_{i=1}^{n-1}\mu_i m_i, \qquad \psi(m_n) = -\frac{1}{\mu_n}\sum_{i=1}^{n-1}\mu_i\psi(m_i).$$

This clearly yields that $\psi(m_n) = \langle y, m_n\rangle$.
(Necessity) If the support of M is infinite then all bounded functions

$$\psi(x), \quad x \in supp\{M\}$$

form a linear space of infinite dimension, while the space of linear functions

$$\langle y, x\rangle, \quad x \in supp\{M\}, \quad y \in U$$

is of dimension no greater than *dim* U. It follows that if (2.3.5) holds then M takes a finite number of values only. Thus let us now consider the case when the set of values of M is $\{m_1, m_2, \ldots, m_n\}$ and $dim(span\{m_1, m_2, \ldots, m_n\}) < n - 1$. We may also assume that the last two vectors are some linear combinations of $m_1, m_2, \ldots, m_{n-2}$, i.e. $dim(span\{m_1, m_2, \ldots, m_n\}) = dim(span\{m_1, m_2, \ldots, m_{n-2}\})$. Let ψ be such that (2.3.6) holds. Then $\psi(m_{n-1})$ and $\psi(m_n)$ are uniquely determined by $\psi(m_i), i = 1, 2, \ldots, n-2$. Now let us define the function $\tilde{\psi}$ as follows $\tilde{\psi}(m_i) = \psi(m_i), i = 1, 2, \ldots, n-2$ and $\tilde{\psi}(m_{n-1}) \neq \psi(m_{n-1}), \tilde{\psi}(m_n) \neq \psi(m_n)$ be such that

$$\mu_{n-1}\tilde{\psi}(m_{n-1}) + \mu_n\tilde{\psi}(m_n) = \mu_{n-1}\psi(m_{n-1}) + \mu_n\psi(m_n).$$

Then $\mathbb{E}[\tilde{\psi}(M)] = 0$ but (2.3.6) is not satisfied. □

Proof of Theorem 2.3.1 Let X be an arbitrary martingale starting from zero. It can be identified with a sequence of functions $\psi_1, \psi_2, \ldots, \psi_{T^*}$ such that

$$X_t = \psi_t(M_1, M_2, \ldots, M_t), \quad t = 1, 2, \ldots, T^*.$$

The integrand in (2.3.3) can also be written in that form, i.e.

$$Y_t = y_t(M_0, M_1, \ldots, M_t), \quad t = 0, 1, \ldots, T^* - 1.$$

We will prove the required distributional property of M inductively. In the first period the problem has the form

$$\psi_1(M_1) = \langle y_0(M_0), M_1\rangle.$$

By Lemma 2.3.4 we see that necessary and sufficient condition for the representation to hold is that the values of M_1 form a set A_0 satisfying (2.3.4).

Now let us assume that (2.3.4) holds for some $t-1$ and show that (2.3.4) holds for t if and only if M has the martingale representation property. We know that

$$\psi_t(M_1, M_2, \ldots, M_t) = \sum_{s=0}^{t-1}\langle y_s(M_0, M_1, \ldots, M_s), M_{s+1} - M_s\rangle$$

for some functions $y_0, y_1, \dots, y_{t-1}$. For any function ψ_{t+1} such that $\mathbb{E}[\psi_{t+1}(M_1, \dots, M_{t+1}) \mid \mathcal{F}_t] = \psi_t(M_0, M_1, \dots, M_t)$, we are looking for a function y_t such that

$$\begin{aligned}\psi_{t+1}(M_1, M_2, \dots, M_{t+1}) &= \sum_{s=0}^{t} \langle y_s(M_0, M_1, \dots, M_s), M_{s+1} - M_s \rangle \\ &= \psi_t(M_1, M_2, \dots, M_t) + \langle y_t(M_0, M_1, \dots, M_t), M_{t+1} - M_t \rangle.\end{aligned} \tag{2.3.7}$$

Let us consider an arbitrary path of M up to time t, i.e.

$$A := \{M_1 = m_1, M_2 = m_2, \dots, M_t = m_t\}.$$

Condition (2.3.7) implies the following

$$[\psi_{t+1}(m_1, \dots, m_t, M_{t+1}) - \psi_t(m_1, \dots, m_t)]\mathbf{1}_A = [\langle y_t(m_0, m_1, \dots, m_t), M_{t+1} - m_t \rangle]\mathbf{1}_A. \tag{2.3.8}$$

From the martingale properties of M and X we obtain

$$\mathbb{E}\big[[\psi_{t+1}(m_1, \dots, m_t, M_{t+1}) - \psi_t(m_1, \dots, m_t)]\mathbf{1}_A\big] = 0, \quad \mathbb{E}\big[(M_{t+1} - m_t)\mathbf{1}_A\big] = 0,$$

hence we can again apply Lemma 2.3.4. This tells that $y_t(m_0, m_1, \dots, m_t)$ satisfying (2.3.8) exists if and only if the values of $(M_{t+1} - m_t)\mathbf{1}_A$ satisfy (2.3.4). This proves the assertion. Uniqueness follows also from Lemma 2.3.4. □

2.3.2 Generalized Martingale Representation Property

To prove the generalized martingale representation property of the martingale

$$Z_0 := 0, \qquad Z_t := \sum_{s=1}^{t} \xi_s, \qquad \mathbb{E}[\xi_t] = 0, \qquad t = 1, 2, \dots, T^*, \tag{2.3.9}$$

where $\{\xi_i\}$ is an i.i.d. sequence living in $U = \mathbb{R}^d$ we need to extend the concept of stochastic integral related to Z. Recall that the classical integral of an adapted process Y over Z is given by

$$I_0^1 := 0, \quad I_t^1 := \sum_{s=0}^{t-1} \langle Y_s, Z_{s+1} - Z_s \rangle = \sum_{s=0}^{t-1} \langle Y_s, \xi_{s+1} \rangle, \quad t = 1, \dots, T^*. \tag{2.3.10}$$

For the definition of the extended stochastic integral we need the concept of *random measure*. Let $\pi : \{1, 2, \dots\} \times U \longrightarrow \{0, 1, \dots\}$ be given by

$$\pi(\{s\}, A) = \mathbf{1}_{\{Z_s - Z_{s-1} \in A\}} = \mathbf{1}_{\{\xi_s \in A\}},$$

where $A \subseteq U$. Then

$$\pi(\{1,2,\ldots,t\},A) = \sum_{s=1}^{t} \mathbf{1}_{\{\xi_s \in A\}}$$

gives the number of increments of Z on the interval $[0,t]$, which take values in the set A, and therefore π is called a *jump measure* or *random measure* of Z. Since

$$\mathbb{E}[\pi(\{s\},A)] = \mathbb{E}[\mathbf{1}_{\{\xi_s \in A\}}] = \mathbb{P}(\xi_s \in A) = \mu(A),$$

where μ stands for the law of ξ_1, we see that the process

$$\tilde{\pi}(\{1,2,\ldots,t\},A) := \sum_{s=1}^{t} \mathbf{1}_{\{\xi_s \in A\}} - t\mu(A), \quad t = 1,2,\ldots,T^*$$

is a martingale. The measure $\tilde{\pi}$ is called a *compensated random measure* of Z. An *extended stochastic integral* we obtain by integrating a process $\psi(\cdot,y)$ over the measure $\tilde{\pi}$ in the following way

$$\begin{aligned} I_0^2 = 0, \quad I_t^2 &:= \sum_{s=0}^{t-1} \int_U \psi(s,y)\tilde{\pi}(\{s+1\},dy), \\ &:= \sum_{s=0}^{t-1} \left\{ \psi(s,\xi_{s+1}) - \int_U \psi(s,y)\mu(dy) \right\} \quad t = 1,2,\ldots,T^*. \end{aligned} \tag{2.3.11}$$

For each y the process $\psi(\cdot,y)$ is assumed to be adapted, i.e. $\psi(s,y)$ is $\mathcal{F}_s$-measurable such that

$$\int_U |\psi(s,y)| \mu(dy) < +\infty, \quad s = 0,1,\ldots,T^*-1.$$

If ψ is of the form $\psi(s,x) = \langle \psi(s), x\rangle$ then

$$I_t^2 = \sum_{s=0}^{t-1} \left\{ \langle \psi(s), \xi_{s+1}\rangle - \psi(s) \int_U y\mu(dy) \right\} = \sum_{s=0}^{t-1} \langle \psi(s), \xi_{s+1}\rangle, \quad t = 1,2,\ldots,T^*$$

because

$$\int_U y\mu(dy) = 0.$$

So, in this case I^2 equals I^1 in (2.3.10) with $Y_s = \psi(s)$.

With the use of the concept of extended stochastic integral one can easily prove the generalized martingale representation property of the process Z.

Proposition 2.3.5 *Any martingale N can be represented in the form*

$$N_t = N_0 + \sum_{s=0}^{t-1} \int_U \psi(s,y)\tilde{\pi}(\{s+1\},dy), \quad t = 1,2,\ldots,T^* \tag{2.3.12}$$

for some adapted process $\psi(s,y)$ satisfying

$$\int_U |\ \psi(s,y)\ |\ \mu(dy) < +\infty, \quad s = 0, 1, \ldots, T^* - 1. \tag{2.3.13}$$

Proof Since $N_s = h(s, \xi_1, \ldots, \xi_s), s = 1, 2, \ldots, T^*$ for some function h, by the martingale property we obtain

$$N_s = \mathbb{E}(N_{s+1} \mid \mathcal{F}_s) = \int_U h(s, \xi_1, \ldots, \xi_s, y)\mu(dy), \quad s = 0, 1, \ldots, T^* - 1.$$

Consequently,

$$\begin{aligned} N_t &= N_0 + \sum_{s=0}^{t-1}(N_{s+1} - N_s) = N_0 + \sum_{s=0}^{t-1}(N_{s+1} - \mathbb{E}(N_{s+1} \mid \mathcal{F}_s)) \\ &= N_0 + \sum_{s=0}^{t-1} h(s, \xi_1, \ldots, \xi_s, \xi_{s+1}) - \sum_{s=0}^{t-1} \int_U h(s, \xi_1, \ldots, \xi_s, y)\mu(dy) \\ &= N_0 + \sum_{s=0}^{t-1} \int_U \psi(s,y)\tilde{\pi}(\{s+1\}, dy) \quad t = 1, 2, \ldots, T^*, \end{aligned}$$

where $\psi(s,y) := h(s, \xi_1, \ldots, \xi_s, y)$. (2.3.13) follows from the definition of ψ. □

2.3.3 Girsanov's Theorems

Let $\mathbb{Q}$ be a measure that is equivalent to $\mathbb{P}$. Girsanov's theorem provides a description of the corresponding density process

$$\rho_t := \frac{d\mathbb{Q}}{d\mathbb{P}}|\mathcal{F}_t, \quad t = 0, 1, \ldots, T^*,$$

where $\{\mathcal{F}_t\}$ is the filtration generated by a sequence of zero mean, independent and identically distributed random variables $\{\xi_i\}, i = 1, 2, \ldots, T^*$. In the classical version of Girsanov's theorem (see Theorem 2.3.6), ρ has an exponential form involving some martingale. Such representation holds, in particular, in the case when the process $Z_t = \xi_1 + \ldots \xi_t$ has the martingale representation property. This case corresponds to the Wiener process in the continuous time setting. In the general version of Girsanov's theorem (see Theorem 2.3.8) ρ has a different form that can be proven under weaker assumptions.

Let φ_ξ be the Laplace exponent of ξ_1, i.e.

$$\mathbb{E}[e^{-\langle u,\xi\rangle}] = e^{\varphi_\xi(u)}, \qquad u \in U,$$

and let us define the set

$$\Gamma := \{\lambda \in U : \varphi_\xi(\lambda) < +\infty\}. \tag{2.3.14}$$

Recall that μ stands for the distribution of ξ_1.

Theorem 2.3.6 [Girsanov's theorem – classical version] *Let $\mathbb{Q} \sim \mathbb{P}$ be a measure with density process ρ satisfying:*

$$\mathbb{E}\Big[\mid \ln(\rho_t) \mid \Big] < +\infty, \quad t = 0, 1, 2, \ldots, T^*. \tag{2.3.15}$$

(a) Then there exists an adapted process $\psi(s, y)$ satisfying

$$\int_U \mid \psi(s, y) \mid \mu(dy) < +\infty, \quad \int_U e^{\psi(s,y)} \mu(dy) < +\infty \quad s = 0, 1, \ldots, T^* - 1, \tag{2.3.16}$$

such that

$$\rho_0 = 1, \quad \rho_t = e^{\sum_{s=0}^{t-1} \int_U \psi(s,y) \tilde{\pi}(\{s+1\},dy) - \sum_{s=0}^{t-1} \{\ln\left(\int_U e^{\psi(s,y)} \mu(dy)\right) - \int_U \psi(s,y) \mu(dy)\}},$$
$$t = 1, 2, \ldots, T^*. \tag{2.3.17}$$

Conversely, if ρ is of the form (2.3.17) *with ψ satisfying* (2.3.16) *then it is a density process of some measure that is equivalent to $\mathbb{P}$.*

(b) If the process

$$Z_0 := 0, \qquad Z_t := \sum_{s=1}^{t} \xi_s, \qquad t = 1, 2, \ldots, T^*$$

has the martingale representation property, then there exists an adapted process δ that $\mathbb{P}$-a.s. takes values in the set Γ such that

$$\rho_0 = 1, \quad \rho_t = e^{\sum_{s=0}^{t-1} \langle \delta_s, \xi_{s+1} \rangle - \sum_{s=0}^{t-1} \varphi_\xi(-\delta_s)}, \quad t = 1, 2, \ldots, T^*. \tag{2.3.18}$$

Conversely, if δ is an adapted process taking values $\mathbb{P}$-a.s. in Γ, then ρ given by (2.3.18) *defines the density process of some measure that is equivalent to $\mathbb{P}$.*

Remark 2.3.7 It follows from Corollary 2.3.3 that (2.3.15) is satisfied if Z has the martingale representation property.

Theorem 2.3.8 [Girsanov's theorem – general version] *Let $\mathbb{Q} \sim \mathbb{P}$ be a measure with density process $\rho_0 = 1$, $\rho_t, t = 1, 2, \ldots, T^*$.*

(a) Then there exists an adapted process $\psi(s, y), s = 0, 1, \ldots, T^* - 1, y \in U$ such that

$$\int_U e^{\psi(s,y)} \mu(dy) < +\infty, \quad \mathbb{P} - a.s., \quad s = 0, 1, \ldots, T^* - 1 \tag{2.3.19}$$

and for $t = 1, 2, \ldots, T^*$,

$$\rho_t = e^{\sum_{s=0}^{t-1} \int_U \psi(s,y) \pi(\{s+1\},dy) - \sum_{s=0}^{t-1} \ln \int_U e^{\psi(s,y)} \mu(dy)}. \tag{2.3.20}$$

(b) Conversely, if ρ is of the form (2.3.20) *with a predictable process ψ satisfying* (2.3.19) *then ρ is the density process of some measure $\mathbb{Q}$ that is equivalent to $\mathbb{P}$.*

Proof of Theorem 2.3.6 By (2.3.15) the process

$$A_0 = 0, \quad A_t := \sum_{s=0}^{t-1} \mathbb{E}[\ln \rho_{s+1} - \ln \rho_s \mid \mathcal{F}_s], \quad t = 1, 2, \ldots, T^*$$

is well defined and A_t is $\mathcal{F}_{t-1}$-measurable for each $t = 1, 2, \ldots, T^*$. Since

$$\begin{aligned}
\mathbb{E}[\ln \rho_{t+1} - A_{t+1} \mid \mathcal{F}_t] &= \mathbb{E}[\ln \rho_{t+1} \mid \mathcal{F}_t] - \sum_{s=0}^{t} \mathbb{E}[\ln \rho_{s+1} - \ln \rho_s \mid \mathcal{F}_s] \\
&= \ln \rho_t - \sum_{s=0}^{t-1} \mathbb{E}[\ln \rho_{s+1} - \ln \rho_s \mid \mathcal{F}_s] \\
&= \ln \rho_t - A_t, \qquad t = 0, 1, \ldots, T^* - 1,
\end{aligned}$$

we see that $\ln \rho - A$ is a martingale.

(a) By Proposition 2.3.5 there exists an adapted process $\psi(s, y)$ satisfying the first condition in (2.3.16) such that

$$\ln \rho_t - A_t = \sum_{s=0}^{t-1} \int_U \psi(s, y)\tilde{\pi}(\{s+1\}, dy), \quad t = 1, 2, \ldots, T^*.$$

Hence

$$\rho_t = e^{\sum_{s=0}^{t-1} \int_U \psi(s,y)\tilde{\pi}(\{s+1\},dy) + A_t}, \quad t = 1, 2, \ldots, T^* \tag{2.3.21}$$

and the martingale property of ρ yields

$$\begin{aligned}
\rho_t &= \mathbb{E}[\rho_{t+1} \mid \mathcal{F}_t] \\
&= e^{\sum_{s=0}^{t-1} \int_U \psi(s,y)\tilde{\pi}(\{s+1\},dy) + A_t} \\
&\quad \cdot \mathbb{E}[e^{\psi(t,\xi_{t+1}) - \int_U \psi(t,y)\mu(dy)} \mid \mathcal{F}_t] \cdot e^{A_{t+1} - A_t} \\
&= \rho_t \mathbb{E}[e^{\psi(t,\xi_{t+1}) - \int_U \psi(t,y)\mu(dy)} \mid \mathcal{F}_t] \cdot e^{A_{t+1} - A_t}, \quad t = 0, 1, \ldots, T^* - 1.
\end{aligned}$$

It follows that

$$\begin{aligned}
e^{-(A_{t+1} - A_t)} &= \mathbb{E}[e^{\psi(t,\xi_{t+1}) - \int_U \psi(t,y)\mu(dy)} \mid \mathcal{F}_t] \\
&= e^{-\int_U \psi(t,y)\mu(dy)} \int_U e^{\psi(t,y)} \mu(dy), \quad t = 0, 1, \ldots, T^* - 1,
\end{aligned}$$

so the second condition in (2.3.16) is satisfied. Thus we can represent A in the form

$$A_t = \sum_{s=0}^{t-1}(A_{s+1} - A_s) = \sum_{s=0}^{t-1}\left\{\int_U \psi(s,y)\mu(dy) - \ln\left(\int_U e^{\psi(s,y)}\mu(dy)\right)\right\}, \quad t = 1,2,\dots,T^*.$$

Consequently, from (2.3.21) we obtain (2.3.17). If ρ is given by (2.3.17), then (2.3.16) ensures that it is a martingale, hence a density process.

(b) By the martingale representation property of Z we have

$$\ln \rho_t - A_t = \sum_{s=0}^{t-1}\langle \delta_s, \xi_{s+1}\rangle, \quad t = 1,2,\dots,T^*$$

for some adapted process δ. This yields

$$\rho_t = e^{\sum_{s=0}^{t-1}\langle \delta_s, \xi_{s+1}\rangle + A_t}, \quad t = 1,2,\dots,T^*.$$

It follows from the martingale property of ρ that

$$\begin{aligned}\rho_t = \mathbb{E}[\rho_{t+1} \mid \mathcal{F}_t] &= \mathbb{E}\left[e^{\sum_{s=0}^{t}\langle \delta_s, \xi_{s+1}\rangle + A_{t+1}} \mid \mathcal{F}_t\right] \\ &= e^{A_{t+1}} e^{\sum_{s=0}^{t-1}\langle \delta_s, \xi_{s+1}\rangle}\mathbb{E}\left[e^{\langle \delta_t, \xi_{t+1}\rangle} \mid \mathcal{F}_t\right] \\ &= \rho_t e^{A_{t+1} - A_t}\mathbb{E}\left[e^{\langle \delta_t, \xi_{t+1}\rangle} \mid \mathcal{F}_t\right], \quad t = 0,1,\dots,T^*-1.\end{aligned}$$

Thus

$$e^{-A_{t+1}+A_t} = \mathbb{E}\left[e^{\langle \delta_t, \xi_{t+1}\rangle} \mid \mathcal{F}_t\right] = e^{\varphi_\xi(-\delta_t)}, \quad t = 0,1,\dots,T^*-1,$$

which shows that δ takes values in Γ and allows determining the process A by summing its increments. This yields $A_t = -\sum_{s=0}^{t-1}\varphi_\xi(-\delta_s)$ and, consequently, (2.3.18).

To see the converse let us notice that ρ given by (2.3.18) is a strictly positive martingale, so the assertion follows. □

Proof of Theorem 2.3.8 If $\mathbb{Q} \sim \mathbb{P}$ then there exists a positive measurable function $h(x_1,\dots,x_{T^*})$, $x_i \in U$, $i = 1,2,\dots,T^*$ such that

$$\frac{d\mathbb{Q}}{d\mathbb{P}} = h(\xi_1,\dots,\xi_{T^*}), \quad \mathbb{P}-a.s.$$

Note that

$$\mathbb{E}\big[h(\xi_1,\dots,\xi_{T^*})\big] = \int_{U^{T^*}} h(x_1,\dots,x_{T^*})\mu(dx_1)\dots\mu(dx_{T^*}) = 1.$$

Let us define for $t = 1, 2, \ldots, T^*$, $(x_1, \ldots, x_t) \in U^t$

$$h_t(x_1, \ldots, x_t) := \int_{U^{T^*-t}} h(x_1, \ldots, x_t, y_1, \ldots, y_{T^*-t})\mu(dy_1)\ldots\mu(dy_{T^*-t}), \quad h_0 = 1.$$

Then

$$\int_U h_t(x_1, \ldots, x_{t-1}, y)\mu(dy) = h_{t-1}(x_1, \ldots, x_{t-1}), \quad t = 2, \ldots, T^*,$$

$$\int_U h_1(y)\mu(dy) = h_0 = 1.$$

Let ρ'_t be given by (2.3.20) with ψ given by the formula:

$$\psi(s, y) = \ln h_{s+1}(\xi_1, \ldots, \xi_s, y), \ \ s = 0, 1, \ldots, T^* - 1, \ \ y \in U. \tag{2.3.22}$$

We have to show that $\rho'_t = \rho_t, t = 1, \ldots, T^*$ where $\rho_t := \frac{d\mathbb{Q}}{d\mathbb{P}} |_{\mathcal{F}_t}$. Since

$$\begin{aligned}
\rho'_{T^*} &= e^{\sum_{s=0}^{T^*-1} \ln h_{s+1}(\xi_1,\ldots,\xi_s,\xi_{s+1}) - \sum_{s=0}^{T^*-1} \ln \int_U h_{s+1}(\xi_1,\ldots,\xi_s,y)\mu(dy)} \\
&= \frac{\prod_{s=0}^{T^*-1} h_{s+1}(\xi_1, \ldots, \xi_{s+1})}{\prod_{s=0}^{T^*-1} \int_U h_{s+1}(\xi_1, \ldots, \xi_s, y)\mu(dy)} \\
&= \frac{\prod_{s=0}^{T^*-1} h_{s+1}(\xi_1, \ldots, \xi_{s+1})}{\prod_{s=0}^{T^*-1} h_s(\xi_1, \ldots, \xi_s)} \\
&= h_{T^*}(\xi_1, \ldots, \xi_{T^*}) = \rho_{T^*}.
\end{aligned}$$

It is easy to check that $\rho'_t, t = 1, 2, \ldots, T^*$ is an $\mathcal{F}_t$-martingale. Since ρ_t, $t = 1, 2, \ldots, T^*$ is an $\mathcal{F}_t$-martingale we get that $\rho'_t = \rho_t, t = 1, 2, \ldots, T^*$ as required. By the latter argument one can prove also (b). □

2.3.4 Application to HJM Models

Now we present results characterizing martingale measures in the HJM model

$$f(t+1, T) = f(t, T) + \alpha(t, T) + \langle\sigma(t, T), \xi_{t+1}\rangle, \quad t, T = 0, 1, \ldots. \tag{2.3.23}$$

This problem was already studied in Section 2.2, but now we do this in an alternative manner involving martingale representation properties of the process

$$Z_0 = 0, \quad Z_t = \xi_1 + \xi_2 + \cdots + \xi_t, \quad t = 1, 2, \ldots, T^*. \tag{2.3.24}$$

This kind of describing martingale measures is commonly used in continuous time HJM models driven by a Lévy process and therefore discussed here in detail. In particular, we derive the drift conditions that correspond to the HJM conditions in continuous time.

Recall that in Section 2.3.1 we showed that Z has the martingale representation property only under very strong distributional restrictions. Therefore the following

Theorem 2.3.9 can be viewed as an alternative characterization of martingale measures to that given by Theorem 2.2.1. This particular situation can be compared to the HJM model in continuous time based on the Wiener process.

Recall that the set Γ is given by

$$\Gamma := \{\lambda \in U : \varphi_\xi(\lambda) < +\infty\},$$

where φ_ξ stands for the Laplace exponent of ξ_1 (see (2.3.14)).

Theorem 2.3.9 *Assume that the martingale Z given by* (2.3.24) *has the martingale representation property. Then the model* (2.3.23) *satisfies (MM) if and only if there exists an adapted process δ that $\mathbb{P}$-a.s. takes values in the set Γ and such that for each $T = 0, 1, \ldots$,*

$$\alpha(t,T) = \varphi_\xi\left(\sum_{s=0}^{T}\sigma(t,s) - \delta_t\right) - \varphi_\xi\left(\sum_{s=0}^{T-1}\sigma(t,s) - \delta_t\right), \quad t = 0, 1, 2, \ldots, T-1. \tag{2.3.25}$$

In particular, $\hat{P}(t,T)$, $T = 0, 1, \ldots$ are martingales under $\mathbb{P}$ if and only if

$$\alpha(t,T) = \varphi_\xi\left(\sum_{s=0}^{T}\sigma(t,s)\right) - \varphi_\xi\left(\sum_{s=0}^{T-1}\sigma(t,s)\right), \quad t = 0, 1, 2, \ldots, T-1. \tag{2.3.26}$$

If Z does not necessarily have the martingale representation property, then we can characterize martingale measures using its generalized martingale representation property. This leads to the following Theorem 2.3.10, which is, of course, equivalent to Theorem 2.2.1.

Theorem 2.3.10 *The model* (2.3.23) *satisfies (MM) if and only if there exists an adapted process $\psi(t,y)$ satisfying*

$$\int_U e^{\psi(s,y)}\mu(dy) < +\infty, \quad s = 0, 1, \ldots, T^* - 1,$$

such that for each $T = 0, 1, \ldots$

$$\alpha(t,T) = \ln\left(\frac{\int_U e^{\psi(t,y) - \langle\sum_{s=0}^{T}\sigma(t,s), y\rangle}\mu(dy)}{\int_U e^{\psi(t,y) - \langle\sum_{s=0}^{T-1}\sigma(t,s), y\rangle}\mu(dy)}\right), \quad t = 0, 1, 2, \ldots, T-1. \tag{2.3.27}$$

In particular, $\hat{P}(t,T)$, $T = 0, 1, \ldots$ are martingales under $\mathbb{P}$ if and only if

$$\alpha(t,T) = \varphi_\xi\left(\sum_{s=0}^{T}\sigma(t,s)\right) - \varphi_\xi\left(\sum_{s=0}^{T-1}\sigma(t,s)\right), \quad t = 0, 1, 2, \ldots, T-1. \tag{2.3.28}$$

Remark 2.3.11 The formulae (2.3.26) and (2.3.28) are the same as (2.2.7) in Theorem 2.2.1.

Example 2.3.12 We show that for a given volatility processes $\sigma(\cdot,\cdot)$ there are models such that the discounted bond prices are martingales under some equivalent measure $\mathbb{Q}$. Let g_t be a function such that $g_t(\sigma(t,\cdot)) := g_t(\sigma(t,t), \sigma(t,t+1), \ldots, \sigma(t,T^*))$ satisfies

$$\mathbb{E}\left[e^{-\langle g_t(\sigma(t,\cdot)),\xi_{t+1}\rangle}\right] < +\infty.$$

If the drift is given by

$$\begin{aligned}\alpha(t,T) &= \varphi_\xi\left(\sum_{s=0}^{T}\sigma(t,s) - g_t(\sigma(t,\cdot))\right)\\ &\quad - \varphi_\xi\left(\sum_{s=0}^{T-1}\sigma(t,s) - g_t(\sigma(t,\cdot))\right), \qquad t = 0,1,2,\ldots,T-1\end{aligned}$$

for $T = 0,1,2,\ldots,T^*$, then the resulting model admits a martingale measure. This follows from Theorem 2.3.9 because $\delta_t = g_t(\sigma(t,\cdot))$ satisfies the required assumptions.

Proof of Theorem 2.3.9 By Remark 2.3.7 and Theorem 2.3.6 (*b*) any measure $\mathbb{Q} \sim \mathbb{P}$ can be identified with an adapted process δ taking values in Γ. Since the density process has the form (2.3.18), for any $T = 0,1,2,\ldots,T^*$ and $t = 0,1,\ldots,T-1$, we have

$$\begin{aligned}\mathbb{E}[\rho_{t+1}\hat{P}(t+1,T) \mid \mathcal{F}_t] &= \mathbb{E}[e^{\sum_{s=0}^{t}\langle\delta_s,\xi_{s+1}\rangle - \sum_{s=0}^{t}\varphi_\xi(-\delta_s)}e^{-\sum_{s=0}^{T-1}f(t+1,s)} \mid \mathcal{F}_t]\\ &= \mathbb{E}[\rho_t e^{\langle\delta_t,\xi_{t+1}\rangle - \varphi_\xi(-\delta_t)}e^{-\sum_{s=0}^{T-1}f(t,s)+\alpha(t,s)+\langle\sigma(t,s),\xi_{t+1}\rangle} \mid \mathcal{F}_t]\\ &= \rho_t\hat{P}(t,T)e^{-\varphi_\xi(-\delta_t)-\sum_{s=0}^{T-1}\alpha(t,s)}\cdot\mathbb{E}[e^{\langle\delta_t,\xi_{t+1}\rangle - \sum_{s=0}^{T-1}\langle\sigma(t,s),\xi_{t+1}\rangle} \mid \mathcal{F}_t].\end{aligned}$$

Taking into account that

$$\mathbb{E}[e^{\langle\delta_t,\xi_{t+1}\rangle - \sum_{s=0}^{T-1}\langle\sigma(t,s),\xi_{t+1}\rangle} \mid \mathcal{F}_t] = \mathbb{E}[e^{-\langle\sum_{s=0}^{T-1}\sigma(t,s)-\delta_t,\xi_{t+1}\rangle} \mid \mathcal{F}_t] = e^{\varphi_\xi(\sum_{s=0}^{T-1}\sigma(t,s)-\delta_t)},$$

we obtain that

$$\rho_t\hat{P}(t,T)e^{-\sum_{s=0}^{T-1}\alpha(t,s)}\cdot e^{-\varphi_\xi(-\delta_t)}\cdot e^{\varphi_\xi(\sum_{s=0}^{T-1}\sigma(t,s)-\delta_t)} = \rho_t\hat{P}(t,T)$$

if and only if

$$\sum_{s=0}^{T-1}\alpha(t,s) + \varphi_\xi(-\delta_t) = \varphi_\xi\left(\sum_{s=0}^{T-1}\sigma(t,s) - \delta_t\right).$$

From that condition we arrive at (2.3.25).

Condition (2.3.26) we obtain by setting $\delta_t \equiv 0$ in (2.3.25). □

Proof of Theorem 2.3.10 In view of the general Girsanov theorem (see Theorem 2.3.8) we can write the density of an equivalent to $\mathbb{P}$ measure $\mathbb{Q}$ in the form

$$\rho_t = e^{\sum_{s=0}^{t-1}\int_U \psi(s,y)\pi(\{s+1\},dy) - \sum_{s=0}^{t-1}\ln\int_U e^{\psi(s,y)}\mu(dy)}, \quad t = 1, 2, \ldots, T^*.$$

Then

$$\mathbb{E}[\rho_{t+1}\hat{P}(t+1,T) \mid \mathcal{F}_t] = \mathbb{E}[e^{\sum_{s=0}^{t}\int_U \psi(s,y)\pi(\{s+1\},dy) - \sum_{s=0}^{t}\ln\int_U e^{\psi(s,y)}\mu(dy)} e^{-\sum_{s=0}^{T-1} f(t+1,s)} \mid \mathcal{F}_t]$$

$$= \mathbb{E}[\rho_t e^{\int_U \psi(t,y)\pi(\{t+1\},dy) - \ln\int_U e^{\psi(t,y)}\mu(dy)} e^{-\sum_{s=0}^{T-1} f(t,s)+\alpha(t,s)+\langle\sigma(t,s),\xi_{t+1}\rangle} \mid \mathcal{F}_t]$$

$$= \rho_t \hat{P}(t,T) e^{-\ln\int_U e^{\psi(t,y)}\mu(dy)} e^{-\sum_{s=0}^{T-1}\alpha(t,s)} \cdot \mathbb{E}[e^{\int_U \psi(t,y)\pi(\{t+1\},dy)} e^{-\sum_{s=0}^{T-1}\langle\sigma(t,s),\xi_{t+1}\rangle} \mid \mathcal{F}_t].$$

Since

$$\begin{aligned}
&\mathbb{E}[e^{\int_U \psi(t,y)\pi(\{t+1\},dy)} e^{-\sum_{s=0}^{T-1}\langle\sigma(t,s),\xi_{t+1}\rangle} \mid \mathcal{F}_t] \\
&= \mathbb{E}\left[\int_U e^{\psi(t,y)-\sum_{s=0}^{T-1}\langle\sigma(t,s),y\rangle}\pi(\{t+1\},dy) \mid \mathcal{F}_t\right] \\
&= \int_U e^{\psi(t,y)-\sum_{s=0}^{T-1}\langle\sigma(t,s),y\rangle}\mu(dy),
\end{aligned}$$

we see that $\mathbb{Q}$ is a martingale measure if and only if

$$\frac{1}{\int_U e^{\psi(t,y)}\mu(dy)} e^{-\sum_{s=0}^{T-1}\alpha(t,s)} \int_U e^{\psi(t,y)-\sum_{s=0}^{T-1}\langle\sigma(t,s),y\rangle}\mu(dy) = 1.$$

It follows that

$$e^{\sum_{s=0}^{T-1}\alpha(t,s)} = \frac{\int_U e^{\psi(t,y)-\sum_{s=0}^{T-1}\langle\sigma(t,s),y\rangle}\mu(dy)}{\int_U e^{\psi(t,y)}\mu(dy)}$$

and one easily comes to (2.3.27).

Condition (2.3.28) follows from (2.3.27) by setting $\psi \equiv 0$. □

Remark 2.3.13 One can also prove Theorem 2.3.9 with the use of Theorem 2.2.1. From the form of the density (2.3.18) we deduce that

$$\psi_{t,t+1}(\xi_1, \ldots, \xi_t, x) = \mathbb{E}\left[e^{-\langle x,\xi_{t+1}\rangle}\frac{\rho_{t+1}}{\rho_t} \mid \mathcal{F}_t\right] = e^{\varphi_\xi(x-\delta_t)-\varphi_\xi(-\delta_t)}.$$

Then (2.2.6) yields (2.3.25).

2.4 Markovian Models under the Martingale Measure

In this section we formulate conditions for the discounted bond prices to be martingales under the original measure. They are based on Markovian properties of the forward rate models introduced in Section 1.2. We follow here, with some modification, the paper of Filipović and Zabczyk [59] and its first version [58].

2.4.1 Models with Markovian Trace

Writing forward rates $r(t) = (r(t,0), r(t,1),\dots)$, $t = 0,1,2,\dots$ in the Musiela parametrization leads to the following form of bond prices

$$P(t,T) = e^{-\sum_{s=0}^{T-t-1} r(t,s)}, \; P(T,T) = 1, \quad t = 0,1,\dots,T-1. \tag{2.4.1}$$

Since $R(t) = r(t,0)$ determines the short rate, the evolution of a savings account is given by

$$B(0) = 1, \quad B(t) = e^{\sum_{s=0}^{t-1} R(s)} = e^{\sum_{s=0}^{t-1} r(s,0)}, \quad t = 1,2,\dots. \tag{2.4.2}$$

Here we will assume that, for given $\gamma \in \mathbb{N}_0$, the γ-trace

$$r^{\gamma}(t) = (r(t,0),\dots,r(t,\gamma)), \quad t = 0,1,\dots$$

of the forward curve is, under the original measure $\mathbb{P}$, a Markov chain on the state space $\mathbb{R}_+^{\gamma+1}$ with some transition operator $\mathcal{P}(\cdot,\cdot)$. Our goal is to characterize models satisfying (MP) in terms of $\mathcal{P}$. It turns out that if $\gamma = 0$ then arbitrary $\mathbb{R}_+^1$-valued Markov chain generates some bond market model satisfying (MP). This case corresponds to the Markovian short rate. In the case when the trace is multidimensional, i.e. $\gamma > 0$, the transition semigroup must satisfy some additional conditions.

To formulate solution of the problem let us set

$$\varphi(x) := e^{-x_0}, \quad x = (x_0, x_1, \dots, x_\gamma),$$

and define inductively functions $\varphi_0, \varphi_1, \dots$ as follows

$$\varphi_0(x) := 1, \qquad \varphi_{k+1}(x) := \int_{\mathbb{R}_+^{\gamma+1}} e^{-y_0}\varphi_k(y_0,\dots,y_\gamma)\mathcal{P}(x,dy), \quad x \in \mathbb{R}_+^{\gamma+1}, \tag{2.4.3}$$

where $y := (y_0, y_1, \dots, y_\gamma)$. In other words,

$$\varphi_{k+1}(x) = \mathcal{P}[\varphi \cdot \varphi_k](x), \quad k = 1,2,\dots, \quad x \in \mathbb{R}_+^{\gamma+1}.$$

Theorem 2.4.1 *Assume that the trace r^γ is a Markov chain in $\mathbb{R}_+^{\gamma+1}$ with the transition operator $\mathcal{P}$.*

(a) If (MP) holds then the forward rate has the representation

$$r(t,k+1) = \ln\left(\frac{\varphi_k}{\varphi_{k+1}}\right)(r^\gamma(t)), \quad t,k \in \mathbb{N}_0, \tag{2.4.4}$$

where $\{\varphi_k\}, k = 0,1,2,\dots$ are given by (2.4.3). Moreover, if $\gamma \geq 1$ and (MP) holds for r^γ starting from an arbitrary state in $\mathbb{R}_+^{\gamma+1}$ then

$$\varphi_k(x_0,x_1,\dots,x_\gamma) = e^{-(x_1+\dots+x_k)}, \quad k = 1,\dots,\gamma, \quad x = (x_0,x_1,\dots,x_\gamma) \in \mathbb{R}_+^{\gamma+1}, \tag{2.4.5}$$

and thus the transition function $\mathcal{P}(\cdot,\cdot)$ satisfies the conditions:

$$\int_{\mathbb{R}_+^{\gamma+1}} e^{-(y_0+\ldots+y_k)}\mathcal{P}(x,dy) = e^{-(x_1+\ldots+x_{k+1})}, \quad k=0,\ldots,\gamma-1,\ x\in\mathbb{R}_+^{\gamma+1}. \tag{2.4.6}$$

(b) *If the forward rate is given by* (2.4.4) *and for* $\gamma \geq 1$ *the transition function satisfies* (2.4.6) *then the bond market with prices*

$$P(T,T)=1, \quad P(t,T)=e^{-\sum_{s=0}^{T-t-1} r(t,s)}, \quad t=0,1,\ldots,T-1$$

satisfies (MP).

Proof (a) From (MP) we have

$$\frac{P(t,T)}{B(t)} = \mathbb{E}\left(\frac{1}{B(T)} \mid \mathcal{F}_t\right), \quad t\leq T,$$

which, in view of (2.4.2), yields

$$P(t,T) = \mathbb{E}\left(e^{-\sum_{s=t}^{T-1} R(s)} \mid \mathcal{F}_t\right), \quad t\leq T-1.$$

Consequently, for $t\leq T-2$,

$$P(t,T) = \mathbb{E}\left(e^{-\sum_{s=t}^{T-2} R(s)}\mathbb{E}(e^{-R(T-1)} \mid \mathcal{F}_{T-2}) \mid \mathcal{F}_t\right)$$

and, by the Markov property,

$$P(t,T) = \mathbb{E}\left(e^{-\sum_{s=t}^{T-2} R(s)}\mathcal{P}\varphi(r^\gamma(T-2)) \mid \mathcal{F}_t\right).$$

Similarly, taking into account that $\mathcal{P}\varphi = \mathcal{P}(\varphi\varphi_0) = \varphi_1$, we obtain for $t\leq T-3$,

$$\begin{aligned} P(t,T) &= \mathbb{E}\left(e^{-\sum_{s=t}^{T-3} R(s)}\mathbb{E}(e^{-R(T-2)}\varphi_1(r^\gamma(T-2)) \mid \mathcal{F}_{T-3}) \mid \mathcal{F}_t\right) \\ &= \mathbb{E}\left(e^{-\sum_{s=t}^{T-3} R(s)}\varphi_2(r^\gamma(T-3)) \mid \mathcal{F}_t\right). \end{aligned}$$

Finally, by induction, we obtain

$$P(t,T) = e^{-R(t)}\varphi_{T-t-1}(r^\gamma(t)), \quad t\leq T-1.$$

In view of (2.4.1), for $t,k=0,1,\ldots,$

$$P(t,t+k+1) = e^{-\sum_{j=0}^{k} r(t,j)} = e^{-R(t)}\varphi_k(r^\gamma(t)).$$

Since $R(t)=r(t,0)$, so finally

$$\sum_{j=1}^{k} r(t,j) = -\ln\varphi_k(r^\gamma(t)),\ k\geq 1, \tag{2.4.7}$$

which yields (2.4.4). Now we show (2.4.5). It follows from (2.4.7) that

$$\varphi_k(r(t,0),\ldots,r(t,\gamma)) = e^{-\sum_{j=1}^{k} r(t,j)}, \quad k=0,1,\ldots,\gamma-1.$$

Taking the initial trace $r^\gamma(0) = (x_0,x_1,\ldots,x_\gamma) \in \mathbb{R}_+^{\gamma+1}$ we get the required representation (2.4.5). The restrictions on the transition function follows now from the definition of φ_k and (2.4.5).

(b) To prove the converse we show that for a fixed $T \in \mathbb{N}_0$ the process $P(t,T)/B(t),\ t \le T$ is a martingale. We can assume that $T \ge 1$. Then

$$\begin{aligned}\frac{P(t,T)}{B(t)} &= e^{-[r(t,0)+\cdots+r(t,T-t-1)]}e^{-\sum_{s=0}^{t-1} r(s,0)}\\ &= e^{-\left[r(t,0)+\left(\ln\frac{\varphi_0}{\varphi_1}+\cdots+\ln\frac{\varphi_{T-t-2}}{\varphi_{T-t-1}}\right)(r^\gamma(t))\right]-\sum_{s=0}^{t-1} r(s,0)},\end{aligned}$$

and consequently

$$\frac{P(t,T)}{B(t)} = e^{-\sum_{u=0}^{t} r(u,0)-\left(\ln\frac{\varphi_0}{\varphi_{T-t-1}}\right)(r^\gamma(t))} = e^{-\sum_{u=0}^{t} r(u,0)}\varphi_{T-t-1}(r^\gamma(t)). \tag{2.4.8}$$

Since

$$\begin{aligned}&\mathbb{E}\left(e^{-\sum_{u=0}^{t} r(u,0)}\varphi_{T-t-1}(r^\gamma(t)) \mid \mathcal{F}_{t-1}\right)\\ &\quad = e^{-\sum_{u=0}^{t-1} r(u,0)} \int_{\mathbb{R}_+^{\gamma+1}} e^{-y_0}\varphi_{T-t-1}(y)\mathcal{P}(r^\gamma(t-1),dy)\end{aligned}$$

and

$$\int_{\mathbb{R}_+^{\gamma+1}} e^{-y_0}\varphi_{T-t-1}(y)\mathcal{P}(x,dy) = \varphi_{T-t}(x),$$

by (2.4.8) we obtain

$$\mathbb{E}\left(\frac{P(t,T)}{B(t)} \mid \mathcal{F}_{t-1}\right) = e^{-\sum_{u=0}^{t-1} r(u,0)}\varphi_{T-t}(r^\gamma(t-1)) = \frac{P(t-1,T)}{B(t-1)}. \qquad \square$$

Remark 2.4.2 Let us stress that if $\gamma = 0$, i.e. the short-rate process is Markovian, then (2.4.4) determines forward rates for which the corresponding bond market satisfies (MP). There are no requirements for the transition operator, and the Markov chain can be chosen freely in this case. If $\gamma > 0$, then the transition function $\mathcal{P}$ must satisfy additional conditions (2.4.6) implied by the fact that $\varphi_k, k = 1,2,\ldots,\gamma$ are of the form (2.4.5) and, on the other hand, are defined inductively by (2.4.3).

2.4.2 Affine Models

Let $r^\gamma(t), t = 0, 1, \ldots$ with $\gamma \geq 0$ be an $\mathbb{R}_+^{\gamma+1}$-valued Markov chain with transition operator $\mathcal{P}$ describing trace of the forward rate $r(t) = r(t,k);\ t, k = 0, 1, 2, \ldots$. If there exist deterministic functions $C(t)$, $D(t)$, $t = 0, 1, \ldots$ with values in $\mathbb{R}_+$, and $\mathbb{R}_+^{\gamma+1}$ respectively, such that

$$P(t,T) = e^{-C(T-t)-\langle D(T-t), r^\gamma(t)\rangle}, \qquad t \leq T < +\infty, \tag{2.4.9}$$

then the bond market with such prices is called *affine*. The positivity of $C(t)$ and $D(t)$ ensures that the model is regular.

We find conditions characterizing (MP) in terms of the transition operator $\mathcal{P}$ and exponential functions of the form

$$f_\lambda(x) := e^{-\langle \lambda, x\rangle} \quad \text{for} \quad \lambda, x \in \mathbb{R}_+^{1+\gamma}.$$

Proposition 2.4.3 *A Markovian model of the trace process $r^\gamma(t), t = 0, 1, \ldots$ governed by the transition function $\mathcal{P}$ together with functions C, D form an affine model* (2.4.9) *satisfying (MP) if and only if*

$$\mathcal{P}f_{D(T)}(x) = e^{-(C(T+1)-C(T))-\langle D(T+1)-e_0, x\rangle}, \ T = 0, 1, \ldots, \ x \in \mathbb{R}_+^{\gamma+1}, \tag{2.4.10}$$

where $e_0 := (1, 0, \ldots, 0) \in \mathbb{R}^{1+\gamma}$.

Proof Note that $\hat{P}(0, T+1) = \mathbb{E}(\hat{P}(1, T+1))$ is satisfied if and only if

$$\begin{aligned}
P(0, T+1) &= e^{-C(T+1)-\langle D(T+1), r^\gamma(0)\rangle} \\
&= \mathbb{E}\left(\frac{P(1, T+1)}{e^{r(0,0)}} \,\Big|\, \mathcal{F}_0\right) \\
&= \mathbb{E}\left(e^{-r(0,0)} e^{-C(T)-\langle D(T), r^\gamma(1)\rangle} | \mathcal{F}_0\right) \\
&= e^{-r(0,0)-C(T)} \cdot \mathcal{P}f_{D(T)}(r^\gamma(0)).
\end{aligned}$$

Consequently, for all $x = (x_0, x_1, \ldots, x_\gamma) \in \mathbb{R}_+^{1+\gamma}$ we have

$$\begin{aligned}
\mathcal{P}f_{D(T)}(x) &= e^{-C(T+1)-\langle D(T+1), x\rangle} e^{x_0 + C(T)} \\
&= e^{-C(T+1)-\langle D(T+1)-e_0, x\rangle + C(T)} \\
&= e^{-(C(T+1)-C(T))-\langle D(T+1)-e_0, x\rangle}.
\end{aligned} \tag{2.4.11}$$

The Markovian property of r^γ ensures that, for any t, the condition $\hat{P}(t, T+1) = \mathbb{E}(\hat{P}(t+1, T+1) \mid \mathcal{F}_t)$ is also equivalent to (2.4.11). □

It is of interest to give a complete characterization of all affine models satisfying (MP). Taking into account Proposition 2.4.3 we pose the following open problem.

Problem: Describe all transition functions $\mathcal{P}(x,\cdot)$, $x \in \mathbb{R}_+^{\gamma+1}$, on $\mathbb{R}_+^{\gamma+1}$, for which there exist a nondecreasing sequence $C(T)$, $T = 0,1,\dots, C(0) = 0$, and a sequence of vectors $D(T) \in \mathbb{R}_+^{\gamma+1}$, $T = 0,1,\dots, D(0) = 0$, such that

$$\int_{\mathbb{R}_+^{\gamma+1}} e^{-\langle D(T),y\rangle}\mathcal{P}(x,dy) = e^{-(C(T+1)-C(T))-\langle D(T+1)-e_0,x\rangle},\ x \in \mathbb{R}_+^{\gamma+1},\ T = 0,1,\dots. \tag{2.4.12}$$

Now we introduce some subclass of transition functions solving the problem. Note that the identity (2.4.12) implies that necessarily the transition function $\mathcal{P}$ transforms some exponential functions f_λ onto multiple of exponential functions, that is,

$$\mathcal{P}f_\lambda(x) = e^{-\langle\psi(\lambda),x\rangle-\varphi(\lambda)}, \tag{2.4.13}$$

where ψ and φ are some functions of λ. We say that a family $\{\mu_x\}, x \in \mathbb{R}_+^{1+\gamma}$ of probability measures on $\mathbb{R}_+^{1+\gamma}$ such that

$$\mu_0 = \delta_0, \quad \mu_x \underset{x\to 0}{\overset{weakly}{\longrightarrow}} \delta_0, \quad \mu_x * \mu_y = \mu_{x+y}$$

is called *infinitely divisible* family or *convolution semigroup* of measures. Let $e_0,\dots,e_\gamma$ be the standard basis in $\mathbb{R}_+^{1+\gamma}$. Since

$$\mu_{(x_0,x_1,\dots,x_\gamma)} = \mu_{x_0e_0} * \mu_{x_1e_1} * \dots * \mu_{x_\gamma e_\gamma},$$

and for each $k = 0,\dots,\gamma$ the family $\mu_{ye_k}, y \geq 0$, is infinite divisible on $\mathbb{R}_+^{1+\gamma}$. The Laplace transform of μ_x defined by

$$\int_{\mathbb{R}_+^{1+\gamma}} f_\lambda(y)\mu_x(dy) = e^{-\langle\psi(\lambda),x\rangle}$$

is characterized by the formula

$$\psi_k(\lambda) = \langle\beta_k,\lambda\rangle + \int_{\mathbb{R}_+^{1+\gamma}} (1 - e^{-\langle\lambda,y\rangle})m_k(dy), \tag{2.4.14}$$

where $\psi_k, k = 0,1,\dots,\gamma$ stand for the components of ψ. Above $\beta_k \in \mathbb{R}_+^{1+\gamma}$ and m_k are nonnegative measures on $\mathbb{R}_+^{1+\gamma}$, without atoms at 0 and such that

$$\int_{\mathbb{R}_+^{1+\gamma}} (1 \wedge |y|)m_k(dy) < +\infty,\ k = 0,\dots,\gamma, \tag{2.4.15}$$

see, e.g. Kallenberg [79, p. 291], for the case $\gamma = 0$, which can be extended easily to arbitrary γ.

In what follows we restrict our considerations to transition functions of the form

$$\mathcal{P}(x,dy) = (\mu_x * \nu)(dy), \tag{2.4.16}$$

where $\{\mu_x\}$ is an infinitely divisible family and ν is a probability measure on $\mathbb{R}_+^{1+\gamma}$ with the Laplace transform

$$\int_{\mathbb{R}_+^{1+\gamma}} e^{-\langle \lambda, y\rangle} \nu(dy) = e^{-\varphi(\lambda)}, \ \lambda \in \mathbb{R}_+^{1+\gamma}. \tag{2.4.17}$$

It is clear that in this case (2.4.13) holds for each $\lambda \in \mathbb{R}_+^{1+\gamma}$. The infinite divisible families of measures will reappear in the part of the book devoted to Lévy processes, see Chapter 5 and Section 5.3.2.

Remark 2.4.4 The family of Markov chains for which (2.4.13) holds is larger than that for which (2.4.16) is satisfied. In the example constructed by Hubalek [68] the condition (2.4.13) is satisfied but $\mu_x(A) < 0$, for some x and a set A, so μ_x is a signed measure. Let us remark, however, that the left side of (2.4.13) is analytic in λ, so it is determined uniquely by its values on some convergent sequence. So, the requirement that (2.4.13) holds for each $\lambda \in \mathbb{R}_+^{1+\gamma}$ is not very restrictive.

Taking into account Proposition 2.4.3 we arrive at the following solution of the preceding problem (see (2.4.12)) for the transition function (2.4.16).

Theorem 2.4.5 *(a) The transition function given by* (2.4.16) *satisfies the constraints* (2.4.6) *if and only if*

$$\varphi(e_0 + \cdots + e_{k-1}) = 0, \ \ \psi(e_0 + \cdots + e_{k-1}) = e_0 + \cdots + e_{k-1}, \ k = 0, \ldots, \gamma, \tag{2.4.18}$$

where $e_0, \ldots e_\gamma$ *is the standard basis in* $\mathbb{R}^{1+\gamma}$.

(b) Assume that the transition function of the Markov chain r^γ *is of the form* (2.4.16) *and condition* (2.4.18) *holds. Then (MP) is satisfied if and only if the functions* C *and* D *are given by*

$$C(T) - C(T-1) = \varphi(D(T-1)), \qquad T \geq 1, \tag{2.4.19}$$

$$D(T) = \psi(D(T-1)) + e_0, \quad T \geq 1. \tag{2.4.20}$$

(c) If $\gamma = 0$*, that is,* $R(t) = r^\gamma(t)$*, and the transition function of the process* $R(t)$ *has the form* (2.4.16) *and the functions* $C(T)$*,* $D(T)$ *are given by* (2.4.19)*,* (2.4.20)*, then (MP) is satisfied.*

Proof Once we prove (a) the part (b) follows directly from (2.4.10). In the present situation the constraint conditions are equivalent to the identity

$$\int_{\mathbb{R}_+^{1+\gamma}} \mathcal{P}(x, dy) e^{-\langle e_0 + \ldots + e_{k-1}, y\rangle} = e^{-\langle e_1 + \ldots + e_k, x\rangle}, \tag{2.4.21}$$

valid for $k = 1, \ldots, \gamma$. It follows from the definition of the convolution of measures that the left side of (2.4.21) is equal to

$$\int_{\mathbb{R}_+^{1+\gamma}} \int_{\mathbb{R}_+^{1+\gamma}} e^{-\langle e_0+\ldots+e_{k-1}, y+z\rangle} \mu_x(dy)\nu(dz),$$

and therefore to the product

$$\int_{\mathbb{R}_+^{1+\gamma}} e^{-\langle e_0+\ldots+e_{k-1}, y\rangle} \mu_x(dy) \int_{\mathbb{R}_+^{1+\gamma}} e^{-\langle e_0+\ldots+e_{k-1}, z\rangle} \nu(dz).$$

Now the definitions of the functions ψ and φ easily lead to the required identities. (c) follows from (a) and (b). □

Example 2.4.6 Assume that $\gamma = 1$. Then the constraints on $\mathcal{P}$ are of the form:

$$\varphi(\epsilon_0) = 0, \ \psi(e_0) = (\psi_0(e_0), \psi_1(e_0)) = (0, 1).$$

Thus the measure ν should be supported by the set $\{(y_0, y_1); y_1 = 0\}$, $\beta_0^0 = 0$, $m_0 = 0$ and

$$\beta_1^0 + \int_{\mathbb{R}_+^2} (1 - e^{-y_0}) m_1(dy_0, dy_1) = 1.$$

For more detailed analysis of the constraints for general γ we refer to Filipović and Zabczyk [59]. □

It is instructive to derive (2.4.19) and (2.4.20) using Theorem 2.4.1 from the previous section. If (2.4.16) holds then the functions φ_k, $k = 0, 1, \ldots$ from (2.4.3) are of the following form

$$\varphi_0 = 1, \quad \varphi_k(x) = e^{-C_k - \langle D_k, x\rangle}, \quad k = 1, 2, \ldots,$$

where (C_k), (D_k) satisfy the recursive relations

$$C_0 = 0, \quad C_{k+1} = C_k + \varphi(D_k + e_0), \tag{2.4.22}$$

$$D_0 = 0, \quad D_{k+1} = \psi(D_k + e_0). \tag{2.4.23}$$

In fact, by (2.4.3), we have

$$\begin{aligned}
\varphi_{k+1}(x) &= \int_{\mathbb{R}_+^{1+\gamma}} e^{-y_0} \varphi_k(y) \mathcal{P}(x, dy) = \int_{\mathbb{R}_+^{1+\gamma}} e^{-y_0} e^{-C_k - \langle D_k, y\rangle} (\mu_x * \nu)(dy) \\
&= e^{-C_k} \int_{\mathbb{R}_+^{1+\gamma}} \int_{\mathbb{R}_+^{1+\gamma}} e^{-\langle D_k + e_0, y+z\rangle} \mu_x(dy)\, \nu(dz) \\
&= e^{-C_k} \int_{\mathbb{R}_+^{1+\gamma}} e^{-\langle D_k + e_0, y\rangle} \mu_x(dy) \cdot \int_{\mathbb{R}_+^{1+\gamma}} e^{-\langle D_k + e_0, z\rangle} \nu(dz) \\
&= e^{-C_k} e^{-\langle \psi(D_k + e_0), x\rangle} e^{-\varphi(D_k + e_0)},
\end{aligned}$$

which leads to (2.4.22) and (2.4.23). Thus, by Theorem 2.4.1, formula (2.4.4)

$$
\begin{aligned}
r(t,k+1) &= \ln e^{-C_k - \langle D_k, r^\gamma(t)\rangle + C_{k+1} + \langle D_{k+1}, r^\gamma(t)\rangle} \\
&= C_{k+1} - C_k + \langle D_{k+1} - D_k, r^\gamma(t)\rangle, \quad t,k \geq 0. \qquad (2.4.24)
\end{aligned}
$$

Taking into account that (C_k), (D_k) satisfy (2.4.22), (2.4.23), we define

$$
\begin{aligned}
&C(0) := 0, \quad D(0) := 0, \\
&C(k) := C_{k-1}, \quad D(k) := D_{k-1} + e_0, \quad k = 1,2,\ldots.
\end{aligned}
$$

Then $(C(k))$, $(D(k))$ satisfy (2.4.19) and (2.4.20). In fact this is true for $k = 0$. Assume that the result is true for $k = 0, 1, \ldots, n$. Then

$$
\begin{aligned}
C(n+1) &= C_n = C_{n-1} + \varphi(D_{n-1} + e_0) \\
&= C(n) + \varphi(D(n))
\end{aligned}
$$

and

$$
D(n+1) = D_n + e_0 = \psi(D_{n-1} + e_0) + e_0 = \psi(D(n)) + e_0.
$$

2.4.3 Dynamics of the Short Rate in Affine Models

It is interesting to find equations of the form

$$
R(t+1) = F(R(t)) + G(R(t))\xi_{t+1}, \quad t = 0, 1, \ldots, \qquad (2.4.25)
$$

with a sequence of independent identically distributed random variables $\xi_1, \xi_2, \ldots$ such that the transition function admits the representation

$$
\mathcal{P}(x, dy) = (\mu_x * \nu)(dy), \quad x > 0, \qquad (2.4.26)
$$

where (μ_x) is an infinitely divisible family on $[0, +\infty)$ and ν is a probability measure on $[0, +\infty)$. Recall that in view of Theorem 2.4.5 (c), short rates with the transition function (2.4.26) generate affine models satisfying (MP). The answer is given by the following theorem, which provides a counterpart to CIR equations in the continuous time setting (see Section 10.2.2).

We say that a nonnegative random variable ξ has the standard α-stable distribution with $\alpha \in (0, 1)$ if

$$
\mathbb{E}(e^{-\lambda\xi}) = e^{-\lambda^\alpha}, \quad \lambda \geq 0.
$$

Theorem 2.4.7 *If either*

$$
R(t+1) = (aR(t) + \tilde{a}) + (bR(t) + \tilde{b})^{\frac{1}{\alpha}}\xi_{t+1}, \quad t = 0, 1, \ldots, \qquad (2.4.27)
$$

where ξ_t is a sequence of independent standard α-stable distributions and $a, \tilde{a}, b, \tilde{b}$ are nonnegative numbers, or

$$
R(t+1) = aR(t) + \xi_{t+1}, \quad t = 0, 1, \ldots, \qquad (2.4.28)
$$

where (ξ_t) *is an arbitrary sequence of independent, identically distributed nonnegative random variables and* $a \geq 0$*, then the transition function of the Markov chain* $R(t)$ *is of the form* (2.4.26).
Conversely, if a Markov chain $R(t)$ *of the form* (2.4.25) *with* F, G *continuous on* $[0, +\infty)$ *and twice differentiable on* $(0, +\infty)$ *has the transition function* (2.4.26) *then it is either of the form* (2.4.27) *or* (2.4.28).

Only the converse of the theorem requires a proof. It will follow easily from the following two propositions of independent interest covering separately the cases $G(0) = 0$ and $G(0) > 0$. The case $G(0) < 0$ can be treated in a similar way. We use the standard notation for the Laplace transforms of the measures ν and μ_x,

$$\int_0^{+\infty} e^{-\lambda y}\nu(dy) = e^{-\varphi(\lambda)}, \qquad \int_0^{+\infty} e^{-\lambda y}\mu_x(dy) = e^{-x\psi(\lambda)},$$

and the Laplace transform of the noise, i.e.

$$\mathbb{E}(e^{-\lambda\xi_1}) = e^{-\varphi_\xi(\lambda)}.$$

Proposition 2.4.8 *The short-rate process with the transition function* (2.4.26) *has the representation* (2.4.25) *with differentiable on* $(0, +\infty)$ *functions* F *and* G *such that* $G(0) = 0$ *and* $G(x_0) > 0$ *for some* $x_0 > 0$*, if and only if*

$$\psi(\lambda) = a\lambda + b\lambda^\alpha, \quad \varphi(\lambda) = c\lambda,$$

$$G(x) = dx^{\frac{1}{\alpha}}, \quad F(x) = c + ax + ex^{\frac{1}{\alpha}},$$

$$\varphi_\xi(\lambda) = -\frac{e}{d}\lambda + bd^{-\alpha}\lambda^\alpha,$$

with constants $a, b, c \geq 0$*,* $d > 0$*,* $0 < \alpha < 1$ *and* $e \in \mathbb{R}$.

Proof It follows from (2.4.26) that

$$\int_0^{+\infty} e^{-\lambda y}\mathcal{P}(x, dy) = e^{-x\psi(\lambda)-\varphi(\lambda)}. \tag{2.4.29}$$

Setting $R(0) = x$ in (2.4.25) we obtain

$$\mathbb{E}[e^{-\lambda R(1)}] = \mathbb{E}[e^{-\lambda F(x)-\lambda G(x)\xi_1}] = e^{-\lambda F(x)}e^{-\varphi_\xi(\lambda G(x))}. \tag{2.4.30}$$

The equality of (2.4.29) and (2.4.30) yields the following basic relation

$$\varphi_\xi(\lambda G(x)) + \lambda F(x) = x\psi(\lambda) + \varphi(\lambda), \quad x \geq 0, \quad \lambda \geq 0. \tag{2.4.31}$$

Since $G(0) = 0$, (2.4.31) yields $\varphi(\lambda) = c\lambda$ with $c := F(0) = R(1) \geq 0$. Consequently,

$$\varphi_\xi(\lambda G(x)) = x\psi(\lambda) - \lambda(F(x) - c). \tag{2.4.32}$$

Differentiation of (2.4.32) over λ and x yields

$$\varphi'_\xi(G(x)\lambda)G(x) = x\psi'(\lambda) - (F(x) - c),$$

$$\varphi'_\xi(G(x)\lambda)G'(x)\lambda = \psi(\lambda) - \lambda F'(x).$$

Dividing the identities by $G(x)$ and $G'(x)\lambda$, respectively, we obtain two formulas for the value $\varphi'_\xi(G(x)\lambda)$. Comparison of them yields

$$\psi'(\lambda) = \frac{\psi(\lambda)}{\lambda} \cdot \frac{G(x)}{G'(x)x} + \left(\frac{F(x) - c}{x} - \frac{G(x)F'(x)}{xG'(x)}\right), \tag{2.4.33}$$

which hold for $\lambda, x > 0$. Choosing a particular x we obtain

$$\psi'(\lambda) = \frac{\psi(\lambda)}{\lambda}A + B, \quad \lambda > 0 \tag{2.4.34}$$

for some constants A, B. Using Lemma 2.4.10 from the sequel we obtain the solution

$$\psi(\lambda) = \left(\frac{\lambda}{\lambda_0}\right)^A \left[\psi(\lambda_0) - \frac{\lambda_0 B}{1 - A}\right] + \lambda\frac{B}{1 - A} \tag{2.4.35}$$

when $A \neq 1$ and

$$\psi(\lambda) = \left(\frac{\psi(\lambda_0)}{\lambda_0} - B \ln \lambda_0\right)\lambda + B\lambda \ln \lambda$$

when $A = 1$, with some $\lambda_0 > 0$. In the case $A = 1$, the positivity of ψ implies that $B = 0$ and, consequently that ψ is linear. So, in each case ψ has the form $\psi(\lambda) = a\lambda + b\lambda^\alpha$ for $\alpha = A$ and relevant constants a, b. Therefore, in view of (2.4.14) we obtain

$$a\lambda + b\lambda^\alpha = \beta\lambda + \int_{\mathbb{R}_+} (1 - e^{-\lambda y})m(dy), \quad \lambda > 0$$

for some $\beta > 0$ and a measure $m(dy)$, which integrates $(1 \wedge y)$. One can justify that the preceding right side is differentiable and obtain

$$a + \alpha b\lambda^{\alpha-1} = \beta + \int_0^{+\infty} e^{-\lambda y} y\, m(dy), \quad \lambda > 0.$$

We see that for $\alpha > 1$ the preceding left side is an increasing function of λ while the right side decreases. The case $\alpha = 1$ can also be excluded because then $G(0) \neq 0$. Thus $\alpha \in (0, 1)$ and $a, b \geq 0$. Since ψ satisfies (2.4.33) for all $x > 0$ and it is given by (2.4.35), we conclude that for all $x > 0$,

$$\frac{G(x)}{G'(x)x} = A = \alpha, \quad \frac{F(x) - c}{x} - \frac{G(x)F'(x)}{xG'(x)} = B.$$

Applying Lemma 2.4.10 again, with $x_0 > 0$ such that $G(x_0) > 0$, we obtain

$$G(x) = (x/x_0)^{1/\alpha}\, G(x_0) =: dx^{\frac{1}{\alpha}},$$

$$F(x) = (x/x_0)^{1/\alpha}\left[F(x_0) - c + x_0\frac{B}{\alpha-1}\right] - x\frac{B}{\alpha-1} + c =: ex^{\frac{1}{\alpha}} + ax + c.$$

To determine the formula for φ_ξ we set $z := \lambda G(x)$ in (2.4.32). This yields

$$\varphi_\xi(z) = x\psi\left(\frac{z}{G(x)}\right) - \frac{z}{G(x)}(F(x) - c)$$
$$= x\left(a\frac{z}{dx^{\frac{1}{\alpha}}} + b\Big(\frac{z}{dx^{\frac{1}{\alpha}}}\Big)^{\alpha}\right) - \frac{z}{dx^{\frac{1}{\alpha}}}\left(ax + ex^{\frac{1}{\alpha}}\right) = bd^{-\alpha}z^{\alpha} - \frac{e}{d}z.$$

The fact that F, G and φ_ξ given by the theorem satisfy really the basic identity (2.4.31) one shows by direct calculation. □

Proposition 2.4.9 *The short-rate process with the transition semigroup* (2.4.26) *has the representation* (2.4.25) *with continuous on* $[0, +\infty)$ *and twice differentiable on* $(0, +\infty)$ *functions F and G such that $G(0) > 0$, if and only if*

$$\psi(\lambda) = a\lambda + b\lambda^{\alpha}, \quad \varphi(\lambda) = \tilde{a}\lambda + \tilde{b}\lambda^{\alpha},$$

$$G(x) = \frac{G(0)}{\tilde{b}^{1/\alpha}}\left(bx + \tilde{b}\right)^{1/\alpha}, \quad F(x) = ax + \tilde{a} - \left(x\frac{b}{\tilde{b}} + 1\right)^{\frac{1}{\alpha}}(\tilde{a} - F(0)),$$

$$\varphi_\xi(\lambda) = \lambda\frac{\tilde{a} - F(0)}{G(0)} + \tilde{b}\left(\frac{\lambda}{G(0)}\right)^{\alpha}$$

with constants $a, b, \tilde{a}, \tilde{b} \geq 0$, $0 < \alpha < 1$, *or*

$$\psi(\lambda) = a\lambda, \quad \varphi(\lambda) - \textit{arbitrary},$$
$$G(x) \equiv 1, \quad F(x) = ax,$$
$$\varphi_\xi(\lambda) = \varphi(\lambda),$$

where $a \geq 0$.

Proof We sketch the proof because it is similar to the proof of Proposition 2.4.8. We start from (2.4.31) by putting into the relation

$$\varphi_\xi(\lambda G(x)) + \lambda F(x) = x\psi(\lambda) + \varphi(\lambda), \quad x \geq 0, \quad \lambda \geq 0 \tag{2.4.36}$$

the value $x = 0$. This yields

$$\varphi_\xi(\lambda G(0)) + \lambda F(0) = \varphi(\lambda), \quad \lambda \geq 0,$$

and, since $G(0) > 0$,

$$\varphi_{\xi}(\lambda) = \varphi\left(\frac{\lambda}{G(0)}\right) - \lambda\frac{F(0)}{G(0)}, \quad \lambda \geq 0. \tag{2.4.37}$$

Using this formula we put $\varphi_{\xi}(\lambda G(x))$ into (2.4.36), which yields

$$\lambda F(x) + \varphi\left(\frac{\lambda G(x)}{G(0)}\right) - \lambda G(x)\frac{F(0)}{G(0)} = x\psi(\lambda) + \varphi(\lambda), \quad \lambda, x \geq 0. \tag{2.4.38}$$

Now one can differentiate (2.4.38) over x and over λ and determine the following two formulas

$$\varphi'\left(\frac{\lambda G(x)}{G(0)}\right) = F(0) - \frac{F'(x)G(0)}{G'(x)} + \frac{\psi(\lambda)G(0)}{\lambda G'(x)}, \quad \lambda, x \geq 0, \tag{2.4.39}$$

$$\varphi'\left(\frac{\lambda G(x)}{G(0)}\right) = F(0) - \frac{F(x)G(0)}{G(x)} + \frac{x\psi'(\lambda)G(0)}{G(x)} + \frac{\varphi'(\lambda)G(0)}{G(x)}, \quad \lambda, x \geq 0. \tag{2.4.40}$$

Comparison of them yields

$$F(x) - F'(x)\frac{G(x)}{G'(x)} + \frac{\psi(\lambda)}{\lambda}\frac{G(x)}{G'(x)} = x\psi'(\lambda) + \varphi'(\lambda), \quad \lambda, x \geq 0.$$

It follows by differentiation over x that

$$\psi'(\lambda) = A\frac{\psi(\lambda)}{\lambda} + B,$$

where

$$A := \frac{d}{dx}\left[\frac{G(x)}{G'(x)}\right]\Big|_{\bar{x}}, \quad B := \frac{d}{dx}\left[F(x) - F'(x)\frac{G(x)}{G'(x)}\right]\Big|_{\bar{x}}$$

and $\bar{x}$ is some point from $[0, +\infty)$. So, ψ satisfies the same equation as in the previous proof (see (2.4.34)) and one concludes that

$$\psi(\lambda) = a\lambda + b\lambda^{\alpha}, \quad \lambda \geq 0$$

with $\alpha = A \in (0, 1]$ and $a, b \geq 0$. By (2.4.39),

$$\varphi(\lambda) = \tilde{a}\lambda + \tilde{b}\lambda^{\alpha}, \quad \lambda \geq 0,$$

where $\tilde{a}, \tilde{b} \geq 0$. Putting the preceding formulas into (2.4.38) yields

$$\lambda\left(F(x) + \frac{G(x)\tilde{a}}{G(0)} - \frac{G(x)F(0)}{G(0)} - ax - \tilde{a}\right) + \lambda^{\alpha}\left(\left(\frac{G(x)}{G(0)}\right)^{\alpha}\tilde{b} - xb - \tilde{b}\right) = 0,$$
$$\lambda, x \geq 0.$$

It follows that the coefficients standing by λ and by λ^{α} disappear, which enables us to determine the functions F and G. Finally, we obtain

$$G(x) = \frac{G(0)}{\tilde{b}^{1/\alpha}}\left(bx + \tilde{b}\right)^{1/\alpha}, \quad x \geq 0,$$

$$F(x) = ax + \tilde{a} - \left(x\frac{b}{\tilde{b}} + 1\right)^{\frac{1}{\alpha}} (\tilde{a} - F(0)).$$

Let us consider the case $\alpha \in (0, 1)$. Let $\tilde{\xi}$ be a random variable with distribution ν. Then

$$\mathbb{E}(e^{-\lambda\tilde{\xi}}) = e^{-\varphi(\lambda)} = e^{-\tilde{a}\lambda - \tilde{b}\lambda^{\alpha}},$$

which means that $\tilde{\xi} - \tilde{a}$ is α-stable distributed. It follows from (2.4.37) that in this case the distribution of the noise in (2.4.25) is given by

$$\xi_1 = \frac{\tilde{\xi}}{G(0)} - \frac{F(0)}{G(0)}.$$

One can check that the condition for positivity of the short rate

$$F(x) + G(x)\xi_1 \geq 0, \quad x \geq 0$$

is satisfied.

If $\alpha = 1$ then

$$\psi(\lambda) = a\lambda, \quad \lambda \geq 0,$$

which means that $\mu_x = \delta_{\{xa\}}$. Let ζ have now an arbitrary distribution ν on $\mathbb{R}_+$. Then

$$\mathcal{P}(x, dy) = \mu_x * \nu$$

if and only if

$$R(t+1) = aR(t) + \zeta, \quad t \geq 0.$$

□

Proof of Theorem 2.4.7 If a random variable ξ is such that

$$\varphi_{\xi}(\lambda) = -\frac{e}{d}\lambda + bd^{-\alpha}\lambda^{\alpha}, \quad \alpha \in (0, 1),$$

then it can be represented in the form

$$\xi = -\frac{e}{d} + \frac{b^{1/\alpha}}{d}\tilde{\xi},$$

where $\tilde{\xi}$ has a standard α-stable distribution. Similarly, if

$$\varphi_{\xi}(\lambda) = \lambda\frac{\tilde{a} - F(0)}{G(0)} + \tilde{b}\left(\frac{\lambda}{G(0)}\right)^{\alpha},$$

then

$$\xi = \frac{\tilde{a} - F(0)}{G(0)} + \frac{\tilde{b}^{1/\alpha}}{G(0)}\tilde{\xi}.$$

Taking into account the formulae for F and G in the cases $G(0) = 0$ and $G(0) > 0$, respectively, one can obtain, after some calculations, the required formula (2.4.27) for R involving the standard α-stable random variables. If φ_ξ is arbitrary, which is the second case in Proposition 2.4.9, one obtains (2.4.28). □

Lemma 2.4.10 *Solution of the equation*

$$y'(x) = \frac{y(x)}{x}\alpha + \gamma, \qquad x > x_0,$$

with given $y(x_0)$ *for* $x_0 > 0$, *is of the form*

$$y(x) = \left(\frac{x}{x_0}\right)^{\alpha}\left[y(x_0) - \frac{x_0\gamma}{1-\alpha}\right] + x\frac{\gamma}{1-\alpha}$$

for $\alpha \neq 1$ *and*

$$y(x) = \left(\frac{y(x_0)}{x_0} - \gamma \ln x_0\right)x + \gamma x \ln x$$

for $\alpha = 1$.

2.4.4 Shape of Forward Curves in Affine Models

We consider now the shapes of forward curves

$$k \mapsto r(t, k)$$

for affine models satisfying (MP) in the case when $\gamma = 0$. The transition function $\mathcal{P}$ has the form introduced earlier, i.e.

$$\mathcal{P}(x, dy) = (\mu_x * \nu)(dy), \qquad x, y \in \mathbb{R}_+.$$

To exclude trivial case we assume that $\nu \neq \delta_{\{0\}}$ and $\mu_x((0, +\infty)) > 0$, for $x > 0$. Since μ_x is a convolution semigroup on $\mathbb{R}_+$, the function ψ given by

$$\int_{\mathbb{R}_+} e^{-\lambda y}\mu_x(dy) = e^{-\psi(\lambda)x}, \qquad \lambda \geq 0 \tag{2.4.41}$$

must have the form

$$\psi(\lambda) = \beta\lambda + \psi_0(\lambda), \quad \text{with} \quad \psi_0(\lambda) := \int_{\mathbb{R}_+}(1 - e^{-\lambda y})m(dy), \qquad \lambda \geq 0, \tag{2.4.42}$$

where $\beta \geq 0$ and the measure m satisfies

$$\int_{\mathbb{R}_+}(1 \wedge y)m(dy) < +\infty$$

(compare (2.4.14), (2.4.15)). Recall that φ is given by

$$\int_{\mathbb{R}_+} e^{-\lambda y}\,\nu(dy) = e^{-\varphi(\lambda)}, \qquad \lambda \geq 0. \tag{2.4.43}$$

Using (2.4.22), (2.4.23) and (2.4.24), we can write forward rates in the form

$$r(t,k) = c_k + d_k R(t), \qquad k \in \mathbb{N}_0, \tag{2.4.44}$$

where

$$c_k := \begin{cases} 0 & \text{for } k = 0, \\ C_k - C_{k-1} & \text{for } k \geq 1, \end{cases} \qquad d_k := \begin{cases} 1 & \text{for } k = 0, \\ D_k - D_{k-1} & \text{for } k \geq 1. \end{cases}$$

The following result describes the behaviour of the sequences (c_k), (d_k), which determine the shapes of forward curves.

Theorem 2.4.11 *The sequence (c_k) is strictly increasing with*

$$\lim_{k\to+\infty} c_k \begin{cases} = +\infty & \text{if } \beta \geq 1, \\ < +\infty & \text{if } \beta < 1. \end{cases}$$

(a) *If $\beta > 1$ then $\{d_k\}$ is strictly increasing and* $\lim_k d_k = +\infty$.
(b) *If $\beta = 1$ then $\{d_k\}$ is nondecreasing and* $\lim_k d_k = 1 + \int_0^{+\infty} m(dy)$.
(c) *If $\beta < 1$ then* $\lim_k d_k = 0$ *and there exists $k^* \in \mathbb{N}_0$ such that d_k, $k \geq k^*$ is strictly decreasing. If $\psi'(1+\psi(1)) \geq 1$ then $\{d_k\}$ has a hump.*

Proof First we show that $\{D_k\}$ is increasing. Since $\psi(\cdot)$ is increasing and, by (2.4.23) we get

$$D_{k+1} - D_k = \psi(D_k + 1) - \psi(D_{k-1} + 1), \tag{2.4.45}$$

we see that $D_{k-1} \leq D_k$ implies that $D_k \leq D_{k+1}$. So, the monotonicity of $\{D_k\}$ follows by induction. Since $c_k = \varphi(D_{k-1} + 1)$ and $\varphi(\cdot)$ is also increasing, we obtain that $c_k \uparrow +\infty$ if and only if $D_k \uparrow +\infty$. We show now that

$$\{D_k\} \quad \text{is bounded} \quad \Longleftrightarrow \quad \beta < 1. \tag{2.4.46}$$

Let us assume that $\beta \geq 1$ and $\{D_k\}$ is bounded. There exists D such that $D_k \uparrow D$ and passing to the limit in the condition $D_{k+1} = \psi(D_k + 1)$ we obtain $D = \psi(D+1)$. Consequently,

$$D = \psi(D+1) = \beta(1+D) + \psi_0(1+D) \geq 1 + D + \psi_0(1+D),$$

which is impossible because $\psi_0(\cdot) \geq 0$. Now let us assume that $\beta < 1$ and $\{D_k\}$ is unbounded. By (2.4.45) and the Lagrange theorem there exists η_k such that

$$D_{k+1} - D_k = \psi'(\eta_k)(D_k - D_{k-1}), \quad D_{k-1} + 1 \leq \eta_k \leq D_k + 1. \tag{2.4.47}$$

It follows that $\eta_k \longrightarrow +\infty$ and, since

$$\psi'(\lambda) = \beta + \int_0^{+\infty} ye^{-\lambda y} m(dy),$$

there exists k^* and γ such that $\psi'(\eta_k) \leq \gamma < 1$ for $k > k^*$. Then $D_{k+1} - D_k \leq \gamma(\psi(D_k+1) - \psi(D_{k-1}+1)), k > k^*$, which means that $\{D_k\}$ converges geometrically to zero, which is a contradiction.

Now we examine the sequence $\{d_k\}$. If $\beta < 1$ then, by (2.4.46), $d_{k+1} = D_{k+1} - D_k \leq D - D_k \longrightarrow 0$, where D stands for the limit of D_k. Writing (2.4.47) in the form

$$d_{k+1} = \psi'(\eta_k) d_k$$

and using the fact that $\psi'(\eta_k)$ is decreasing we see that a sufficient condition for $\{d_k\}$ to have a hump is that $\psi'(\eta_1) \geq 1$. Since $1 \leq \eta_1 \leq D_1 + 1$ and $D_1 = \psi(1)$, we need $\psi'(1 + \psi(1)) \geq 1$.

If $\beta = 1$ then

$$\begin{aligned} d_{k+1} = D_{k+1} - D_k &= [(D_k + 1) + \psi_0(D_k + 1)] - D_k \\ &= 1 + \psi_0(D_k + 1) \longrightarrow 1 + \int_0^{+\infty} m(dy), \end{aligned}$$

where the latter convergence holds because $D_k \uparrow +\infty$.

Since $\psi'(\cdot) > \beta$, (2.4.47) yields

$$D_{k+1} - D_k \geq \beta(D_k - D_{k-1}).$$

Hence, $d_k \longrightarrow +\infty$ geometrically fast if $\beta > 1$. □

Example 2.4.12 We examine the sufficient condition $\psi'(1 + \psi(1)) \geq 1$, from Theorem 2.4.11, for $\{d_k\}$ to have a hump in the case $m(dy) = \frac{c}{y^{1+\delta}} dy$, where $0 < \delta < 1$ and c is a constant. Then $\psi(\lambda) = \beta\lambda + \bar{c}c\lambda^\delta$, where $\bar{c}$ is another constant, which is independent of c, see Example 5.3.3 in the sequel for detailed calculations. Since $\psi'(\lambda) = \beta + \bar{c}c\delta \frac{1}{\lambda^{1-\delta}}$ and $\psi(1) = \beta + c\bar{c}$, it follows that

$$\psi'(1 + \psi(1)) = \beta + c\bar{c}\delta \frac{1}{(\beta + c\bar{c})^{1-\delta}}.$$

Consequently,

$$\psi'(1 + \psi(1)) \geq 1 \quad \Longleftrightarrow \quad \frac{\delta c\bar{c}(\beta + c\bar{c})^\delta}{\beta + c\bar{c}} \geq 1 - \beta.$$

It is clear now that we can find c such that the preceding right side is satisfied.

2.4.5 Factor Models

In the Markovian setting one can incorporate into the model concrete shapes of bond and forward curves.

Let us consider a factor model in which forward rate is of the form

$$f(t,T) = G(T-t, X_t), \quad t,T = 0,1,\dots,T \geq t, \tag{2.4.48}$$

where G is some deterministic positive function and X a Markov chain on $(E,\mathcal{E})$, with transition operator $\mathcal{P}(\cdot,\cdot)$, defined on a probability space $(\Omega,\mathcal{F},\{\mathcal{F}_t\},\mathbb{P})$. For a fixed value x of the factor $X(t)$, the function $v \mapsto G(v,x)$ describes thus the shape of the *forward curve* in terms of time to maturity $v := T-t$. Since

$$P(t,T) = e^{-\sum_{s=t}^{T-1} f(t,s)} = e^{-\sum_{s=t}^{T-1} G(s-t,X_t)} = e^{-\sum_{u=0}^{T-t-1} G(u,X_t)}, \quad t,T = 0,1,\dots,T \geq t,$$

it follows that

$$P(t,T) = F(T-t, X(t)), \tag{2.4.49}$$

where F is given by

$$F(0,x) = 1, \quad F(v,x) := e^{-\sum_{s=0}^{v-1} G(s,x)}, \quad v = 1,2,\dots. \tag{2.4.50}$$

In particular,

$$F(1,x) = e^{-G(0,x)}, \ x \in E.$$

Similarly, for a given value x of $X(t)$ the function $v \mapsto F(v,x)$ describes the *bond curve*. In the case when $X(t)$ is equal to the short rate $R(t)$ on $E = [0,+\infty)$, we have

$$R(t) = f(t,t) = G(0,R(t)),$$

and consequently G must be such that

$$G(0,x) = x, \quad x \geq 0.$$

More generally, if $X(t)$ is the γ-trace of the forward rate $r^{\gamma}(t) = (r(t,0), r(t,1),\dots, r(t,\gamma))$ with $\gamma > 0$, then

$$r(t,k) = f(t,t+k) = G(k, r(t,0),\dots r(t,k),\dots,r(t,\gamma)), \quad k = 0,1,\dots,\gamma,$$

so G must satisfy

$$G(k,x_0,\dots,x_\gamma) = x_k, \quad x_0,x_1,\dots,x_\gamma \geq 0, \quad k = 0,1,\dots,\gamma.$$

In factor models satisfying (MM) or (MP) the function G, or equivalently, F, and the transition operator $\mathcal{P}$ must satisfy certain conditions that we deduce in the following. If the underlying filtration $\{\mathcal{F}_t\}$ is generated by a sequence $\xi_1,\xi_2,\dots$ of independent identically distributed random variables, then the Markov process X can be represented in the form

$$X(t+1) = K(X(t), \xi_{t+1}), \quad t = 0, 1, \dots \tag{2.4.51}$$

for some function K. The existence of the representation (2.4.51) for an arbitrary Markov chain is a direct conclusion from Theorem 1.2 in Peszat and Zabczyk [100]. Let μ stand for the distribution of ξ_1. Recall that for a measure $\mathbb{Q} \sim \mathbb{P}$ one can write the density process in the form

$$\rho_0 = 1, \quad \rho_t = \psi_t(\xi_1, \dots \xi_t), \quad t = 1, 2, \dots T^* \tag{2.4.52}$$

for some functions ψ_t.

Theorem 2.4.13 *Let $\mathbb{Q} \sim \mathbb{P}$ be a measure with density* (2.4.52). *Then $\mathbb{Q}$ is a martingale measure if and only if*

$$\begin{aligned}\int_U & \psi_{t+1}(\xi_1, \dots, \xi_t, y) F(T-t-1, K(X(t), y)) \mu(dy) \\ &= \psi_t(\xi_1, \dots, \xi_t) e^{G(0,X(t))} F(T-t, X(t))\end{aligned} \tag{2.4.53}$$

for each $T = 1, 2, \dots$ and $t = 0, 1, \dots, T-1$.

Proof The process $\hat{P}(t,T)$ is a $\mathbb{Q}$-martingale if and only if

$$\mathbb{E}(\rho_{t+1}\hat{P}(t+1,T) \mid \mathcal{F}_t) = \rho_t \hat{P}(t,T), \quad t = 0, 1, \dots, T-1.$$

Since the bank account is given by

$$B(t) = e^{-\sum_{s=0}^{t-1} R(s)} = e^{-\sum_{s=0}^{t-1} G(0,X(s))},$$

by (2.4.49) we obtain

$$\rho_t \hat{P}(t,T) = \psi_t(\xi_1, \dots, \xi_t) e^{-\sum_{s=0}^{t-1} G(0,X(s))} F(T-t, X(t)). \tag{2.4.54}$$

In view of (2.4.51) we have

$$\begin{aligned}&\mathbb{E}(\rho_{t+1}\hat{P}(t+1,T) \mid \mathcal{F}_t) \\ &\quad = e^{-\sum_{s=0}^{t} G(0,X(s))} \mathbb{E}\left(\psi_{t+1}(\xi_1, \dots, \xi_{t+1}) F(T-t-1, X(t+1)) \mid \mathcal{F}_t\right) \\ &\quad = e^{-\sum_{s=0}^{t} G(0,X(s))} \mathbb{E}\left(\psi_{t+1}(\xi_1, \dots, \xi_{t+1}) F(T-t-1, K(X(t), \xi_{t+1})) \mid \mathcal{F}_t\right) \\ &\quad = e^{-\sum_{s=0}^{t} G(0,X(s))} \int_U \psi_{t+1}(\xi_1, \dots, \xi_t, y) F(T-t-1, K(X(t), y)) \mu(dy).\end{aligned} \tag{2.4.55}$$

From the equality of (2.4.55) and (2.4.54) we obtain the required formula. □

It follows from (2.4.51) that the transition operator of X is given by

$$\mathcal{P}h(x) = \mathbb{E}(h(K(x,\xi_1))) = \int_U h(K(x,y))\mu(dy).$$

This implies, that in the case when $\psi = \psi_t \equiv 1$, (2.4.53) boils down to the condition

$$\int_U F(T-t-1, K(X(t),y))\mu(dy) = \mathcal{P}(F(T-t-1, X(t))) = e^{G(0,X(t))}F(T-t, X(t)).$$

This allows formulating conditions for models to satisfy (MP). As a consequence of Theorem 2.4.13, we obtain the following result.

Theorem 2.4.14 *The factor model satisfies (MP) if and only if*

$$F(k+1,x) = F(1,x)\mathcal{P}(F(k,\cdot))(x), \quad F(0,x) = 1, \ x \in E \quad k = 0,1,2,\ldots.$$

In particular, any factor model satisfying (MP) is determined by the function $F(1,x)$ and the transition operator $\mathcal{P}$ of the Markov process X.

With the use of Theorem 2.4.14 one can characterize models satisfying (MP) with multiplicative factor X.

Proposition 2.4.15 *Let the factor X be given by*

$$X_{t+1} = aX_t + bX_t\xi_{t+1}, \quad X_0 = x > 0, \quad t = 1,2,\ldots,$$

where $\{\xi_t\}$ is an i.i.d. sequence and a, b are constants. Then the model (2.4.48) *with $F(1,x) := x^\gamma$, $x > 0, \gamma \in \mathbb{R}$, satisfies (MP) if and only if*

$$F(k,x) = c_\gamma c_{2\gamma}\ldots c_{(k-1)\gamma}x^{k\gamma}, \quad k = 1,2,3,\ldots, \tag{2.4.56}$$

where $c_\gamma := \mathbb{E}[(a+b\xi_1)^\gamma]$. Consequently, under (MP), the bond prices are given by

$$P(t,T) = F(T-t, X(t)) = c_\gamma c_{2\gamma}\ldots c_{(T-t-1)\gamma}X_t^{(T-t)\gamma}.$$

If, additionally,

$$\gamma > 0, E = (0,1), \quad 0 < a + b\xi_1 < 1, \quad \mathbb{P}-a.s.$$

or

$$\gamma < 0, E = (1,+\infty), \quad a + b\xi_1 > 1, \quad \mathbb{P}-a.s.,$$

then the forward rates are positive. In particular, the model is then regular.

Proof Let us notice that the class of power functions $h_\alpha(x) = x^\alpha$, $x > 0$, is invariant for the transition operator $\mathcal{P}$. That is

$$\mathcal{P}h_\alpha(x) = x^\alpha\mathbb{E}[(a+b\xi_1)^\alpha] = c_\alpha x^\alpha, \quad \alpha \neq 0, \ x > 0.$$

It follows from Theorem 2.4.14 that for $F(1,x) = x^{\gamma},\ x > 0$,

$$F(2,x) = F(1,x)c_{\gamma}x^{\gamma} = c_{\gamma}x^{2\gamma}$$

and by the inductive argument

$$F(k,x) = \Big(\prod_{j=1}^{k-1} c_{j\gamma}\Big)x^{k\gamma},$$

which is the required formula (2.4.56).

Let us formulate conditions for the positivity of $f(t,T) = G(T-t,X_t)$. Since

$$e^{-G(k,x)} = \frac{F(k+1,x)}{F(k,x)} = \frac{c_{\gamma}c_{2\gamma}\ldots c_{k\gamma}x^{(k+1)\gamma}}{c_{\gamma}c_{2\gamma}\ldots c_{(k-1)\gamma}x^{k\gamma}}$$
$$= c_{k\gamma}x^{\gamma}, \quad k = 1,2,\ldots,$$

we obtain

$$G(0,x) = -\gamma \ln x, \quad G(k,x) = -\ln c_{k\gamma} - \gamma \ln x, \quad k = 1,2,\ldots.$$

It follows that forward rates are positive if X evolves in $E = (0,1)$ for $\gamma > 0$ or in $E = (1,+\infty)$ for $\gamma < 0$ and, in both cases, $c_{k\gamma} \in (0,1)$ for $k = 1,2,\ldots$. This leads to the required conditions. □

3

Completeness

In this chapter we study the completeness problem. It turns out that the majority of market models are not complete but, nevertheless, may still be approximately complete. The crucial role here is played by portfolios with an infinite number of bonds. Without them, even approximate completeness may fail. The first sections discuss completeness with respect to the minimal filtration. Models with martingale discounted bond prices are covered next. The final section is about the interplay between completeness and martingale measures.

3.1 Concepts of Completeness

Let X be an $\mathcal{F}_t$-measurable claim, with $t > 0$, on a probability space $(\Omega, \mathcal{F}, \mathbb{P})$ with filtration $\{\mathcal{F}_s\}, s = 0, 1, \ldots$. Recall that X is attainable at t if there exists a self-financing strategy $\{(b(s), \varphi_s), s = 0, 1, \ldots, t\}$, where φ_s takes values in l^1, such that the corresponding portfolio wealth

$$X(s) = b(s) + \langle \varphi_s, P(s) \rangle, \quad s = 0, 1, \ldots, t$$

satisfies $X(t) = X$. As explained in Section 1.4 (see Proposition 1.4.1), X is attainable if and only if the discounted claim admits the representation

$$\hat{X} = x + \sum_{s=0}^{t-1} \langle \varphi_s, \Delta \hat{P}(s+1) \rangle, \quad \mathbb{P} - \text{a.s.} \tag{3.1.1}$$

for some $x \in \mathbb{R}$ and some l^1-valued adapted strategy φ_s. Here $\Delta \hat{P}(s+1) = \Delta \hat{P}(s+1, j), j = s+2, s+3, \ldots$ stands for the increments of discounted bond prices.

Loosely speaking, the market is called *complete* if, for any $t > 0$, each claim is attainable. A precise definition of completeness requires specific conditions for the class of claims that are to be replicated. Typically X or $\hat{X}$ is supposed to be bounded or p-th power integrable with $p \geq 1$. A specification of the underlying filtration is also important. Often, but not always, $\{\mathcal{F}_s\}$ is assumed to be the *minimal filtration*, i.e.

$$\mathcal{F}_s = \sigma\{ P(u,T);\ u = 0,1,\dots,s,\ T = 0,1,2,\dots\}, \quad s = 0,1,\dots.$$

In this case each claim X at time $t > 0$, can be written as some function of bond prices observed up to time t, i.e.

$$X = X(P(0), P(1), \dots, P(t)), \tag{3.1.2}$$

where $P(s) = (P(s,j), j = 0,1,2,\dots)$ stands for the bond curve at time $s = 0, 1,\dots,t$. The class of claims of the preceding form is wide enough to comprise financial contracts that are traded in practice. If $\{\mathcal{F}_s\}$ is required to be the minimal filtration, one can write the replicating condition (3.1.1) in the more convenient form, which will be used in the sequel. Let us notice that (3.1.1) involves increments of the discounted bond prices. Therefore it is natural to expect that one can replace in (3.1.1) the minimal filtration $\{\mathcal{F}_s\}$ by the filtration generated by the discounted bond prices, i.e.

$$\mathcal{F}_s^{\hat{P}} := \sigma\{\hat{P}(u,T);\ u \le s,\ T = 0,1,\dots\}, \quad s = 0,1,\dots$$

The following result confirms this conjecture.

Proposition 3.1.1 *For every $t = 0, 1, \dots$,*

$$\mathcal{F}_t = \mathcal{F}_t^{\hat{P}} = \mathcal{F}_t^f,$$

where

$$\mathcal{F}_t^f := \sigma\{f(s,T);\ s \le t,\ T = 0,1,\dots\}.$$

The filtration $\{\mathcal{F}_t^f\}$ allows us to deal also with claims depending on forward curves.

Proof Let us recall basic relations between f, P and $\hat{P}$:

$$P(t,T) := \begin{cases} e^{-\sum_{s=t}^{T-1} f(t,s)} & \text{for } t = 0,1,\dots,T-1; \\ 1 & \text{for } t = T; \\ e^{\sum_{s=T}^{t-1} f(s,s)} & \text{for } t = T+1, T+2,\dots; \end{cases} \tag{3.1.3}$$

$$f(t,s) = f(s,s), \quad s = 0,1,\dots,t; \tag{3.1.4}$$

$$\hat{P}(t,T) := \begin{cases} e^{-\sum_{s=0}^{T-1} f(t,s)} & \text{for } t = 0,1,\dots; T = 1,2,\dots; \\ 1 & \text{for } t = 0,1,\dots; T = 0. \end{cases} \tag{3.1.5}$$

If $s \ge t$ then, by (3.1.3),

$$f(t,s) = \ln P(t,s) - \ln P(t,s+1),$$

and consequently $\sigma\{f(t,s)\} \subseteq \mathcal{F}_t$. If $s < t$, then, by (3.1.4), $\sigma\{f(t,s)\} = \sigma\{f(s,s)\}$ and therefore $\sigma\{f(t,s)\} \subseteq \mathcal{F}_t$. Consequently, $\sigma\{f(t,s)\} \subseteq \mathcal{F}_t$ for all $s = 0, 1, \ldots$. Thus

$$\mathcal{F}_t^f \subseteq \mathcal{F}_t.$$

Since

$$\sigma\{P(t,T); T = 0, 1, \ldots\} \subseteq \sigma\{f(s,T); s = 0, 1, \ldots, t; T = 0, 1, \ldots\},$$

we have that

$$\mathcal{F}_t \subseteq \mathcal{F}_t^f,$$

and consequently

$$\mathcal{F}_t^f = \mathcal{F}_t.$$

By (3.1.5),

$$\mathcal{F}_t^{\hat{P}} \subseteq \mathcal{F}_t^f.$$

On the other hand, for $t = 0, 1, \ldots, s = 0, 1, \ldots$

$$f(t,s) = \ln \hat{P}(t,s) - \ln \hat{P}(t,s+1),$$

so

$$\mathcal{F}_t^f = \mathcal{F}_t^{\hat{P}}.$$

The proof is complete. □

Let us denote by $\eta_s := (\eta_s(s+1), \eta_s(s+2), \ldots)$ the sequence describing increments of the discounted bond prices, i.e.

$$\eta_s(j) := \hat{P}(s,j) - \hat{P}(s-1,j), \quad j = s+1, s+2, \ldots$$

and $\eta_0 := (P(0,1), P(0,2), \ldots)$. Then, obviously,

$$\mathcal{F}_s^{\hat{P}} = \sigma\{\eta_0, \eta_1, \ldots, \eta_s\}, \quad s = 0, 1, \ldots. \tag{3.1.6}$$

In view of (3.1.1), (3.1.6) and Proposition 3.1.1 the completeness problem with the minimal filtration boils down to representing each bounded claim $\hat{X}(\eta_0, \eta_1, \ldots, \eta_t)$ in the form

$$\hat{X}(\eta_0, \eta_1, \ldots, \eta_t) = x + \sum_{s=0}^{t-1} \langle \varphi_s(\eta_0, \eta_1, \ldots, \eta_s), \eta_{s+1} \rangle, \quad \mathbb{P} - \text{a.s.} \tag{3.1.7}$$

As we will see in the sequel, completeness is rather a rare feature of the market as it enforces very restrictive conditions for the bond price process. Therefore one considers also weaker concepts of completeness, where claims are approximated in a certain sense. The market is *weakly complete* at time $t > 0$ if for each bounded $\mathcal{F}_t$-measurable claim $\hat{X}$ and an arbitrary level of accuracy $\varepsilon > 0$, there exists a strategy such that

$$| \hat{X}(t) - \hat{X} | < \varepsilon, \quad \mathbb{P} - a.s.$$

The interpretation of the preceding condition is that the replication error can be made arbitrarily close to zero. The market is *L^p-approximately complete* a time t, where $p \geq 1$, if for any $t > 0$, $\hat{X} \in L^p(\Omega, \mathcal{F}_t, \mathbb{P})$ and $\varepsilon > 0$ one can find a strategy such that

$$\mathbb{E}\left(| \hat{X}(t) - \hat{X} |^p\right) < \varepsilon.$$

If the properties above hold for all $t = 1, 2, \ldots$, then one says that the markets are, respectively, weakly complete or L^p-approximately complete.

In the sequel we characterize complete, weakly complete and approximately complete models and discuss the link with the existence and uniqueness of the martingale measure. In Section 3.5 and Section 3.6 we study the completeness problem with other filtrations than the minimal one. To avoid ambiguities, in the formulation of each result we will indicate the class of claims in which the completeness problem is considered. Recall that $L^\infty(\Omega, \mathcal{F}_t, \mathbb{P})$ stands for the set of all bounded $\mathcal{F}_t$-measurable random variables.

3.2 Necessary Conditions for Completeness

The representation (3.1.7) required for completeness under the minimal filtration $\{\mathcal{F}_t\}$ depends on properties of the process (η_t). First we show that weak completeness, like completeness, fails if at least one η_t takes an infinite number of values with positive probability.

If the distribution of a random variable η is concentrated on a finite set, one says that η takes a finite number of values. In the opposite case η is said to take an infinite number of values.

Theorem 3.2.1 *If, for some $t \geq 1$, η_t takes an infinite number of values, then the market is not weakly complete at t in the class $\hat{X} \in L^\infty(\Omega, \mathcal{F}_t, \mathbb{P})$.*

Proof Assume to the contrary, that η_t takes an infinite number of values and the market is weakly complete at time t. Without loss of generality we can require that the random variables $\eta_1, \ldots, \eta_{t-1}$ take a finite number of values, say η_s takes values $a_1^s, a_2^s, \ldots, a_{K_s}^s$, $s = 1, 2, \ldots, t-1$ and $K_1, \ldots, K_s$ are some natural numbers.

Let U be the space of infinite sequences $x = \{x_m\}$ taking values in $[-1, 1]$ equipped with the metric

$$\rho(x, \bar{x}) = \sum_{m=1}^{\infty} \frac{1}{2^m} \cdot \frac{| x_m - \bar{x}_m |}{1 + | x_m - \bar{x}_m |}, \qquad x = \{x_m\}, \bar{x} = \{\bar{x}_m\}.$$

With the use of this metric one can decompose U into the form

$$U = \bigcup_{j=1}^{\infty} U_j,$$

where U_j are disjoint Borel subsets of U such that

$$\mathbb{P}(\eta_t \in U_j) > 0, \quad j = 1, 2, \ldots .$$

If $h, \bar{h}$ are two different functions on U taking values 0 or 1 and constant on each of the sets U_j, then

$$\| h(\eta_t) - \bar{h}(\eta_t) \| := \operatorname{ess\,sup} \mid h(\eta_t(\omega)) - \bar{h}(\eta_t(\omega)) \mid = 1.$$

Weak completeness at time t implies that for any function h taking values 0 or 1 on the sets $U_j, j = 1, 2, \ldots$, there exist x and portfolios $\varphi_0, \varphi_1, \ldots, \varphi_{t-1}$ such that

$$\varphi_s = \varphi_s(\eta_1, \ldots, \eta_s) = \{\varphi_s^j(\eta_1, \ldots, \eta_s), j = 1, 2, \ldots\},$$

$$\mid \varphi_s \mid_{l^1} = \sum_{j=1}^{+\infty} \mid \varphi_s^j \mid < +\infty, \quad s = 0, 1, \ldots, t-1,$$

and

$$\mathbb{P}\Big(\mid h(\eta_t) - (x + \sum_{s=0}^{t-1} \langle \varphi_s, \eta_{s+1} \rangle) \mid < \varepsilon \Big) = 1.$$

Note that

$$\langle \varphi_s, \eta_{s+1} \rangle = \sum_{j=1}^{+\infty} \varphi_s^j(\eta_1, \ldots, \eta_s) \eta_{s+1}^j, \quad s = 0, 1, \ldots, t-1, \tag{3.2.1}$$

where η_{s+1}^j are coordinates of η_{s+1} and satisfy $\mid \eta_{s+1}^j \mid \leq 1$. The series in (3.2.1) converges uniformly. Moreover,

$$\sum_{j=1}^{+\infty} \varphi_s^j(\eta_1, \ldots, \eta_s) \eta_{s+1}^j = \sum_{j=1}^{+\infty} \sum_{\substack{k_1 = 1, 2, \ldots, K_1 \\ \cdots \\ k_s = 1, 2, \ldots, K_s}} \varphi_s^j(a_{k_1}^1, \ldots, a_{k_s}^s) \eta_{s+1}^{j, k_1, \ldots, k_s},$$

where

$$\eta_{s+1}^{j, k_1, \ldots, k_s} := \mathbf{1}_{\{\eta_1 = a_{k_1}^1, \ldots, \eta_s = a_{k_s}^s\}} \eta_{s+1}^j$$

and

$$\sum_{j=1}^{+\infty} \sum_{\substack{k_1 = 1, 2, \ldots, K_1 \\ \cdots \\ k_s = 1, 2, \ldots, K_s}} \mid \varphi_s^j(a_{k_1}^1, \ldots, a_{k_s}^s) \mid < +\infty, \quad \mid \eta_{s+1}^{j, k_1, \ldots, k_s} \mid \leq 1.$$

For each s, the family $\{\eta_{s+1}^{j,k_1,\dots,k_s}; j = 1,2,\dots,k_i = 1,2,\dots,K_i, i = 1,2,\dots,s\}$ is countable and the series is convergent in the *essential supremum*-norm. Thus each random variable $h(\eta_t)$ can be approximated in this norm by linear combinations of finite sums of random variables $\{\eta_{s+1}^{j,k_1,\dots,k_s}\}$ which is a contradiction with the fact that for a continuum of random variables $h(\eta_t)$ the mutual distances are equal to 1.

Assume that for each $h(\eta_t)$ there exist an approximating linear finite combination with rational coefficients of elements of the set

$$\{\eta_{s+1}^{j,k_1,\dots,k_s}, \quad j = 1,2,\dots;\ k_i = 1,2,\dots,K_i;\ i = 1,\dots,s;\ s = 1,2,\dots,t-1\},$$

with the distance to $h(\eta_t)$ less than $1/2$. Then for different random variables $h(\eta_t)$ the approximating sequnces are different. This is, however, impossible because the set of random variables $h(\eta_t)$ is uncountable. □

3.3 Sufficient Conditions for Completeness

It follows from Theorem 3.2.1 that a necessary condition for the market to be complete under the natural filtration $\{\mathcal{F}_t\}$ is that η_t takes a finite number of values for each $t = 1,2,\dots$. Now we formulate sufficient conditions in this case. We begin by showing that the concepts of completeness and weak completeness coincide. Since each claim $\hat{X}$ takes a finite number of values, we obviously have that $\hat{X} \in L^\infty(\Omega, \mathcal{F}_t, \mathbb{P})$ with some $t \geq 1$.

Proposition 3.3.1 *Let $\{\mathcal{F}_t\}$ be the minimal filtration. If, for each $t > 0$, η_t takes a finite number of values then the market is complete if and only if it is weakly complete.*

Proof Let $d_t < +\infty$ stand for the number of values taken by η_t, where $t = 1,2,\dots$ and $d_0 = 1$ by definition. The set of all portfolios at time $t \geq 0$ of the form

$$\varphi_t(\eta_0, \eta_1, \dots, \eta_t)$$

can be identified with the space $(l^1)^{d_1 \cdot d_2 \cdot \dots \cdot d_t}$. Consequently, the discounted wealth process at time t corresponding to the pair $(x, (\eta_t))$ given by

$$\mathcal{K}_t(x, \varphi_0, \varphi_1, \dots, \varphi_{t-1}) := x + \sum_{s=0}^{t-1} \langle \varphi_s, \eta_{s+1} \rangle$$

is a linear transformation from

$$\mathbb{R} \times l^1 \times (l^1)^{d_1} \times (l^1)^{d_1 \cdot d_2} \times \dots \times (l^1)^{d_1 \cdot d_2 \cdot \dots \cdot d_{t-1}} \quad \text{into} \quad \mathbb{R}^{d_1 \cdot d_2 \cdot \dots \cdot d_t}.$$

Since $\mathbb{R}^{d_1 \cdot d_2 \cdot \dots \cdot d_t}$ describes the set of all discounted claims paid at time t of the form

$$\hat{X} = \hat{X}(\eta_0, \eta_1, \dots, \eta_t),$$

the market is weakly complete if and only if, for each $t > 0$, the image of $\mathcal{K}_t$ truncated to the ball $B := \{y \in \mathbb{R}^{d_1 \cdot d_2 \cdot \ldots \cdot d_t} :| y | \leq 1\}$ is dense in B. This is, however, possible if and only if

$$Im(\mathcal{K}_t) = \mathbb{R}^{d_1 \cdot d_2 \cdot \ldots \cdot d_t}, \quad t > 0,$$

which means that the market is complete. □

Now we formulate specific conditions for the market to be complete. For a finite sequence of vectors $\{a_1, a_2, \ldots, a_d\}$ in m and their linear span

$$G_d := span\{a_1, a_2, \ldots, a_d\},$$

let us introduce the following nondegeneracy conditions:

- $(ND1)$ $\quad dim\ G_d = d$;
- $(ND2)$ $\quad dim\ G_d = d - 1$ and the d-dimensional vector $(1, 1, \ldots, 1)$ does not belong to the image of the operator $A : l^1 \longrightarrow \mathbb{R}^d$ defined by

$$Ab := \begin{pmatrix} \langle b, a_1 \rangle \\ \langle b, a_2 \rangle \\ \vdots \\ \langle b, a_d \rangle \end{pmatrix}, \quad b \in l^1. \tag{3.3.1}$$

Theorem 3.3.2 *Let $\{\mathcal{F}_t\}$ be the minimal filtration and the number of values of η_t be finite for any $t > 0$. Then the market is complete if and only if, for each $t > 0$, the conditional distribution of η_t with respect to $\eta_0, \eta_1, \ldots, \eta_{t-1}$ is concentrated on the set $\{a_1^t, a_2^t, \ldots, a_{d_t}^t\}$, with $d_t < +\infty$, of vectors in m which satisfies $(ND1)$ or $(ND2)$. If this is the case then each replicating strategy can be replaced by a replicating strategy with finite portfolios at any time.*

In the preceding formulation the set $\{a_1^t, a_2^t, \ldots, a_{d_t}^t\}$ and d_t depend, of course, on $\eta_0, \eta_1, \ldots, \eta_{t-1}$. The second part of Theorem 3.3.2 follows from the fact that any l^1-valued strategy φ_s can be reduced to some finite one $\tilde{\varphi}$, so that the final wealth remains the same, i.e.

$$\hat{X}(t) = x + \sum_{s=0}^{t-1} \langle \varphi_s, \eta_{s+1} \rangle = x + \sum_{s=0}^{t-1} \langle \tilde{\varphi}_s, \eta_{s+1} \rangle.$$

This is possible because η_t takes a finite number of values in m for each $t \geq 0$. We prove this in Proposition 3.3.4. The proof of the first part of Theorem 3.3.2 is based on the following auxiliary result.

Proposition 3.3.3 *Let $a_1, a_2, \ldots, a_d$ be a finite set of vectors in m. Any sequence of reals $\gamma_1, \gamma_2, \ldots, \gamma_d$ can be represented in the form*

$$\gamma_i = a + \langle b, a_i \rangle, \qquad i = 1, 2, \ldots, d, \tag{3.3.2}$$

where $a \in \mathbb{R}$ and $b \in l^1$ if and only if (ND1) or (ND2) is satisfied. If (ND2) holds then a in the representation (3.3.2) is unique while if (ND1) then for each a there exists b such that (3.3.2) holds.

Proof Let $m \leq d$ denote the dimension of G_d and assume for simplicity that the first m-vectors $a_1, a_2, \dots, a_m$ are linearly independent. Then for any sequence $\gamma_1, \gamma_2, \dots, \gamma_m$ there exists $b \in l^1$ such that $\langle b, a_i \rangle = \gamma_i, i = 1, 2, \dots, m$, which means that the image of the operator A given by (3.3.1) is of dimension m. Writing (3.3.2) in the form

$$\begin{pmatrix} \gamma_1 \\ \gamma_2 \\ \vdots \\ \gamma_d \end{pmatrix} = Ab + a \begin{pmatrix} 1 \\ 1 \\ \vdots \\ 1 \end{pmatrix}, \qquad b \in l^1, \tag{3.3.3}$$

we see that the representation of an arbitrary sequence $\gamma_1, \gamma_2, \dots, \gamma_d$ is possible if and only if the image of the right side of (3.3.3) equals $\mathbb{R}^d$. This may happen in two situations only. Either $m = d$ or $m = d-1$ and $(1, 1, \dots, 1) \notin ImA$, which correspond to (*ND*1) and (*ND*2), respectively. If (*ND*1) holds then clearly for any $a \in \mathbb{R}$ we can find $b \in l^1$ such that (3.3.3) is satisfied. If (*ND*2) holds then $\gamma = (\gamma_1, \dots, \gamma_d)$ has a unique decomposition of the form $\gamma = \gamma' + \gamma''$, where $\gamma' \in ImA$ and $\gamma'' \in (ImA)^\perp$. Since $(ImA)^\perp = span\{(1, 1, \dots, 1)\}$, the constant a in (3.3.3) is unique. In general, there are many vectors $b \in l^1$ solving $Ab = \gamma''$. □

Proposition 3.3.4 *Let $a_1, \dots, a_d$ be a finite set of vectors in m. Then there exists $n \in \mathbb{N}$ such that for any $\varphi \in l^1$ there exists $\tilde{\varphi} \in \mathbb{R}^n$ such that*

$$\langle \varphi, a_i \rangle = \langle \tilde{\varphi}, a_i^{(n)} \rangle_{\mathbb{R}^n}, \quad i = 1, 2, \dots, d.$$

Above $\langle \cdot, \cdot \rangle_{\mathbb{R}^n}$ stands for the scalar product in $\mathbb{R}^n$ and $a_i^{(n)}$ is the truncation of $a_i = (a_i(1), a_i(2), \dots)$ restricted to the first n coordinates, i.e.

$$a_i^{(n)} := (a_i(1), a_i(2), \dots, a_i(n)), \ i = 1, 2, \dots, d.$$

Proof Let $dim\ G_d = m \leq d$, where $G_d := span\{a_1, a_2, \dots, a_d\}$ and assume for simplicity that $a_1, a_2, \dots, a_m$ are linearly independent. We prove first that for some $n \in \mathbb{N}$ also $a_1^{(n)}, a_2^{(n)}, \dots, a_m^{(n)}$ are linearly independent. Otherwise, for each $n \geq 1$, we can find non-vanishing $\alpha_1^n, \alpha_2^n, \dots, \alpha_m^n$ such that

$$\alpha_1^n a_1^{(n)} + \cdots \alpha_m^n a_m^{(n)} = 0,$$

and without loss of generality we can assume that $\sum_{i=1}^m \mid \alpha_i^n \mid = 1$. Then we can find a subsequence $\{n_k\}$ such that

$$(\alpha_1^{n_k}, \dots, \alpha_m^{n_k}) \longrightarrow (\tilde{\alpha}_1, \dots, \tilde{\alpha}_m),$$

where $(\tilde{\alpha}_1, \ldots, \tilde{\alpha}_m)$ is some vector in $\mathbb{R}^n$ such that $\sum_{i=1}^m \mid \tilde{\alpha}_i \mid= 1$. Since for each $l > 1$ we have

$$\tilde{\alpha}_1 a_1^{(l)} + \cdots + \tilde{\alpha}_m a_m^{(l)} = \lim_{k\to+\infty} \alpha_1^{n_k} a_1^{(l)} + \cdots + \alpha_m^{n_k} a_m^{(l)} = 0,$$

it follows that

$$\tilde{\alpha}_1 a_1 + \cdots + \tilde{\alpha}_m a_m = 0,$$

which contradicts the linear independence of $a_1, a_2, \ldots, a_m$.

It follows from the linear independence of $a_1^{(n)}, a_2^{(n)}, \ldots, a_m^{(n)}$, that for any $\varphi \in l^1$, we can find $\tilde{\varphi} \in \mathbb{R}^n$ such that

$$\langle \varphi, a_i \rangle = \langle \tilde{\varphi}, a_i^{(n)} \rangle_{\mathbb{R}^n}, \quad i = 1, 2, \ldots, m.$$

Since $a_k = \sum_{j=1}^m \beta_j a_j, k = d+1, \ldots, m$ for some $\{\beta_j\}$, we have

$$\langle \varphi, a_k \rangle = \sum_{j=1}^m \beta_j \langle \varphi, a_j \rangle = \sum_{j=1}^m \beta_j \langle \tilde{\varphi}, a_j^{(n)} \rangle_{\mathbb{R}^n} = \langle \tilde{\varphi}, a_k^{(n)} \rangle_{\mathbb{R}^n}, \quad k = d+1, \ldots, m,$$

and the assertion follows. □

Proof of Theorem 3.3.2 For $\hat{X} = \hat{X}(\eta_0, \eta_1, \eta_2, \ldots, \eta_t)$, $t > 0$, we are looking for $a \in \mathbb{R}$ and a process $\varphi_s = \varphi_s(\eta_0, \eta_1, \ldots, \eta_s)$ such that

$$\hat{X} = a + \sum_{s=0}^{t-1} \langle \varphi_s, \eta_{s+1} \rangle. \tag{3.3.4}$$

(Sufficiency) For fixed $t > 0$ let us consider the path of the discounted price process

$$\eta_0 = x_0, \ldots, \eta_{t-1} = x_{t-1},$$

which appears with positive probability. Given this path, the random variable $\hat{X}(x_0, x_1, \ldots, x_{t-1}, \eta_t)$ is a sequence of numbers, i.e.

$$\hat{X}(x_0, x_1, \ldots, \eta_t) = \hat{X}(x_0, x_1, \ldots, x_{t-1}, a_i^t), \qquad i = 1, 2, \ldots, d_t,$$

and in view of Proposition 3.3.3 there exist a real number $a_{t-1} = a_{t-1}(x_0, x_1, \ldots, x_{t-1})$ and a vector $\varphi_{t-1} = \varphi_{t-1}(x_0, x_1, \ldots, x_{t-1})$ in l^1 such that

$$\begin{aligned}\hat{X}(x_0, x_1, \ldots, x_{t-1}, a_i^t) &= a_{t-1}(x_0, x_1, \ldots, x_{t-1}) \\ &\quad + \langle \varphi_{t-1}(x_0, x_1, \ldots, x_{t-1}), a_i^t \rangle, \quad i = 1, 2, \ldots, d_t.\end{aligned}$$

Since the preceding relation holds for every path of positive probability, we obtain

$$\hat{X}(\eta_0, \eta_1, \ldots, \eta_t) = a_{t-1}(\eta_0, \eta_1, \ldots, \eta_{t-1}) + \langle \varphi_{t-1}(\eta_0, \eta_1, \ldots, \eta_{t-1}), \eta_t \rangle. \tag{3.3.5}$$

By induction, we obtain analogous representations for $a_{t-1}(\eta_0, \eta_1, \ldots, \eta_{t-1})$, i.e.

$$a_{t-1}(\eta_0, \eta_1, \ldots, \eta_{t-1}) = a_{t-2}(\eta_0, \eta_1, \ldots, \eta_{t-2}) + \langle \varphi_{t-2}(\eta_0, \eta_1, \ldots, \eta_{t-2}), \eta_{t-1} \rangle, \tag{3.3.6}$$

and continue the procedure till the final formula for $a_1(\eta_0, \eta_1)$:

$$a_1(\eta_0, \eta_1) = a_0(\eta_0) + \langle \varphi_0(\eta_0), \eta_1 \rangle. \tag{3.3.7}$$

Combining (3.3.5), (3.3.6) and (3.3.7) yields

$$\hat{X}(\eta_0, \eta_1, \ldots, \eta_t) = a_0(\eta_0) + \sum_{s=0}^{t-1} \langle \varphi_s(\eta_0, \eta_1, \ldots, \eta_s), \eta_{s+1} \rangle,$$

and thus (3.3.4) holds.

(Necessity) We show that completeness implies that for any $t > 0$ the set $\{a_1^t, a_2^t, \ldots, a_{d_t}^t\}$ satisfies $(ND1)$ or $(ND2)$. Since the market is complete, it follows from (3.3.4) that for any function $\hat{X}(\eta_0, \eta_1, \ldots, \eta_t)$ and a fixed trajectory that occurs with positive probability

$$\eta_0 = x_0, \ldots, \eta_{t-1} = x_{t-1},$$

we have the representation

$$\hat{X}(x_0, x_1, \ldots, x_{t-1}, a_i^t) = a_{t-1}(x_0, x_1, \ldots, x_{t-1}) + \langle \varphi(x_0, x_1, \ldots, x_{t-1}), a_i^t \rangle,$$
$$i = 1, 2, \ldots, d_t.$$

In view of Proposition 3.3.3 the set $\{a_1^t, a_2^t, \ldots, a_{d_t}^t\}$ satisfies $(ND1)$ or $(ND2)$. □

3.4 Approximate Completeness

In this section we characterize bond markets that are L^p-approximately complete, with $p \geq 1$, under the minimal filtration $\{\mathcal{F}_t\}$. Recall that for any discounted claim $\hat{X} = \hat{X}(\eta_0, \eta_1, \ldots, \eta_t) \in L^p(\Omega, \mathcal{F}_t, \mathbb{P})$ and any $\varepsilon > 0$ we are looking for an initial capital $x \in \mathbb{R}$ and l^1-valued strategy (φ_s) such that

$$\mathbb{E}\left(\mid x + \sum_{s=0}^{t-1} \langle \varphi_s, \eta_{s+1} \rangle - \hat{X} \mid^p \right) < \varepsilon. \tag{3.4.1}$$

The strategy (φ_s) is assumed to be adapted to $\{\mathcal{F}_s\}$ such that the corresponding discounted wealth is p-integrable, i.e.

$$\mathbb{E}(\mid \hat{X}(t) \mid^p) = \mathbb{E}\left(\mid x + \sum_{s=0}^{t-1} \langle \varphi_s, \eta_{s+1} \rangle \mid^p \right) < +\infty.$$

For the case $p = 2$ we will assume that

$$\mathbb{E}(\mid \varphi_s \mid_{l^1}^2) < +\infty, \quad s = 0, 1, \ldots, \tag{3.4.2}$$

that is, $\varphi_s \in L^2(\Omega, \mathcal{F}_s, \mathbb{P}; l^1)$. Since, for each $s > 0$, η_s takes values in m and its coordinates are in the interval $[0, 1]$, (3.4.2) implies that

$$\mathbb{E}\Big(\mid \hat{X}(t) \mid^2 \Big) \le 2^{t+1}\Big(x^2 + \sum_{s=0}^{t-1} \mathbb{E}(\mid \varphi_s \mid_{l^1}^2)\Big) < +\infty.$$

Hence, the requirement (3.4.2) makes the problem of L^2-approximate completeness well posed.

3.4.1 General Characterization

Here we characterize L^2-approximate completeness under the minimal filtration in the case when (η_t) is a general process taking values in m.

Theorem 3.4.1 *The bond market is L^2-approximately complete in the class of claims $\hat{X} \in L^2(\Omega, \mathcal{F}_t, \mathbb{P})$ and strategies satisfying (3.4.2) if and only if for any $t = 0, 1, 2, \ldots$ and $Y_{t-1} \in L^2(\Omega, \mathcal{F}_{t+1}, \mathbb{P})$ the following implication holds:*

$$\mathbb{E}[Y_{t+1} \mid \mathcal{F}_t] = 0, \quad \mathbb{E}[Y_{t+1}\eta_{t+1} \mid \mathcal{F}_t] = 0 \quad \Rightarrow \quad Y_{t+1} = 0. \tag{3.4.3}$$

Remark 3.4.2 If $\mathbb{P}$ is a martingale measure then the condition (3.4.3) is equivalent to the implication

$$X_t\hat{P}_t \text{ is a martingale} \quad \Rightarrow \quad X_t = 0,\ t = 1, 2, \ldots$$

for a square integrable martingale X_t with $X_0 = 0$. To see this note that the left side of (3.4.3) can be written with the use of the square integrable martingale X_t defined by

$$X_0 := 0, \quad X_t := \sum_{s=1}^{t} Y_s, \quad t = 1, 2, \ldots$$

in the following way:

$$\mathbb{E}[(X_{t+1} - X_t)(\hat{P}_{t+1} - \hat{P}_t) \mid \mathcal{F}_t] = 0, \quad t = 0, 1, \ldots. \tag{3.4.4}$$

Since $\hat{P}_t$ is a martingale, (3.4.4) is equivalent to the condition

$$\mathbb{E}[X_{t+1}\hat{P}_{t+1} \mid \mathcal{F}_t] = X_t\hat{P}_t, \quad t = 0, 1, \ldots,$$

which clearly means that the product $X_t\hat{P}_t$ is a martingale.

Proof of Theorem 3.4.1 We simplify notation in the proof by writing $L^2(\mathcal{F}_t) := L^2(\Omega, \mathcal{F}_t, \mathbb{P})$ and $L^2(\mathcal{F}_t; l^1) := L^2(\Omega, \mathcal{F}_t, \mathbb{P}; l^1)$, $L^2(\mathcal{F}_t; m) := L^2(\Omega, \mathcal{F}_t, \mathbb{P}; m)$.

First we show that the market is L^2-approximately complete if and only if for each $t = 0, 1, \ldots$ the operator $\mathcal{K}_t \colon L^2(\mathcal{F}_t) \times L^2(\mathcal{F}_t; l^1) \longrightarrow L^2(\mathcal{F}_{t+1})$ defined by

$$\mathcal{K}_t(Z, \varphi) = Z + \langle \varphi, \eta_{t+1} \rangle$$

has a dense image in $L^2(\mathcal{F}_{t+1})$. If the market is L^2-approximately complete then for any $\hat{X} \in L^2(\mathcal{F}_t)$ and $\varepsilon > 0$ there exist $x \in \mathbb{R}$ and $\{\varphi_s\}$ such that (3.4.1) holds with $p = 2$. Consequently, for $Z := x + \sum_{s=0}^{t-2} \langle \varphi_s, \eta_{s+1} \rangle$ and $\varphi := \varphi_{t-1}$ we obtain

$$\mathbb{E}\Big(\mid \hat{X} - \mathcal{K}_{t-1}(Z, \varphi) \mid^2 \Big) < \varepsilon,$$

which means that the image of $\mathcal{K}_{t-1}$ is dense in $L^2(\mathcal{F}_t)$.

Conversely, if the image of $\mathcal{K}_t$ is dense in $L^2(\mathcal{F}_{t+1})$ for any t, then for $\hat{X} \in L^2(\mathcal{F}_t)$ and $\varepsilon > 0$ we can find $Z_{t-1} \in L^2(\mathcal{F}_{t-1})$ and $\varphi_{t-1} \in L^2(\mathcal{F}_{t-1}; l^1)$ such that

$$\mathbb{E}\Big(\mid \hat{X} - (Z_{t-1} + \langle \varphi_{t-1}, \eta_t \rangle) \mid^2 \Big) < \varepsilon.$$

Since the image of $\mathcal{K}_{t-1}$ is also dense, for Z_{t-1} we can find $Z_{t-2} \in L^2(\mathcal{F}_{t-2})$ and $\varphi_{t-2} \in L^2(\mathcal{F}_{t-2}, l^1)$ such that

$$\mathbb{E}\Big(\mid Z_{t-1} - (Z_{t-2} + \langle \varphi_{t-2}, \eta_{t-1} \rangle) \mid^2 \Big) < \varepsilon,$$

which yields

$$\mathbb{E}\Big(\mid \hat{X} - (Z_{t-2} + \langle \varphi_{t-2}, \eta_{t-1} \rangle + \langle \varphi_{t-1}, \eta_t \rangle) \mid^2 \Big) < 4\varepsilon.$$

Repetition of the procedure provides $x \in \mathbb{R}$ and $\{\varphi_s\}$ such that

$$\mathbb{E}\Big(\mid \hat{X} - (x + \sum_{s=0}^{t-1} \langle \varphi_s, \eta_{s+1} \rangle) \mid^2 \Big) < k(t)\varepsilon,$$

where $k(t)$ is some constant. Hence, the market is L^2-approximately complete.

Now we show that (3.4.3) is equivalent to the fact that $\mathcal{K}_t$ has a dense image in $L^2(\mathcal{F}_{t+1})$. For $Y_{t+1} \in L^2(\mathcal{F}_{t+1})$ we have

$$\begin{aligned} \mathbb{E}[\mathcal{K}_t(Z, \varphi) \cdot Y_{t+1}] &= \mathbb{E}[Z Y_{t+1}] + \mathbb{E}[\langle \varphi, \eta_{t+1} \rangle Y_{t+1}] \\ &= \mathbb{E}\Big(Z \cdot \mathbb{E}[Y_{t+1} \mid \mathcal{F}_t]\Big) + \mathbb{E}\Big(\langle \varphi, \mathbb{E}[\eta_{t+1} Y_{t+1} \mid \mathcal{F}_t] \rangle\Big). \end{aligned} \tag{3.4.5}$$

Let us notice that $\mathbb{E}[Y_{t+1} \mid \mathcal{F}_t] \in L^2(\mathcal{F}_t)$ and $\mathbb{E}[\eta_{t+1} Y_{t+1} \mid \mathcal{F}_t] \in L^2(\mathcal{F}_t, m)$, so the operator

$$\mathcal{H}_t(Y_{t+1}) := \Big(\mathbb{E}[Y_{t+1} \mid \mathcal{F}_t], \mathbb{E}[\eta_{t+1} Y_{t+1} \mid \mathcal{F}_t]\Big) \tag{3.4.6}$$

acts from $L^2(\mathcal{F}_{t+1})$ into $L^2(\mathcal{F}_t) \times L^2(\mathcal{F}_t, m)$. But $L^2(\mathcal{F}_t) \times L^2(\mathcal{F}_t, m) \subseteq (L^2(\mathcal{F}_t) \times L^2(\mathcal{F}_t, l^1))^*$, which follows from Remark 3.4.3. It follows from (3.4.5) and (3.4.6) that

$$\langle \mathcal{K}_t(Z, \varphi), Y_{t+1} \rangle_{L^2(\mathcal{F}_{t+1}); L^2(\mathcal{F}_{t+1})} = \Big((Z, \varphi), \mathcal{H}_t(Y_{t+1})\Big)_{L^2(\mathcal{F}_t) \times L^2(\mathcal{F}_t, l^1); (L^2(\mathcal{F}_t) \times L^2(\mathcal{F}_t, l^1))^*},$$

so $\mathcal{K}_t^* = \mathcal{H}_t$. The image of $\mathcal{K}_t$ is dense in $L^2(\mathcal{F}_{t+1})$ if and only if $Ker\ \mathcal{K}_t^* = \{0\}$ that leads to (3.4.3). □

Remark 3.4.3 Let E be a Banach space with dual E^* and V be a measurable space with a measure μ. Then

$$L^2(\mu, E^*) \subseteq \left(L^2(\mu, E)\right)^*.$$

For $g \in L^2(\mu, E^*)$, let us define $G_g(f) := \int_V \left(g(v), f(v)\right)_{E^*;E}\mu(dv)$ for $f \in L^2(\mu, E)$. Then

$$\begin{aligned}
| G_g | &= \sup_{|f|\leq 1} \left| \int_V \left(g(v), f(v)\right)_{E^*;E}\mu(dv) \right| \\
&\leq \sup_{|f|\leq 1} \int_V | g(v) |_{E^*} | f(v) |_E \, \mu(dv) \\
&\leq | g |,
\end{aligned}$$

so G_g belongs to $\left(L^2(\mu, E)\right)^*$.

Remark 3.4.4 The general bond market is not L^2-approximately complete if trading strategies are restricted to finite portfolios only. To see this, let us consider the operator $\mathcal{K}_0 : \mathbb{R} \times l^1 \longrightarrow L^2(\Omega, \mathcal{F}_1, \mathbb{P})$ given by

$$\mathcal{K}_0(x, \varphi_0) = x + \langle \varphi_0, \eta_1 \rangle, \tag{3.4.7}$$

which we used in the proof of Theorem 3.4.1. Portfolios with a number of bonds bounded by n restrict the domain of $\mathcal{K}_0$ to $\mathbb{R} \times \mathbb{R}^n$ and then the image of $\mathcal{K}_0$ cannot be dense in $L^2(\Omega, \mathcal{F}_1, \mathbb{P})$ because it is a finite dimensional space.

3.4.2 Bond Curves in a Finite Dimensional Space

Let us now consider the L^p-approximate completeness problem in the particular case when the process (η_s) takes values is a finite dimensional subspace of m. The underlying filtration $\{\mathcal{F}_t\}$ is still assumed to be the minimal one.

Theorem 3.4.5 *Assume that η_t, for each $t > 0$, takes values in a finite dimensional subspace of m. The following conditions are equivalent.*

(a) *The market is L^p-approximately complete in the class $\hat{X} \in L^p(\Omega, \mathcal{F}_t, \mathbb{P})$ for some $p \in [1, +\infty)$.*
(b) *The market is L^p-approximately complete in the class $\hat{X} \in L^p(\Omega, \mathcal{F}_t, \mathbb{P})$ for each $p \in [1, +\infty)$.*
(c) *For each $t > 0$, η_t takes a finite number of values and conditional values of η_t given $\eta_0, \eta_1, \dots \eta_{t-1}$ satisfy* (ND1) *or* (ND2).

Proof Let η_t have a support K_t in some finite dimensional subspace H_t of m. We show that if the market is L^p-approximately complete for some $p \geq 1$ then K_t is finite for each $t > 0$ and the market is complete. The rest of the assertion follows from Theorem 3.3.2.

First we show the assertion for $t = 1$. Let μ be the distribution of η_1 and $\mathcal{H}_1 \subseteq L^p(m, \mu)$ be the linear space of functions of the form

$$\mathcal{H}_1 := \{h(y) : h(y) := x + \langle \varphi_0, y \rangle, \quad x \in \mathbb{R},\ \varphi_0 \in l^1\}.$$

Since H_1 is finite dimensional, so is $\mathcal{H}_1$. Since an arbitrary claim $\hat{X} \in L^p(m, \mu)$ can be approximated by elements from $\mathcal{H}_1$ it follows that $\mathcal{H}_1 = L^p(m, \mu)$. In particular $L^p(m, \mu)$ is finite dimensional, say $\dim L^p(m, \mu) = d$. We claim that the cardinality of K_1 must be no greater than d. Assume to the contrary that $\#K_1 \geq d + 1$. Then there exist disjoint sets $A_1, A_2, \ldots, A_{d+1}$ such that $\mu(A_j) > 0, j = 1, 2, \ldots, d + 1$. The functions

$$\mathbf{1}_{A_1}, \mathbf{1}_{A_2}, \ldots, \mathbf{1}_{A_{d+1}}$$

are linearly independent and all belong to $L^p(m, \mu)$. It follows that the dimension of $L^p(m, \mu)$ is greater than or equal to $d + 1$, which is a contradiction. Consequently K_1 is finite and the measure μ is a sum of a finite number of atoms.

By fixing $\eta_1 = x_1$ we can use the preceding arguments to show that the conditional distribution of η_2 is finite, which implies that K_2 is finite as well. By induction K_t is finite for any $t > 0$. □

3.4.3 Bond Curves in Hilbert Spaces

In this section we characterize L^2-approximately complete markets under the minimal filtration $\{\mathcal{F}_t\}$ in the case when the process (η_t) is a square integrable process taking values in some Hilbert space H. Since H should be a subspace of m, a good choice of H can be, for instance, the space of square summable sequences l^2 or its weighted version l^2_ρ. Let $\{e_k\}$ be an orthonormal complete basis in H. Then

$$\eta_t = \sum_{n=1}^{+\infty} \alpha_n(t) e_n, \quad t = 1, 2, \ldots \tag{3.4.8}$$

for some adapted square integrable functions $\{\alpha_n(t)\}$ such that

$$\mathbb{E}(\mid \eta_t \mid_H^2) = \mathbb{E}\Big(\sum_{n=1}^{+\infty} \mid \alpha_n(t) \mid^2 \Big) < +\infty, \quad t = 1, 2, \ldots. \tag{3.4.9}$$

This model setting allows us to construct some useful examples presented in the second part of this section. They are based on the following result.

Theorem 3.4.6 *Let (η_t) be an H-valued process with the representation* (3.4.8) *satisfying* (3.4.9).

(a) *Then the market is L^2-approximately complete in the class $\hat{X} \in L^2(\Omega, \mathcal{F}_t, \mathbb{P})$ if and only if, for each $t \geq 0$, the set*

$$\bar{\mathcal{A}}_1(t+1) := \left\{\mathbf{1}_{A_0}, \alpha_1(t+1)\mathbf{1}_{A_1}, \alpha_2(t+1)\mathbf{1}_{A_2}, \ldots; \quad A_0, A_1, \ldots \in \mathcal{F}_t\right\}$$

is linearly dense in $L^2(\Omega, \mathcal{F}_{t+1}, \mathbb{P})$. In particular, if, for any $t = 1, 2, \ldots$, the functions

$$\mathcal{A}_1(t) := \{1, \alpha_1(t), \alpha_2(t), \ldots\}$$

are linearly dense in $L^2(\Omega, \mathcal{F}_t, \mathbb{P})$, then the market is L^2-approximately complete.

(b) *If there exists a martingale measure $\mathbb{Q}$ with square integrable density then, for each $t \geq 0$, the set*

$$\bar{\mathcal{A}}(t+1) := \left\{\alpha_1(t+1)\mathbf{1}_{A_1}, \alpha_2(t+1)\mathbf{1}_{A_2}, \ldots; \quad A_0, A_1, \ldots \in \mathcal{F}_t\right\}$$

is not linearly dense in $L^2(\Omega, \mathcal{F}_{t+1}, \mathbb{P})$.

Proof (a) We characterize condition (3.4.3) form Theorem 3.4.1 in terms of the sequence $\{\alpha_n(t)\}$. For any $Y_{t+1} \in L^2(\Omega, \mathcal{F}_{t+1}, \mathbb{P})$, we have

$$\begin{aligned}\mathbb{E}(\eta_{t+1} Y_{t+1} \mid \mathcal{F}_t) &= \mathbb{E}\Big(\Big(\sum_{n=1}^{+\infty} \alpha_n(t+1) e_n\Big) Y_{t+1} \mid \mathcal{F}_t\Big) \\ &= \mathbb{E}\Big(\Big(\sum_{n=1}^{+\infty} \alpha_n(t+1) Y_{t+1} e_n\Big) \mid \mathcal{F}_t\Big),\end{aligned}$$

which, in view of (3.4.9), yields

$$\mathbb{E}(\eta_{t+1} Y_{t+1} \mid \mathcal{F}_t) = \sum_{n=1}^{+\infty} \mathbb{E}\Big(\alpha_n(t+1) Y_{t+1} \mid \mathcal{F}_t\Big) e_n.$$

Hence condition (3.4.3) can be reformulated to the form

$$\begin{aligned}\mathbb{E}[Y_{t+1} \mid \mathcal{F}_t] = 0, \qquad &\mathbb{E}(Y_{t+1}\alpha_n(t+1) \mid \mathcal{F}_t) = 0, \\ n = 1, 2, \ldots \quad &\Rightarrow \quad Y_{t+1} = 0,\end{aligned}$$

which means that if for all sets $A_0, A_1, \ldots \in \mathcal{F}_t$

$$\mathbb{E}[Y_{t+1}\mathbf{1}_{A_0}] = 0, \quad \mathbb{E}(Y_{t+1}\mathbf{1}_{A_n}\alpha_n(t+1)) = 0,$$

then $Y_{t+1} = 0$. This, however, clearly means that $\bar{\mathcal{A}}_1(t+1)$ is linearly dense in $L^2(\Omega, \mathcal{F}_{t+1}, \mathbb{P})$.

Since $\mathcal{A}_1(t) \subseteq \bar{\mathcal{A}}_1(t)$, it follows that the linear density of $\mathcal{A}_1(t)$ in $L^2(\Omega, \mathcal{F}_t, \mathbb{P})$ implies L^2-completeness.

(b) If $\mathbb{Q}$ is a martingale measure then

$$\mathbb{E}^{\mathbb{Q}}[\eta_{t+1} \mid \mathcal{F}_t] = 0, \quad t = 0, 1, \ldots. \tag{3.4.10}$$

By the Bayes rule,

$$\mathbb{E}^{\mathbb{Q}}[\eta_{t+1} \mid \mathcal{F}_t] = \frac{\mathbb{E}[\rho_{t+1}\eta_{t+1} \mid \mathcal{F}_t]}{\mathbb{E}[\rho_{t+1} \mid \mathcal{F}_t]}, \quad t = 0, 1, \ldots, \tag{3.4.11}$$

where (ρ_t) stands for the density process. By (3.4.10) and (3.4.11) we obtain

$$\mathbb{E}[\rho_{t+1}\eta_{t+1} \mid \mathcal{F}_t] = 0, \quad t = 0, 1, \ldots. \tag{3.4.12}$$

If (ρ_t) is square integrable, then, arguing as in (*a*), we see that (3.4.12) is equivalent to the condition

$$\mathbb{E}\Big(\rho_{t+1}\alpha_n(t+1) \mid \mathcal{F}_t\Big) = 0, \quad n = 1, 2, \ldots \quad t = 0, 2, \ldots.$$

Hence, for any $A_1, A_2, \ldots \in \mathcal{F}_t$,

$$\mathbb{E}[\rho_{t+1}\alpha_n(t+1)\mathbf{1}_{A_n}] = 0, \quad n = 1, 2, \ldots,$$

which means that ρ_{t+1} is orthogonal to all elements of $\bar{\mathcal{A}}(t+1)$. Thus $\bar{\mathcal{A}}(t+1)$ is not linearly dense in $L^2(\Omega, \mathcal{F}_{t+1}, \mathbb{P})$. □

It follows from Theorem 3.4.6 that each model of the form

$$\eta_t = \sum_{n=1}^{+\infty} \alpha_n(t) e_n,$$

where, for each $t > 0$, $\{1, \alpha_1(t), \alpha_2(t), \ldots\}$ are linearly dense in $L^2(\Omega, \mathcal{F}_t, \mathbb{P})$, is L^2-approximately complete. Let us recall that the necessary condition for the market with finite dimensional noise to be complete was that η_t, for each $t \geq 0$, takes a finite number of values only (see Theorem 3.4.5). This may suggest that L^2-approximate completeness can take place only if η_t is discretely distributed. In Example 3.4.7 we show that this conjecture is not true. We construct a one-period model that is L^2-approximately complete and η_1 has a non-atomic distribution. On the other hand, discrete distribution of η_1 does not imply L^2-approximate completeness. This we show in Example 3.4.8.

Example 3.4.7 The distribution of η_1 has an atom $a \in H$ if and only if

$$\mathbb{P}(\eta_1 = a) = c > 0.$$

If $a = \sum_{n=1}^{+\infty} a_n e_n$ then

$$\mathbb{P}(\eta_1 = a) = \mathbb{P}\Big(\alpha_n(1) = a_n, \ n = 1, 2, \ldots\Big).$$

Now, let $\{1, f_1, f_2, \ldots\}$ be an orthonormal complete basis in $L^2([0,1], dx)$ such that the distribution of at least one f_n has no atoms. Then for any sequence $\{\gamma_n\}$ such that $\gamma_n \neq 0, n = 1, 2, \ldots$ and $\sum \gamma_n^2 < +\infty$ the random variable

$$\eta_1 := \sum_{n=1}^{+\infty} \gamma_n f_n e_n$$

has a non-atomic distribution. Since the set $\{1, \gamma_1 f_1, \gamma_2 f_2, \ldots\}$ is linearly dense in $L^2([0,1], dx)$, by Theorem 3.4.6, the market is L^2-approximately complete.

Example 3.4.8 Let $\{e_i\}$ be an orthonormal basis in H and the distribution of η_1 be given by the following

$$\mathbb{P}(\eta_1 = e_i) = p_i > 0, \quad \sum_{i=1}^{+\infty} p_i < 1, \quad \mathbb{P}(\eta_1 = a) = p_a := 1 - \sum_{i=1}^{+\infty} p_i,$$

where $H \ni a := \sum a_i e_i$ and the sequence $\{a_i\}$ satisfies the conditions

$$\sum_{i=1}^{+\infty} a_i^2 < +\infty, \quad \sum_{i=1}^{+\infty} \frac{a_i^2}{p_i} < +\infty, \quad \sum_{i=1}^{+\infty} a_i = 1.$$

For the random variable Y defined by

$$Y := -p_a \sum_{i=1}^{+\infty} \frac{a_i}{p_i} \mathbf{1}_{\{\eta = e_i\}} + \mathbf{1}_{\{\eta = a\}},$$

we have

$$\begin{aligned}
\mathbb{E}(Y^2) &= p_a^2 \sum \left(\frac{a_i}{p_i}\right)^2 p_i + p_a = p_a^2 \sum \frac{a_i^2}{p_i} + p_a < +\infty, \\
\mathbb{E}(Y) &= -p_a \sum \frac{a_i}{p_i} p_i + p_a = p_a(1 - \sum a_i) = 0, \\
\mathbb{E}(\eta_1 Y) &= -p_a \sum \frac{a_i}{p_i} e_i p_i + p_a a = -p_a \sum a_i e_i + p_a a = 0.
\end{aligned}$$

In view of Theorem 3.4.1 the market is not L^2-approximately complete.

We close this section with an example showing that the uniqueness of the martingale measure with square integrable density does not imply L^2-approximate completeness.

Example 3.4.9 Let η_1 be given by (3.4.8) satisfying (3.4.9) such that

$$\mathbb{E}[\alpha_n(1)] = 0, \quad n = 1, 2, \ldots.$$

It follows from Theorem 3.4.6 that such a market is L^2-approximately complete if and only if the orthogonal complement of $\bar{\mathcal{A}}(1) = \{\alpha_1(1), \alpha_2(1), \ldots\}$ satisfies

$$\left(\text{span}\{\bar{\mathcal{A}}(1)\}\right)^{\perp} = \text{span}\{1\}. \tag{3.4.13}$$

To see this, assume that (3.4.13) is not satisfied. Then $(\text{span}\{\bar{\mathcal{A}}(1)\})^{\perp}$ is at least two dimensional and hence there exists $Y \in (\text{span}\{\bar{\mathcal{A}}(1)\})^{\perp}$, which is orthogonal to the vector 1, i.e. $\mathbb{E}[Y] = 0$. This implies, however, that the set $\{1, \alpha_1(1), \alpha_2(1), \ldots\}$ is not linearly dense in $L^2(\Omega, \mathcal{F}_1, \mathbb{P})$.

Now let us assume that

$$\left(\text{span}\{\alpha_1(1), \alpha_2(1), \ldots\}\right)^{\perp} = \text{span}\{1, Y\},$$

where $Y \in L^2(\Omega, \mathcal{F}_1, \mathbb{P})$ is a random variable which is unbounded from below and from above. This market is not L^2-approximately complete. Since each element in $\text{span}\{1, Y\}$ is of the form

$$Z(a, b) = a + bY, \qquad a, b \in \mathbb{R},$$

it follows that it takes negative values with positive probability providing that $b \neq 0$. Thus all positive elements are of the form $Z(a, 0), a > 0$. It follows that there exists a unique martingale measure with density in $L^2(\Omega, \mathcal{F}_1, \mathbb{P})$ and its density is given by $\rho = Z(1, 0) = 1$.

3.5 Models with Martingale Prices

Contingent claims do not necessarily have to be functions of bond curves. One can consider claims depending on some economical factors describing randomness in the bond market. Replicating strategies for such claims are constructed also with the use of information about evolution of the factors. Therefore we introduce here a new filtration $\{\mathcal{G}_t\}, t = 0, 1, \ldots$, which can be greater than the minimal one $\{\mathcal{F}_t\}$ generated by the bond curves, i.e.

$$\mathcal{F}_t \subseteq \mathcal{G}_t, \quad t = 0, 1, \ldots,$$

and consider the completeness problem under $\{\mathcal{G}_t\}$. So, for a $\mathcal{G}_t$-measurable discounted claim $\hat{X}$ paid at time $t > 0$ we are looking for the representation

$$\hat{X} = x + \sum_{s=0}^{t-1} \langle \varphi_s, \eta_{s+1} \rangle, \quad \mathbb{P} - a.s., \tag{3.5.1}$$

where $x \in \mathbb{R}$ and (φ_s) is an l^1-valued process adapted to $\{\mathcal{G}_t\}$. Recall that η_t is a sequence $\{\eta_t(T); T = t+1, t+2, \ldots\}$, belonging to the space m, defined by $\eta_t(T) = \hat{P}(t, T) - \hat{P}(t-1, T)$.

In the particular case, when $\{\mathcal{G}_t\}$ is the natural filtration of the martingale

$$Z(t) = \xi_1 + \xi_2 + \cdots + \xi_t, \quad t = 1, 2, \ldots,$$

where $\{\xi_i\}$ are zero-mean independent and identically distributed random variables in $U = \mathbb{R}^d$, (3.5.1) reads as

$$\hat{X}(\xi_0, \xi_1, \ldots, \xi_t) = x + \sum_{s=0}^{t-1} \langle \varphi_s(\xi_0, \xi_1, \ldots, \xi_s), \eta_{s+1} \rangle, \quad \mathbb{P} - \text{a.s.} \tag{3.5.2}$$

By definition $\xi_0 = 0$. Recall that by the generalized martingale representation property of Z, any martingale N adapted to $\{\mathcal{G}_t\}$ can be written in the form

$$N(t) = N_0 + \sum_{s=0}^{t-1} \int_U \psi(s, y) \tilde{\pi}(\{s+1\}, dy) \tag{3.5.3}$$

(see Section 2.3.2 for details), where $\psi(\cdot, y)$ is an adapted process, that is,

$$\psi(s, y) = h(s, \xi_1, \ldots, \xi_s, y)$$

for some function h and

$$\int_U \psi(s, y) \tilde{\pi}(\{s+1\}, dy) := h(s, \xi_1, \ldots, \xi_s, \xi_{s+1}) - \int_U h(s, \xi_1, \ldots, \xi_s, y) \mu(dy).$$

Here μ stands for the distribution of ξ_1. If $\hat{X} \in L^1(\Omega, \mathcal{G}_t, \mathbb{P})$ we can apply (3.5.3) with $N_s = \mathbb{E}[\hat{X} \mid \mathcal{F}_s]$ to obtain

$$\hat{X} = \mathbb{E}[\hat{X} \mid \mathcal{G}_t] = \mathbb{E}[\hat{X}] + \sum_{s=0}^{t-1} \int_U \psi_{\hat{X}}(s, y) \tilde{\pi}(\{s+1\}, dy) \tag{3.5.4}$$

for some adapted process $\psi_{\hat{X}}$. With the use of (3.5.4) one can formulate sufficient conditions for determining (3.5.2) but one needs to know that

$$\mathbb{E}[\eta_{s+1} \mid \mathcal{G}_s] = 0, \quad s = 0, 1, \ldots, t-1,$$

i.e. the discounted bond price process $\hat{P}(s, T)$ is a martingale for each $T = 1, 2, \ldots$. Therefore we assume in the present section that the condition (MP) is satisfied. The method of determining replicating strategy for $\hat{X}$ from (3.5.4) leads to the so-called *hedging equation*, which is commonly used in the continuous time setting. Therefore we discuss it in the sequel for concrete models in a detailed way.

3.5.1 HJM Models

Recall that in the HJM model

$$f(t+1, T) = f(t, T) + \alpha(t, T) + \langle \sigma(t, T), \xi_{t+1} \rangle, \quad 0 \le t < T \tag{3.5.5}$$

(see Section 2.3.4) the measure $\mathbb{P}$ is a martingale measure if and only if, for each $T = 1, 2, \ldots$,

$$\sum_{v=s}^{T-1} \alpha(s,v) = \varphi_\xi \left(\sum_{v=s}^{T-1} \sigma(s,v) \right), \quad s = 0,1,\ldots,T-1 \tag{3.5.6}$$

(see Theorem 2.2.1 in Section 2.2). Here φ_ξ stands for the Laplace exponent of ξ_1 and, as always, $\xi_1, \xi_2, \ldots$ is an i.i.d. sequence taking values in $U = \mathbb{R}^d$ with distribution of ξ_1 equal μ.

We consider the completeness problem under the filtration

$$\mathcal{G}_t := \sigma\{\xi_1, \xi_2, \ldots, \xi_t\}, \quad t = 1, 2, \ldots.$$

For an integrable claim $\hat{X} = \hat{X}(\xi_1, \xi_2, \ldots, \xi_t)$, let $\psi_{\hat{X}}$ be the process appearing in (3.5.4).

Theorem 3.5.1 *A claim $\hat{X} \in L^1(\Omega, \mathcal{G}_t, \mathbb{P})$ is attainable at t if the hedging equation*

$$\psi_{\hat{X}}(s,y) = \sum_{T \geq s+2} \varphi_s(T)\Gamma_s(T,y), \quad s = 0,1,\ldots,t-1, \quad y \in \text{supp}\{\xi_1\} \tag{3.5.7}$$

has an l^1-valued solution φ. Above $\Gamma_s(T,y)$ stands for

$$\Gamma_s(T,y) := \hat{P}(s,T)[e^{-\varphi_\xi\left(\sum_{v=s}^{T-1} \sigma(s,v)\right) - \langle \sum_{v=s}^{T-1} \sigma(s,v), y \rangle} - 1].$$

Consequently, if the hedging equation has a solution for any $\{\mathcal{G}_t\}$-adapted process $\psi(s,y)$, then the market is complete in the class $\hat{X} \in L^1(\Omega, \mathcal{G}_t, \mathbb{P})$.

Proof The discounted wealth at time t of a strategy φ with initial capital $x \in \mathbb{R}$ equals

$$\hat{X}(t) = x + \sum_{s=0}^{t-1} \langle \varphi_s, \eta_{s+1} \rangle. \tag{3.5.8}$$

We will write (3.5.8) in an alternative form. First we determine the formula for η_{s+1}. It follows from (3.5.12) that

$$\begin{aligned}\hat{P}(s+1,T)/\hat{P}(s,T) &= e^{-\sum_{v=0}^{T-1} f(s+1,v)} / e^{-\sum_{v=0}^{T-1} f(s,v)} \\ &= e^{-\{\sum_{v=0}^{T-1} (f(s+1,v) - f(s,v))\}} = e^{-\{\sum_{v=s}^{T-1} (\alpha(s,v) + \langle \sigma(s,v), \xi_{s+1} \rangle)\}},\end{aligned}$$

and consequently,

$$\begin{aligned}\eta_{s+1}(T) &= \hat{P}(s+1,T) - \hat{P}(s,T) = \hat{P}(s,T)[\hat{P}(s+1,T)/\hat{P}(s,T) - 1] \\ &= \hat{P}(s,T)[e^{-\{\sum_{v=s}^{T-1} (\alpha(s,v) + \langle \sigma(s,v), \xi_{s+1} \rangle)\}} - 1].\end{aligned}$$

By (3.5.6),

$$\eta_{s+1}(T) = \Gamma_s(T, \xi_{s+1}). \tag{3.5.9}$$

It follows from (3.5.8) and (3.5.9) that

$$\hat{X}(t) = x + \sum_{s=0}^{t-1} \sum_{T \geq s+2} \varphi_s(T)\Gamma_s(T, \xi_{s+1}),$$

which can be written in the form involving the jump measure $\pi(\{s\}, dy)$ of the process $Z(s) := \xi_1 + \cdots \xi_s$, that is,

$$\hat{X}(t) = x + \sum_{s=0}^{t-1} \sum_{T \geq s+2} \varphi_s(T) \int_U \Gamma_s(T, y)\pi(\{s+1\}, dy).$$

Since $\hat{P}(t, T)$ are martingales and therefore

$$\int_U \Gamma_s(T, y)\mu(dy) = 0,$$

we can replace the integration over $\pi(\{s+1\}, dy)$ by $\tilde{\pi}(\{s+1\}, dy)$. This yields

$$\hat{X}(t) = x + \sum_{s=0}^{t-1} \sum_{T \geq s+2} \varphi_s(T) \int_U \Gamma_s(T, y)\tilde{\pi}(\{s+1\}, dy). \tag{3.5.10}$$

Now we see from (3.5.10) and (3.5.4) that if (3.5.7) holds then $\hat{X}(t) = \hat{X}$ and $x = \mathbb{E}[\hat{X}]$. □

Corollary 3.5.2 *Using Theorem 3.5.1 we can formulate sufficient conditions for completeness in the case when ξ_1 takes a finite number of values $\{y_1, y_2, \ldots, y_K\}$. Indeed, if, for any $s = 0, 1, \ldots, t-1$, there exist maturities $T_1 < T_2 < \cdots < T_K$ such that*

$$\det([\Gamma_s(T_k, y_k)]_{k=1,2,\ldots,K}) \neq 0,$$

then the hedging equation has a solution for any $\psi_{\hat{X}}$ and the market is complete in the class $\hat{X} \in L^1(\Omega, \mathcal{G}_t, \mathbb{P})$.

Proposition 3.5.3 *If ξ_1 takes an infinite number of values in $U = \mathbb{R}^d$ and, for each $s = 0, 1, \ldots$, there exist maturities $T_1, T_2, \ldots, T_d$ such that*

$$\sigma(s, T_1), \sigma(s, T_2), \ldots, \sigma(s, T_d) \quad \text{are linearly independent in } \mathbb{R}^d, \tag{3.5.11}$$

then the market is not complete in the class $\hat{X} \in L^\infty(\Omega, \mathcal{G}_t, \mathbb{P})$.

Proof First we show that the increments of the discounted bond prices η_s take also an infinite number of values. Since in the proof of Theorem 3.5.1 we have shown that

$$\eta_{s+1}(T) = \Gamma_s(T, \xi_{s+1}),$$

it is enough to prove that

$$\Gamma_s(T, z) = \Gamma_s(T, y), \quad T = 1, 2, \ldots \quad \Longrightarrow \quad z = y.$$

But

$$\Gamma_s(T,z) = \Gamma_s(T,y) \quad \Longleftrightarrow \quad \langle \sum_{v=s}^{T-1} \sigma(s,v), z-y \rangle = 0.$$

The last condition is satisfied by (3.5.11). Moreover, (3.5.11) implies also the identity of $\{\mathcal{G}_t\}$ and the minimal filtration, i.e.

$$\mathcal{G}_t = \mathcal{F}_t = \sigma\{\eta_1, \eta_2, \ldots, \eta_s\}.$$

To see this, recall from Proposition 3.1.1 that $\{\mathcal{F}_t\}$ equals to the filtration $\{\mathcal{F}_t^f\}$ generated by the forward rate process and notice, by (3.5.12), that

$$\mathcal{F}_t^f = \mathcal{G}_t$$

if and only if (3.5.11) is satisfied. Thus it follows from Theorem 3.2.1 that the market is not complete. □

Approximate Completeness

We pass now to the approximate completeness of the HJM model. The following result proven as theorem 6.5 in Borwein and Erdélyi [21] will play an essential role. It is a version of the classical Müntz theorem published in Müntz [96].

Theorem 3.5.4 [Müntz's theorem] *Let $\lambda_0 = 0 < \lambda_1 < \lambda_2 \ldots$ and $\mu \neq 0$ be a finite measure concentrated on a bounded subset of $(0,+\infty)$ and absolutely continuous with respect to the Lebesgue measure. Then the set*

$$\text{span}\{1, x^{\lambda_1}, x^{\lambda_2}, \ldots\}$$

is dense in $L^p((0,+\infty),\mu)$, $p \in (0,+\infty)$ if and only if $\sum_{i=1}^{+\infty} \frac{1}{\lambda_i} = +\infty$.

Theorem 3.5.5 *Let in the HJM model*

$$f(t+1,T) = f(t,T) + \alpha(t,T) + \sigma(t,T)\xi_{t+1}, \quad 0 \le t < T \tag{3.5.12}$$

the distribution of ξ_1 be absolutely continuous with respect to the Lebesgue measure and concentrated on a bounded set. Let the volatility be positive, i.e.

$$\sigma(t,T) > 0, \quad 0 \le t < T, \quad T > 0. \tag{3.5.13}$$

Then the model is L^2-approximately complete under minimal filtration if and only if for each $t \ge 0$, almost surely

$$\sum_{T=t+2}^{+\infty} \frac{1}{\sum_{v=t}^{T-1} \sigma(t,v)} = +\infty. \tag{3.5.14}$$

Proof Condition (3.5.13) implies that $\{\mathcal{G}_t\}$ is identical with the minimal filtration. Taking into account this and also the fact that

$$\eta_{t+1}(T) = \hat{P}(t,T)[e^{-\varphi_\xi(\sum_{v=t}^{T-1}\sigma(t,v))-\sum_{v=t}^{T-1}\sigma(t,v)\xi_{t+1}} - 1],$$

the conditions in (3.4.3) in Theorem 3.4.1 take the form

$$\mathbb{E}[Y_{t+1} \mid \mathcal{G}_t] = 0$$

and

$$\mathbb{E}[Y_{t+1}\hat{P}(t,T)[e^{-\varphi_\xi(\sum_{v=t}^{T-1}\sigma(t,v))-\sum_{v=t}^{T-1}\sigma(t,v)\xi_{t+1}} - 1] \mid \mathcal{G}_t] = 0, \quad T = t+2, t+3, \ldots.$$

Recall that $\eta_{t+1}(T) = 0$ for $T \leq t+1$, so these values could be omitted in the preceding equation. Hence L^2-approximate completeness takes place if and only if for $Y_{t+1} \in L^2(\Omega, \mathcal{G}_{t-1}, \mathbb{P})$ the following conditions

$$\mathbb{E}[Y_{t+1} \mid \mathcal{G}_t] = 0, \quad \mathbb{E}[Y_{t+1}e^{-\sum_{v=t}^{T-1}\sigma(t,v)\xi_{t+1}} \mid \mathcal{G}_t] = 0, \quad T = t+2, t+3, \ldots \tag{3.5.15}$$

imply that $Y_{t+1} = 0$. Since Y_{t+1} can be written in the form

$$Y_{t+1} = h(\xi_1, \xi_2, \ldots, \xi_t, e^{-\xi_{t+1}}),$$

where h is a deterministic function, we can write (3.5.15) in the form

$$\int_I h(\xi_1, \ldots, \xi_t, y)\mu(dy) = 0, \quad \int_I h(\xi_1, \ldots, \xi_t, y)y^{\sum_{v=t}^{T-1}\sigma(t,v)}\mu(dy) = 0, \quad T = t+2, t+3, \ldots, \tag{3.5.16}$$

where $\mu(dy)$ stands for the distribution of $e^{-\xi_1}$ and I for its bounded support in $\mathbb{R}_+$. So, the market is L^2-approximately complete if and only if (3.5.16) implies that $h \equiv 0$. This means, however, that the functions

$$1, \quad y^{\sum_{v=t}^{T-1}\sigma(t,v)}, \quad T = t+2, t+3, \ldots$$

are linearly dense in $L^2(I, \mu)$. In view of Theorem 3.5.4 this is possible if and only if (3.5.14) is satisfied. □

Remark 3.5.6 Let us assume that in the preceding HJM model the volatility satisfies

$$\sum_{s=t}^{+\infty} \mid \sigma(t,s) \mid < +\infty, \quad t = 0, 1, \ldots. \tag{3.5.17}$$

Then, since (MP) holds, it follows from Theorem 2.2.3 that $\mathbb{P}$ is the unique martingale measure. Since (3.5.17) implies (3.5.14), the market is L^2-approximately complete in this case.

3.5.2 Multiplicative Factor Model

We focus now on the factor model from Proposition 2.4.15 where bond prices are given by

$$P(t,T) = F(T-t,X_t) \tag{3.5.18}$$

with $F(1,x) = x^{\gamma},\ x > 0$ and

$$X_{t+1} = aX_t + bX_t\xi_{t+1}, \quad X_0 = x > 0, \quad t = 1,2,\ldots. \tag{3.5.19}$$

Above $a, b \in \mathbb{R}, b \neq 0$ and $\xi_1, \xi_2, \ldots$ are independent and identically distributed. By Proposition 2.4.15 the model satisfies (MP) if and only if

$$F(k,x) = \Big(\prod_{j=1}^{k-1} c_{j\gamma}\Big)x^{k\gamma}, \tag{3.5.20}$$

where $c_\gamma := \mathbb{E}[(a+b\xi_1)^{\gamma}]$ and forward rates are positive if

$$\gamma > 0, \quad x \in (0,1), \quad 0 < a + b\xi_1 < 1 \tag{3.5.21}$$

or

$$\gamma < 0, \quad x > 1, \quad a + b\xi_1 > 1. \tag{3.5.22}$$

In what follows, we consider only models satisfying (3.5.20) and (3.5.21) or (3.5.22). Then

$$P(t,T) = c_\gamma c_{2\gamma} \ldots c_{(T-t-1)\gamma} X_t^{(T-t)\gamma} \tag{3.5.23}$$

and

$$B(t) = e^{\sum_{s=0}^{t-1} R(s)} = e^{\sum_{s=0}^{t-1} G(0,X_s)} = \prod_{s=0}^{t-1} X_s^{-\gamma}. \tag{3.5.24}$$

The completeness problem is considered under the filtration

$$\mathcal{G}_t := \sigma\{\xi_1, \xi_2, \ldots, \xi_t\}, \quad t = 1,2,\ldots,$$

which, clearly satisfies

$$\mathcal{G}_t = \mathcal{G}_t^X := \sigma\{X_1, X_2, \ldots, X_t\}, \quad t = 1,2,\ldots.$$

Theorem 3.5.7 *A claim $\hat{X} \in L^1(\Omega, \mathcal{G}_t, \mathbb{P})$ with the representation* (3.5.4) *is attainable if the hedging equation*

$$\psi_{\hat{X}}(s,y) = \sum_{T \geq s+2} \varphi_s(T)\Gamma_s(T,y), \quad s = 0,1,\ldots,t-1,\ y \in \operatorname{supp}\{\xi_1\} \tag{3.5.25}$$

has a solution φ taking values in l^1. Where

$$\Gamma_s(T,y) := \hat{P}(s,T)\left[\frac{1}{c_{(T-s-1)\gamma}}\,(a+by)^{(T-s-1)\gamma} - 1\right].$$

If the hedging equation has a solution for any $\{\mathcal{G}_t\}$-adapted process $\psi(s,y)$, then the market is complete in the class of claims $\hat{X} \in L^1(\Omega, \mathcal{G}_t, \mathbb{P})$.

Proof First let us determine the process (η_t). It follows from (3.5.23) and (3.5.24) that

$$\hat{P}(t,T) = P(t,T)/B(t) = \Big(\prod_{j=1}^{T-t-1} c_{j\gamma}\Big)\Big(\prod_{s=0}^{t-1} X_s^{\gamma}\Big)X_t^{(T-t)\gamma},$$

and consequently

$$\hat{P}(t+1,T)/\hat{P}(t,T) = \frac{1}{c_{(T-t-1)\gamma}}\,X_t^{\gamma}\,\frac{X_{t+1}^{(T-t-1)\gamma}}{X_t^{(T-t)\gamma}} = \frac{1}{c_{(T-t-1)\gamma}}\left(\frac{X_{t+1}}{X_t}\right)^{(T-t-1)\gamma}$$
$$= \frac{1}{c_{(T-t-1)\gamma}}\,(a+b\xi_{t+1})^{(T-t-1)\gamma}.$$

Hence

$$\eta_{t+1}(T) = \hat{P}(t+1,T) - \hat{P}(t,T) = \hat{P}(t,T)[\hat{P}(t+1,T)/\hat{P}(t,T) - 1]$$
$$= \hat{P}(t,T)\left[\frac{1}{c_{(T-t-1)\gamma}}\,(a+b\xi_{t+1})^{(T-t-1)\gamma} - 1\right]$$
$$= \Gamma_t(T,\xi_{t+1}).$$

The discounted wealth of a pair (x,φ) is given thus by

$$\hat{X}(t) = x + \sum_{s=0}^{t-1}\langle \varphi_s, \eta_{s+1}\rangle = x + \sum_{s=0}^{t-1}\sum_{T\ge s+2} \varphi_s(T)\eta_{s+1}(T)$$
$$= x + \sum_{s=0}^{t-1}\sum_{T\ge s+2} \varphi_s(T)\Gamma_s(T,\xi_{s+1})$$

and can be written in the form

$$\hat{X}(t) = x + \sum_{s=0}^{t-1}\sum_{T\ge s+2} \varphi_s(T)\int_U \Gamma_s(T,y)\tilde{\pi}(\{s+1\},dy). \tag{3.5.26}$$

If (φ_s) is such that (3.5.25) holds, then, by (3.5.4) and (3.5.26), $\hat{X}(t) = \hat{X}$ and $x = \mathbb{E}[\hat{X}]$. □

Corollary 3.5.8 *If ξ_1 takes a finite number of values $\{y_1, y_2, \dots, y_K\}$ and if, for any $s = 0, 1, \dots, t-1$, there exist maturities $T_1 < T_2 < \cdots < T_K$ such that*

$$\det([\Gamma_s(T_k, y_k)]_{k=1,2,\dots,K}) \neq 0,$$

then the hedging equation has a solution for any $\psi_{\hat{X}}$ and the market is complete in the class $\hat{X} \in L^1(\Omega, \mathcal{G}_t, \mathbb{P})$.

Corollary 3.5.9 *It is clear that*

$$b \neq 0 \quad \Longrightarrow \quad \Gamma_s(T, y) \neq \Gamma_s(T, z) \quad \text{for } y \neq z.$$

Consequently, if ξ_1 takes an infinite number of values then so does $\eta_{s+1}(T) = \Gamma_s(T, \xi_{s+1})$. If $b \neq 0$ then $\{\mathcal{G}_t\}$ is identical with the minimal filtration and, consequently, by Theorem 3.2.1, the market is not complete in the class $\hat{X} \in L^\infty(\Omega, \mathcal{G}_t, \mathbb{P})$.

Approximate Completeness

Now we pass to the L^2-approximate completeness of the multiplicative factor model in the class $\hat{X} \in L^2(\Omega, \mathcal{G}_t, \mathbb{P})$.

Theorem 3.5.10 *If the distribution of ξ_1 is absolutely continuous with respect to the Lebesgue measure, then the model is L^2-approximately complete in the class of claims $\hat{X} \in L^2(\Omega, \mathcal{G}_t, \mathbb{P})$.*

In the proof we use again the Müntz theorem 3.5.4.

Proof By (3.5.23) and (3.5.24) we obtain

$$\hat{P}(t, T) = P(t, T)/B(t) = \prod_{s=0}^{t-1} X_s^\gamma c_\gamma c_{2\gamma} \dots c_{(T-t-1)\gamma} X_t^{(T-t)\gamma}$$

and consequently

$$\begin{aligned} \eta_{t+1}(T) &= \hat{P}(t+1, T) - \hat{P}(t, T) \\ &= c_\gamma c_{2\gamma} \dots c_{(T-t-2)\gamma} x^\gamma X_1^\gamma \dots X_{t-1}^\gamma \left(X_t^\gamma X_{t+1}^{(T-t-1)\gamma} - c_{(T-t-1)\gamma} X_t^{(T-t)\gamma} \right), \\ & T = t+2, t+3, \dots \end{aligned}$$

for $t = 0, 1, \dots$. Since, for $b \neq 0$, the filtration $\{\mathcal{G}_t\}$ is identical with the minimal one, for the characterization of L^2-approximate completeness we can apply Theorem 3.4.1. One can check that condition (3.4.3) has the form

$$\begin{gathered} \mathbb{E}(Y_{t+1} \mid \mathcal{F}_t) = 0, \quad \mathbb{E}(Y_{t+1} X_{t+1}^{(T-t-1)\gamma} \mid \mathcal{F}_t) = 0, \\ T = t+2, t+3, \dots \Longrightarrow Y_{t+1} = 0. \end{gathered}$$

Since $Y_{t+1} = h(X_1, X_2, \dots, X_{t+1})$, for a deterministic function h, the Markovianity of X implies that the left side of the preceding implication can be written as

$$\int_{\mathbb{R}_+} h(X_1, X_2, \ldots, X_t, y)\ \mathcal{P}(X_t, dy) = 0,$$

$$\int_{\mathbb{R}_+} h(X_1, X_2, \ldots, X_t, y) y^{n\gamma}\ \mathcal{P}(X_t, dy) = 0, \quad n = 1, 2, \ldots. \tag{3.5.27}$$

In (3.5.27) $\mathcal{P}(x, \cdot)$ stands for the transitions function of the process X. Denoting, $\mu(dy) := \mathcal{P}(X_t, dy)$ and $\psi(y) := h(X_1, X_2, \ldots, X_t, y)$ we see that (3.5.27) has the form

$$\int_{\mathbb{R}_+} \psi(y) y^{\gamma n}\ \mu(dy) = 0, \quad n = 0, 1, 2, \ldots. \tag{3.5.28}$$

It follows that the model is L^2-approximately complete if and only if the functions

$$1, y^{\gamma}, y^{2\gamma}, \ldots \tag{3.5.29}$$

are linearly dense in $L^2(\mathbb{R}, \mu(dy))$. Let us consider two cases when $\gamma > 0$ and $\gamma < 0$. If $\gamma > 0$ then the measure μ is concentrated on $(0, 1)$. Since

$$\sum_{n=1}^{+\infty} \frac{1}{n\gamma} = +\infty, \quad \gamma > 0,$$

it follows from Lemma 3.5.4 that (3.5.29) are linearly dense $L^2((0, 1), \mu)$.

If $\gamma < 0$ then μ is concentrated on $(1, +\infty)$ and (3.5.28) can be written in the form

$$\int_1^{+\infty} \psi(y) \frac{1}{y^{|\gamma| n}}\ \mu(dy) = 0, \quad n = 0, 1, 2, \ldots. \tag{3.5.30}$$

Let us consider the transformation $T(y) = \frac{1}{y}, y \in (1, +\infty)$ and the measure $\nu := T \circ \mu$ on $(0, 1)$ given by

$$\nu(A) = (T \circ \mu)(A) := \mu\{z \in (1, +\infty) : T(z) \in A\}, \quad A \subseteq (0, 1).$$

Then

$$\int_1^{+\infty} \psi(T(y))\ \mu(dy) = \int_0^1 \psi(z)\ (T \circ \mu)(dz).$$

Since $\mu = T^{-1} \circ \nu$, (3.5.30) can be reformulated to

$$\begin{aligned} \int_1^{+\infty} \psi(y) \frac{1}{y^{|\gamma| n}}\ \mu(dy) &= \int_1^{+\infty} \psi(y) \frac{1}{y^{|\gamma| n}}\ (T^{-1} \circ \nu)(dy) \\ &= \int_0^1 \psi\left(\frac{1}{z}\right) z^{|\gamma| n}\ [T \circ (T^{-1} \circ \nu)](dz) \\ &= \int_0^1 \psi\left(\frac{1}{z}\right) z^{|\gamma| n}\ \nu(dz) = 0. \end{aligned}$$

To show that (3.5.29) is linearly dense we can use the same arguments as in the case $\gamma > 0$. □

3.5.3 Affine Models

Let us now consider the affine model

$$P(t,T) = e^{-C(T-t)-D(T-t)R(t)}, \quad t \leq T < +\infty, \tag{3.5.31}$$

where the short-rate R is a Markov chain with transition function

$$\mathcal{P}(x,dy) = (\mu_x * \nu)(dy), \quad x, y \geq 0. \tag{3.5.32}$$

Here μ_x is a convolution semigroup and ν a probability measure, both on $\mathbb{R}_+$. In Section 2.4.2 we described functions $C(\cdot), D(\cdot)$ for which the model satisfies (MP). Recall that functions φ, ψ given by

$$\int_{\mathbb{R}_+} e^{-\lambda y}\mu_x(dy) = e^{-\psi(\lambda)x}, \quad \int_{\mathbb{R}_+} e^{-\lambda y}\nu(dy) = e^{-\varphi(\lambda)},$$

provide the following formulas

$$C(T) - C(T-1) = \varphi(D(T-1)), \quad D(T) = \psi(D(T-1)) + 1, \quad T = 1, 2, \ldots. \tag{3.5.33}$$

and $C(0) = D(0) = 0$. Moreover, ψ admits the representation

$$\psi(\lambda) = \beta\lambda + \int_{\mathbb{R}_+} (1 - e^{-\lambda y})m(dy), \quad \lambda \geq 0, \tag{3.5.34}$$

where $\beta \geq 0$ and the measure m satisfies

$$\int_{\mathbb{R}_+} (1 \wedge y)m(dy) < +\infty.$$

The completeness problem is considered under the filtration

$$\mathcal{G}_t := \sigma\{R(0), R(1), \ldots, R(t)\}, \quad t \geq 0,$$

which, however, in view of (3.5.31) is identical with the minimal filtration, i.e.

$$\mathcal{G}_t = \mathcal{F}_t, \quad t = 0, 1, \ldots. \tag{3.5.35}$$

First we argue that non-trivial affine models are not complete. Let us notice that $R(t)$ takes a finite number of values if and only if the convolution $(\mu_x * \nu)(dy)$ has finite support. This is possible only when the support of ν is finite and μ_x is concentrated on the point βx, i.e. $\mu_x = \mathbf{1}_{\{\beta x\}}$. In the opposite case $R(t)$ takes infinite many values and, by Theorem 3.2.1 and (3.5.35), we obtain the following corollary.

Corollary 3.5.11 *If in the transition function* (3.5.32) *of the short-rate process the support of* ν *is infinite or* $\mu_x \neq \mathbf{1}_{\{\beta x\}}$ *then the affine model is not complete in the class* $\hat{X} \in L^\infty(\Omega, \mathcal{G}_t, \mathbb{P})$.

Approximate Completeness

The following result characterizes the L^2-approximate completeness of the affine models.

Theorem 3.5.12 *Let the affine model (3.5.31) driven by a Markovian short rate with the transition function (3.5.32) satisfy (MP) and β be given by (3.5.34). Let, in addition, the transition function be absolutely continuous with respect to Lebesgue measure. Then the following statements are true.*

(a) If $\beta < 1$ then the market is L^2-approximately complete.
(b) If $\beta > 1$ then the market is not L^2-approximately complete.
(c) If $\beta = 1$ and the measure in (3.5.34) is γ-stable, i.e.

$$m(dy) = \frac{1}{y^{1+\gamma}} \mathbf{1}_{[0,+\infty)}(y)dy, \quad \gamma \in (0,1),$$

then the market is not L^2-approximately complete.

Proof By (3.5.35) we can use Theorem 3.4.1. Since

$$\begin{aligned}\eta_{t+1}(T) &= \hat{P}(t+1,T) - \hat{P}(t,T)\\ &= e^{-\sum_{s=0}^{t} R(s)} e^{-C(T-t-1)-D(T-t-1)R(t+1)} - e^{-\sum_{s=0}^{t-1} R(s)} e^{-C(T-t)-D(T-t)R(t)},\end{aligned}$$

we can write (3.4.3) in the form

$$\begin{aligned}\mathbb{E}(Y_{t+1} \mid \mathcal{F}_t) = 0, \quad \mathbb{E}(Y_{t+1} e^{-D(T-t-1)R(t+1)} \mid \mathcal{F}_t) = 0,\\ T = t+2, t+3, \ldots \implies Y_{t+1} = 0.\end{aligned} \tag{3.5.36}$$

Since $Y_{t+1} = h(R(1), \ldots, R(t), e^{-R(t+1)})$, for some function h, we can write the left side of the preceding implication in the form

$$\int_0^1 h(R(1), \ldots, R(t), y)\, \mu(dy) = 0,$$

$$\int_0^1 h(R(1), \ldots, R(t), y) y^{D(T-t-1)}\, \mu(dy) = 0, \quad T = t+2, t+3, \ldots,$$

where $\mu(dy)$ stands for the conditional distribution of $e^{-R(t+1)}$ given $R(t)$. Denoting $g(y) := h(R(1), \ldots, R(t), y)$ we can thus write (3.5.3) in the form

$$\int_0^1 g(y)\, \mu(dy) = 0, \quad \int_0^1 g(y) y^{D(n)}\, \mu(dz) = 0, \quad n = 1, 2, \ldots \quad \implies \quad h = 0. \tag{3.5.37}$$

So, it follows that the market is L^2-approximately complete if and only if the functions $z^{D(n)}, n = 1, 2, 3, \ldots$ are linearly dense in $L^2((0,1), \mu)$. By Theorem 3.5.4 this is the case if and only if

$$\sum_{n=1}^{+\infty} \frac{1}{D(n)} = +\infty. \tag{3.5.38}$$

In the proof of Theorem 2.4.11 we have shown that $D(n)$ is increasing and

$$\lim_{n \longrightarrow +\infty} D(n) = +\infty \quad \Longleftrightarrow \quad \beta \geq 1. \tag{3.5.39}$$

It follows from (3.5.39) that (3.5.38) is satisfied for $\beta < 1$. For the case $\beta > 1$, we obtain from (3.5.34) that $\psi(\lambda) \geq \beta\lambda, \lambda \geq 0$. This together with (3.5.33) yields

$$D(n) \geq 1 + \beta + \beta^2 + \cdots + \beta^{n-1}, \quad n \geq 1. \tag{3.5.40}$$

Consequently, $D(n) \geq \beta^{n-1}$ and (3.5.38) clearly fails for $\beta > 1$. So, the model is L^2-approximately complete for $\beta < 1$ and is not for $\beta > 1$. Hence (a) and (b) follow.

Now we prove (c). One can check (see Example 5.3.7 for details) that

$$\int_0^{+\infty} (1 - e^{-\lambda y}) \frac{1}{y^{1+\gamma}} dy = c\lambda^\gamma,$$

where $c = \frac{\Gamma(1-\gamma)}{\gamma}$ and Γ stands for the gamma function. Consequently,

$$\psi(\lambda) = \beta\lambda + c\lambda^\gamma.$$

By (2.4.20) we have that

$$D(n+1) = D(n) + c(D(n))^\gamma + 1, \quad D(0) = 0. \tag{3.5.41}$$

It is clear that $D(n) \geq n, n = 0, 1, \ldots$, so (3.5.41) yields

$$D(n+1) = (n+1) + c\sum_{k=1}^{n} (D(k))^\gamma \geq (n+1) + c\sum_{k=1}^{n} k^\gamma. \tag{3.5.42}$$

Now we show that

$$\sum_{k=1}^{n} k^\gamma \geq \left(\frac{n}{2}\right)^{1+\gamma}. \tag{3.5.43}$$

If n is even, i.e. $n = 2l$, $l = 1, 2, \ldots$, then

$$\sum_{k=1}^{n} k^\gamma \geq \sum_{k=l+1}^{2l} k^\gamma \geq l \cdot l^\gamma = l^{1+\gamma} = \left(\frac{n}{2}\right)^{1+\gamma}.$$

If $n = 2l + 1$, $l = 1, 2, \ldots$ then

$$\sum_{k=1}^{n} k^\gamma \geq \sum_{k=l+1}^{2l+1} k^\gamma \geq (l+1)(l+1)^\gamma = (l+1)^{\gamma+1}$$

$$\geq \left(\frac{n-1}{2} + 1\right)^{1+\gamma} = \left(\frac{n+1}{2}\right)^{1+\gamma} \geq \left(\frac{n}{2}\right)^{1+\gamma}.$$

From (3.5.42) and (3.5.43) we obtain that (3.5.38) fails, so the market is not L^2-approximately complete. □

Remark 3.5.13 The transition functions $\mathcal{P}(x,\cdot), x > 0$ are absolutely continuous with respect to Lebesgue measue if, for instance, ν or $\mu_x, x > 0$ are. Specific condition for the latter can be found in Grzywacz, Leżaj and Trajan [65].

We apply Theorem 3.5.12 to the affine model for which the short-rate process $R(t), t \geq 0$ solves the stochastic equation

$$R(t+1) = F(R(t)) + G(R(t))\xi_{t+1}, \quad R(0) = R_0. \tag{3.5.44}$$

We know from Section 2.4.3 that (3.5.44) defines a model satisfying (MP) if it is of the form

$$R(t+1) = (aR(t) + \tilde{a}) + (bR(t) + \tilde{b})^{\frac{1}{\alpha}}\xi_{t+1}, \quad t = 0, 1, \dots, \tag{3.5.45}$$

where ξ_t is a sequence of independent standard α-stable distributions and $a, \tilde{a}, b, \tilde{b}$ are nonnegative numbers, or

$$R(t+1) = aR(t) + \xi_{t+1}, \quad t = 0, 1, \dots, \tag{3.5.46}$$

where (ξ_t) is an arbitrary sequence of independent, identically distributed nonnegative random variables and $a \geq 0$.

Theorem 3.5.14 *The affine model with short rate given by* (3.5.45) *or* (3.5.46) *is* L^2*-approximately complete if and only if* $a < 1$.

Proof By Proposition 2.4.8 and Proposition 2.4.9 we know that

$$\psi(\lambda) = a\lambda + b\lambda^{\alpha}, \text{ or } \quad \psi(\lambda) = a\lambda, \quad \lambda \geq 0.$$

Thus

$$\psi(\lambda) = a\lambda + c\int_{\mathbb{R}_+}(1 - e^{-\lambda y})\frac{1}{y^{1+\alpha}}dy$$

with some $c \geq 0$. The result follows from Theorem 3.5.12. □

3.6 Replication with Finite Portfolios

In this section we show how martingale measures and martingale representation property can be used to characterize replicating portfolios. Our analysis is concerned with contingent claims at time t_0 of the form

$$X = h\Big(P(t_0, T_1), P(t_0, T_2), \dots, P(t_0, T_n)\Big), \tag{3.6.1}$$

which are some function of prices of bonds with future maturities $T_1 < T_2 < \cdots < T_n$, where $t_0 < T_1$. Many important contingent claims can be represented in this way. In particular, the so called LIBOR spot rate (London Interbank Offered Rate) for $[t_0, T_1]$ defined by

$$L(t_0, T_1) := \frac{1 - P(t_0, T_1)}{(T_1 - t_0)P(t_0, T_1)},$$

and the related caplet, respectively floorlet, given by $(L(t_0, T_1) - K)^+$, respectively $(K - L(t_0, T_1))^+$, with some $K > 0$ (see Filipović [52]).

Our aim is to formulate conditions allowing to replicate claims of the form (3.6.1) only with the use of bonds with maturities $t_0, T_1, \ldots, T_n$ and the bank account, of course. The problem may seem to be embraced by the classical model setting with a finite number of trading assets, where the uniqueness of martingale measure implies completeness. There is, however, a subtle difference concerning the bank account process that makes the problem more involved. Recall that by (1.1.5) the bank account process is given by

$$B(t) = \frac{1}{P(0,1)P(1,2) \cdot \ldots \cdot P(t-1,t)}, \qquad t = 1, 2, \ldots.$$

This means that for the determination of the bank account process on the interval $[0, t_0]$ we need all bonds with maturities lying in $[0, t_0]$. In other words, the bank account process is not predictable with respect to the σ-field generated merely by the bonds with maturities $t_0, T_1, \ldots, T_n$, in general. The predictability of $B(t)$ with respect to the σ-field generated by risky assets is, however, necessary for the implication of completeness by the uniqueness of martingale measure (see Shiryaev [116, p. 482 and p. 495]). Hence, to solve our problem it is not sufficient to find conditions for the uniqueness of a martingale measure for the bonds with maturities $t_0, T_1, \ldots, T_n$.

Our solution of the problem formulated above provides conditions for the increments of discounted prices of bonds with maturities $t_0, T_1, \ldots, T_n$. The analysis is based on the condition (MM), i.e. we assume here that there exists a measure $\mathbb{Q}$ such that the discounted bond prices $\hat{P}(t, T), t \in [0, T]$ are $\mathbb{Q}$-martingales for any $T = 1, 2, \ldots$.

Theorem 3.6.1 *Any contingent claim of the form* (3.6.1) *can be replicated by a strategy involving bank account and bonds with maturities* $t_0, T_1, \ldots, T_n$ *if and only if the* $\mathbb{Q}$*-martingale*

$$S(t) := (\hat{P}(t, t_0), \hat{P}(t, T_1), \ldots, \hat{P}(t, T_n)), \quad t = 0, 1, \ldots, t_0$$

is such that, for each $t = 1, 2, \ldots, t_0$*, its increment* $\Delta S(t) = S(t) - S(t-1)$ *for a given path* $S(0), S(1), \ldots, S(t-1)$*, is concentrated on a finite set* A_t *in* $\mathbb{R}^{n+1}$ *such that*

$$dim(span\ A_t) = m(t) - 1, \tag{3.6.2}$$

where $m(t) := \sharp A_t$ *stands for the number of elements of* A_t*. If this is the case then the price of* X *equals* $\mathbb{E}^{\mathbb{Q}}[X/B(t_0)]$ *and the replicating strategy is unique providing that* $m(t) = n + 2$ *for each* $t = 1, 2, \ldots, t_0$.

Proof In view of Proposition 1.4.1, the contingent claim X can be replicated by a strategy involving bonds with maturities $t_0, T_1, \ldots, T_n$ if and only if the discounted claim $\hat{X} = X/B(t_0)$ admits the representation

$$\hat{X} = x + \sum_{s=0}^{t_0-1} \langle \varphi_s, \triangle S(s+1) \rangle \tag{3.6.3}$$

for some $x \in \mathbb{R}$ and $\mathcal{F}^S$-adapted process φ. First we show that the discounted value of the claim (3.6.1) admits the representation

$$\hat{X} = g\Big(\hat{P}(t_0-1, t_0), \hat{P}(t_0, T_1), \ldots, \hat{P}(t_0, T_n)\Big) \tag{3.6.4}$$

for some function g. Due to the specific form of the discount factor

$$1/B(t) = P(0,1)P(1,2) \cdot \ldots \cdot P(t-1,t), \quad t = 1, 2, \ldots \tag{3.6.5}$$

(see (1.1.5)), we obtain the following formula

$$\hat{P}(t,T) = P(0,1)P(1,2) \cdot \ldots \cdot P(t-1,t)P(t,T), \quad t \leq T. \tag{3.6.6}$$

Recursive application of (3.6.6) with the initial condition $\hat{P}(0,T) = P(0,T)$ yields

$$P(t,T) = \hat{P}(t,T)/\hat{P}(t-1,t), \quad t \leq T. \tag{3.6.7}$$

As a consequence of (3.6.5) and (3.6.7) we obtain

$$1/B(t) = \hat{P}(0,1) \cdot \hat{P}(1,2)/\hat{P}(0,1) \cdot \ldots \cdot \hat{P}(t-1,t)/\hat{P}(t-2,t-1) = \hat{P}(t-1,t) \tag{3.6.8}$$

for $t = 1, 2, \ldots$. By (3.6.1), (3.6.7) and (3.6.8) we obtain

$$\hat{X} = \hat{P}(t_0-1,t_0) \cdot h\Big(\hat{P}(t_0,T_1)/\hat{P}(t_0-1,t_0), \ldots, \hat{P}(t_0,T_n)/\hat{P}(t_0-1,t_0)\Big),$$

so (3.6.4) holds.

By Theorem 2.3.1, condition (3.6.2) is necessary and sufficient for S to have the martingale representation property. Since the process

$$M(t) = \mathbb{E}^{\mathbb{Q}}[\hat{X} \mid \mathcal{F}^S(t)], \quad t = 0, 1, \ldots, t_0$$

is a $\mathbb{Q}$-martingale, it can be represented in the form

$$M(t) = M(0) + \sum_{s=0}^{t-1} \langle \varphi(s), \triangle S(s+1) \rangle, \quad t = 0, 1, \ldots, t_0$$

for some $\mathcal{F}^S$-adapted process φ. Since $\hat{X}$ is $\mathcal{F}^S_{t_0}$-measurable, we have that $M(t_0) = \hat{X}$ and consequently (3.6.3) is satisfied with $x = M(0)$. Taking $\mathbb{Q}$-expectations in (3.6.3) yields that the price of X equals $\mathbb{E}^{\mathbb{Q}}[\hat{X}]$. The uniqueness of φ in the case when $m(t) = n+2$ follows from Theorem 2.3.1. □

Proposition 3.6.2 *Let us consider the HJM model*

$$f(t+1,T) = f(t,T) + \alpha(t,T) + \sigma(t,T)\xi_{t+1}, \quad t,T = 0,1,\dots,T^*-1,$$

where $\{\xi_t\}$ is a sequence of i.i.d. real valued random variables taking k values, where $k \geq 1$. Let us assume that

$$\alpha(t,T) = G\Big(t,T,\sigma(t,t+1),\sigma(t,t+2),\dots,\sigma(t,T)\Big) \tag{3.6.9}$$

for some function G, and, for given constants $t_0 < T_1 < \cdots < T_n \leq T^*$,

$$\sigma(t,T) = g\Big(t,T,\hat{P}(t,t_0),\hat{P}(t,T_1),\dots,\hat{P}(t,T_n)\Big) \tag{3.6.10}$$

for some function g. Let the model admit a martingale measure $\mathbb{Q}$. Then the property of replicating claims of the form

$$X = h\Big(P(t_0,T_1),P(t_0,T_2),\dots,P(t_0,T_n)\Big) \tag{3.6.11}$$

with the use of strategies involving bonds with maturities $t_0, T_1, \dots, T_n$ is preserved for $k \leq n+1$ and fails for $k \geq n+3$.

Remark 3.6.3 Recall that in Proposition 2.2.2 we constructed a market satisfying (3.6.9) that admits a martingale measure. Condition (3.6.10) can be viewed as an additional requirement specifying volatility of the model.

Proof of Proposition 3.6.2 We examine condition (3.6.2). By (2.2.2) we obtain the following formula for $\triangle\hat{P}(t+1,T) = \hat{P}(t+1,T) - \hat{P}(t,T)$, with $T = 1,2,\dots,T^*$,

$$\begin{aligned}\triangle\hat{P}(t+1,T) &= e^{-\sum_{s=0}^{T-1} f(t+1,s)} - e^{-\sum_{s=0}^{T-1} f(t,s)} \\ &= e^{-\sum_{s=0}^{T-1} f(t,s)}\Big[e^{-\sum_{s=0}^{T-1}\{\alpha(t,s)+\sigma(t,s)\xi_{t+1}\}} - 1\Big] \\ &= \hat{P}(t,T)\Big[e^{A(t,T)+\Sigma(t,T)\xi_{t+1}} - 1\Big],\end{aligned} \tag{3.6.12}$$

with $A(t,T) := -\sum_{s=t}^{T-1}\alpha(t,s)$, $\Sigma(t,T) := -\sum_{s=t}^{T-1}\sigma(t,s)$. Consequently, the increments of the $\mathbb{Q}$-martingale

$$S(t) := \Big(\hat{P}(t,t_0),\hat{P}(t,T_1),\dots,\hat{P}(t,T_n)\Big)$$

have the form

$$\triangle S(t+1) := \begin{pmatrix} \hat{P}(t,t_0)\Big[e^{A(t,t_0)+\Sigma(t,t_0)\xi_{t+1}} - 1\Big] \\ \hat{P}(t,T_1)\Big[e^{A(t,T_1)+\Sigma(t,T_1)\xi_{t+1}} - 1\Big] \\ \vdots \\ \hat{P}(t,T_n)\Big[e^{A(t,T_n)+\Sigma(t,T_n)\xi_{t+1}} - 1\Big] \end{pmatrix},$$

and by (3.6.9) and (3.6.10) we obtain that for given values $S(0), S(1), \ldots, S(t)$, the number of values of $\triangle S(t+1)$ equals the number of values of ξ_{t+1}, providing that $\Sigma(t,t_0), \Sigma(t,T_1), \ldots, \Sigma(t,T_n)$ do not vanish. Let $a_1, a_2, \ldots, a_k$ stand for the values of ξ_1. Then the values of $\triangle S(t+1)$ are $v_1, v_2, \ldots, v_k$ with

$$v_i := \begin{pmatrix} \hat{P}(t,t_0)\left[e^{A(t,t_0)+\Sigma(t,t_0)a_i} - 1\right] \\ \hat{P}(t,T_1)\left[e^{A(t,T_1)+\Sigma(t,T_1)a_i} - 1\right] \\ \vdots \\ \hat{P}(t,T_n)\left[e^{A(t,T_n)+\Sigma(t,T_n)a_i} - 1\right] \end{pmatrix}, \quad i = 1,2,\ldots,k.$$

It follows from the existence of a martingale measure that the vectors $v_1, v_2, \ldots, v_k$ are linearly dependent. We argue now that in the set $v_1, \ldots, v_k$ there are $k-1$ linearly independent vectors providing that $k \leq n+1$. By Corollary 2.2.6 we see that the vectors

$$\begin{pmatrix} e^{A(t,t_0)+\Sigma(t,t_0)a_i} \\ e^{A(t,T_1)+\Sigma(t,T_1)a_i} \\ \vdots \\ e^{A(t,T_n)+\Sigma(t,T_n)a_i} \end{pmatrix}, \quad i = 1,2,\ldots,k$$

are linearly independent. It follows from Lemma 3.6.4 that the set

$$\begin{pmatrix} e^{A(t,t_0)+\Sigma(t,t_0)a_i} - 1 \\ e^{A(t,T_1)+\Sigma(t,T_1)a_i} - 1 \\ \vdots \\ e^{A(t,T_n)+\Sigma(t,T_n)a_i} - 1 \end{pmatrix}, \quad i = 1,2,\ldots,k$$

contains $k-1$ linearly independent vectors. The same result is clearly true for the original set $\{v_1, \ldots, v_k\}$. Consequently, (3.6.2) is satisfied if $k \leq n+1$ and fails if $k \geq n+3$. □

Lemma 3.6.4 *Let the vectors $w_1, w_2, \ldots, w_n$ be linearly independent. Then in the set*

$$\{w_1 - z, w_2 - z, \ldots, w_n - z\}, \tag{3.6.13}$$

where z is some vector, there exist $n-1$ linearly independent vectors.

Proof Let us assume to the contrary that each arbitrarily chosen $n-1$ vectors from (3.6.13) are linearly dependent. Then there exist at least constants $c_1, \ldots, c_{n-1}$, not all equal zero, such that

$$c_1(w_1 - z) + c_2(w_2 - z) + \cdots + c_{n-1}(w_{n-1} - z) = 0.$$

Equivalently,

$$c_1 w_1 + c_2 w_2 + \cdots + c_{n-1} w_{n-1} - (c_1 + \cdots + c_{n-1}) z = 0.$$

Since $w_1, \ldots, w_{n-1}$ are linearly independent, we have that $c_1 + \cdots + c_{n-1} \neq 0$ and consequently

$$z = \tilde{c}_1 w_1 + \cdots + \tilde{c}_{n-1} w_{n-1} \tag{3.6.14}$$

for some constants $\tilde{c}_1, \ldots, \tilde{c}_{n-1}$. We may assume that $\tilde{c}_1 \neq 0$. Repeating the same procedure with the set $\{w_2, \ldots, w_n\}$, we obtain

$$z = \tilde{\tilde{c}}_2 w_2 + \cdots + \tilde{\tilde{c}}_n w_n. \tag{3.6.15}$$

Subtracting (3.6.15) from (3.6.14) yields

$$0 = \tilde{c}_1 w_1 + (\tilde{c}_2 - \tilde{\tilde{c}}_2) w_2 + \cdots + (\tilde{c}_{n-1} - \tilde{\tilde{c}}_{n-1}) w_{n-1} - \tilde{\tilde{c}}_n w_n$$

with the non-vanishing sequence $\tilde{c}_1, \tilde{c}_2 - \tilde{\tilde{c}}_2, \ldots, \tilde{c}_{n-1} - \tilde{\tilde{c}}_{n-1}, \tilde{\tilde{c}}_n$. This contradicts the linear independence of $w_1, \ldots, w_n$. □

3.7 Completeness and Martingale Measures

In the previous section we showed that completeness of the market may hold only if increments of the discounted bond prices $\eta_1, \eta_2, \ldots$ take a finite number of values. Such a market, with a finite time horizon $t > 0$, can be treated as a classical stock market with a finite number of shares, which additionally take a finite number of values. To see this, let us notice that all possible values of the random variables

$$\eta_0, \eta_1, \ldots, \eta_t$$

form also a finite set in m and, by Proposition 3.3.4, one can find $n > 0$ such that for any $\varphi \in l^1$,

$$\langle \varphi, \eta_i \rangle = \langle \varphi^{(n)}, \eta_i^{(n)} \rangle_{\mathbb{R}^n}, \quad i = 1, 2, \ldots, t, \quad \mathbb{P} - a.s.,$$

where $b^{(n)}$ stands for the truncation of $b \in l^1$ to the first n coordinates. This means that, without losing generality, one can trade with bonds with maturities from the set $1, 2, \ldots, n$ only. In addition to that, for each $T \in \{0, 1, \ldots, n\}$ the discounted bond price $\hat{P}(s, T)$ takes a finite number of values for each $s = 1, 2, \ldots, t$. This simple model setting allows us to use the classical Fundamental Theorems of Asset Pricing describing properties of the model in terms of martingale measures.

In the sequel $\{\mathcal{F}_t\}$ stands for the minimal filtration.

Theorem 3.7.1 *Assume that, for each $t > 0$, the random variable η_t takes a finite number of values.*

(a) Then the market is arbitrage free if and only if there exists a martingale measure.

(b) Assume that the market admits a martingale measure. Then the market is complete under the minimal filtration if and only if the martingale measure is unique.

In view of Theorem 3.3.2 and Theorem 3.7.1 we obtain the following result, which, in particular, shows that completeness does not imply existence of a martingale measure.

Proposition 3.7.2 *Let the bond market be complete, i.e. for any t the conditional distribution of η_t with respect to $\eta_0, \eta_1, \dots, \eta_{t-1}$ is concentrated on the set $\{a_1^t, a_2^t, \dots, a_{d_t}^t\}$, with $d_t < +\infty$, of vectors in m which satisfies (ND1) or (ND2). Then the following statements are true.*

(a) If $\{a_1^t, a_2^t, \dots, a_{d_t}^t\}$ satisfy (ND1) for some $t > 0$ then the market does not admit a martingale measure.

(b) Let, for each $t > 0$, the vectors $\{a_1^t, a_2^t, \dots, a_{d_t}^t\}$ satisfy (ND2). Assume for simplicity that $a_1^t, a_2^t, \dots, a_{d_t-1}^t$ are linearly independent. Define the positive cone of the set $a_1^t, a_2^t, \dots, a_{d_t-1}^t$ by

$$Con_t^+ := \{u \in m : u = \sum_{i=1}^{d_t-1} \alpha_i a_i^t;\ \alpha_i > 0\}.$$

Then

i) *If, for each $t > 0$, $-a_{d_t} \in Con_t^+$ then the market admits a unique martingale measure.*

ii) *If, for some $t > 0$, $-a_{d_t} \notin Con_t^+$ then the market does not allow a martingale measure.*

Proof By definition, a measure $\mathbb{Q}$ is a martingale measure if

$$\mathbb{E}^{\mathbb{Q}}[\hat{P}(t,T) \mid \mathcal{F}_{t-1}] = \hat{P}(t-1,T), \quad T > t,$$

or, equivalently,

$$\mathbb{E}^{\mathbb{Q}}[\eta_t \mid \mathcal{F}_{t-1}] = \sum_{i=1}^{d_t} a_i^t q_i = 0, \tag{3.7.1}$$

where $q_i := \mathbb{Q}(\eta_t = a_i^t \mid \eta_0 = x_0, \eta_1 = x_1, \dots, \eta_{t-1} = x_{t-1}) > 0$ for $i = 1, 2, \dots, d_t$.

(a) If a martingale measure exists then condition (3.7.1) clearly means that vectors $a_1^t, \dots, a_{d_t}^t$ are linearly dependent which contradicts (*ND*1).

(b) If $-a_{d_t} \in Con_t^+$ then from the unique representation of $-a_{d_t}$,

$$-a_{d_t} = \sum_{i=1}^{d_t-1} \alpha_i a_i^t$$

we can construct a martingale measure in a unique way via

$$q_i := \frac{\alpha_i}{\sum_{j=1}^{d_t-1}\alpha_j + 1}, \; i = 1, 2, \ldots, d_t - 1, \qquad q_{d_t} := \frac{1}{\sum_{j=1}^{d_t-1}\alpha_j + 1}.$$

If $-a_{d_t} \notin Con_t^+$ then at least one q_i defined above is nonpositive. □

A Counterexample

It turns out that when $\eta_t, t = 1, 2, \ldots$ take an infinite number of values, uniqueness of the martingale measure does not imply completeness. Now we construct an example, in the spirit of Proposition 1.5.3, of a one period regular bond market for which there exists a unique martingale measure but the model is not complete.

Proposition 3.7.3 *There exists a one period regular bond market model that admits a unique martingale measure and is not complete in the class $\hat{X} \in L^\infty(\Omega, \mathcal{F}_1, \mathbb{P})$, where $\mathcal{F}_1$ is the minimal σ-field at time* 1.

Proof We construct the required market on the probability space $\Omega = \{\omega_1, \omega_2, \ldots\}$ consisting of natural numbers, i.e. $\omega_k = k, k = 1, 2, \ldots$, with the measure $\mathbb{P}(\{\omega_k\}) = \frac{1}{2^k}, k = 1, \ldots$. Let the initial prices be given by

$$P(0, T) = \frac{1}{2^T}, \quad T = 0, 1, \ldots.$$

If the market is regular then the prices at time 1 satisfy

$$P(1, 1) = 1, \quad 0 < P(1, T) \le 1, \quad P(1, T) \ge P(1, T + 1), \quad T = 1, 2, \ldots. \tag{3.7.2}$$

Using

$$\eta(T) = \eta_1(T) := \hat{P}(1, T) - \hat{P}(0, T) = P(1, T)P(0, 1) - P(0, T), \quad T = 1, 2, \ldots,$$

one can check, as we did in the proof of Proposition 1.5.3, that (3.7.2) is equivalent to

$$\begin{aligned} \eta(1) = 0, \quad \eta(T) + P(0, T) > 0, \\ \eta(T + 1) - \eta(T) \le P(0, T) - P(0, T + 1), \quad T = 1, 2, \ldots. \end{aligned} \tag{3.7.3}$$

Let us define $\eta(T)$ by

$$\eta(1) \equiv 0, \quad \eta(T)(\omega_k) := \begin{cases} \frac{1}{2^{T+3}} & \text{for } k = T - 1, \\ -\frac{1}{2^{T+2}} & \text{for } k = T, \\ 0 & \text{elsewhere.} \end{cases} \qquad T = 2, 3, \ldots,$$

One can check directly that (3.7.3) is satisfied, so the market is regular. Let $\mathbb{Q}$ be a measure equivalent to $\mathbb{P}$ and set $q_i := \mathbb{Q}(\omega_i), i = 1, 2, \ldots$. The martingale condition

$$\mathbb{E}^{\mathbb{Q}}[\eta] = 0$$

reads as

$$0 = \mathbb{E}^{\mathbb{Q}}[\eta(T)] = q_{T-1}\frac{1}{2^{T+3}} - q_T\frac{1}{2^{T+2}}, \quad T = 2, 3, \ldots.$$

This yields $q_T = \frac{1}{2}q_{T-1}$ for $T = 2, 3, \ldots$ and consequently

$$\sum_{i=1}^{+\infty} q_i = q_1 \sum_{i=0}^{+\infty} \frac{1}{2^i} = 2q_1 = 1$$

if and only if $q_1 = \frac{1}{2}$. This means that there exists exactly one martingale measure $\mathbb{Q}$ and it is equal to $\mathbb{P}$. Since η takes an infinite number of values, by Theorem 3.2.1, we see that the market is not complete. □

Remark 3.7.4 The bond market from Proposition 3.7.3 is L^2-approximately complete. To see this, we use Theorem 3.4.1. For $Y \in L^2(\Omega, \mathcal{F}_1, \mathbb{P})$, we have

$$\mathbb{E}[Y] = \sum_{k=1}^{+\infty} Y(\omega_k)\frac{1}{2^k}$$

and for $T = 2, 3, \ldots$,

$$\begin{aligned}
\mathbb{E}[Y\eta(T)] &= \sum_{k=1}^{+\infty} Y(\omega_k)\eta(T)(\omega_k)\frac{1}{2^k} \\
&= Y(\omega_{T-1})\eta(T)(\omega_{T-1})\frac{1}{2^{T-1}} + Y(\omega_T)\eta(T)(\omega_T)\frac{1}{2^T} \\
&= Y(\omega_{T-1})\frac{1}{2^{T+3}}\frac{1}{2^{T-1}} - Y(\omega_T)\frac{1}{2^{T+2}}\frac{1}{2^T} \\
&= \frac{1}{2^{2T+2}}(Y(\omega_{T-1}) - Y(\omega_T)).
\end{aligned}$$

Hence

$$\mathbb{E}[Y] = 0, \quad \mathbb{E}[Y\eta(T)] = 0, \quad T = 2, 3, \ldots$$

if and only if $Y \equiv 0$. So, condition (3.4.3) is satisfied and the assertion follows.

Part II

Fundamentals of Stochastic Analysis

4

Stochastic Preliminaries

This chapter is devoted to general stochastic analysis and covers such topics as square integrable martingales, Doob–Meyer decomposition, semimartingales and their characteristics, random measures, compensators and compensating measures. Stochastic integration with respect to semimartingales and random measures is presented as well. The material is compiled with almost no proofs and is based on Protter [102], Jacod and Shiryaev [75], where more detailed presentations are available.

4.1 Generalities

Let us recall that a measurable space $(E, \mathcal{E})$ consists of a set E and a σ-field $\mathcal{E}$ of its subsets. If E is a metric space then $\mathcal{B}(E)$ stands for the Borel σ-field, the smallest σ-field containing all open subset of E. If $(E_1, \mathcal{E}_1)$, $(E_2, \mathcal{E}_2)$ are two measurable spaces then the product of $\mathcal{E}_1$ and $\mathcal{E}_2$ is denoted by $\mathcal{E}_1 \times \mathcal{E}_2$ and it is the smallest σ-field containing all sets of the form $A \times B$, where $A \in \mathcal{E}_1$, $B \in \mathcal{E}_2$. If μ_1, μ_2 are measures on $(E_1, \mathcal{E}_1)$, $(E_2, \mathcal{E}_2)$, respectively, then $\mu_1 \times \mu_2$ denotes the product measure on $(E_1 \times E_2, \mathcal{E}_1 \times \mathcal{E}_2)$. Sometimes we simplify the notation and write also $\mu_1(de_1)\mu(de_2)$ for the product measure.

Let $(\Omega, \mathcal{F}, \mathbb{P})$ be a probability space with filtration $\mathcal{F}_t, t \geq 0$, satisfying $\mathcal{F} = \mathcal{F}_\infty$. The filtration is assumed to be right continuous, i.e. $\mathcal{F}_t = \bigcap_{s>t} \mathcal{F}_s$ for $t \geq 0$, and to satisfy the *usual conditions*, which means that $\mathcal{F}$ is *complete*, i.e.

$$B \subseteq A,\ A \in \mathcal{F},\ \mathbb{P}(A) = 0 \quad \Longrightarrow \quad B \in \mathcal{F},$$

and $\mathcal{F}_t$ contains all $\mathbb{P}$-null sets of $\mathcal{F}$ for each $t \geq 0$. A transformation $X \colon \Omega \times \mathbb{R}_+ \mapsto U$, where U is a separable Hilbert space with the Borel σ-field $\mathcal{B}(U)$, is called a *stochastic process* if it is a measurable function from $(\Omega \times \mathbb{R}_+, \mathcal{F} \times \mathcal{B}(\mathbb{R}_+)$ to $(U, \mathcal{B}(U))$. It will be denoted by $X_t, t \geq 0$, (X_t), $X(t)$ or just X for short. For a fixed time point $t \geq 0$, the function

$$\omega \mapsto X_t(\omega)$$

is a U-valued random variable and for any $\omega \in \Omega$,

$$t \mapsto X_t(\omega)$$

is a U-valued function. It is called a *path* or *trajectory* of X. The paths of X are *càdlàg* if they are right continuous on $[0, +\infty)$ and have left limits on $(0, +\infty)$ in U. If this is the case then the process of *left limit* X_-:

$$X_{t-} = \lim_{s \uparrow t} X_s, \quad t > 0, \tag{4.1.1}$$

and the process of *jumps* $\triangle X$:

$$\triangle X_t := X_t - X_{t-}, \quad t > 0, \tag{4.1.2}$$

are well defined in U. If $\triangle X_t = 0, t \geq 0$, then X is called *continuous*. Two processes X and Y are *indistinguishable* if their paths are identical almost surely, i.e.

$$\mathbb{P}(X(t) = Y(t), t \geq 0) = 1. \tag{4.1.3}$$

If the laws of X_t and Y_t are equal for each $t \geq 0$, i.e.

$$\mathbb{P}(X_t = Y_t) = 1, \quad t \geq 0, \tag{4.1.4}$$

then Y is a *modification* or *version* of X. It is clear that (4.1.3) implies (4.1.4) and it is easy to construct an example showing that the converse implication is not true. However, under the assumption that both X and Y have càdlàg paths, (4.1.3) and (4.1.4) are equivalent.

The following boundedness properties of càdlàg processes can be proven by application of elementary arguments to their paths.

Proposition 4.1.1 *Let X be a process with càdlàg paths in U. Then for any $t > 0$*

(a) and any $\varepsilon > 0$, we have

$$\mathbb{P}\Big(\sharp\{s \in [0,t] :\mid \triangle X(s) \mid > \varepsilon\} < +\infty\Big) = 1,$$

(b) we have

$$\mathbb{P}\Big(\sup_{s\in[0,t]} \mid X(s) \mid < +\infty\Big) = 1, \quad t \geq 0.$$

Above $| \cdot |$ stands for the norm in U. The result tells us that, on each finite time interval, the number of jumps of a càdlàg process exceeding a positive constant is finite and that the process is bounded. The second property is the same as for continuous processes, but a càdlàg process does not have to take its extreme values on compacts.

Recall that if the process X is such that X_t is $\mathcal{F}_t$-measurable for any $t \geq 0$ then it is called *adapted*. Adapted processes generate two σ-fields on the space $\Omega \times \mathbb{R}_+$, which we introduce now.

Definition 4.1.2 The σ-field on $\tilde{\Omega} = \Omega \times \mathbb{R}_+$ generated by all adapted processes with càdlàg (resp. continuous) paths is called optional (predictable) and denoted by $\mathcal{O}$, $(\mathcal{P})$.

A stochastic processes is called optional (predictable) if it is $\mathcal{O}$ $(\mathcal{P})$-measurable. It is clear that $\mathcal{P} \subseteq \mathcal{O}$, so each predictable process is also optional. It can be proved that any optional process is adapted and that the σ-field $\mathcal{P}$ is equal to that generated by all adapted left continuous processes. Since every process X with continuous paths can be pointwise approximated by the sequence

$$X^n(t) := X_0 \mathbf{1}_{\{0\}}(t) + \sum_{k=1}^{+\infty} X_{\frac{k-1}{n}} \mathbf{1}_{(\frac{k-1}{n}, \frac{k}{n}]}(t), \qquad t \geq 0, \qquad n = 1, 2, \ldots,$$

we obtain the following alternative characterization of $\mathcal{P}$.

Proposition 4.1.3 *The σ-field $\mathcal{P}$ is generated by the following family of sets*

$$A \times \{0\}, A \in \mathcal{F}_0 \quad \textit{and} \quad A \times (s, t],\ A \in \mathcal{F}_s,\ s < t.$$

4.2 Doob–Meyer Decomposition

In this section we present some basic facts concerning martingales and submartingales taking values in $U = \mathbb{R}^d$ including the Doob–Meyer decomposition, which is of prime importance for the sequel.

Definition 4.2.1 An adapted process X taking values in U satisfying $\mathbb{E}[\mid X_t \mid] < +\infty, t \geq 0$ is a *martingale* if

$$\mathbb{E}[X(t) \mid \mathcal{F}_s] = X(s), \quad 0 \leq s \leq t.$$

If $U = \mathbb{R}$ and $\mathbb{E}(X_t \mid \mathcal{F}_s) \geq X_s$, $(\mathbb{E}(X_t \mid \mathcal{F}_s) \leq X_s)$ then X is called a submartingale (supermartingale).

A martingale M is square integrable if

$$\sup_{t \geq 0} \mathbb{E}(\mid M(t) \mid^2) < +\infty.$$

The space of all square integrable U-valued martingales with càdlàg paths will be denoted by $\mathcal{M}^2(U)$ while $\mathcal{M}^{2,c}(U)$ stands for all elements of $\mathcal{M}^2(U)$ with continuous paths. For $U = \mathbb{R}$ we abbreviate the notation by setting $\mathcal{M}^2 := \mathcal{M}^2(\mathbb{R})$ and $\mathcal{M}^{2,c} := \mathcal{M}^{2,c}(\mathbb{R})$.

Recall that a $[0, +\infty]$-valued random variable τ is a *stopping time* with respect to the filtration $\mathcal{F}_t$ if $\{\tau \leq t\} \in \mathcal{F}_t$ for each $t \geq 0$.

Definition 4.2.2 An adapted process M is called a *local martingale* if there exists a sequence of stopping times $\{\tau_n\}$ such that

$$\tau_n \leq \tau_{n+1}, \qquad \lim_{n\to+\infty} \tau_n = +\infty, \quad \mathbb{P} - a.s. \tag{4.2.1}$$

and the stopped process $M^{\tau_n}(t) := M(\tau_n \wedge t)$ is a martingale for each n.

The sequence $\{\tau_n\}$ satisfying (4.2.1) will be called a *localizing sequence* for M. In the definition of local martingale one can, additionally, require that M^{τ_n} is a uniformly integrable martingale. In fact both definitions are equivalent. If M is a martingale then $\tau_n := n, n = 1, 2, \ldots$ defines its localizing sequence and hence M is also a local martingale. There are local martingales that fail to be martingales. By $\mathcal{M}^2_{loc}(U)$ we denote the class of local martingales M such that there exist localizing sequences $\{\tau_n\}$ such that $M^{\tau_n} \in \mathcal{M}^2(U)$. Likewise we define the classes $\mathcal{M}^{2,c}_{loc}(U)$, $\mathcal{M}^2_{loc}$ and $\mathcal{M}^{2,c}_{loc}$.

For the path regularity of martingales and submartingales we need the concept of stochastic continuity. Recall that a U-valued process is *stochastically continuous* if

$$\lim_{s\to t} \mathbb{P}(| X_t - X_s |> \varepsilon) = 0, \qquad \varepsilon > 0,\ t \geq 0.$$

Clearly, any process with continuous paths is stochastically continuous but the opposite implication is not true. For stochastically continuous submartingales we have, however, the following regularity result.

Theorem 4.2.3 *Any stochastically continuous submartingale has a modification with càdlàg paths.*

It follows from Theorem 4.2.3 that any stochastically continuous submartingale X the processes X_-, $\triangle X$, (see (4.1.1), (4.1.2)) are well defined. This holds, in particular, if X is a real-valued martingale.

Theorem 4.2.4 (Optional Sampling) *Let X be a càdlàg submartingale and τ a bounded stopping time. Then $X(\tau)$ is integrable and*

$$\mathbb{E}[X(\tau) \mid \mathcal{F}_\sigma] \geq X(\tau \wedge \sigma)$$

for any stopping time σ.

It follows immediately that

$$\mathbb{E}[X(t \wedge \tau) \mid \mathcal{F}_s] \geq X(s \wedge \tau), \qquad 0 \leq s \leq t$$

for any stopping time τ and a submartingale X from Theorem 4.2.4, which means that the *stopped process* $X^\tau(t) := X(\tau \wedge t)$ is also a càdlàg submartingale.

The following result, which is due to Doob and Meyer, describes decomposition of submartingales and plays a central role in further study of the properties of square integrable martingales and compensators of increasing processes in the sequel. We need an auxiliary definition.

Definition 4.2.5 A process X is of class (D) if the set of random variables

$$\{X_\tau : \tau - \text{finite valued stopping time}\}$$

is uniformly integrable.

Recall, that a family of random variables $(Y_\alpha)_{\alpha\in A}$ is *uniformly integrable* if

$$\lim_{n\longrightarrow+\infty} \sup_{\alpha\in A} \mathbb{E}[\mid Y_\alpha \mid \mathbf{1}_{\{|Y_\alpha|\geq n\}}] = 0.$$

Theorem 4.2.6 (Doob–Meyer) *For any càdlàg submartingale X of class (D) there exists a unique predictable, increasing, integrable process X^p with $X^p_0 = 0$ such that the process*

$$X_t - X^p_t, \quad t \geq 0$$

is a uniformly integrable martingale.

4.2.1 Predictable Quadratic Variation of Square Integrable Martingales

The following result concerned with a $U = \mathbb{R}^d$-valued square integrable martingale is a consequence of Theorem 4.2.6.

Theorem 4.2.7 *For any $M \in \mathcal{M}^2_{loc}(U)$ there exists a unique $\mathbb{R}$-valued process $\langle M,M\rangle_t; t\geq 0$ that is predictable, increasing, $\langle M,M\rangle_0 = 0$ and such that*

$$\mid M_t \mid^2 - \langle M,M\rangle_t = \langle M_t,M_t\rangle - \langle M,M\rangle_t; \qquad t\geq 0, \tag{4.2.2}$$

is a local martingale. If $M \in \mathcal{M}^2(U)$ then $\langle M,M\rangle_t; t\geq 0$ is integrable and (4.2.2) *is a martingale.*

The process $\langle M,M\rangle$ is called a *predictable quadratic variation* or an *angle bracket* of M. Notice a hardly noticeable difference in notation between the angle bracket $\langle\cdot,\cdot\rangle$ and the scalar product $\langle\cdot,\cdot\rangle$.

The definition of quadratic variation can be extended. For any $M,N \in \mathcal{M}^2_{loc}(U)$ we can write

$$\langle M_t,N_t\rangle = \frac{1}{4}\Big[\mid M_t+N_t \mid^2 - \mid M_t - N_t \mid^2\Big], \qquad t\geq 0,$$

which immediately yields, that

$$\langle M_t,N_t\rangle - \frac{1}{4}\Big[\langle M_t+N_t, M_t+N_t\rangle - \langle M_t-N_t, M_t-N_t\rangle\Big], \qquad t\geq 0 \tag{4.2.3}$$

is a local martingale. The process

$$\langle M,N\rangle_t := \frac{1}{4}\Big[\langle M+N,M+N\rangle_t - \langle M-N,M-N\rangle_t\Big], \qquad t\geq 0 \tag{4.2.4}$$

is called a *predictable quadratic covariation* or an *angle bracket* of M and N. Clearly, $\langle M,N\rangle$ inherits the properties of quadratic variations of $M+N$ and $M-N$, i.e. it is unique, predictable and starts from zero. If, additionally, $M,N\in\mathcal{M}^2(U)$, then $\langle M,N\rangle$ is integrable and (4.2.3) is a martingale. If M^i are coordinates of the martingale M then a *predictable operator quadratic variation* is defined by

$$\langle\langle M,M\rangle\rangle_t := (\langle M^i,M^j\rangle_t)_{i,j}.$$

Moreover, there exists a predictable process Q_t taking values in the space of positive symmetric $d\times d$ matrices such that

$$\langle\langle M,M\rangle\rangle_t = \int_0^t Q_s d\langle M,M\rangle_s \tag{4.2.5}$$

(see Theorem 8.2 in Peszat and Zabczyk [100]).

4.2.2 Compensators of Finite Variation Processes

The aim of this section is to introduce the notion of compensator of an increasing process. Since increasing and finite variation processes are closely related we extend our discussion on finite variation processes.

The class of all $\mathbb{R}$-valued adapted increasing process with càdlàg paths starting from 0 will be denoted $\mathcal{V}^+$. For any $A\in\mathcal{V}^+$ there exists an $\mathbb{R}\cup\{+\infty\}$-valued random variable

$$A_\infty := \lim_{t\to+\infty} A_t.$$

If $\mathbb{E}[A_\infty]<+\infty$ then the process A is called *integrable* and the corresponding class is denoted by $\mathcal{A}^+$, that is,

$$\mathcal{A}^+ := \Big\{A\in\mathcal{V}^+, \quad \mathbb{E}[A_\infty]<+\infty\Big\}.$$

If there exists an increasing sequence of stopping times $\{\tau_n\}_{n=1,2,\ldots}$ that diverges to $+\infty$ and such that the stopped process A^{τ_n} belongs to $\mathcal{A}^+$ for each n, then A will be called *locally integrable*. $\mathcal{A}^+_{loc}$ stands for the class of all locally integrable increasing processes.

An $\mathbb{R}$-valued adapted process X is called of *finite variation* if its (total) *variation* defined by

$$TV(X)_t := \lim_{n\to+\infty}\sum_{k=1}^{n-1} \mid X_{\frac{(k+1)t}{n}} - X_{\frac{kt}{n}} \mid$$

is finite for any $t\geq 0$ and any path. By $\mathcal{V}$ we denote the family of $\mathbb{R}$-valued adapted processes starting from 0 with càdlàg paths of finite variation. Clearly, any process from $\mathcal{V}^+$ is of finite variation. On the other hand, any $\mathbb{R}$-valued process of finite variation X is a difference of two increasing processes because it can be decomposed to the form

$$X_t = X_t^1 - X_t^2, \quad t \geq 0, \tag{4.2.6}$$

with $X^1 := \frac{1}{2}(TV(X)+X)$ and $X^2 := \frac{1}{2}(TV(X)-X)$. If $A \in \mathcal{V}$ and $E[TV(A)_\infty] < +\infty$ then A is said to be of *integrable variation* and

$$\mathcal{A} := \left\{A \in \mathcal{V},\ \mathbb{E}[TV(A)_\infty] < +\infty\right\}$$

stands for the corresponding class. By $\mathcal{A}_{loc}$ we denote the localized class.

It is clear that a process A from $\mathcal{V}$ does not have to belong to $\mathcal{A}_{loc}$ nor to $\mathcal{A}$. The situation changes, however, if A is predictable.

Proposition 4.2.8 *Let $A \in \mathcal{V}$ be a predictable process. Then $A \in \mathcal{A}_{loc}$.*

The variation of a local martingale turns out to be locally integrable providing that it is finite.

Proposition 4.2.9 *Let M be an $\mathbb{R}$-valued local martingale that belongs to $\mathcal{V}$. Then $M \in \mathcal{A}_{loc}$.*

The variation of local martingales is, however, typically infinite. This happens always if the martingale is continuous or, more generally, predictable.

Proposition 4.2.10 *Let M be an $\mathbb{R}$-valued predictable local martingale that belongs to $\mathcal{V}$. Then $M(t) = M(0), t \geq 0$.*

Further stochastic properties of increasing and finite variation processes arise from Theorem 4.2.6.

Theorem 4.2.11 *For $A \in \mathcal{A}_{loc}^+$ there exists a unique predictable process $A^p \in \mathcal{A}_{loc}^+$ such that*

$$A(t) - A^p(t), \qquad t \geq 0$$

is a local martingale.

The process A^p in Theorem 4.2.11 is called the *compensator* of A. It can be alternatively characterized as follows.

Proposition 4.2.12 *Let $A \in \mathcal{A}_{loc}^+$. Then $A^p \in \mathcal{A}_{loc}^+$ is the compensator of A if and only if*

$$\mathbb{E}[A(\tau)] = \mathbb{E}[A^p(\tau)] \tag{4.2.7}$$

for any stopping time τ.

Since every real valued process of finite variation can be represented as a difference of two increasing processes (see (4.2.6)), as a consequence of Theorem 4.2.11 we obtain the following.

Theorem 4.2.13 *For $A \in \mathcal{A}_{loc}$ there exists a unique predictable process $A^p \in \mathcal{A}_{loc}$ such that $A - A^p$ is a local martingale.*

Also in this case A^p will be called the *compensator* of A.

4.3 Semimartingales

A real valued process of the form

$$X_t = X_0 + M_t + A_t, \qquad t \geq 0, \tag{4.3.1}$$

where $X_0 \in \mathbb{R}$, M is a local martingale with $M_0 = 0$ and càdlàg paths, A adapted process with $A_0 = 0$ and càdlàg paths of finite variation, is called a *semimartingale*. The sum in (4.3.1) will be referred to as a *decomposition* of the semimartingale X. It is, in general, not unique unless the process A is predictable. If A is predictable then Z is called a *special semimartingale*. If X is a special semimartingale then its decomposition with predictable A is called *canonical* and its uniqueness one can easily prove with the use of Proposition 4.2.10. The class of special semimartingales can be characterized as follows.

Theorem 4.3.1 *Let X be a semimartingale. Then X is a special semimartingale if and only if the process*

$$X_t^* := \sup_{s \in [0,t]} | X_s |, \qquad t \geq 0$$

belongs to $\mathcal{A}_{loc}^+$.

Application of Theorem 4.3.1 yields a useful criteria for identifying special semimartingales.

Corollary 4.3.2 *Every semimartingale X with bounded jumps, i.e. such that $| \Delta X | \leq a$ with $a \geq 0$, is special. In particular, every continuous semimartingale is special.*

It turns out that the property of bounded jumps is inherited by the canonical decomposition of a semimartingale. More precisely, the following result can be proven.

Theorem 4.3.3 *Let X be a semimartingale such that $| \Delta X | \leq a$, where $a \geq 0$, and $X = X_0 + M + A$ be its canonical decomposition. Then $| \Delta A | \leq a$ and $| \Delta M | \leq 2a$. Specifically, both components of the canonical decomposition of a continuous semimartingale are continuous.*

As already mentioned, a general semimartingale admits many decompositions with a non-predictable finite variation part. A wide and useful class of decompositions arises from the following result dealing with decompositions of local martingales.

Theorem 4.3.4 *Let M be a local martingale. For any constant $\alpha > 0$ there exist two local martingales M^1 and M^2 such that $\mid \triangle M_1 \mid \leq \alpha$, M^2 is of finite variation and $M = M^1 + M^2$.*

Corollary 4.3.5 *It follows from Theorem 4.3.4 that any decomposition of a semimartingale can be modified in such a way that the local martingale part has bounded jumps, hence is locally square integrable. So, any semimartingale X admits the representation*

$$X(t) = X(0) + M(t) + A(t), \quad t > 0,$$

where $M \in \mathcal{M}^2_{loc}$ and $A \in \mathcal{V}$.

Theorem 4.3.4 can, in a sense, be extended by proving that any local martingale M can be written as a sum of two unique local martingales $M = M^1 + M^2$ such that M^1 is continuous and M^2 is such that the product $M^2 \cdot N$ is again a local martingale for any continuous local martingale N. In contrast to the decomposition in Theorem 4.3.4, M^2 doesn't have to be of finite variation, in general. What may seem surprising is that M^1 is robust for the martingale part of a given semimartingale. Let us formulate this fact precisely.

Theorem 4.3.6 *For any semimartingale X there exists a unique continuous local martingale X^c such that any decomposition $X = X_0 + M + A$ satisfies $M^c = X^c$.*

X^c defined in Theorem 4.3.6 is called a *continuous martingale part* of X.

Recall that for any semimartingale X there exists a *quadratic variation* process $[X,X]$ defined by

$$[X,X]_t := \lim_{n \to +\infty} \sum_{k=1}^{n-1} \mid X_{\frac{(k+1)t}{n}} - X_{\frac{kt}{n}} \mid^2, \quad t \geq 0,$$

where the convergence holds in probability uniformly on every compact interval. Clearly, $[X,X]$ is an adapted, increasing process with càdlàg paths, $[X,X]_0 = 0$ and $\triangle[X,X]_t = (\triangle X_t)^2$. If X is a local martingale then its quadratic variation has the following property.

Theorem 4.3.7 *If M is an $\mathbb{R}$-valued local martingale then the process*

$$M_t^2 - [M,M]_t, \quad t \geq 0$$

is a local martingale.

Let us notice that Theorem 4.3.7 is, in a sense, similar to Theorem 4.2.7 if considered in the class $\mathcal{M}^2_{loc}$. In this case both of the processes

$$M^2 - [M,M], \qquad M^2 - \langle M,M \rangle,$$

are local martingales, but the quadratic variation $[M,M]$ and predictable quadratic variation $\langle M,M \rangle$ differ in general. This fact does not contradict the uniqueness of $\langle M,M \rangle$ formulated in Theorem 4.2.7 because $[M,M]$ is, usually, not predictable. Since $[M,M]$ is increasing one may suppose that it meets the requirements allowing to construct its compensator. This is the case that shows the next result.

Theorem 4.3.8 *If $M \in \mathcal{M}^2_{loc}$ then $[M,M]$ is locally integrable, i.e. $[M,M] \in \mathcal{A}^+_{loc}$ and $\langle M,M \rangle$ is the compensator of $[M,M]$.*

If $M \in \mathcal{M}^{2,c}_{loc}$ then $[M,M]$ has continuous paths, hence is predictable, and thus both quadratic variations are equal, i.e.

$$[M,M]_t = \langle M,M \rangle_t, \quad t \geq 0, \quad M \in \mathcal{M}^{2,c}_{loc}. \tag{4.3.2}$$

If the paths of $M \in \mathcal{M}^2_{loc}$ are not continuous then the relation between quadratic variations is

$$[M,M]_t = \langle M^c, M^c \rangle_t + \sum_{s \in [0,t]} | \triangle M_s |^2, \quad t \geq 0.$$

Notice, that $\langle M^c, M^c \rangle$ is well defined because M^c has continuous paths and hence is a local square integrable martingale. The preceding formula can be extended also for semimartingales.

Theorem 4.3.9 *Let X be a semimartingale and X^c its continuous martingale part. Then*

$$[X,X]_t = \langle X^c, X^c \rangle_t + \sum_{s \in [0,t]} | \triangle X_s |^2, \quad t \geq 0. \tag{4.3.3}$$

The process $[X,X]$ is increasing with $\triangle[X,X] = | \triangle X |^2$, so its jumps are summable. Moreover, one defines $[X,X]^c$ by

$$[X,X]^c_t = [X,X]_t - \sum_{s \in [0,t]} \triangle[X,X]_s = [X,X]_t - \sum_{s \in [0,t]} | \triangle X_s |^2, \quad t \geq 0,$$

which is the pathwise continuous part of $[X,X]$.

The quadratic variation is a useful tool for estimating the supremum of a local martingale (see Kallenberg [79], Theorem 26.12).

Theorem 4.3.10 *(Burkholder–Davis–Gundy) For each $p \geq 1$ there exists a constant $0 < c_p < +\infty$ such that for any local martingale M with $M(0) = 0$ and any $t \geq 0$,*

$$\frac{1}{c_p} \mathbb{E}[M,M]_t^{\frac{p}{2}} \leq \mathbb{E}\Big(\sup_{s \in [0,t]} | M_s |^p \Big) \leq c_p \mathbb{E}[M,M]_t^{\frac{p}{2}}. \tag{4.3.4}$$

If the local martingale M in the preceding result has continuous paths, then (4.3.2) takes the form

$$\frac{1}{c_p}\,\mathbb{E}\langle M,M\rangle_t^{\frac{p}{2}} \le \mathbb{E}\Big(\sup_{s\in[0,t]} |M_s|^p\Big) \le c_p\,\mathbb{E}\langle M,M\rangle_t^{\frac{p}{2}}. \qquad (4.3.5)$$

4.4 Stochastic Integration

Here we recall construction of the stochastic integrals

$$\int_0^t g(s)dX(s), \qquad \int_0^t\int_U h(s,y)\pi(ds,dy),$$

with respect to a semimartingale X and its jump measure π.
We consider first stochastic integrals

$$\int_0^t g(s)dX(s), \quad t>0, \qquad (4.4.1)$$

where the integrator X is a semimartingale with values in $U=\mathbb{R}^d$ and the integrand g is a predictable process taking values in the set of matrices $\mathcal{M}^{k\times d}$.
The definitions and results extend easily to the case when U,H are infinite dimensional separable Hilbert spaces and g is an operator valued process.

4.4.1 Bounded Variation Integrators

In view of Corollary 4.3.5, any semimartingale X can be represented in the form

$$X(t)=X(0)+M(t)+A(t), \quad t>0, \qquad (4.4.2)$$

where $M\in\mathcal{M}^2_{loc}(U)$ and the coordinates of A belong to $\mathcal{V}$. Then (4.4.1) can be defined by

$$\int_0^t g(s)dX(s) := \int_0^t g(s)dM(s) + \int_0^t g(s)dA(s), \quad t>0. \qquad (4.4.3)$$

The second integral in (4.4.3) can be viewed as the classical Lebesgue integral. In fact, $A(t)$ admits the representation

$$A(t)=\Big(A_1^1(t),\dots,A_d^1(t)\Big)-\Big(A_1^2(t),\dots,A_d^2(t)\Big), \quad t>0,$$

where A_l^1, A_l^2, $l=1,\dots,d$, are increasing, right continuous functions. The functions define measures μ_l^1, μ_l^2, $l=1,\dots,d$ on $[0,+\infty)$ and if $g(s)=(g_{j,l}(s))_{j,l}$, then one sets,

$$\int_0^t g(s)dA(s) = \left(\sum_{l=1}^d\int_0^t g_{j,l}(s)(\mu_l^1-\mu_l^2)(ds)\right)_j, \quad j=1,2,\dots,k. \qquad (4.4.4)$$

If g is locally bounded then the integral (4.4.4) is well defined.

4.4.2 Square Integrable Martingales as Integrators

Let M be a square integrable martingale taking values in $U = \mathbb{R}^d$ and the integrand g be a process taking values in the set of matrices $\mathcal{M}^{k\times d}$, i.e.

$$g(s) = g_{j,l}(s), \quad j = 1, 2, \ldots, k, \quad l = 1, 2, \ldots, d.$$

For the columns of $g(s)$ we use a single index, i.e.

$$g_l(s) = [g_{1,l}(s), g_{2,l}(s), \ldots, g_{k,l}(s)], \quad l = 1, 2, \ldots, d.$$

Similarly, the rows of $g(s)$ are denoted by

$$g^l(s) = [g_{l,1}(s), g_{l,2}(s), \ldots, g_{l,d}(s)], \quad l = 1, 2, \ldots, k.$$

Then the stochastic integral

$$I(g)_t := \int_0^t g(s)dM(s), \quad t > 0 \tag{4.4.5}$$

will take values in $H := \mathbb{R}^k$. The scalar valued integral (4.4.5), i.e. when $k = 1$, will be denoted by

$$\int_0^t \langle g(s), dM(s)\rangle, \quad t > 0.$$

Since $M \in \mathcal{M}^2(U)$, its coordinates $M^j, j = 1, 2, \ldots, d$ belong to $\mathcal{M}^2$. Consequently, quadratic covariations are well defined and

$$M^j(t)M^l(t) - \langle M^j, M^l\rangle_t, \quad t > 0, \quad j, l = 1, 2, \ldots, d$$

are martingales.

For a compact formulation of the so-called isometric formula we need some more notation. If A is in $\mathcal{M}^{k\times d}$ then its Hilbert–Schmidt $\mid A \mid_2$ is given by

$$\mid A \mid_2 := \Big(\sum_{i,j} a_{i,j}^2\Big)^{1/2}.$$

More generally, a linear operator A acting from a Hilbert space U into a Hilbert space H is called Hilbert–Schmidt if, for an orthonormal basis (e_k) in U:

$$\mid A \mid_2 = \Big(\sum_{k=1}^{+\infty} |Ae_k|_H^2\Big)^{1/2} < +\infty.$$

Recall that Q_t is a $d \times d$ positive symmetric matrix given by (4.2.5).

Theorem 4.4.1 *Let $g(s)$ be a predictable process satisfying*

$$\mathbb{E}\left(\int_0^{T^*} \mid g(s)Q_s^{\frac{1}{2}} \mid_2^2 \, d\langle M, M\rangle_s\right) < +\infty \tag{4.4.6}$$

for some $T^* > 0$. Then the integral (4.4.5) is a well-defined H-valued square integrable martingale on $[0, T^*]$ satisfying the following isometric formula

$$\mathbb{E}(| I(g)_t |_H^2) = \mathbb{E}\left(\int_0^t | g(s)Q_s^{\frac{1}{2}} |_2^2 \langle M, M\rangle_s\right), \quad t \in [0, T^*]. \tag{4.4.7}$$

Remark 4.4.2 One can check by direct calculation that

$$\begin{aligned}\mathbb{E}\left(\int_0^t | g(s)Q_s^{\frac{1}{2}} |_2^2 \langle M, M\rangle_s\right) &= \mathbb{E}\left(\int_0^{T^*} Trace(g(s)Q_s g^*(s))d\langle M, M\rangle_s\right)\\ &= \mathbb{E}\left(\int_0^{T^*} \sum_{j,l=1}^{d} Q_s^{j,l}\langle g_j(s), g_l(s)\rangle d\langle M, M\rangle_s\right)\\ &= \mathbb{E}\left(\sum_{j,l=1}^{d} \int_0^{T^*} \langle g_j(s), g_l(s)\rangle d\langle M^j, M^l\rangle_s\right),\end{aligned}$$

which enables us to write the isometric identity (4.4.7) in several equivalent forms.

Proof of Theorem 4.4.1 We show the assertion only for simple processes of the form

$$g(s) = a(0)\mathbf{1}_{\{s=0\}} + \sum_{i=0}^{n-1} a(t_i)\mathbf{1}_{(t_i,t_{i+1}]}(s),$$

where $0 = t_0 < t_1 < \cdots < t_n = T^*$ and $a(t_i)$ is $\mathcal{F}_{t_i}$-measurable, bounded and takes values in $\mathcal{M}^{k\times d}$. Then

$$I(g)_t = \sum_{i=0}^{n-1} a(t_i)(M(t_{i+1} \wedge t) - M(t_i \wedge t)), \quad t \in [0, T^*].$$

To avoid technical difficulties, we check the martingale condition and the isometric formula for the partition points only. For $0 < m < n$ we have

$$\begin{aligned}\mathbb{E}[I(g)_{t_{m+1}} \mid \mathcal{F}_{t_m}] &= \mathbb{E}\left[\sum_{i=0}^{m} a(t_i)(M(t_{i+1}) - M(t_i)) \mid \mathcal{F}_{t_m}\right]\\ &= \sum_{i=0}^{m-1} a(t_i)(M(t_{i+1}) - M(t_i)) + a(t_m)\,\mathbb{E}[M(t_{m+1}) - M(t_m) \mid \mathcal{F}_{t_m}]\\ &= \sum_{i=0}^{m-1} a(t_i)(M(t_{i+1}) - M(t_i)) = I(g)_{t_m}.\end{aligned}$$

Since

$$\mathbb{E}\left[\mid \int_0^t g(s)dM(s) \mid^2\right] = \mathbb{E}\left[\mid \int_0^t \langle g^1(s), dM(s)\rangle \mid^2\right] + \cdots + \mathbb{E}\left[\mid \int_0^t \langle g^k(s), dM(s)\rangle \mid^2\right], \tag{4.4.8}$$

we determine first

$$\mathbb{E}\left[\mid \int_0^t \langle g^l(s), dM(s)\rangle \mid^2\right], \quad l = 1, 2, \ldots, k.$$

We have

$$\begin{aligned}
\mathbb{E}\left[\mid \int_0^{t_m} \langle g^l(s), dM(s)\rangle \mid^2\right] &= \mathbb{E}\left[\mid \sum_{i=0}^{m-1} \langle a(t_i)^l, M(t_{i+1}) - M(t_i)\rangle \mid^2\right] \\
&= \mathbb{E}\left[\sum_{i=0}^{m-1} \mid \langle a(t_i)^l, M(t_{i+1}) - M(t_i)\rangle \mid^2\right] \\
&= \mathbb{E}\left[\sum_{i=0}^{m-1} \left(\sum_{j=1}^{d} a(t_i)_{l,j}(M^j(t_{i+1}) - M^j(t_i))\right)^2\right] \\
&= \mathbb{E}\left[\sum_{i=0}^{m-1} \sum_{j,r=1}^{d} a(t_i)_{l,j} a(t_i)_{l,r} (M^j(t_{i+1}) - M^j(t_i))(M^r(t_{i+1}) - M^r(t_i))\right].
\end{aligned}$$

Changing the order of the sums and using the quadratic covariation of M^j and M^r we obtain finally

$$\begin{aligned}
\mathbb{E}\left[\mid \int_0^{t_m} \langle g^l(s), dM(s)\rangle \mid^2\right] &= \mathbb{E}\left[\sum_{j,r=1}^{d} \sum_{i=1}^{m-1} a(t_i)_{l,j} a(t_i)_{l,r} (\langle M^j, M^r\rangle_{t_{i+1}} - \langle M^j, M^r\rangle_{t_i})\right] \\
&= \mathbb{E}\left[\sum_{j,r=1}^{d} \int_0^{t_m} g(s)_{l,j} g(s)_{l,r} d\langle M^j, M^r\rangle_s\right].
\end{aligned} \tag{4.4.9}$$

In view of (4.4.8) and (4.4.9) we obtain

$$\begin{aligned}
\mathbb{E}\left[\mid \int_0^{t_m} g(s)dM(s) \mid^2\right] &= \mathbb{E}\left[\sum_{l=1}^{k} \sum_{j,r=1}^{d} \int_0^{t_m} g(s)_{l,j} g(s)_{l,r} d\langle M^j, M^r\rangle_s\right] \\
&= \mathbb{E}\left[\sum_{j,r=1}^{d} \int_0^{t_m} g(s)_j g(s)_r d\langle M^j, M^r\rangle_s\right].
\end{aligned}$$

□

The definition of the integral can be extended to integrands g satisfying the condition

$$\mathbb{P}\left(\int_0^{T^*} | g(s)Q_s^{\frac{1}{2}} |_2^2 \, d\langle M, M\rangle_s < +\infty\right) = 1, \tag{4.4.10}$$

by the usual technique of localization. As a localizing sequence $\{\tau_n\}$ one may take

$$\inf_{s\in[0,T^*]} \int_0^s | g(s)Q_s^{\frac{1}{2}} |_2^2 \, d\langle M, M\rangle_s \geq n, \quad n = 1, 2, \ldots.$$

Then, for each n, the integrand $g(t \wedge \tau_n)$ satisfies the condition (4.4.6) from the theorem and one sets:

$$\int_0^t g(s)dM(s) = \lim_{n\to+\infty} \int_0^t g(s \wedge \tau_n)dM(s).$$

In particular, if g is locally bounded then the integral is well defined.

4.4.3 Integration over Random Measures

Here we discuss the construction of the integral

$$\int_0^t \int_U h(s, y)\pi(ds, dy),$$

where $\pi(dt, dy)$ is the so-called *jump measure* of an $U = \mathbb{R}^d$-valued semimartingale X and $h(\cdot, \cdot)$ takes values in some Hilbert space H.

Since X has càdlàg paths, one may count its jumps along a set $I \in \mathcal{B}(\mathbb{R}_+)$ which belong to $A \in \mathcal{B}(U)$. This number is denoted by $\pi(I \times A)$, i.e.

$$\pi(I \times A) := \sharp\{s \in I : \Delta X(s) \in A\}, \qquad I \in \mathcal{B}(\mathbb{R}_+), \quad A \in \mathcal{B}(U). \tag{4.4.11}$$

If A is *separated from zero*, that is, $0 \notin \bar{A}$, where $\bar{A}$ stands for the closure of A, and I is bounded, then it follows from Proposition 4.1.1 that $\pi(I \times A)$ is finite. If A is not separated from zero or I is unbounded then $\pi(I \times A)$ may be infinite. For a sequence $\{I_i \times A_i\}$ of disjoint sets in $\mathbb{R}_+ \times U$

$$\pi\left(\bigcup_{i=1}^{+\infty} I_i \times A_i\right) = \sum_{i=1}^{+\infty} \pi(I_i \times A_i),$$

hence $\pi(\cdot)$ can be viewed as a measure on $\mathcal{B}(\mathbb{R}_+) \times \mathcal{B}(U)$. In fact,

$$\pi = \sum_{s\geq 0} \mathbf{1}_{\Delta X(s)} \delta_{s,\Delta X(s)},$$

and it can be regarded as a measure valued random variable. It will be called the *jump measure* of X. It is clear that the jump measure takes values in the set of natural numbers, is σ-finite and random, as it depends on paths of X.

For $h\colon \Omega \times \mathbb{R}_+ \times U = \tilde{\Omega} \times U \longrightarrow H$ one defines an integral over π in the natural way

$$(h * \pi)_t := \int_0^t \int_U h(s,y)\pi(ds,dy) := \sum_{s\in[0,t]} h(s,\triangle X(s)).$$

If the function h is measurable with respect to the product σ-field $\mathcal{P} \times \mathcal{B}(U)$ then it is called *predictable* process or *predictable* random field.

With $\pi(\cdot)$ we can associate another random measure, the so-called *(predictable) compensating measure* π^p, which we characterize in the following proposition. For the sake of brevity we write $\pi(t,A) := \pi([0,t] \times A)$ in the sequel.

Proposition 4.4.3 *Let π be a jump measure of an adapted, U-valued process with càdlàg paths. There exists a unique random measure $\pi^p(\cdot,\cdot)$ on $\mathcal{B}(\mathbb{R}) \times \mathcal{B}(U \setminus \{0\})$, such that*

$$\pi(t,A) - \pi^p(t,A),$$

is a local martingale for any set A separated from zero.

Moreover, if $h\colon \Omega \times \mathbb{R}_+ \times U \longrightarrow \mathbb{R}$ is a predictable process such that

$$(\mid h \mid *\pi).\in \mathcal{A}^+_{loc},$$

then

$$(\mid h \mid *\pi).\in \mathcal{A}^+_{loc}, \tag{4.4.12}$$

and $(h * \pi^p)$ is the compensator of $(h * \pi)$.

Proof If $A \in \mathcal{B}(U)$ is separated from zero then the process

$$t \longrightarrow \pi(t,A), \qquad t \geq 0$$

is adapted, increasing and càdlàg with jump sizes equal to 1, hence locally integrable. It follows from Theorem 4.2.11 that there exists its compensator, which one denotes by $Y(A)$, i.e. $\pi(t,A) - Y_t(A)$, $\;t \geq 0$, is a local martingale. Since for any disjoint, sets A, B separated from zero $\in \mathcal{B}(U)$ we have

$$\pi(t, A\cup B) = \pi(t,A) + \pi(t,B), \quad t \geq 0,$$

it follows from the uniqueness of the compensator that

$$Y_t(A\cup B) = Y_t(A) + Y_t(B), \quad t\geq 0, \quad Y_t(\emptyset) = 0,$$

so $Y_t(\cdot)$ is a finitely additive measure defined on the ring of separated from zero Borel sets of U. By the Carathéodory theorem there exists an extension of $Y_t(\cdot)$ to the measure defined on the σ-field generated by the ring that is identical with $\mathcal{B}(U\setminus\{0\})$. Moreover, since $U \setminus \{0\}$ is a union of separated from zero sets, i.e.

$$U\setminus\{0\} = \bigcup_{i=1}^{\infty} \Big\{u \in U :\mid u \mid\geq 1/n\Big\},$$

it follows that $Y_t(\cdot)$ is σ-finite and thus the extension is unique. We set

$$\pi^P(t,A) := Y_t(A), \quad t \geq 0, \quad A \in \mathcal{B}(U \setminus \{0\}),$$

and extend $\pi_X^P(\cdot,\cdot)$ to the measure

$$\pi^P(I \times A), \quad I \in \mathcal{B}(\mathbb{R}), \quad A \in \mathcal{B}(U \setminus \{0\})$$

in a standard way.

Let us consider the function

$$h(\omega, w, x) := \mathbf{1}_{A\times(s,v]\times B}(\omega, w, x), \qquad A \in \mathcal{F}_s,\ B \in \mathcal{B}(U),\ 0 \notin \bar{B},\ s < v. \quad (4.4.13)$$

Since $A \times (s,t]$, where $A \in \mathcal{F}_s, s < t$ belongs to $\mathcal{P}$, h is predictable. Integration of (4.4.13) over π and π^P yields the following expressions

$$\begin{aligned}(h * \pi)_t &= \mathbf{1}_{A\times(s,+\infty)}(\omega,t)\pi((s, t\wedge v] \times B)\\ &= \mathbf{1}_{A\times(s,+\infty)}(\omega,t)(\pi(t\wedge v, B) - \pi(s,B)), \qquad (4.4.14)\\ (h * \pi^P)_t &= \mathbf{1}_{A\times(s,+\infty)}(\omega,t)\pi^P((s, t\wedge v] \times B)\\ &= \mathbf{1}_{A\times(s,+\infty)}(\pi^P(t\wedge v, B) - \pi^P(s,B)). \qquad (4.4.15)\end{aligned}$$

It is clear that (4.4.14) is optional, that (4.4.15) is predictable and also that

$$(h * \pi)_t - (h * \pi^P)_t, \qquad t \geq 0$$

is a local martingale. Hence, it follows that (4.4.15) is the compensator of (4.4.14). If h is a general predictable function, then, by Proposition 4.1.3, it can be approximated by finite combinations of functions of the form (4.4.13). Hence, if only $(h*\pi)$ is of locally integrable variation then $(h * \pi^P)$ is its compensator. □

Remark 4.4.4 The measure $\tilde{\pi}$ defined by

$$\tilde{\pi}(t,A) := \pi(t,A) - \pi^P(t,A)$$

is called a *compensated jump measure* of X. The integral

$$(h * \tilde{\pi}) := (h * \pi) - (h * \pi^P)$$

is a local martingale for predictable h such that $(\mid h \mid *\pi)_\cdot \in \mathcal{A}^+_{loc}$.

4.4.4 Itô's Formula

Let us start with the classical Itô formula for a real valued semimartingale X (see for instance Protter [102]).

Theorem 4.4.5 *Let X be a semimartingale and $f\colon \mathbb{R} \longrightarrow \mathbb{R}$, $f \in \mathcal{C}^2(\mathbb{R})$. Then $f(X)$ admits the representation*

$$f(X_t) = f(X_0) + \int_0^t f'(X_{s-})dX_s + \frac{1}{2}\int_0^t f''(X_{s-})d[X,X]_s^c$$
$$+ \sum_{s\in[0,t]} \Big\{ f(X_s) - f(X_{s-}) - f'(X_{s-})\Delta X_s \Big\}, \quad t \geq 0. \tag{4.4.16}$$

In particular, $f(X)$ is a semimartingale.

Let us notice that the integral over X in formula (4.4.16) is well defined because the process $f'(X(s-))$ is predictable and locally bounded. In (4.4.16) $[X,X]^c$ stands for the continuous part of the quadratic variation process. One can show that

$$[X,X]_t^c = [X^c, X^c]_t, \quad t \geq 0,$$

where X^c stands for the continuous martingale part of X (see Theorem 4.3.6). Using the fact that $[X^c, X^c] = \langle X^c, X^c \rangle$ (see (4.3.2)) or the relation $[X,X] = \langle X^c, X^c \rangle + \sum \mid \Delta X_s \mid^2$ (see (4.3.3)) we can obtain several equivalent formulations for (4.4.16). To see that $f(X)$ is a semimartingale, we can use (4.4.2) and write (4.4.16) in the form $f(X_t) = f(X_0) + N_t + B_t$, where, for $t \geq 0$,

$$N_t := \int_0^t f'(X_{s-})dM_s,$$

$$B_t := \int_0^t f'(X_{s-})dA_s + \frac{1}{2}\int_0^t f''(X_{s-})d[X,X]_s^c + \sum_{s\in[0,t]} \Big\{ f(X_s) - f(X_{s-}) - f'(X_{s-})\Delta X_s \Big\}.$$

Moreover, N is a local martingale and B is of finite variation.

Now, let X be an $\mathbb{R}^n$-valued semimartingale, that is, each of its coordinate is a real valued semimartingale. Then for a twice differentiable function $f\colon \mathbb{R}^n \longrightarrow \mathbb{R}$ the following multidimensional version of Itô's formula can be proven

$$f(X_t) = f(X_0) + \sum_{i=1}^n \int_0^t f'_{x_i}(X_{s-})dX_s^i + \frac{1}{2}\sum_{i,j=1}^n \int_0^t f''_{x_i x_j}(X_{s-})d[X^i, X^j]_s^c$$
$$+ \sum_{s\in[0,t]} \Big\{ f(X_s) - f(X_{s-}) - \sum_{i=1}^n f'_{x_i}(X_{s-})\Delta X_s^i \Big\}, \quad t \geq 0. \tag{4.4.17}$$

A direct application of (4.4.17) for two real-valued semimartingales X, Y and the function $f(x,y) = xy$ provides the *Itô product formula*, also called the *integration by parts formula*

$$X_tY_t = X_0Y_0 + \int_0^t X_{s-}dY_s + \int_0^t Y_{s-}dX_s + [X, Y]_t^c + \sum_{s\in[0,t]} \triangle X_s \triangle Y_s$$

$$= X_0Y_0 + \int_0^t X_{s-}dY_s + \int_0^t Y_{s-}dX_s + [X, Y]_t, \qquad t \geq 0. \qquad (4.4.18)$$

As a direct application of the Itô formula we can prove the following.

Theorem 4.4.6 *Let X be a semimartingale. There exists a unique semimartingale Y solving the equation*

$$Y_t = Y_0 + \int_0^t Y_{s-}dX_s, \quad t \geq 0,$$

and it has the form

$$Y_t = Y(0)e^{X_t - X_0 - \frac{1}{2}[X,X]_t^c} \prod_{s\in[0,t]} (1 + \triangle X_s)e^{-\triangle X_s}, \quad t \geq 0. \qquad (4.4.19)$$

The solution Y is called the *Doléans-Dade exponential* or *stochastic exponential* of the semimartingale X. Again, similar to the Itô formula, using the relations between $[X, X]$, $[X, X]^c$, $[X^c, X^c]$ and $\langle X^c, X^c\rangle$ we can write Y in several equivalent forms.

5

Lévy Processes

This chapter is devoted to Lévy processes, a basic tool used in this book. We start from general definitions and describe specific classes of Lévy processes. For the missing proofs, the readers are referred to Applebaum [2], Sato [114] and Peszat and Zabczyk [100].

5.1 Basics on Lévy Processes

Lévy processes are of basic importance for the present book. We will be mainly, but not solely, concerned with Lévy processes taking values in $U = \mathbb{R}^d$, but the definition and several properties are the same if the processes take values in separable Hilbert spaces U.

Definition 5.1.1 A U-valued Lévy process $\{Z(t), t \geq 0\}$ is an adapted process satisfying the following conditions:

(a) $Z_0 = 0$,

(b) its increments are independent, i.e. the random variables $Z(t_0), Z(t_1) - Z(t_0), Z(t_2) - Z(t_1), \ldots, Z(t_n) - Z(t_{n-1})$ are independent for any sequence of times $0 \leq t_0 \leq t_1 \leq \cdots \leq t_n$,

(c) the law of $Z(t+h) - Z(t), t \geq 0, h > 0$, does not depend on t, that is, the increments of Z are stationary,

(d) Z is stochastically continuous, i.e. for each $t \geq 0$:

$$\forall \varepsilon > 0, \qquad \mathbb{P}(\mid Z_{t+h} - Z_t \mid > \varepsilon) \underset{h \to 0}{\longrightarrow} 0.$$

Lévy processes have càdlàg modifications. We will work solely with such versions. Let $\mu_t, t \geq 0$ be the laws of the random variables $Z_t, t \geq 0$ of a Lévy process Z. Then the family (μ_t) is *infinitely divisible*, that is,

$$\mu_0 = \delta_{\{0\}}, \qquad \mu_t * \mu_s = \mu_{t+s}, t, s \geq 0, \qquad \mu_t(\{u : |u| \leq r\}) \to 1 \text{ as } t \downarrow 0, \forall r > 0.$$

If (μ_t) is an arbitrary infinitely divisible family of measures on U then there exists a Lévy process Z such that μ_t is the law of Z_t. This is a direct consequence of the Kolmogorov's existence theorem. In fact the family

$$
\begin{aligned}
& p_{t_1,\dots,t_n}, 0 < t_1 < \dots < t_n, \quad n = 1, 2, \dots, \\
& p_{t_1,t_2,\dots,t_n}(\Gamma_1, \Gamma_2, \dots, \Gamma_n) \\
& = \int \mathbb{1}_{\Gamma_1}(x_1)\mathbb{1}_{\Gamma_2}(x_1 + x_2)\dots\mathbb{1}_{\Gamma_n}(x_1 + \dots + x_n)\mu_{t_1}(dx_1)\mu_{t_2-t_1}(dx_2)\dots \\
& \quad \mu_{t_n-t_{n-1}}(dx_n),
\end{aligned}
$$

where $\Gamma_1, \dots, \Gamma_n$ are arbitrary Borel subsets of U, satisfies the consistency conditions and therefore there exists a stochastic process Z on the probability space $(\Omega, \mathcal{F}, \mathbb{P})$ such that

$$
\mathbb{E}(\varphi(Z_{t_1}, \dots, Z_{t_n})) = \int \varphi(x_1, x_1 + x_2, \dots, x_1 + \dots + x_n)\mu_{t_1}(dx_1)\dots\mu_{t_n-t_{n-1}}(dx_n),
$$

for arbitrary bounded Borel function φ. Setting $\varphi(x_1, \dots, x_n) = \mathbb{1}_{\Gamma_1}(x_1)\mathbb{1}_{\Gamma_2}(x_2 - x_1)$ $\dots\mathbb{1}_{\Gamma_n}(x_n - x_{n-1})$ we get

$$
\begin{aligned}
& \mathbb{P}(Z_{t_1} \in \Gamma_1, Z_{t_2} - Z_{t_1} \in \Gamma_2, \dots, Z_{t_n} - Z_{t_{n-1}} \in \Gamma_n) \\
& = \prod_{k=1}^{n} \mu_{t_k - t_{k-1}}(\Gamma_k) = \prod_{k=1}^{n} \mathbb{P}(Z_{t_k - t_{k-1}} \in \Gamma_k) \\
& = \prod_{k=1}^{n} \mathbb{P}(Z_{t_k} - Z_{t_{k-1}} \in \Gamma_k),
\end{aligned}
$$

where $t_0 := 0$, and the independence of increments follows.

Basic examples of Lévy processes are the deterministic function $Z_t := at$, with $a \in U$, the Wiener process, the Poisson process, the compound Poisson process and the finite activity process.

Wiener process: If $Q = [q_{ij}]_{i,j=1,2,\dots,d}$ is a symmetric and positive definite matrix in $\mathbb{R}^{d\times d}$ then the U-valued Q-Wiener process is characterized by the following properties: $W_0 = 0$, W has independent and stationary increments, W_t has Gaussian distribution with mean 0 and covariance tQ, W has continuous trajectories. The matrix Q is called the *covariance operator* of $W = (W^1, \dots, W^d)$ as it describes covariance of its coordinates, i.e.

$$
Cov(W_t^i, W_t^j) = tq_{i,j}, \quad i,j = 1, 2, \dots, d, \quad t > 0.
$$

A real valued Wiener process is called *standard* if $Q = q = 1$.

Compound Poisson process: Let $\tau_n, n = 1, 2, \dots$ be a sequence of independent identically distributed (i.i.d.) random variables with exponential distribution with parameter $\lambda > 0$, that is,

$$
\mathbb{P}(\tau_n > t) = e^{-\lambda t}, \quad t \geq 0.
$$

A *Poisson process* $\{N_t, t \geq 0\}$ with *intensity* λ is defined by

$$N_t = 0 \quad \text{if} \quad t < \tau_1, \qquad N_t = n \quad \text{if} \quad \sum_{i=1}^{n} \tau_i \leq t < \sum_{i=1}^{n+1} \tau_i \quad \text{for} \quad n = 1, 2, \ldots.$$

Hence, the sequence $\{\tau_n\}$ describes time intervals between consecutive jumps of the Poisson process. The paths of N are piecewise constant càdlàg increasing functions with jumps equal to 1 only.

Let N be a Poisson process with intensity $\lambda > 0$ and $Y_1, Y_2, \ldots$ a sequence of i.i.d. U-valued random variables that are also independent on N. The *compound Poisson process* is defined by

$$Z_0 = 0 \quad \text{if} \quad N_t = 0, \qquad Z_t = \sum_{i=1}^{N_t} Y_i, \quad \text{if} \quad N_t > 0, \qquad t \geq 0. \tag{5.1.1}$$

Its paths are piecewise constant càdlàg functions with jump moments governed by the Poisson process and jump magnitudes determined by the sequence $\{Y_i\}$.

Finite activity processes: By a *finite activity process* we mean a process that admits only a finite number of jumps on any bounded time interval. If Z is a sum of independent components: linear function, Wiener process and compound Poisson process, i.e.

$$Z_t := at + W_t + \sum_{i=1}^{N_t} Y_i = at + W_t + \sum_{s \in [0,t]} \Delta Z_s, \tag{5.1.2}$$

then it is clearly of finite activity. It turns out that each Lévy process of finite activity is of the form (5.1.2).

Other examples of Lévy processes will be discussed in Section 5.3.

5.2 Lévy–Itô Decomposition

To describe the general form of a Lévy process let us start from the description of its jump measure. Since Z has càdlàg paths we can define, as in (4.4.11), the jump measure of Z by

$$\pi(I \times A) := \sharp\{s \in I : \Delta Z(s) \in A\}, \qquad I \in \mathcal{B}(\mathbb{R}_+),\ A \in \mathcal{B}(U),$$

which counts all jumps of Z, over the time set I, which lie in A. If I is bounded and A separated from zero, then $\pi(I \times A) < +\infty$. We briefly write $\pi(t, A) := \pi([0,t] \times A)$. The jump measure of a Lévy process is called a *Poisson random measure* because for any set A separated from zero the function

$$t \mapsto \pi(t, A), \quad t \geq 0,$$

is a Poisson process. Its intensity is given by the formula

$$\nu(A) := \mathbb{E}[\pi(1,A)]. \tag{5.2.1}$$

Since the processes $\pi(t,A)$ and $\pi(t,B)$ are independent if $A \cap B = \emptyset$, it follows that $\nu(\cdot)$ is a finitely additive measure defined on the ring of Borel subsets of U separated from zero. Hence, by the Carathéodory theorem, it can be extended in a unique way to the measure on $\mathcal{B}(U \setminus \{0\})$. This measure is called the *Lévy measure* or also the *intensity measure* of the process Z. It can be shown that each Lévy measure $\nu(dy)$ satisfies

$$\int_U (\mid y \mid^2 \wedge 1)\, \nu(dy) < +\infty. \tag{5.2.2}$$

Conversely, if $\nu(dy)$ is a measure satisfying (5.2.2), then it is a Lévy measure of some Lévy process. Condition (5.2.2) has a direct interpretation for jumps of Z, that is, for any $\varepsilon > 0$ and $t > 0$

$$\sum_{s\in[0,t]} \mathbf{1}_{\{|\triangle Z(s)|>\varepsilon\}} < +\infty, \quad \mathbb{P} - a.s.,$$

and

$$\sum_{s\in[0,t]} \mid \triangle Z(s) \mid^2 \mathbf{1}_{\{|\triangle Z(s)|\leq\varepsilon\}} < +\infty, \quad \mathbb{P} - a.s.$$

This means that the number of large jumps of Z on a finite interval is finite while small jumps of Z are square summable.

Example 5.2.1 Let Z be a compound Poisson process (5.1.1) with intensity $\lambda > 0$ of the underlying Poisson process and $g(dy)$ be the distribution of Y_i. Then the Lévy measure of Z is given by

$$\nu(dy) = \lambda g(dy).$$

Indeed, for a set A separated from zero we have

$$\pi(t,A) = \sum_{i=1}^{N_t} \hat{Y}_i, \quad \text{where} \quad \hat{Y}_i := \mathbf{1}_A(Y_i), \quad t \geq 0.$$

Since $P(\hat{Y}_i = 1) = g(A) = 1 - P(\hat{Y}_i = 0)$, we see that $\pi(t,A), t \geq 0$ is a Poisson process with intensity $E[\pi(1,A)] = \lambda g(A)$. Here ν is a finite measure.

Since for any separated set A from zero the process

$$\pi(t,A) - t\nu(A), \quad t \geq 0,\ A \in \mathcal{B}(U)$$

is a compensated Poisson process, hence a martingale, it follows that $dt\nu(dy)$ is the compensating measure of $\pi(dt,dy)$. Consequently, the *compensated jump measure* of Z is of the form

$$\tilde{\pi}(dt, dy) := \pi(dt, dy) - dt\, \nu(dy).$$

Let us consider the integrals

$$I^{\varepsilon}(t) := \int_0^t \int_{\{\varepsilon<|y|\leq 1\}} y\tilde{\pi}(ds, dy) := \int_0^t \int_{\{\varepsilon<|y|\leq 1\}} y\pi(ds, dy) - \int_0^t \int_{\{\varepsilon<|y|\leq 1\}} y\, ds\nu(dy),$$

with $t > 0$ and $\varepsilon \in (0, 1)$. It can be shown with the use of (5.2.2) that each I^{ε} is a square integrable martingale and that I^{ε} converges in the mean square sense to some square integrable martingale that will be denoted by

$$\int_0^t \int_{\{|y|\leq 1\}} y\tilde{\pi}(ds, dy) := \lim_{\varepsilon \longrightarrow 0} \int_0^t \int_{\{\varepsilon<|y|\leq 1\}} y\tilde{\pi}(ds, dy).$$

Each I^{ε} as well as the limit can be shown to be Lévy processes.

Theorem 5.2.2 (Lévy–Itô decomposition) *Any Lévy process Z admits the following decomposition*

$$Z(t) := at + W(t) + Z_0(t) + Z_1(t), \quad t \geq 0, \tag{5.2.3}$$

where $a \in U$, W is a Q-Wiener processes with some covariance operator Q and

$$Z_0(t) := \int_0^t \int_{\{|y|\leq 1\}} y\tilde{\pi}(ds, dy), \quad Z_1(t) := \int_0^t \int_{\{|y|>1\}} y\pi(ds, dy), \quad t > 0.$$

Moreover, all components in (5.2.3) *are independent.*

Formula (5.2.3) is called the *Lévy–Itô decomposition* of Z. The preceding Z_0 is a square integrable martingale with jumps of norm less than 1, while Z_1 is a compound Poisson process with jumps of norm greater than one (big jumps part). Since Q uniquely determines W and the Lévy measure uniquely determines the jump measure, the quantities (a, Q, ν) uniquely determine the Lévy process Z. They are called *characteristics* or a *characteristic triplet* of Z. On the other hand, it can be proven that for any $a \in U$, matrix Q, which is symmetric and positive definite and a measure ν satisfying (5.2.2), there is a Lévy process with characteristic triplet (a, Q, ν).

It follows from (5.2.3) that Z is a sum of four Lévy processes, two of them are square integrable martingales and one is a process of finite activity. Hence Lévy processes are semimartingales. Moreover, a Lévy process is a special semimartingale if and only if the process Z_1, consisting of large jumps, is integrable, that is, if

$$\int_{|y|>1} |y|\nu(dy) < +\infty$$

(see Kallenberg [79, p. 536]).

It turns out that the quadratic variation of a Lévy process has the form

$$[Z,Z]_t = tQ + \Big(\sum_{s\le t}(\Delta Z^i(s)Z^j(s))\Big)_{i,j} = tQ + \int_0^t\int_U y\otimes y\;\pi(ds,dy), \qquad (5.2.4)$$

where $y\otimes y := (y_iy_j)_{i,j=1,\dots,d}$ for $y=(y_1,\dots,y_d)\in\mathbb{R}^d$.

Another consequence of (5.2.3) is the form of the characteristic function of $Z(t)$. Using the independence of all components in (5.2.3) we obtain

$$\mathbb{E}[e^{i\langle u,Z_t\rangle}] = e^{t\psi(u)}, \quad t\ge 0,\ u\in U, \qquad (5.2.5)$$

with

$$\psi(u) := i\langle a,u\rangle - \frac{1}{2}\langle Qu,u\rangle + \int_U \big(e^{i\langle u,y\rangle} - 1 - i\langle u,y\rangle \mathbf{1}_B(y)\big)\nu(dy), \qquad (5.2.6)$$

where $B := \{y :\mid y\mid\le 1\}$. (5.2.6) is called the *characteristic exponent* of Z and (5.2.5) the *Lévy–Khinchin formula*.

In much the same way we can obtain the direct formula for the *Laplace transform* of Z defined by

$$\mathbb{E}[e^{-\langle u,Z_t\rangle}], \quad u\in U,\ t\ge 0. \qquad (5.2.7)$$

It can be shown that (5.2.7) is well defined for $u\in U$ for any $t\ge 0$ if and only if

$$\int_{\{|y|>1\}} e^{-\langle u,y\rangle}\nu(dy) < +\infty \qquad (5.2.8)$$

(see Theorem 25.17 in Sato [114] and Theorem 4.30 in Peszat and Zabczyk [100]). If (5.2.8) holds then

$$\mathbb{E}[e^{-\langle u,Z_t\rangle}] = e^{tJ(u)}, \quad t\ge 0,$$

where the *Laplace exponent* $J(u)$ of Z is given by

$$J(u) = -\langle a,u\rangle + \frac{1}{2}\langle Qu,u\rangle + \int_U \Big(e^{-\langle u,y\rangle} - 1 + \mathbf{1}_B(y)\langle u,y\rangle\Big)\,\nu(dy). \qquad (5.2.9)$$

5.3 Special Classes

5.3.1 Finite Variation Processes

Here we characterize Lévy processes of finite variation with the use of their characteristic triplets.

Proposition 5.3.1 *A real-valued Lévy process Z with characteristic triplet (a,q,ν) is a process of finite variation if and only if*

$$q=0, \quad \textit{and} \quad \int_{\{|y|\le 1\}} \mid y\mid \nu(dy) < +\infty. \qquad (5.3.1)$$

Proof If (5.3.1) is satisfied then, in view of (5.2.3), Z can be represented in the form

$$\begin{aligned} Z(t) &= at + Z_0(t) + Z_1(t) \\ &= \Big(a - \int_B y\nu(dy)\Big)t + \int_0^t \int_B y\pi(ds,dy) + \int_0^t \int_{B^c} y\pi(ds,dy), \quad t \geq 0, \end{aligned} \tag{5.3.2}$$

with $B := \{y : | y | \leq 1\}$. It is clear that the first and the third component in (5.3.2) are processes of finite variation. The variation of the second satisfies

$$\mathbb{E}\Big[TV\Big(\int_0^t \int_B y\pi(ds,dy)\Big)\Big] = t\int_B | y | \nu(dy) < +\infty, \quad t \geq 0.$$

Now assume that Z is of finite variation. For any $\varepsilon > 0$

$$TV(Z(t)) \geq \sum_{s\in[0,t]} | \Delta Z(s) | \mathbf{1}_{\{\varepsilon<|\Delta Z_s|\leq 1\}} = \int_0^t \int_{\{\varepsilon<|y|\leq 1\}} | y | \pi(ds,dy), \quad t \geq 0,$$

and letting $\varepsilon \downarrow 0$ we obtain

$$\int_0^t \int_B | y | \pi(ds,dy) < +\infty, \quad t \geq 0. \tag{5.3.3}$$

Assume that $\int_B | y | \nu(dy) = +\infty$ and consider a disjoint partition $\{A_k\}$ of B such that $\int_{A_k} | y | \nu(dy) \geq 1, k = 1,2,\ldots$. Then the random variables

$$\int_0^t \int_{A_k} | y | \pi(ds,dy), \tag{5.3.4}$$

are independent and the sum of their means diverges, because

$$\sum_{k=1}^{+\infty} \mathbb{E}\Big(\int_0^t \int_{A_k} | y | \pi(ds,dy)\Big) = t\sum_{k=1}^{+\infty} \int_{A_k} | y | \nu(dy) = t\int_B | y | \nu(dy) = +\infty.$$

But this implies that the sums of (5.3.4) also diverge, which leads to contradiction with (5.3.3). So, we proved that the second condition in (5.3.1) holds. To see that $q = 0$ notice that it follows from the decomposition of Z that

$$W(t) = Z(t) - \Big(a - \int_B y\nu(dy)\Big)t - \int_0^t \int_{\mathbb{R}} y\pi(ds,dy), \quad t \geq 0,$$

and all the processes on the right side are of finite variation. Since W is of infinite variation, it must disappear, that is, $q = 0$. □

Since all jumps of processes of finite variation are summable on finite time intervals, using Proposition 5.3.1 we can represent them in an alternative way.

Corollary 5.3.2 *If Z is a Lévy process of finite variation with characteristic triplet* $(a, 0, \nu)$ *then it admits the representation*

$$Z_t = \hat{a}t + \sum_{s\in[0,t]} \triangle Z_s, \quad t \geq 0,$$

and its characteristic function is given by

$$\mathbb{E}[e^{iuZ_t}] = e^{t(i\hat{a}u+\int_{\mathbb{R}}(e^{iuy}-1)\nu(dy))}, \quad u \in \mathbb{R},\ t \geq 0, \tag{5.3.5}$$

where $\hat{a} := a - \int_{\{|y|\leq 1\}} y\nu(dy)$. *In particular, the variation of Z equals*

$$TV(Z(t)) = \mid \hat{a} \mid t + \sum_{s\in[0,t]} \mid \triangle Z(s) \mid, \quad t > 0.$$

5.3.2 Subordinators

Subordinators are Lévy processes with nondecreasing trajectories. If Z is a subordinator then its (total) variation

$$TV(Z(t)) = Z(t), \quad t \geq 0,$$

is finite, and by Proposition 5.3.1 the Wiener part in the Lévy–Itô decomposition of Z disappears, i.e. $q = 0$. Further, Z may not have negative jumps that is $supp\{\nu\} \subseteq [0, +\infty)$ and its jumps must be summable, which means that

$$\int_{\{|y|\leq 1\}} y\nu(dy) < +\infty.$$

Finally, each path between each two consecutive big jumps, which is the linear function $t \longrightarrow at$, must also be nondecreasing, that is, $a \geq 0$. Summarizing, a Lévy process is a subordinator if and only if its characteristic triplet (a, q, ν) satisfies

$$a \geq 0, \quad q = 0, \quad supp\{\nu\} \subseteq [0, +\infty) \quad \text{and} \quad \int_0^{+\infty} (y \wedge 1)\ \nu(dy) < +\infty. \tag{5.3.6}$$

In view of (5.3.6) and (5.3.5) the characteristic exponent of a subordinator has the form

$$\psi(u) = i\hat{a}u + \int_0^{+\infty} (e^{iuy} - 1)\nu(dy), \quad u \in \mathbb{R},$$

with $\hat{a} := a - \int_{\{|y|\leq 1\}} y\nu(dy)$. Further, it is clear that the Laplace transform (5.2.7) of a subordinator is well defined for positive arguments and the Laplace exponent (5.2.9) has the form

$$J(u) = -\hat{a}u + \int_0^{+\infty} (e^{-uy} - 1)\nu(dy), \quad u \in \mathbb{R}. \tag{5.3.7}$$

Example 5.3.3 **(Stable subordinator with index $\alpha \in (0,1)$)** An α-stable subordinator Z, with $\alpha \in (0,1)$, is given by

$$Z(t) := \int_0^t \int_0^{+\infty} y\pi(ds, dy), \quad t \geq 0,$$

with the Lévy measure $\nu(dy) := \frac{1}{y^{1+\alpha}}dy$ concentrated on $(0,+\infty)$. Since $\alpha \in (0,1)$, one can check directly that (5.3.6) is satisfied. By (5.3.7) the Laplace exponent is given by

$$J(u) = \int_0^{+\infty} \left(e^{-uy} - 1\right) \frac{1}{y^{1+\alpha}} dy, \quad u > 0.$$

Putting $z = uy$ we obtain

$$J(u) = u^\alpha \int_0^{+\infty} \left(e^{-z} - 1\right) \frac{1}{z^{1+\alpha}} dz = u^\alpha J(1).$$

Integration by part yields

$$\begin{aligned} J(1) &= \int_0^{+\infty} \left(e^{-y} - 1\right) \frac{1}{y^{1+\alpha}} dy \\ &= -\frac{1}{\alpha}(e^{-y} - 1)y^{-\alpha}\Big|_0^{+\infty} - \frac{1}{\alpha}\int_0^{+\infty} y^{-\alpha}e^{-y}dy \\ &= -\frac{1}{\alpha}\Gamma(1-\alpha), \end{aligned}$$

where $\Gamma(z)$ stands for the Gamma function, i.e. $\Gamma(z) := \int_0^{+\infty} x^{z-1}e^{-x}dx, z > 0$. So, finally we obtain

$$J(u) = -\frac{1}{\alpha}\Gamma(1-\alpha)u^\alpha, \quad u > 0.$$

5.3.3 Lévy Martingales

Here we characterize Lévy processes that are martingales and square integrable martingales. The following equivalence is valid for $\alpha > 0$

$$\mathbb{E}\big[\mid Z_t \mid^\alpha \big] < +\infty, \quad t \geq 0 \quad \Longleftrightarrow \quad \int_{\{|y|>1\}} \mid y \mid^\alpha \nu(dy) < +\infty \tag{5.3.8}$$

(see Sato [114], Theorem 25.3 and Proposition 25.4).

Using the Lévy–Itô decomposition and (5.2.9) one can deduce the following.

Proposition 5.3.4 *Let Z be a U-valued Lévy process with characteristic triplet (a, Q, ν). Then the following conditions are equivalent:*

(a) Z is a martingale,

(b)

$$\int_{\{|y|>1\}} \mid y \mid \nu(dy) < +\infty, \quad \textit{and} \quad a = -\int_{\{|y|>1\}} y\nu(dy), \tag{5.3.9}$$

(c) Z admits the representation

$$Z(t) = W(t) + \int_0^t \int_U y\, \tilde{\pi}(ds, dy), \qquad t \geq 0, \tag{5.3.10}$$

(d) the Laplace exponent of Z equals

$$J(u) = \frac{1}{2}\langle Qu, u\rangle + \int_U \left(e^{-\langle u,y\rangle} - 1 + \langle u, y\rangle\right) \nu(dy). \tag{5.3.11}$$

Moreover, if a Lévy process Z is a martingale, then it is a square integrable martingale if and only if

$$\int_{\{|y|>1\}} |\, y\,|^2\, \nu(dy) < +\infty. \tag{5.3.12}$$

If Z is a square integrable Lévy martingale, then its predictable quadratic variation equals

$$\langle Z, Z\rangle_t = t\left(TraceQ + \int_U |\, y\,|^2\, \nu(dy)\right); \qquad t \geq 0$$

and predictable operator quadratic variation

$$\langle\langle Z, Z\rangle\rangle_t = tQ + t\int_U y \otimes y\ \nu(dy),$$

compare to (4.2.5). Above $a \otimes b := (a_i b_j)_{i,j}$ where $a, b \in \mathbb{R}^d$.

It is rather surprising that Lévy processes are local or local square integrable martingales if and only if they are martingales resp. square integrable martingales (see Kallenberg [79, p. 518, 536]). We prove this fact under the square integrability condition.

Proposition 5.3.5 *If Z is a real valued local square integrable Lévy martingale then it is a square integrable martingale.*

Proof The following direct proof was communicated to us by Słomiński [118]. In view of the Lévy–Itô decomposition (5.2.3) of Z we can omit the Wiener part and compensated small jumps, as they are both square integrable martingales, and consider Z of the form

$$Z(t) = at + Z_1(t) = at + \int_0^t \int_{|y|>1} y\pi(ds, dy).$$

Then there exists an increasing sequence of stopping times $\{\tau_n\}$ such that $\mathbb{E}Z^2(t \wedge \tau_n) < +\infty, n = 1, 2, \ldots$. However, for each n,

$$\mathbb{E}Z^2(t \wedge \tau_n) = \mathbb{E}[Z, Z]_{t\wedge\tau_n} < +\infty$$

(see Section 4.3 or Protter [102, p. 66]). But

$$\begin{aligned}\mathbb{E}[Z,Z]_{t\wedge\tau_n} &= \mathbb{E}\Big(\int_0^{t\wedge\tau_n}\int_{|y|>1} y\,\pi(ds,dy)\Big)^2\\ &= \mathbb{E}\int_0^{t\wedge\tau_n}\int_{|y|>1} y^2 ds\nu(dy) = \Big(\int_{|y|>1} y^2\nu(dy)\Big)\mathbb{E}(t\wedge\tau_n),\end{aligned}$$

so

$$\int_{|y|>1} y^2\nu(dy) < +\infty.$$

Therefore

$$\mathbb{E}[Z,Z]_t \leq t\int_{|y|>1} y^2\nu(dy) < +\infty,$$

and by Theorem 4.3.10 Z is a square integrable martingale. □

The following *α-stable martingale* will frequently appear in the sequel.

Example 5.3.6 (Stable martingale with index $\alpha\in(1,2)$**)** A real valued α-stable martingale, with $\alpha\in(1,2)$, is a process

$$Z(t) := \int_0^t\int_0^{+\infty} y\tilde{\pi}(ds,dy), \quad t\geq 0,$$

where the related Lévy measure equals $\nu(dy) = \frac{1}{y^{1+\alpha}}dy$ on $(0,+\infty)$. As in Example 5.3.3 we can determine the Laplace exponent of Z. By direct calculation we obtain

$$J(u) = \int_0^{+\infty}\left(e^{-uy}-1+uy\right)\frac{1}{y^{1+\alpha}}dy = u^\alpha J(1),$$

and

$$J(1) = \frac{1}{\alpha(\alpha-1)}\Gamma(2-\alpha).$$

This yields

$$J(u) = \frac{1}{\alpha(\alpha-1)}\Gamma(2-\alpha)u^\alpha, \quad u>0.$$

5.4 Stochastic Integration

In this section we are concerned with stochastic integrals when the integrator Z is a U-valued Lévy martingale. We allow the space H, where the integrands take values, to be infinite dimensional Hilbert spaces. More specific properties and formulae will be derived for

$$\int_0^t g(s)dZ(s), \quad \int_0^t \int_U g(s,y)\tilde{\pi}(ds,dy), \quad t \in [0,T^*],\ T^* < +\infty, \tag{5.4.1}$$

where the integrands are predictable processes taking values in the space of linear operators from U into H and H -valued predictable random fields, respectively. The second integral is with respect to the compensated jump measure $\tilde{\pi}$ determined by Z. If Z is a martingale then both integrals are local martingales.

5.4.1 Square Integrable Integrators

Recall that by Proposition 5.3.4, a Lévy process Z with characteristic triplet (a, Q_W, ν) belongs to $\mathcal{M}^2(U)$ if and only if

$$\int_{\{|y|>1\}} |\, y\,|^2\ \nu(dy) < +\infty, \qquad a = -\int_{\{|y|>1\}} y\nu(dy).$$

If this is the case then Z admits the representation

$$Z(t) = W(t) + \int_0^t \int_U y\tilde{\pi}(ds,dy) = W(t) + Z_0(t), \quad t \geq 0.$$

Moreover,

$$\langle\langle Z, Z\rangle\rangle_t = t\left(Q_W + \int_U y \otimes y\ \nu(dy)\right) = t\ Q,$$

where Q is the covariance operator of $Z(1)$, equals to the sum of the covariance operators of W and Z_0. In addition, the angle bracket of Z is equal to

$$\langle Z, Z\rangle_t = t\ TrQ = t\ \mathbb{E}|Z(1)|^2$$

and $Q_t = Q/TrQ$ (see the definition (4.2.5)). Therefore the general isometric formula (4.4.7) takes the following transparent form:

$$\mathbb{E}\Big(\,\Big|\int_0^t g(s)dZ(s)\,\Big|_H^2\,\Big) = \mathbb{E}\left(\int_0^t |\,g(s)Q^{\frac{1}{2}}\,|_2^2\ ds\right), \quad t \in [0,T^*]. \tag{5.4.2}$$

In view of Theorem 4.4.1 we obtain the following characterization.

Theorem 5.4.1 *If $g(s)$ is a predictable process satisfying*

$$\mathbb{E}\left(\int_0^{T^*} |\,g(s)Q^{\frac{1}{2}}\,|_2^2\ ds\right) < +\infty, \tag{5.4.3}$$

then the integral

$$\int_0^t g(s)dZ(s), \quad t \in [0,T^*],$$

is a well-defined H-valued square integrable martingale on $[0, T^*]$ *and the following isometric formula holds*

$$\mathbb{E}\Big(\mid \int_0^t g(s)dZ(s) \mid_H^2 \Big) = \mathbb{E}\left(\int_0^t \mid g(s)Q^{\frac{1}{2}} \mid_2^2 ds \right), \quad t \in [0, T^*]. \tag{5.4.4}$$

Moreover, if

$$\mathbb{P}\left(\int_0^{T^*} \mid g(s)Q^{\frac{1}{2}} \mid_2^2 ds < +\infty \right) = 1,$$

then the stochastic integral is a well-defined local martingale.

Remark 5.4.2 For arbitrary linear operators A, B, $\mid AB \mid_2 \leq \mid A \mid \cdot \mid B \mid_2$, therefore the sufficient condition for the integral to be local martingale can be formulated in a simpler way as

$$\mathbb{P}\left(\int_0^{T^*} \mid g(s) \mid^2 ds < +\infty \right) = 1.$$

5.4.2 Integration over Compensated Jump Measures

In the case when the jump measure π is related to a Lévy process, then the integral

$$\int_0^t \int_U g(s, y)\tilde{\pi}(ds, dy), \quad t \in [0, T^*]$$

can be defined for more general integrands than those discussed in Section 4.4.3.

The process $g = g(t, y)$ is *simple* if it has the form

$$g(s, y) = g(0, y)\mathbf{1}_{\{s=0\}} + \sum_{i=0}^{n-1} \left(\sum_{j=1}^{m_i} g_{ij}\mathbf{1}_{(t_i,t_{i+1}]}(s)\mathbf{1}_{A_{ij}} \right), \qquad s \in [0, T^*],\ y \in U, \tag{5.4.5}$$

where $0 = t_0 < t_1 < \cdots < t_n = T^*$ is a partition of $[0, T^*]$ and A_{ij} is a family of sets in U that are separated from zero, i.e. $0 \notin \bar{A}_{ij}$. For a given subinterval $(t_i, t_{i+1}]$ the process g is a linear combination of terms $g_{ij}\mathbf{1}_{(t_i,t_{i+1}]}(s)\mathbf{1}_{A_{ij}}$, where g_{ij} are H-valued bounded $\mathcal{F}_{t_i}$- measurable random variables and $A_{ij}, j = 1, 2, \ldots, m_i$ are disjoint. Obviously, it is a predictable process (random field). Denote the class of simple processes by $\mathcal{S}$. It is clear that

$$(g * \tilde{\pi})_t = \int_0^t \int_U g(s, y)\tilde{\pi}(ds, dy) := \sum_{i=0}^{n} \sum_{j=1}^{m_i} g_{ij}\tilde{\pi}((t_i \wedge t, t_{i+1} \wedge t] \times A_{ij}), \quad t \in [0, T^*].$$

It is a simple consequence of the following lemma that the introduced stochastic integral is a square integrable martingale.

Lemma 5.4.3 *For sets $A, B \in U$ separated from zero and $s < t, s, t \in [0, T^*]$ hold*

$$\mathbb{E}[\tilde{\pi}^2 ((s,t] \times A) \mid \mathcal{F}_s] = (t-s)\nu(A),$$

$$\mathbb{E}[\tilde{\pi} ((s,t] \times A) \cdot \tilde{\pi} ((s,t] \times B) \mid \mathcal{F}_s] = 0, \quad \text{if } A \cap B = \emptyset,$$

$$\mathbb{E}[\tilde{\pi} ((s,t] \times A) \cdot \tilde{\pi} ((u,v] \times B) \mid \mathcal{F}_u] = 0, \quad \text{for } t \le u < v \le T^*.$$

We have the following result.

Proposition 5.4.4 *The stochastic integral*

$$(g * \tilde{\pi})_t = \int_0^t \int_U g(s,y)\tilde{\pi}(ds,dy), \quad t \in [0, T^*]$$

can be extended from $\mathcal{S}$ to all predictable random fields g, as a local martingale, such that

$$\mathbb{P}\left[\int_0^{T^*} \int_U \mid g(s,y) \mid_H^2 ds\nu(dy)) < +\infty\right] = 1.$$

Moreover, if

$$\mathbb{E}\left[\int_0^{T^*} \int_U \mid g(s,y) \mid_H^2 ds\nu(dy))\right] < +\infty,$$

then $(g * \tilde{\pi})_t$, $t \in [0, T^*]$, is an H-valued square integrable martingale and the following isometric formula holds

$$\mathbb{E}\Big[\mid (g * \tilde{\pi})_t \mid_H^2\Big] = \mathbb{E}\left[\int_0^t \int_U \mid g(s,y) \mid_H^2 ds\nu(dy))\right], \quad t \in [0, T^*]. \tag{5.4.6}$$

The proof of the result is rather standard. First, with the help of the lemma, one shows the isometric formula for simple processes and using it one establishes the second part of the proposition. The first part is established by a localization technique.

Denote by $\Psi_2(H)$ the class of all predictable processes satisfying

$$\Psi_2(H): \qquad \int_0^{T^*} \int_U \mid g(s,y) \mid_H^2 ds\nu(dy) < +\infty, \quad \mathbb{P}-a.s.,$$

and by $\Psi_1(H)$ the class all predictable processes satisfying

$$\Psi_1(H): \qquad \int_0^{T^*} \int_U \mid g(s,y) \mid_H ds\nu(dy) < +\infty, \qquad \mathbb{P}-a.s.$$

Using localizing arguments one can show that for $g \in \Psi_2(H)$ the integral

$$\int_0^t \int_U g(s,y)\tilde{\pi}(ds,dy), \quad t \in [0, T^*]$$

is a well-defined H-valued locally square integrable martingale and for $g \in \Psi_1(H)$ a local martingale. For $H = \mathbb{R}$ the corresponding classes will be denoted briefly by Ψ_1, Ψ_2.

5.4.3 Stochastic Fubini's Theorem

In some situations the integrands depend on a parameter a from a measurable space $(E, \mathcal{E})$ with a finite positive measure $\lambda(dx)$ and the problem of changing the order of integration arises. Results of this type are usually called Fubini's theorems. It turns out that, under rather weak assumptions, the following formulae hold for $t \in [0, T^*]$:

$$\int_E \left(\int_0^t g(s,a)dZ(s) \right) \lambda(da) = \int_0^t \left(\int_E g(s,a)\lambda(da) \right) dZ(s), \tag{5.4.7}$$

$$\int_E \left(\int_0^t \int_U g(s,y,a)\tilde{\pi}(ds,dy) \right) \lambda(da) = \int_0^t \int_U \left(\int_E g(s,y,a)\lambda(da) \right) \tilde{\pi}(ds,dy), \tag{5.4.8}$$

$$\int_E \left(\int_0^t \int_U g(s,y,a)\pi(ds,dy) \right) \lambda(da) = \int_0^t \int_U \left(\int_E g(s,y,a)\lambda(da) \right) \pi(ds,dy). \tag{5.4.9}$$

The following theorem formulates sufficient conditions for the validity of (5.4.7), (5.4.8) and (5.4.9). The identity (5.4.7) is a special case of Fubini's theorem from Protter's monograph [102, p. 160]. One can arrive at (5.4.8) using similar methods as in [102, p. 157–162]. Formula (5.4.9) is a corollary from the classical Fubini theorem applied pathwise.

Theorem 5.4.5 (Stochastic Fubini) *(a) Let Z be a Lévy process and $g(t,\omega,a)$, $(t,\omega,a) \in [0,T^*] \times \Omega \times E$ be measurable with respect to σ-field $\mathcal{P} \times \mathcal{E}$ and*

$$\mathbb{P}\left(\int_0^{T^*} \int_E \mid g(s,a) \mid^2 ds\, \lambda(da) < +\infty \right) = 1.$$

If $X(t,a) = \int_0^t g(s,a)dZ(s)$, $(t,a) \in [0,T^*] \times E$, is $\mathcal{B}([0,T^*]) \times \mathcal{E} \times \mathcal{F}$-measurable and $X(\cdot, a)$ is càdlàg for each a, then (5.4.7) holds.

(b) Let $\tilde{\pi}$ be the compensated random measure corresponding to a Lévy process Z and $g(t,\omega,y,a)$, $(t,\omega,y,a) \in [0,T^*] \times \Omega \times U \times E$ be measurable with respect to σ-field $\mathcal{P} \times \mathcal{U} \times \mathcal{E}$ and

$$\mathbb{P}\left(\int_0^{T^*} \int_U \int_E \mid g(s,y,a) \mid^2 ds\, \nu(dy)\lambda(da) < +\infty \right) = 1.$$

If $X(t,a) = \int_0^t \int_U g(s,y,a)\tilde{\pi}(ds,dy)$, $(t,a) \in [0,T^*] \times E$, *is* $\mathcal{B}([0,T^*]) \times \mathcal{E} \times \mathcal{F}$ *measurable and* $X(\cdot,a)$ *is càdlàg for each a,then* (5.4.8) *holds. The result is true if g is a Hilbert space valued process as well.*

(c) *Let* π *be the random measure corresponding to a Lévy process Z and* $g(t,\omega,y,a)$, $(t,\omega,y,a) \in [0,T^*]\times\Omega\times U\times E$ *be measurable with respect to* σ*-field* $\mathcal{P}\times\mathcal{U}\times\mathcal{E}$ *and*

$$\mathbb{P}\left(\int_0^{T^*}\int_U\int_E \mid g(s,y,a) \mid ds\, \nu(dy)\lambda(da) < +\infty\right) = 1.$$

If $X(t,a) = \int_0^t \int_U g(s,y,a)\tilde{\pi}(ds,dy)$, $(t,a) \in [0,T^*] \times E$ *is* $\mathcal{B}([0,T^*]) \times \mathcal{E} \times \mathcal{F}$*-measurable and* $X(\cdot,a)$ *is càdlàg for each a, then* (5.4.9) *holds. The result is true if g is a Hilbert space valued process as well.*

5.4.4 Ito's Formula for Lévy Processes

From Theorem 4.4.5 one derives easily the stochastic representation of the process $f(X(t))$ for a regular function f and semimartingale X of the form

$$X_t = \int_0^t g_1(s)ds + \int_0^t \langle g_2(s), dW_s\rangle + \int_0^t\int_U g_3(s,y)\tilde{\pi}(ds,dy)$$
$$+ \int_0^t\int_U g_4(s,y)\pi(ds,dy), \quad t \geq 0.$$

Here W is a Wiener process with covariance operator Q and π a jump measure – both related to the Lévy process Z. The integrals are assumed to exist. Moreover,

$$[X,X]_t^c = \left\langle \int_0^\cdot \langle g_2(s), dW_s\rangle\right\rangle_t = \int_0^t g_2(s)Qg_2^*(s)ds = \int_0^t \mid Q^{\frac{1}{2}}g_2(s) \mid^2 ds, \quad t \in [0,T^*],$$

and

$$\triangle X_s = \Big(g_3(s,\triangle Z_s) + g_4(s,\triangle Z_s)\Big)\mathbf{1}_{\{\triangle Z_s \neq 0\}}.$$

Thus (4.4.16), for a twice continuously differentiable function f from U into $\mathbb{R}$ yields

$$f(X_t) = f(X_0) + \int_0^t \left(f'(X_s)g_1(s) + \frac{1}{2}f''(X_s) \mid Q^{\frac{1}{2}}g_2(s) \mid^2\right)ds$$
$$+ \int_0^t f'(X_{s-})\langle g_2(s), dW_s\rangle + \int_0^t\int_U f'(X_{s-})g_3(s,y)\tilde{\pi}(ds,dy)$$
$$+ \int_0^t\int_U f'(X_{s-})g_4(s,y)\pi(ds,dy)$$
$$+ \int_0^t\int_U \Big\{f(X_s) - f(X_{s-}) - f'(X_{s-})(g_3(s,y) + g_4(s,y))\Big\}\pi(ds,dy), \quad t \in [0,T^*]. \tag{5.4.10}$$

6

Martingale Representation and Girsanov's Theorems

We present the theorem on representation of local martingales adapted to the filtration generated by a Lévy process as a sum of stochastic integrals over the Wiener process and compensated random measure. The representation theorem is applied in the proof of the Girsanov formula for densities of equivalent probability measures.

6.1 Martingale Representation Theorem

Let Z be a Lévy process with characteristic triplet (a, Q, ν) and let $\mathcal{F}_t = \sigma\{Z(s), s \leq t\}$ be its natural filtration. Let W denote its Wiener part and $\tilde{\pi}$ its compensated random measure.

The class of predictable U-valued processes ϕ such that

$$\int_0^{T^*} \mid \phi(s) \mid^2 ds < +\infty$$

is denoted by $\Phi(U)$. For such processes the stochastic integral

$$\int_0^t \langle \phi(s), dW(s) \rangle, \quad t \in [0, T^*]$$

is well defined. The class of predictable processes $g(s, y), s \in [0, T^*], y \in U$, satisfying

$$\int_0^{T^*} \int_U (\mid g(s,y) \mid^2 \wedge \mid g(s,y) \mid) ds\nu(dy) < +\infty, \quad \mathbb{P} - a.s.$$

will be denoted by $\Psi_{1,2}$. For $g \in \Psi_{1,2}$ one defines the integral

$$\int_0^t \int_U g(s,y)\tilde{\pi}(ds, dy) := \int_0^t \int_U g\mathbf{1}_{\{|g| \leq 1\}}\tilde{\pi}(ds, dy) + \int_0^t \int_U g\mathbf{1}_{\{|g| > 1\}}\tilde{\pi}(ds, dy),$$

which is a local martingale on $[0, T^*]$.

Now we formulate the martingale representation theorem.

Theorem 6.1.1 *Let M be a real valued $\mathbb{P}$-local martingale on $[0,T^*]$ adapted to $(\mathcal{F}_t)$. Then there exist predictable processes ϕ and ψ such that, $\mathbb{P}-a.s.$,*

$$\phi \in \Phi(U), \quad \psi \in \Psi_{1,2}, \tag{6.1.1}$$

for which

$$M_t = M_0 + \int_0^t \langle \phi(s), dW(s)\rangle + \int_0^t \int_U \psi(s,y)\tilde{\pi}(ds,dy), \qquad t \in [0,T^*]. \tag{6.1.2}$$

Moreover, the pair (ϕ,ψ) is unique i.e. if (ϕ',ψ'), $\phi' \in \Phi(U), \psi' \in \Psi_{1,2}$, satisfy (6.1.2) *then*

$$Q\phi = Q\phi', \quad d\mathbb{P}\times dt - \ a.s. \quad \text{and} \quad \psi = \psi', \quad d\mathbb{P}\times dt \times d\nu - \ a.s.$$

Moreover, if M is a square integrable martingale then

$$\mathbb{E}M_t^2 = \mathbb{E}M_0^2 + \mathbb{E}\left(\int_0^t \mid Q^{1/2}\phi(s)\mid^2 ds + \int_0^t\int_U \mid \psi(s,y)\mid^2 ds\nu(dy)\right) < +\infty.$$

The problem of representing martingales as stochastic integrals is of fundamental importance in financial modelling and has a very long history. Its importance was noticed in the 1960s in connection with stochastic filtering (Fujisaki–Kallianpur–Kunita filtering equation) and the general theory was developed in the book of Jacod [74]. For early papers dealing with jump processes see Dellacherie [38] and [39], and Davis [34]. The general version of the theorem in a semimartingale framework, can be found in Jacod and Shiryaev [75], but it is not directly applicable to specific cases. The presented version is due to Kunita [84] who used some results of Itô [73].

The result is well known if Z is a Wiener process, and we present in Appendix A a detailed proof of Theorem 6.1.1 in the case when Z is a jump process without the Wiener part.

6.2 Girsanov's Theorem and Equivalent Measures

For a U-valued Lévy process Z, with $U = \mathbb{R}^d$, defined on $(\Omega,\mathcal{F},\mathbb{P})$ let us consider an equivalent to $\mathbb{P}$ measure $\mathbb{Q}$ on $(\Omega,\mathcal{F}_{T^*})$. Then there exists a density process

$$\rho_t := \frac{d\mathbb{Q}}{d\mathbb{P}}\bigg|_{\mathcal{F}_t} \qquad t \in [0,T^*],$$

which is adapted to the filtration $(\mathcal{F}_t)$ generated by Z. The aim of this section is to prove the Girsanov theorem that provides an explicit form of ρ and characterizes the process Z under the measure $\mathbb{Q}$.

Theorem 6.2.1 (Girsanov) *Let $\mathbb{Q}\sim\mathbb{P}$ and Z be a U-valued Lévy process with characteristic triplet (a,Q,ν) under $\mathbb{P}$. The following statements are true.*

(a) *There exists a pair of processes* (ϕ, ψ) *such that* $\phi \in \Phi(U)$ *and* $e^{\psi} - 1 \in \Psi_{1,2}$ *such that the density process* ρ *has the form*

$$d\rho(t) = \rho(t-)\left[\langle \phi(t), dW(t)\rangle + \int_U (e^{\psi(t,y)} - 1)\tilde{\pi}(dt, dy)\right],$$
$$\rho(0) = 1, \quad t \in [0, T^*], \tag{6.2.1}$$

with $\mathbb{E}[\rho_t] = 1, t \in [0, T^*]$.

(b) *Under the measure* $\mathbb{Q}$ *the process*

$$\tilde{W}(t) := W(t) - \int_0^t \phi(s)ds, \quad t \in [0, T^*]$$

is a Q-Wiener process in U and the measure

$$\nu_{\mathbb{Q}}(dt, dy) := e^{\psi(t,y)}dt\nu(dy), \quad t \in [0, T^*], y \in U$$

is the compensating measure for the jump measure $\pi(dt, dy)$ *of Z. Moreover, it satisfies*

$$\int_0^{T^*}\int_U (|\, y\,|^2 \wedge 1)\nu_{\mathbb{Q}}(ds, dy) < +\infty. \tag{6.2.2}$$

(c) *Under* $\mathbb{Q}$ *the process Z admits the representation*

$$Z(t) = \tilde{a}_t + \tilde{W}(t) + \int_0^t \int_{\{|y|\le 1\}} y\, \tilde{\pi}_{\mathbb{Q}}(ds, dy) + \int_0^t \int_{\{|y|>1\}} y\, \pi(ds, dy), \tag{6.2.3}$$

with

$$\tilde{a}_t := at + \int_0^t \phi(s)ds + \int_0^t \int_{\{|y|\le 1\}} y(e^{\psi(s,y)} - 1)ds\nu(dy), \tag{6.2.4}$$

and $\tilde{\pi}_{\mathbb{Q}}(ds, dy) := \pi(ds, dy) - \nu_{\mathbb{Q}}(ds, dy)$.

(d) *Under* $\mathbb{Q}$ *the process Z is still a Lévy process if and only if* ϕ *is a deterministic constant and* ψ *is a deterministic function independent of time, i.e.*

$$\phi(\omega, t) = \phi, \quad \psi(\omega, t, y) = \psi(y), \qquad t \in [0, T^*],\ y \in U. \tag{6.2.5}$$

The pair (ϕ, ψ) appearing in the theorem will be called a *generating pair* of the measure $\mathbb{Q}$.

First let us comment on Theorem 6.2.1. The Doléans-Dade equation (6.2.1) can be solved explicitly to see that ρ has the form $\rho_t = e^{Y_t}$ with

$$Y(t) = \int_0^t \langle \phi(s), dW(s)\rangle - \frac{1}{2}\int_0^t |\, Q^{\frac{1}{2}}\phi(s)\,|^2\, ds + \int_0^t \int_U (e^{\psi(s,y)} - 1)\tilde{\pi}(ds, dy)$$
$$- \int_0^t \int_U (e^{\psi(s,y)} - 1 - \psi(s, y))\pi(ds, dy), \quad t \in [0, T^*]. \tag{6.2.6}$$

It follows from Theorem 6.2.1 that if $\mathbb{Q}$ is a measure equivalent to $\mathbb{P}$ with the generating pair (ϕ, ψ) then Z is not a Lévy process under $\mathbb{Q}$, in general. In particular, the measure $\mathbb{Q}$ changes stochastic properties of the jumps of Z because the new compensating measure $\nu_{\mathbb{Q}}$ is random and time dependent. Hence π is no longer a Poisson random measure. It is clear that under $\mathbb{Q}$, small jumps are square summable on compacts and since $\mathbb{Q} \sim \mathbb{P}$, there are only a finite number of big jumps. However, (6.2.2) does not imply that their expectations are finite just as it was under $\mathbb{P}$. By (6.2.2) there exists an increasing sequence of stopping times $\{\tau_n, n = 1, 2, \ldots\}$ such that

$$\mathbb{E}^{\mathbb{Q}}\Big[\int_0^{\tau_n}\int_U (|\, y\,|^2 \wedge 1)\nu_{\mathbb{Q}}(ds, dy)\Big] < +\infty, \qquad n = 1, 2, \ldots,$$

which implies that

$$\mathbb{E}^{\mathbb{Q}}\Big[\sum_{s\in[0,\tau_n]} |\,\Delta Z_s\,|^2\, \mathbf{1}_{\{|\Delta Z_s|\le 1\}}\Big] < +\infty, \quad \mathbb{E}^{\mathbb{Q}}\Big[\sum_{s\in[0,\tau_n]} \mathbf{1}_{\{|\Delta Z_s|>1\}}\Big] < +\infty, \quad n = 1, 2, \ldots.$$

Moreover, in view of (6.2.9), for any set $A \in U$ with $0 \notin \bar{A}$ holds

$$\nu_{\mathbb{Q}}([0, t], A) = \int_0^t \int_A e^{\psi(s,y)} ds\nu(dy) < +\infty, \tag{6.2.7}$$

and using similar arguments as in the preceding text one can argue that the process

$$\tilde{\pi}_{\mathbb{Q}}(t, A) := \pi(t, A) - \int_0^t \int_A \nu_{\mathbb{Q}}(ds, dy), \qquad t \ge 0,$$

is a $\mathbb{Q}$-local martingale and not a $\mathbb{Q}$-martingale, in general.

For the proof of Theorem 6.2.1 we need two auxiliary results, which we recall now.

Lemma 6.2.2 *Let $\mathbb{Q}$ be equivalent to $\mathbb{P}$ and have the density process ρ_t, $t \in [0, T^*]$. Then the process $M(t)$ is a $\mathbb{Q}$-local martingale if and only if $M(t)\rho(t)$ is a $\mathbb{P}$-local martingale.*

Proof Assume that $M(t)$ is a $\mathbb{Q}$-local martingale and let $\{\tau_n\}, n = 1, 2, \ldots$ be a localizing sequence. Then

$$\mathbb{E}^{\mathbb{Q}}[\mathbf{1}_\Gamma M(t \wedge \tau_n)] = \mathbb{E}^{\mathbb{Q}}[\mathbf{1}_\Gamma M(s \wedge \tau_n)], \qquad \Gamma \in \mathcal{F}_s, \quad s < t,$$

and thus

$$\mathbb{E}^{\mathbb{P}}[\rho(t)\mathbf{1}_\Gamma M(t \wedge \tau_n)] = \mathbb{E}^{\mathbb{P}}[\rho(s)\mathbf{1}_\Gamma M(s \wedge \tau_n)], \qquad \Gamma \in \mathcal{F}_s, \quad s < t,$$

which implies that for each n the process $\rho(t)M(t \wedge \tau_n)$ is a $\mathbb{P}$-martingale. For a bounded stopping time $\hat{\tau}_n := \tau_n \wedge n$ the stopped process

$$(\rho_t M(t \wedge \tau_n))^{\hat{\tau}_n} = \rho(t \wedge \hat{\tau}_n) M(t \wedge \hat{\tau}_n)$$

is also a $\mathbb{P}$-martingale and thus $\{\hat{\tau}_n\}, n = 1, 2, \ldots$ forms a localizing sequence for the process $\rho(t)M(t)$. Moreover, since $\hat{\tau}_n$ is bounded, we have

$$\lim_{t \to +\infty} \rho(t \wedge \hat{\tau}_n) M(t \wedge \hat{\tau}_n) = \rho(\tau_n) M(\tau_n),$$

and uniform integrability of $\rho(t \wedge \hat{\tau}_n) M(t \wedge \hat{\tau}_n)$ for each n follows. Hence $\rho(t)M(t)$ is a $\mathbb{P}$-local martingale.

The opposite implication can be shown analogously using the density process $\frac{1}{\rho(t)}$. □

Proposition 6.2.3 *Let X be a semimartingale under $\mathbb{P}$ and $\mathbb{Q}$ be a measure equivalent to $\mathbb{P}$. Then X is a semimartingale under $\mathbb{Q}$ and the quadratic variation of X under $\mathbb{Q}$ is the same as under $\mathbb{P}$.*

Proof The proof that X is a $\mathbb{Q}$-semimartingale can be found in Protter [102, Theorem 2, p. 53]. By definition of quadratic variation we have

$$\mathbb{P}(\sup_{s \in [0,t]} \mid Z_s^n - [X, X]_s \mid > \varepsilon) \underset{n}{\longrightarrow} 0, \quad t > 0, \varepsilon > 0,$$

with $Z_t^n := \sum_{k=1}^{n-1} \mid X_{\frac{(k+1)t}{n}} - X_{\frac{kt}{n}} \mid^2$. Let $\rho = \frac{d\mathbb{Q}}{d\mathbb{P}}$ be the corresponding density of $\mathbb{Q}$. Since ρ is $\mathbb{P}$-integrable it follows that

$$\mathbb{Q}(\sup_{s \in [0,t]} \mid Z_s^n - [X, X]_s \mid > \varepsilon) = \int_\Omega \mathbf{1}_{\{\sup_{s \in [0,t]} | Z_s^n - [X,X]_s | > \varepsilon\}} \rho d\mathbb{P} \underset{n}{\longrightarrow} 0,$$

and consequently that $[X, X]$ is the quadratic variation of X under $\mathbb{Q}$. □

Proof of Theorem 6.2.1 *(a)* To prove (6.2.1) one can use elements of the proof of the martingale representation theorem presented in Appendix A. The positivity of the martingale ρ allows representing it in the exponential form that leads to the Doléans-Dade equation (6.2.1). Details are presented in Lemma A.1.8 and Remark A.1.10 in the pure jump case, but the analysis containing the Wiener part is similar.

(b) First we prove that $\tilde{W}$ is a Q-Wiener process in U under $\mathbb{Q}$. We show that, for any $u \in U$, the process

$$\tilde{W}_t^u := \langle u, \tilde{W}_t \rangle = \langle u, W(t) \rangle - \int_0^t \langle u, \phi(s) \rangle ds, \quad t \in [0, T^*], \tag{6.2.8}$$

is, under $\mathbb{Q}$, a real valued Wiener process with variance $\langle Qu, u \rangle$. Application of the product Itô formula (4.4.18) to the product $\rho_t \cdot \tilde{W}_t^u$ yields

$$\rho_t \cdot \tilde{W}_t^u = \int_0^t \rho_{s-} d\tilde{W}_s^u + \int_0^t \tilde{W}_s^u d\rho_s + [\rho, \tilde{W}^u]_t.$$

Since stochastic integration of a locally bounded process over a local martingale yields local martingale, so $\int_0^t \tilde{W}_s^u d\rho_s$ is a $\mathbb{P}$-local martingale. Moreover,

$$
\begin{aligned}
[\rho, \tilde{W}^u]_t &= \Big[\int_0^t \rho_{s-}\langle \phi_s, dW_s\rangle, W_t^u\Big] \\
&= \Big[\int_0^t \rho_{s-}\langle \phi_s, dW_s\rangle, \int_0^t \langle u, dW_s\rangle\Big] = \int_0^t \rho_{s-}\langle \phi_s, u\rangle ds,
\end{aligned}
$$

where $W_t^u := \langle u, W_t\rangle$, so

$$
\begin{aligned}
\int_0^t \rho_{s-} d\tilde{W}_s^u + [\rho, \tilde{W}^u]_t &= \int_0^t \rho_{s-} dW_s^u - \int_0^t \rho_{s-}\langle u, \phi_s\rangle ds + \int_0^t \rho_{s-}\langle u, \phi_s\rangle ds \\
&= \int_0^t \rho_{s-} dW_s^u,
\end{aligned}
$$

and, consequently, $\rho \cdot \tilde{W}^u$ is a $\mathbb{P}$-local martingale. Hence $\tilde{W}^u$ is a continuous $\mathbb{Q}$-local martingale. By (6.2.8) the quadratic variations of $\tilde{W}^u$ and W^u are the same. By Proposition 6.2.3 the quadratic variation of W^u under $\mathbb{Q}$ and $\mathbb{P}$ remains the same and equals $t\langle Qu, u\rangle$. So, we obtain

$$
[\tilde{W}^u, \tilde{W}^u]_t = [W^u, W^u]_t = t\langle Qu, u\rangle.
$$

Since the quadratic variation is continuous, it is equal to the predictable quadratic variation of $\tilde{W}^u$ under $\mathbb{Q}$, i.e.

$$
\langle \tilde{W}^u, \tilde{W}^u\rangle_t = t\langle Qu, u\rangle.
$$

It follows that $\tilde{W}^u$ is a Wiener process under $\mathbb{Q}$. With the use of (4.2.4) one can determine the predictable quadratic covariation of $\tilde{W}^u$ and $\tilde{W}^v$, which is

$$
\langle \tilde{W}^u, \tilde{W}^v\rangle_t = t\langle Qu, v\rangle_U, \qquad t > 0, \quad u, v \in U.
$$

This proves that $\tilde{W}$ is a Q-Wiener process under $\mathbb{Q}$.

Now we will show that $\nu_{\mathbb{Q}}$ is a compensating measure of π under $\mathbb{Q}$. It follows from the condition $e^{\psi} - 1 \in \Psi_{1,2}$ that for any set A separated from zero

$$
\begin{aligned}
\int_0^{T^*}\int_A |\, e^{\psi(s,y)} - 1 \,|\, ds\nu(dy) &= \int_0^{T^*}\int_A |\, e^{\psi(s,y)} - 1 \,|\, \mathbf{1}_{\{|e^{\psi}-1|\le 1\}} ds\nu(dy) \\
&\quad + \int_0^{T^*}\int_A |\, e^{\psi(s,y)} - 1 \,|\, \mathbf{1}_{\{|e^{\psi}-1|> 1\}} ds\nu(dy) \\
&\le T^*\nu(A) + \int_0^{T^*}\int_A |\, e^{\psi(s,y)} - 1 \,|\, \mathbf{1}_{\{|e^{\psi}-1|> 1\}} ds\nu(dy) < +\infty. \qquad (6.2.9)
\end{aligned}
$$

Hence the integral $\int_0^t \int_A e^{\psi(s,y)} ds\nu(dy),\ t \in [0, T^*]$ is well defined and

$$X_t^A := \pi(t,A) - \int_0^t \int_A e^{\psi(s,y)} ds\nu(dy) = \tilde{\pi}(t,A) + \int_0^t \int_A (1 - e^{\psi(s,y)}) ds\nu(dy), \quad t \in [0, T^*].$$

It follows from the Itô product formula that

$$X_t^A \rho_t = \int_0^t X_{s-}^A d\rho_s + \int_0^t \rho_{s-} dX_s^A + [\rho, X^A]_t, \quad t \in [0, T^*].$$

Since

$$\begin{aligned}
&\int_0^t \rho_{s-} dX_s^A + [\rho, X^A]_t \\
&= \int_0^t \int_A \rho_{s-} \tilde{\pi}(ds, dy) + \int_0^t \int_A \rho_{s-}(1 - e^{\psi(s,y)}) ds\nu(dy) \\
&\quad + \int_0^t \int_A \rho_{s-}(e^{\psi(s,y)} - 1)\pi(ds, dy) \\
&= \int_0^t \int_A \rho_{s-} \tilde{\pi}(ds, dy) + \int_0^t \int_A \rho_{s-}(e^{\psi(s,y)} - 1)\tilde{\pi}(ds, dy), \quad t \in [0, T^*],
\end{aligned}$$

we conclude that $X_t^A = \pi(t,A) - \int_0^t \int_A e^{\psi(s,y)} ds\nu(dy)$ is a $\mathbb{Q}$-local martingale. So, the measure $e^{\psi(s,y)} ds\nu(dy)$ is the compensating measure of π under $\mathbb{Q}$. To prove (6.2.2) let us notice that

$$\begin{aligned}
&\int_0^{T^*} \int_U (|\, y\,|^2 \wedge 1) e^{\psi(s,y)}\, ds\nu(dy) < +\infty \\
&\iff \quad \int_0^{T^*} \int_U (|\, y\,|^2 \wedge 1) \mid e^{\psi(s,y)} - 1 \mid\, ds\nu(dy) < +\infty.
\end{aligned}$$

Using the fact that $e^{\psi} - 1 \in \Psi_{1,2}$ we obtain

$$\begin{aligned}
&\int_0^{T^*} \int_U (|\, y\,|^2 \wedge 1) \mid e^{\psi(s,y)} - 1 \mid ds\nu(dy) \\
&\leq \int_0^{T^*} \int_U (|\, y\,|^2 \wedge 1) \mathbf{1}_{\{|e^{\psi} - 1| \leq 1\}} ds\nu(dy) \\
&+ \int_0^{T^*} \int_U \mid e^{\psi(s,y)} - 1 \mid \mathbf{1}_{\{|e^{\psi} - 1| > 1\}} ds\nu(dy) < +\infty.
\end{aligned}$$

(c) The decomposition (6.2.3) follows immediately from the Lévy–Itô decomposition (5.2.3) by adding and subtracting the terms $\int \phi ds$ and $\int \int_{\{|y| \leq 1\}} y e^{\psi} ds\nu(dy)$. We need, however, to show that all terms in (6.2.3) are well defined. It follows from the estimation

$$
\begin{aligned}
&\int_0^{T^*}\int_{\{|y|\le 1\}} | y(e^{\psi(s,y)}-1) | \, ds\nu(dy) \\
&= \int_0^{T^*}\int_{\{|y|\le 1\}} | y(e^{\psi(s,y)}-1) | \, \mathbf{1}_{\{|e^{\psi}-1|\le 1\}} ds\nu(dy) \\
&\quad + \int_0^{T^*}\int_{\{|y|\le 1\}} | y(e^{\psi(s,y)}-1) | \, \mathbf{1}_{\{|e^{\psi}-1|> 1\}} ds\nu(dy) \\
&\le \left(\int_0^{T^*}\int_{\{|y|\le 1\}} | y |^2 ds\nu(dy)\right)^{\frac{1}{2}} \left(\int_0^{T^*}\int_{\{|y|\le 1\}} | e^{\psi(s,y)}-1 |^2 \mathbf{1}_{\{|e^{\psi}-1|\le 1\}} ds\nu(dy)\right)^{\frac{1}{2}} \\
&\quad + \int_0^{T^*}\int_{\{|y|\le 1\}} | e^{\psi(s,y)}-1 | \, \mathbf{1}_{\{|e^{\psi}-1|> 1\}} ds\nu(dy) < +\infty
\end{aligned}
$$

that $\tilde{a}_t$ is well defined. In view of (6.2.2) the integral $\int_0^t\int_{\{|y|\le 1\}} y\tilde{\pi}(ds,dy)$ is also well defined.

(*d*) Let us assume that (6.2.5) is satisfied. Then the measure $\nu_{\mathbb{Q}}(dt,dy)$ splits into the product of $e^{\psi(y)}\nu(dy)$ and dt. It follows from (6.2.2) that

$$\int_U (| y |^2 \wedge 1) e^{\psi(y)}\nu(dy) < +\infty,$$

so $e^{\psi(y)}\nu(dy)$ is a Lévy measure and consequently $\pi(dt,dy)$ becomes a Poisson random measure under $\mathbb{Q}$. Since

$$at + \int_0^t \phi(s)ds + \int_0^t\int_{\{|y|\le 1\}} y(e^{\psi(y)}-1) ds\nu(dy) = ct,$$

with $c := a+\phi+\int_{\{|y|\le 1\}} y(e^{\psi(y)}-1)\nu(dy)$, formula (6.2.3) provides the Lévy–Itô decomposition of Z under $\mathbb{Q}$. This shows that Z is a Lévy process under $\mathbb{Q}$.

Let us now assume that Z is a Lévy process under $\mathbb{Q}$. Then $\nu_{\mathbb{Q}}(dt,dy)$ splits into the product of dt and some Lévy measure. Consequently, for any set A separated from zero there exists a constant c_A such that

$$\nu_{\mathbb{Q}}([0,t]\times A) = c_A\cdot t, \quad t\in[0,T^*].$$

Since

$$\nu_{\mathbb{Q}}([0,t]\times A) = \int_0^t\int_A e^{\psi(s,y)} ds\nu(dy), \quad t\in[0,T^*],$$

we obtain that

$$\int_A e^{\psi(t,y)}\nu(dy) = c_A, \quad t\in[0,T^*],$$

and, consequently, that $\psi(\omega, t, y) = \psi(y)$. It follows from (6.2.3) that

$$\tilde{a}_t := Z(t) - \tilde{W}(t) - \int_0^t \int_{\{|y|\leq 1\}} y\, \tilde{\pi}_{\mathbb{Q}}(ds, dy) - \int_0^t \int_{\{|y|>1\}} y\, \pi(ds, dy),$$

is a Lévy process. However, by (6.2.4), $\tilde{a}_t$ is a finite variation process, so it must be a linear function. Hence there exists c such that

$$\tilde{a}_t = at + \int_0^t \phi(s)ds + t \int_{\{|y|\leq 1\}} y(e^{\psi(y)} - 1)\nu(dy) = ct, \quad t \in [0, T^*].$$

This implies that

$$\int_0^t \phi(s)ds = \Big(c - a - \int_{\{|y|\leq 1\}} y(e^{\psi(y)} - 1)\nu(dy)\Big)t, \quad t \in [0, T^*],$$

which yields that $\phi(\omega, t) = \phi$ for $t \in [0, T^*]$. □

Part III

Bond Market in Continuous Time

7

Fundamentals

The chapter deals with the mathematical description of bond market models and their elementary properties. We introduce three basic models that are at the centre of attention in further parts of the book: the Heath–Jarrow–Morton forward rate model, the Markovian factor forward curves model and the affine term structure model. Relation to modelling of forward rates in function spaces is presented as well.

7.1 Prices and Rates

A bond market can be regarded as a collection of two random fields

$$P(t,T), \quad f(t,T), \quad T \in \mathbb{R}_+, \quad 0 \leq t \leq T, \tag{7.1.1}$$

and two stochastic processes

$$B(t), \quad R(t), \quad t \in \mathbb{R}_+, \tag{7.1.2}$$

called respectively, *bond prices*, *forward rates*, *bank account* and *short rate*. They are defined on a probability space $(\Omega, \mathcal{F}, (\mathcal{F}_t), \mathbb{P})$ and adapted. The value $P(t,T)$ is interpreted as the price, at time t, of the bond that *matures* at time T. That is, the owner of this T-bond will receive at time T the so-called *nominal value*, specified on the bond, which we assume to be 1. Hence it is natural to assume that

$$P(T,T) = 1.$$

The relation between bond prices and forward rates is given by the formula

$$P(t,T) = e^{-\int_t^T f(t,s)ds}, \quad 0 \leq t \leq T. \tag{7.1.3}$$

If one spends 1 at fixed time t on buying bonds that mature at time T then the number of bonds one receives is equal to

$$e^{\int_t^T f(t,s)ds} = (P(t,T))^{-1},$$

which is identical with the amount of money one gets at T. So, $f(t,T)$ as a function of T plays a role of rates for bond prices with respect to maturity times T.

Empirical observations show that for fixed $T > 0$, the function

$$t \mapsto P(t,T), \qquad t \geq 0, \quad \textit{is chaotic}, \tag{7.1.4}$$

and for any fixed $t \geq 0$, the function

$$T \mapsto P(t,T), \qquad T \geq t, \quad \textit{is regular}, \tag{7.1.5}$$

so they may be assumed to be differentiable. Moreover, in typical situations bond prices satisfy

$$0 < P(t,T) \leq 1, \qquad 0 \leq t \leq T, \tag{7.1.6}$$

and are monotone functions of T, i.e.

$$P(t,S) \geq P(t,T), \qquad 0 \leq t \leq S \leq T, \tag{7.1.7}$$

which means that higher gains require more time. Models satisfying (7.1.6) and (7.1.7) will be called *regular*. In view of (7.1.3) it is clear that models with positive forward rates are regular. By (7.1.3) we have also the following obvious formula for the forward rates in term of the bond prices

$$T \mapsto f(t,T) := -\frac{\partial}{\partial T} \ln P(t,T), \qquad t \leq T. \tag{7.1.8}$$

More specific assumptions concerned with forward rates and bond prices are motivated by further empirical observations. These are discussed in Section 7.1.2.

The value $B(t)$ defined by

$$B(t) := e^{\int_0^t R(s)ds}, \quad B(0) = 1, \tag{7.1.9}$$

is interpreted as the amount of money on the bank account at time t from depositing 1 at time 0. It satisfies the equation

$$dB(t) = R(t)B(t)dt, \quad t > 0.$$

One requires that the short rate is directly related to the forward rates by

$$R(t) := f(t,t), \quad t \geq 0. \tag{7.1.10}$$

The relation (7.1.10) might look surprising and therefore is discussed in detail in the next section. Since R given by (7.1.10) is adapted, it follows from (7.1.9) that B is a *predictable* process.

With the use of the bank account and (7.1.10) one can extend the definition of bond prices also for $t \geq T$ by "cashing" the bond at its maturity T. We assume thus that 1 is put at T on the bank account. This yields

$$P(t,T) = e^{\int_T^t R(s)ds} = e^{\int_T^t f(s,s)ds}, \quad t \geq T. \tag{7.1.11}$$

If, additionally, forward rates satisfy

$$f(t,T) = f(T,T) = R(T); \quad t > T, \tag{7.1.12}$$

which will often be the case, then (7.1.3) and (7.1.11) yield

$$P(t,T) = e^{-\int_t^T f(t,u)du}, \quad t,T \geq 0, \tag{7.1.13}$$

which extends the definition of the bond prices on $\mathbb{R}_+ \times \mathbb{R}_+$.

7.1.1 Bank Account and Discounted Bond Prices

Here we give a financial justification of the relation

$$R(t) := f(t,t), \quad t \geq 0, \tag{7.1.14}$$

by showing that the bank can offer such rate by constructing a "risk-free strategy". Let us recall that having capital a at time t one can buy

$$a \cdot \left[P(t,T)\right]^{-1} = a \cdot e^{\int_t^T f(t,s)ds}, \tag{7.1.15}$$

T-bonds with the risk free gain at time T equal exactly to (7.1.15). *A roll-over strategy* consists of frequently buying bonds maturing in short times, selling them at their maturity times and investing the gain again into bonds. Let us fix time t and a natural number n. Starting with the capital $B_n(0) = 1$ at time 0, one invests it into bonds maturing at time t/n and reinvests the gain

$$B_n(t/n) = e^{\int_0^{t/n} f(0,s)ds},$$

at time t/n, in bonds maturing at time $2t/n$ and so on. After n operations, at time t, the gain will be equal to

$$B_n(nt/n) = \left[P(0,t/n)P(t/n,2t/n)\dots P((n-1)t/n, tn/n)\right]^{-1}, \tag{7.1.16}$$

or, more explicitly,

$$B_n(t) = e^{\sum_{k=0}^{n-1}\int_{kt/n}^{(k+1)t/n} f(kt/n,s)ds}. \tag{7.1.17}$$

Under very mild conditions on the forward rates one has that

$$\lim_{n\to+\infty} B_n(t) = e^{\int_0^t f(u,u)du} = B(t), \tag{7.1.18}$$

for $t > 0$ with $B(t)$ given by (7.1.9), so (7.1.14) holds. In particular, we have the following result.

Proposition 7.1.1 *Assume that the forward rate $f(s,T)$, $s,T \in [0,T^*]$, with $T^* < +\infty$, is bounded and one of the following two conditions holds:*

i) *for almost all $T \in [0,T^*]$,*

$$\lim_{s \uparrow T}[f(s,T) - f(T,T)] = 0,$$

ii) *for almost all $T \in [0,T^*]$,*

$$\lim_{s \uparrow T}[f(s,T) - f(s,s)] = 0,$$

and the points of discontinuity of $f(s,s)$, $s \in [0,T^*]$ are of measure 0.

Then the convergence (7.1.18) *holds for $t \in [0,T^*]$.*

Proof It is enough to show that

$$\lim_{n \to +\infty} \sum_{k=0}^{n-1} \int_{kt/n}^{(k+1)t/n} f(kt/n,T)dT = \int_0^t f(T,T)dT, \quad t \in [0,T^*]. \tag{7.1.19}$$

Let us define the sequence of functions $F_n(T), T \in [0,t]$ by

$$F_n(T) = f(kt/n,T) - f(T,T), \quad \text{if } kt/n \le T < (k+1)t/n.$$

Then

$$\sum_{k=0}^{n-1} \int_{kt/n}^{(k+1)t/n} f(kt/n,T)dT - \int_0^t f(T,T)dT = \int_0^t F_n(T)dT,$$

and by *i*) and the Lebesgue dominated convergence theorem (7.1.19) holds. Similarly, define

$$G_n(T) = f(kt/n,T) - f(kt/n,kt/n), \quad \text{if } kt/n \le T < (k+1)t/n.$$

Since $f(s,s), s \in [0,T^*]$ is Riemann integrable, by the assumption on points of discontinuity, one has:

$$\lim_{n \to +\infty} \sum_{k=0}^{n-1} f(kt/n,kt/n)t/n = \int_0^t f(T,T)dT.$$

By the first part of *ii*),

$$\lim_{n \to +\infty} \int_0^t G_n(T)dT = 0,$$

so the result follows under *ii*) as well. □

The regularity required in the latter part of *ii*) in Proposition 7.1.1, is satisfied, for instance, if the short rate is right continuous with finite left limits. This condition will be satisfied in models considered in the sequel.

An important concept is the so-called *discounted* bond price $\hat{P}(t,T)$ defined by the formula

$$\hat{P}(t,T) = B^{-1}(t)P(t,T) = P(t,T)e^{-\int_0^t R(s)ds}, \quad t,T \geq 0. \tag{7.1.20}$$

If $t > T$ then, by (7.1.20) and (7.1.11),

$$\hat{P}(t,T) = e^{-\int_0^t R(s)ds}e^{\int_T^t R(s)ds} = e^{-\int_0^T R(s)ds},$$

so the discounted price of the T-bond is constant after T. If, additionally, (7.1.12) holds, then

$$\hat{P}(t,T) = e^{-\int_0^t f(u,u)du}e^{-\int_t^T f(t,u)du} = e^{-\int_0^T f(t,u)du}, \quad t,T \geq 0. \tag{7.1.21}$$

If $P(t,T)$ is a semimartingale for each $T > 0$, then the Itô product formula applied in (7.1.20) yields,

$$d\hat{P}(t,T) = P(t-,T)d\left(e^{-\int_0^t f(s,s)ds}\right) + e^{-\int_0^t f(s,s)ds}dP(t,T) \tag{7.1.22}$$

and thus we obtain a simple but important relation

$$\frac{d\hat{P}(t,T)}{\hat{P}(t-,T)} = -f(t,t)dt + \frac{dP(t,T)}{P(t-,T)}, \quad t,T \geq 0. \tag{7.1.23}$$

7.1.2 Prices and Rates in Function Spaces

Bond prices and forward rates can be regarded as function-valued processes. For fixed running time $t \in [0,T^*], T^* < +\infty$, they are functions of maturities on $[0,+\infty)$, i.e.

$$T \mapsto f(t,T), \quad T \mapsto P(t,T), \quad T \in [0,+\infty),$$

and they can be elements of some Hilbert spaces. We alternatively use a modified notation to stress that forward rates, bond prices and discounted bond prices are treated as function valued processes. We denote them by

$$f_t := f_t(T) = f(t,T), \quad P_t := P_t(T) = P(t,T), \quad \hat{P}_t := \hat{P}_t(T) = \hat{P}(t,T),$$

where $t \in [0,T^*], T \in [0,+\infty)$ and call them *forward curves, bond curves* and *discounted bond curves*. Under the assumption that $f_t(u) = f_u(u)$ for $t > u$, we clearly obtain by (7.1.13) and (7.1.21) that

$$P_t(T) = e^{-\int_t^T f_t(u)du}, \qquad \hat{P}_t(T) = e^{-\int_0^T f_t(u)du}, \quad t \in [0,T^*], \; T \in [0,+\infty). \tag{7.1.24}$$

Working with function valued forward rates is suggested by financial practice. Central banks of many countries determine forward curves in the class of linear combinations of exponential-polynomial functions of *time to maturity* $u := T - t$, i.e. $f_t(u) = G(u)$, where:

$$G(u) = p_1(u)e^{-\alpha_1 u} + \cdots + p_n(u)e^{-\alpha_n u}, \quad u \geq 0. \tag{7.1.25}$$

Here $p_i(\cdot)$ are some polynomials and $\alpha = (\alpha_1, \ldots, \alpha_n)$ are parameters in $\mathbb{R}^n_+$. In particular, the *Nelson–Siegel* family of functions

$$\beta_1 + \Big[\beta_2 + \beta_3 u\Big]e^{-\alpha_1 u}, \tag{7.1.26}$$

and the *Svensson* family of functions

$$\beta_1 + \Big[\beta_2 + \beta_3 u\Big]e^{-\alpha_1 u} + \beta_4 u e^{-\alpha_2 u}, \tag{7.1.27}$$

are widely used in Europe.

Since working with forward curves taking values in Hilbert spaces is mathematically preferable, we will use the following state space

$$H = L^{2,\gamma} := \left\{h\colon [0,+\infty) \to \mathbb{R} : |h|^2_{L^{2,\gamma}} := \int_0^{+\infty} h^2(x)e^{\gamma x}dx < +\infty\right\}, \tag{7.1.28}$$

indexed with positive γ. This choice preserves exponential decay of forward curves at infinity suggested by (7.1.26) and (7.1.27). To obtain positive limits at infinity we can enlarge H to the space

$$\hat{H} = \hat{L}^{2,\gamma} := \left\{g : [0,+\infty) \to \mathbb{R} : g = c + h,\ c \text{ a constant},\ h \in L^{2,\gamma}\right\}. \tag{7.1.29}$$

Both spaces are separable Hilbert spaces with norms

$$|h|_{L^{2,\gamma}} = \left(\int_0^{+\infty} h^2(x)e^{\gamma x}dx\right)^{1/2}, \quad |h+c|_{\hat{L}^{2,\gamma}} = (|c|^2 + |h|^2_{L^{2,\gamma}})^{1/2}.$$

Bond prices, as functions of maturities, are usually differentiable, so the following Sobolev space is often taken as a state space for them:

$$G = W^{1,\gamma} := \left\{h\colon [0,+\infty) \to \mathbb{R} : |h|^2_{H^{1,\gamma}} := | h(0) |^2 + \int_0^{+\infty} | h'(x) |^2 e^{\gamma x}dx < +\infty\right\}. \tag{7.1.30}$$

The following proposition is concerned with forward rates taking values in $L^{2,\gamma}$. Its extension to the space $\hat{L}^{2,\gamma}$, containing constant functions, is discussed in the corollary following the proof.

Proposition 7.1.2 *If $f_t, t \in [0,T^*]$ is a càdlàg process in $H = L^{2,\gamma}$ then the processes P_t, $\hat{P}_t$, $t \in [0,T^*]$, are càdlàg processes in $G = W^{1,\gamma}$.*

The proof of Proposition 7.1.2 for $\hat{P}_t$ follows from the local Lipschitz property of the transformation

$$F(h)(T) := e^{-\int_0^T h(s)ds}, \quad T \geq 0, \quad h \in H, \tag{7.1.31}$$

which we prove in Proposition 7.1.3. Then P_t is also càdlàg because

$$P_t(T) = \hat{P}_t(T)e^{\int_0^t R(u)du} = \hat{P}_t(T)e^{\int_0^t f_t(u)du},$$

and $t \mapsto e^{\int_0^t f_t(u)du}$ is càdlàg (see Proposition 7.1.5 and Lemma 7.1.6).

Proposition 7.1.3 *For arbitrary $M > 0$ there exists $C > 0$ such that if*

$$| h |_H \leq M, \quad | g |_H \leq M,$$

then F given (7.1.31) *satisfies*

$$| F(h) - F(g) |_G \leq C | h - g |_H .$$

Proof Since $F(h)(0) = F(g)(0) = 1$, so

$$\begin{aligned} I &:= | F(h) - F(g) |_G^2 \\ &= \int_0^{+\infty} e^{\gamma T}\Big[e^{-\int_0^T h(s)ds}h(T) - e^{-\int_0^T g(s)ds}g(T)\Big]^2 dT \\ &= \int_0^{+\infty} e^{\gamma T}\Big[e^{-\int_0^T h(s)ds}(h(T) - g(T)) + g(T)(e^{-\int_0^T h(s)ds} - e^{-\int_0^T g(s)ds})\Big]^2 dT. \end{aligned}$$

Consequently,

$$I \leq 2(I_1 + I_2),$$

with

$$I_1 := \int_0^{+\infty} e^{\gamma T}(h(T) - g(T))^2 e^{-2\int_0^T h(s)ds}dT,$$

$$I_2 := \int_0^{+\infty} e^{\gamma T}g^2(T)\Big(e^{-\int_0^T h(s)ds} - e^{-\int_0^T g(s)ds}\Big)^2 dT.$$

We estimate I_2. For some constants c_1, c_2

$$| e^{-\int_0^T h(s)ds} - e^{-\int_0^T g(s)ds} |^2 \leq c_1 \left| \int_0^T (h(s) - g(s))ds \right|^2 \leq c_2 | h - g |_H^2,$$

and thus

$$\begin{aligned} I_2 &\leq \int_0^{+\infty} e^{\gamma T}g^2(T)dT \cdot c_2 | h - g |_H^2 \\ &\leq c_2 | g |_H^2 \cdot | h - g |_H^2 \leq c_2 M | h - g |_H^2 . \end{aligned}$$

In the estimate we used the fact that if $| g |_H \le M$ then

$$| \int_0^T g(s)ds | \le \Big(\int_0^T e^{-\gamma s} ds \Big)^{\frac{1}{2}} \Big(\int_0^T e^{\gamma s} | g(s) |^2 ds \Big)^{\frac{1}{2}} \le \frac{M}{\sqrt{\gamma}} | g |_H .$$

To estimate I_1 notice that

$$| e^{-2\int_0^T h(s)ds} | \le e^{2\int_0^{+\infty} |h(s)|ds} \le e^{2\frac{M}{\sqrt{\gamma}}}.$$

Therefore

$$I_1 \le \int_0^{+\infty} e^{\gamma T} (h(T) - g(T))^2 e^{\frac{2M}{\sqrt{\gamma}}} dT \le | h - g |_H^2 \, e^{\frac{2M}{\sqrt{\gamma}}}. \qquad \square$$

Corollary 7.1.4 *If f_t is a càdlàg process in $\hat{H}$ then $f_t = c_t + f_t^0$, where f_t^0 is a càdlàg process in H and c_t is a càdlàg function of time. Thus if c_t is a nonnegative process then the processes P_t, $\hat{P}_t$, $t \in [0, T^*]$, are càdlàg processes in G as well.*

Forward rates taking values in H also generate a regular bank account process.

Proposition 7.1.5 *If f_t is a càdlàg process in H then*

$$e^{\int_0^t f_t(u)du} = e^{\int_0^t R(u)du},$$

is càdlàg as well. In particular, the bank account process is càdlàg.

The proof is a consequence of the following lemma.

Lemma 7.1.6 *If, for $0 \le s \le t \le T^*$, $| f_s |_H, | f_t |_H \le M$, then for a constant c*

$$I := | e^{\int_0^t f_t(T)dT} - e^{\int_0^s f_s(T)dT} | \le c\Big(| f_t - f_s |_H + \sqrt{t-s} \Big). \tag{7.1.32}$$

Proof Let k be such that

$$| e^x - e^y | \le k | x - y |, \quad \text{for } | x |, | y | \le M.$$

The assertion follows from the following estimation

$$\begin{aligned} I &\le k\Big(| \int_0^s (f_t(T) - f_s(T))dT | + | \int_s^t f_t(T)dT | \Big) \\ &\le k\frac{M}{\sqrt{\gamma}} | f_t - f_s |_H + \int_0^t | e^{-\frac{\gamma}{2}T} e^{\frac{\gamma}{2}T} f_t(T) | \, dT \\ &\le k\frac{M}{\sqrt{\gamma}} | f_t - f_s |_H + \Big(\int_s^t e^{-\gamma T} dT \Big)^{\frac{1}{2}} | f_t |_H . \end{aligned} \qquad \square$$

7.2 Portfolios and Strategies

7.2.1 Portfolios

Investor trades on the bond market by changing her/his portfolio in time. By a *simple portfolio* one usually means a sequence:

$$p = (b, (T_1, a_1), \ldots, (T_n, a_n)),$$

where b is a real number, interpreted as an amount of money kept on the bank account, positive numbers $T_1 < T_2 < \cdots < T_n$ are maturities of bonds possessed by the investor in quantities $a_1, a_2, \ldots, a_n$. The *value $X(t)$ of the portfolio* at time t, when the bond prices are $P(t, T)$, $T \geq 0$, is then equal to:

$$X(t) = b + \sum_{k=1}^{n} a_k P(t, T_k).$$

Note that if $T_k \leq t$ then a_k is the number of T_k-bonds that have already matured and its value $P(t, T_k)$, according to the convention introduced earlier, is

$$P(t, T_k) = e^{\int_{T_k}^{t} R(s)ds} = e^{\int_{T_k}^{t} f(s,s)ds}.$$

In fact, the investment in bank account can be interpreted in terms of bonds that matured at time 0 and have value $B(t) = P(t, 0)$. Their number, at time t, is equal to a_0, where

$$b = a_0 e^{\int_0^t R(s)ds} = a_0 P(t, 0).$$

Thus the simple portfolio can be redefined as

$$p = ((0, a_0), (T_1, a_1), \ldots, (T_n, a_n)).$$

From now on we use this interpretation of the investment in the bank account. Let us notice that if, at time $t > 0$, there are in the portfolio T_i-bonds with $T_i \leq t$ then the amount of money held on the bank account equals

$$\sum_{T_i \leq t} e^{\int_{T_i}^{t} R(s)ds} a_i,$$

so is described not only by a_0 but also in terms of bonds that matured before t. Of course, we could modify a_0 such that it would contain also money invested in these bonds and simply neglect them. This idea, however, would make the concept of portfolio dependent on the running time, which we would like to avoid.

If we identify the simple portfolio p with the discrete measure φ

$$\varphi = a_0 \delta_0 + \sum_{k=1}^{n} a_k \delta_{T_k},$$

then its value can be written in a compact way as the integral

$$X(t) := (\varphi, P(t,\cdot)) = \int_0^{+\infty} P(t,T)\varphi(dT). \tag{7.2.1}$$

The concept of simple portfolios can be extended to portfolios being general finite signed measures on $[0,+\infty)$. The space of such measures equipped with weak topology will be denoted by $\mathcal{M}$ or $\mathcal{M}([0,+\infty))$. For $\varphi \in \mathcal{M}$,

$$\varphi([a,b]), \quad 0 \leq a \leq b,$$

can be interpreted as the number of bonds in the portfolio with maturities in the interval $[a,b]$. The value of the portfolio, at time t, is then given by (7.2.1). Writing (7.2.1) in the form

$$\begin{aligned} X(t) &= (\varphi, P(t,\cdot)\mathbf{1}_{[0,t]}(\cdot)) + (\varphi, P(t,\cdot)\mathbf{1}_{(t,+\infty]}(\cdot)) \\ &= \int_0^t e^{\int_T^t R(s)ds}\varphi(dT) + \int_t^{+\infty} P(t,T)\varphi(dT), \end{aligned}$$

allows us to separate money really saved on the bank account from money invested in bonds that mature in the future.

Let us stress that the expression (μ, h), where h is a function and μ a measure, will always denote the integral of the function h with respect to the measure μ.

7.2.2 Strategies and the Wealth Process

To some extent, mathematical finance is concerned with the problem of how to choose trading strategies, i.e. dynamically select portfolios, to achieve financial aims. Let us start with the concept of simple *self-financing strategies*. Suppose that an investor, at time $t_0 = 0$, starts with an initial capital $X(0) = x$ and forms a portfolio

$$\varphi_s = \left((0, a_0^0), (T_1^0, a_1^0), \ldots, (T_{N_0}^0, a_{N_0}^0)\right),$$

which is constant for $s \in [0, t_1)$ with $t_1 > 0$ and positive natural number N_0. At the starting point the portfolio necessarily should satisfy the budget condition:

$$x = (\varphi_0, P(0,\cdot)) = \sum_{k=0}^{N_0} a_k^0 P(0, T_k^0),$$

where $T_0^0 := 0$. More generally, the investor is selecting, at times $t_1, \ldots, t_M$ new portfolios

$$\varphi_{t_m} = ((0, a_0^m), (T_1^m, a_1^m), \ldots, (T_{N_m}^m, a_{N_m}^m)), \qquad m = 1, \ldots, M,$$

which are constant on the intervals $[t_m, t_{m+1})$, and their values are equal to

$$X(t_m) = (\varphi_{t_m}, P(t_m, \cdot)) = \sum_{k=0}^{N_m} a_k^m P(t_m, T_k^m), \quad m = 1, 2, \ldots, M, \tag{7.2.2}$$

with $T_0^m := 0$. The new portfolios are required to be completely financed from the capitals arrived at t_m. In particular, at t_1 we have

$$X(t_1) = (\varphi_0, P(t_1, \cdot)),$$

which, in view of (7.2.2), yields the budget constraint

$$(\varphi_0, P(t_1, \cdot)) = (\varphi_{t_1}, P(t_1, \cdot)),$$

at time t_1. In general, for $k = 1, 2, \ldots, M$, we obtain that

$$(\varphi_{t_{k-1}}, P(t_k, \cdot)) = (\varphi_{t_k}, P(t_k, \cdot)). \tag{7.2.3}$$

Consequently, by (7.2.2) and (7.2.3) in each point of the trading grid the capital can be written in the form

$$\begin{aligned} X(t_m) &= X(0) + \sum_{k=1}^{m-1} (X(t_k) - X(t_{k-1})) \\ &= X(0) + \sum_{k=1}^{m-1} ((\varphi_{t_{k-1}}, P(t_k, \cdot)) - ((\varphi_{t_{k-1}}, P(t_{k-1}, \cdot))) \\ &= X(0) + \sum_{k=1}^{m-1} (\varphi_{t_{k-1}}, \triangle P(t_k, \cdot)), \end{aligned} \tag{7.2.4}$$

where $\triangle P(t_k, T) := P(t_k, T) - P(t_{k-1}, T)$. Notice that the sum in (7.2.4) can be interpreted as an integral over the interval $[0, t_m)$:

$$X(t_m) = X(0) + \int_0^{t_m} (\varphi_s, dP(s, \cdot)).$$

Since the strategy is constant on the interval $[t_m, t_{m+1})$, the *wealth process* $X(t)$, $t \in [t_m, t_{m+1})$, can be written in two ways

$$\begin{aligned} X(t) &= (\varphi_t, P(t, \cdot)) \\ &= X(0) + \int_0^t (\varphi_s, dP(s, \cdot)). \end{aligned} \tag{7.2.5}$$

In the second identity, the integrator is the bond price process and the integrand, the trading strategy. It can be also written in the differential form

$$dX(t) = (\varphi_t, dP(t, \cdot)), \quad t \geq 0, \tag{7.2.6}$$

which means financially that the change of the investor's capital is a result of the movements of the bond prices only.

By a *general trading strategy*, called simply a strategy in the sequel, one understands a predictable process $\varphi_t, t \geq 0$ with values in the space $\mathcal{M}([0, +\infty))$. A strategy is *self-financing* if the wealth process satisfies

$$X(t) = (\varphi_t, P(t, \cdot)) = X(0) + \int_0^t (\varphi_s, dP(s, \cdot)), \quad t \geq 0, \tag{7.2.7}$$

where in the stochastic integral in (7.2.7) strategy φ and bond prices P are integrand and integrator, respectively. General stochastic integration will be discussed in Section 7.2.3.

For a given strategy φ with initial capital $x = (\varphi_0, P(0, \cdot))$ define by Δ the so-called *discrepancy* function of φ by the formula

$$\Delta(t) = x + \int_0^t (\varphi_s, dP(s, \cdot)) - (\varphi_t, P(t, \cdot)), \quad t \geq 0,$$

which measures how much the self-financing condition is violated. It is obvious that $\Delta(0) = 0$ and that φ is self-financing if and only if $\Delta(t) = 0$ for all $t \geq 0$. For theoretical considerations it is important that, by modifying investment on the bank account, one can arrive at a self-financing strategy with the same initial capital.

Proposition 7.2.1 *If, for a given strategy* (φ_t)*, taking values in* $\mathcal{M}$*, with initial capital* $x = (\varphi_0, P(0, \cdot))$*, the discrepancy is locally integrable, then the strategy*

$$\tilde{\varphi}_t := \left(\frac{\Delta(t)}{B(t)} + \int_0^t \frac{\Delta(s)}{B(s)} R(s) ds \right) \delta_0 + \varphi_t, \quad t \geq 0 \tag{7.2.8}$$

is self-financing and $(\tilde{\varphi}_0, P(0, \cdot)) = x$.

Proof Define

$$b(t) := \frac{\Delta(t)}{B(t)} + \int_0^t \frac{\Delta(s)}{B(s)} R(s) ds, \quad t \geq 0. \tag{7.2.9}$$

One has to check that

$$b(t)B(t) + (\varphi_t, P(t, \cdot)) = x + \int_0^t b(s) dB(s) + \int_0^t (\varphi_s, dP(s, \cdot)) \tag{7.2.10}$$

for $t \geq 0$. Set $y(t) := B(t)b(t)$. Then (7.2.10), is equivalent to:

$$y(t) = \int_0^t b(s)B(s) \cdot \frac{1}{B(s)} dB(s) + \Delta(t), \tag{7.2.11}$$

$$= \int_0^t y(s)R(s) ds + \Delta(t). \tag{7.2.12}$$

Thus, for the function $z(t) := y(t) - \Delta(t)$, one gets the equation

$$z'(t) = R(t)z(t) + \Delta(t)R(t), \quad z(0) = 0.$$

The preceding initial condition follows from (7.2.9), which yields $b(0) = \Delta(0) = 0$. Consequently,

$$z(t) = B(t) \int_0^t \frac{\Delta(s)}{B(s)} R(s) ds,$$

and the required formula easily follows. □

Self-financing strategies can be characterized with the use of discounted bond prices. It is easy to check that the *discounted wealth* process $\hat{X}$:

$$\hat{X}(t) = X(t)/B(t) = X(t)/P(t,0), \quad t \geq 0,$$

for a simple self-financing strategy, can be written as the integral with respect to the discounted bond prices:

$$\hat{X}(t) = \hat{X}(0) + \int_0^t (\varphi_s, d\hat{P}(s,\cdot)), \quad t \geq 0. \tag{7.2.13}$$

This important identity extends to general trading strategies.

Proposition 7.2.2 *Let (φ_t) be an $\mathcal{M}$-valued strategy such that the wealth process is a semimartingale. Then (φ_t) is self-financing if and only if*

$$d\hat{X}(t) = (\varphi_t, d\hat{P}(t,\cdot)), \quad t \geq 0. \tag{7.2.14}$$

Proof By the Itô product formula we obtain

$$(\varphi_t, d\hat{P}(t,\cdot)) = \left(\frac{d}{dt}B^{-1}(t)\right)(\varphi_t, P(t,\cdot))dt + B^{-1}(t)(\varphi_t, dP(t,\cdot)).$$

However, since $\hat{X}(t) = B^{-1}(t)(\varphi_t, P(t,\cdot))$, so

$$d\hat{X}(t) = \left(\frac{d}{dt}B^{-1}(t)\right)(\varphi_t, P(t,\cdot))dt + B^{-1}(t)d(\varphi_t, P(t,\cdot)).$$

Hence (7.2.14) is satisfied if and only if

$$d(\varphi_t, P(t,\cdot)) = (\varphi_t, dP(t,\cdot)),$$

which means that φ is self-financing. □

Corollary 7.2.3 *It follows from Proposition 7.2.1 and Proposition 7.2.2 that for any strategy φ with locally integrable discrepancy there exists a self-financing strategy $\tilde{\varphi}$ such that*

$$\int_0^t (\varphi_s, d\hat{P}(s,\cdot)) = \int_0^t (\tilde{\varphi}_s, d\hat{P}(s,\cdot)), \quad t \geq 0.$$

This follows from the fact that, for any $t \geq 0$, the measure $\tilde{\varphi}_t$ given by Proposition 7.2.1 may differ from φ_t only at zero and that $\hat{P}(t,0) \equiv 1$.

7.2.3 Wealth Process as Stochastic Integral

The wealth process corresponding to an $\mathcal{M}$-valued strategy φ starting from an initial capital x was defined by

$$X(t) = x + \int_0^t (\varphi_s, dP(s,\cdot)), \quad t \geq 0. \tag{7.2.15}$$

Here we discuss this definition in more detail. If a strategy

$$\varphi_t = \sum_{k=0}^{M} a_k(t)\delta_{T_k},$$

is based on a finite number of bonds with maturities $0 = T_0 < T_1 < \cdots < T_M$, then (7.2.15) simplifies to

$$\int_0^t (\varphi_s, dP(s,\cdot)) = \sum_{k=0}^{M} \int_0^t a_k(s)dP(s,T_k), \quad t \geq 0.$$

The preceding integral has a well-defined meaning provided that for each $k = 0,1,\ldots,M$ the processes $P(s,T_k)$, $s \in [0,T^*]$ are semimartingales, the processes $a_k, k = 1,2,\ldots,M$ are locally bounded and predictable and a_0 is adapted and locally integrable (see Protter [102]).

If now φ is a general strategy that involves an infinite number of bonds it is convenient to assume that the bond price process takes values in the Hilbert space $G = W^{1,\gamma}$ introduced in Section 7.1.2, and φ takes values in its dual G^*. We refer to Proposition 7.2.6 for important properties of G^*. If (φ_t) is a piecewise constant strategy, say $\varphi_s = \varphi_{s_k}$, $s \in (s_k, s_{k+1}]$, where $s_0 = 0 < s_1 < \cdots < s_m$, and $\varphi(s_k)$ are $\mathcal{F}_{s_k}$-measurable random variables, then one has

$$\int_0^t (\varphi_s, dP(s,\cdot)) = \sum_k (\varphi(s_k), P(t \wedge s_{k+1}) - P(t \wedge s_k)), \quad t \geq 0.$$

The concept of stochastic integral for more general integrands one extends by following classical procedure (see e.g. Protter [102, pp. 134–135] and Peszat and Zabczyk [100], chapter 8). One first defines the integral under additional conditions on the integrator and integrands. They allow the extension due to convenient isometric-type formulas. The final step is based on localization.

The stochastic integration in infinite dimension is well developed especially when the integrator is a square integrable martingale (see Métivier [92], or also Peszat and Zabczyk [100] where the largest class of admissible integrands is described). However, for models studied in the book the general theory is not very convenient, as for instance, the description of the operator Q_t (see (4.2.5)) is rather cumbersome. Therefore we use a direct approach taking into account that the bond prices, studied in the book, can be represented in the following way

$$\begin{aligned} P(t,T) &= P(0,T) - \int_0^t \Gamma_0(s,T)ds + \int_0^t \langle \Gamma_1(s,T), dW(s)\rangle \\ &\quad + \int_0^t \int_U \Gamma_2(s,T,y)\tilde{\pi}(ds,dy) + \int_0^t \int_U \Gamma_3(s,T,y)\pi(ds,dy), \end{aligned} \tag{7.2.16}$$

where $t \in [0,T^*]$ with fixed $T^* > 0$ and $T > 0$. In the decomposition W is a Q-Wiener process, π and $\tilde{\pi}$ are the jump measure and the compensated jump measure with intensity ν corresponding to the process Z. The coefficients in (7.2.16) viewed as functions of T, i.e.

$$\Gamma_0(s) = \Gamma_0(s,T), \quad \Gamma_1(s) = \Gamma_1(s,T), \quad \Gamma_2(s,y) = \Gamma_2(s,T,y), \quad \Gamma_3(s,y) = \Gamma_3(s,T,y),$$

with $s \in [0,T^*]$ are predictable G-valued processes such that

$$\int_0^{T^*} \left(|\Gamma_0(s)|_G + |\Gamma_1(s)|_G^2\right) ds < +\infty, \quad \mathbb{P} - \text{a.s.}, \tag{7.2.17}$$

$$\int_0^{T^*} \int_U \left(|\Gamma_2(s,y)|_G^2 + |\Gamma_3(s,y)|_G\right) ds\, \nu(dy) < +\infty, \quad \mathbb{P} - \text{a.s.} \tag{7.2.18}$$

Under those conditions the stochastic integrals in (7.2.16) are well defined and the bond price process has càdlàg trajectories in G. Now we can define in a rigorous way the stochastic integral (7.2.15) for $\mathcal{M}$-valued integrands by formulating conditions for their G^*-norm.

Theorem 7.2.4 *Assume that conditions* (7.2.17), (7.2.18) *hold and* φ *is an* $\mathcal{M}$*-valued predictable process such that*

$$\int_0^{T^*} |\varphi(s)|_{G^*} \left(|\Gamma_0(s)|_G + \int_U |\Gamma_3(s,y)|_G \nu(dy)\right) ds < +\infty \quad \mathbb{P} - \text{a.s.}, \tag{7.2.19}$$

$$\int_0^{T^*} |\varphi(s)|_{G^*}^2 \left(|\Gamma_1(s)|_G^2 + \int_U |\Gamma_2(s,y)|_G^2 \nu(dy)\right) ds < +\infty \quad \mathbb{P} - \text{a.s.} \tag{7.2.20}$$

Then the following statements are true.

(i) *The process* (φ_t) *is stochastically integrable over* (P_t) *and, for* $t \in [0,T^*]$,

$$\begin{aligned} \int_0^t (\varphi_s, dP(s,\cdot)) &= \int_0^t (\varphi_s, \Gamma_0(s,\cdot))ds + \int_0^t \langle (\varphi_s, \Gamma_1(s,\cdot)), dW(s)\rangle \\ &\quad + \int_0^t \int_U (\varphi_s, \Gamma_2(s,\cdot,y))\tilde{\pi}(ds,dy) + \int_0^t \int_U (\varphi_s, \Gamma_3(s,\cdot,y))\pi(ds,dy). \end{aligned} \tag{7.2.21}$$

(ii) *All predictable processes* φ *whose trajectories are bounded, almost surely in total variation, are integrable.*

We establish first the result under more restrictive conditions.

Proposition 7.2.5 *Assume that*

$$\mathbb{E}\left(\int_0^{T^*}\left(|\Gamma_0(s)|_G+|\Gamma_1(s)|_G^2\right)ds\right)<+\infty, \tag{7.2.22}$$

$$\mathbb{E}\left(\int_0^{T^*}\int_U\left(|\Gamma_2(s,y)|_G^2+|\Gamma_3(s,y)|_G\right)ds\,\nu(dy)\right)<+\infty. \tag{7.2.23}$$

If (φ_t) *is an* $\mathcal{M}$*-valued, predictable process for which the conditions*

$$\mathbb{E}\left(\int_0^{T^*}|\varphi(s)|_{G^*}\left(|\Gamma_0(s)|_G+\int_U|\Gamma_3(s,y)|_G\nu(dy)\right)ds\right)<+\infty, \tag{7.2.24}$$

$$\mathbb{E}\left(\int_0^{T^*}|\varphi(s)|_{G^*}^2\left(|\Gamma_1(s)|_G^2+\int_U|\Gamma_2(s,y)|_G^2\nu(dy)\right)ds\right)<+\infty, \tag{7.2.25}$$

are satisfied, then the stochastic integral $\int_0^t(\varphi_s,dP(s,\cdot)), t\in[0,T^*]$ *is well defined and the identity* (7.2.21) *holds.*

Moreover, if $\Gamma_0=0$ *and* $\Gamma_3=0$ *then the isometric formula holds*

$$\mathbb{E}\left(|\int_0^t(\varphi_s,dP(s,\cdot))|^2\right)=\mathbb{E}\left(\int_0^t|(\varphi_s,\Gamma_1(s,\cdot)Q^{\frac12}|_2^2\,ds\right)$$
$$+\mathbb{E}\left(\int_0^t\int_U|(\varphi_s,\Gamma_2(s,\cdot,y))|^2\,ds\,\nu(dy)\right),\quad t\in[0,T^*]$$

(*compare* (5.4.4) *and* (5.4.6)).

Proof We consider only the case when

$$P(t,T)=P(0,T)+\int_0^t\int_U\Gamma_2(s,T,y)\tilde{\pi}(ds,dy),\ t\in[0,T^*], \tag{7.2.26}$$

as the other cases can be checked similarly. If (7.2.23) is satisfied then an arbitrary simple integrand with deterministic φ^k satisfies condition (7.2.25). For (φ_t) satisfying

$$\mathbb{E}\left(\int_0^{T^*}|\varphi(s)|_{G*}^2\left(\int_U|\Gamma_2(s,y)|_G^2\nu(dy)\right)ds\right)<+\infty,$$

one can construct a sequence φ^n of simple integrands such that

$$\mathbb{E}\left(\int_0^{T^*}|\varphi(s)-\varphi^n(s)|_{G*}^2\left(\int_U|\Gamma_2(s,y)|_G^2\nu(dy)\right)ds\right)\to 0, \tag{7.2.27}$$

as $n\to+\infty$. To do it, let us choose a countable set $\mathcal{M}_0=\{\mu_1,\mu_2,\ldots\}$ of measures from $\mathcal{M}$ dense in G^* and define $\varphi_n(s)$ for every natural n and $s\in[0,T^*]$ as the closest to $\varphi(s)$ element in $\{\mu_1, \mu_2, \ldots, \mu_n\}$. If there are several such elements one chooses the one with the smallest index. The process $\varphi_n(s), s \in [0, T^*]$ is predictable, takes a finite number of values and for every $s \in [0, T^*]$ the sequence $|\varphi(s) - \varphi_n(s)|_{G*}$ tends monotonically to 0. Therefore (7.2.27) holds. Although the constructed processes take a finite number of values and are predictable they are not, in general, simple ones. It is, however, well known (see e.g. Da Prato and Zabczyk [33, p. 98]) that for an arbitrary predictable set $A \in \Omega \times [0, T^*]$ and arbitrary $\epsilon > 0$ there exists a finite number of disjoint predictable rectangles $S_1, \ldots, S_k$ of the form $B \times (s, t]$, $B \in \mathcal{F}_s$ such that their sum S approximates A up to $\epsilon > 0$, i.e.

$$\mathbb{P}_{T^*}((A \setminus S) \cup (S \setminus A)) < \epsilon.$$

Here $\mathbb{P}_{T^*}$ denotes the product of the measure $\mathbb{P}$ and the Lebesgue measure on $[0, T^*]$. This easily allows approximating the sequence φ^n by simple integrands and complete the required construction. By Doob's inequality, as $m, n \to +\infty$:

$$\mathbb{E}\left(\sup_{t \le T^*} \left| \int_0^t (\varphi_s^m, dP(s,\cdot)) - \int_0^t (\varphi_s^m, dP(s,\cdot)) \right|^2\right) \tag{7.2.28}$$

$$\le 4\mathbb{E}\left(\int_0^{T^*} |\varphi_s^m - \varphi_s^n|^2_{G*}\left(\int_U |\Gamma_2(s,y)|^2_G \nu(dy)\right) ds\right) \to 0. \tag{7.2.29}$$

Therefore for a subsequence n_k the integrals $\int_0^t (\varphi_s^{n_k}, dP(s,\cdot))$ converge, almost surely, uniformly on $[0, T^*]$ to a stochastic process, which is the required integral

$$\int_0^t (\varphi_s, dP(s,\cdot)) = \lim_{k \to +\infty} \int_0^t (\varphi_s^{n_k}, dP(s,\cdot)).$$

It is a càdlàg process. It is also clear that the identity (7.2.21) is true. □

Proof of the Theorem 7.2.4 (i) By Proposition 7.2.5 it is enough to define the integral by localization. To simplify notation we restrict again to processes P_t of the form (7.2.26). Thus let us define, for each natural n, m, the stopping times T_n, S_m as follows

$$T_n = \inf\{t \in [0, T^*] : \int_0^t |\varphi_s|^2_{G^*} \int_U |\Gamma_2(s,y)|^2_G ds\, \nu(dy) \ge n\}, \tag{7.2.30}$$

$$S_m = \inf\{t \in [0, T^*] : \int_0^t \int_U |\Gamma_2(s,y)|^2_G ds\, \nu(dy) \ge m\}, \tag{7.2.31}$$

where the infimum of the empty set is, by definition, equal to T^*. Then both sequences converge to T^*. Moreover, the processes $\varphi^n(s) := \varphi(t \wedge T_n)$ and $\Gamma_2^m(t, T, y) := \Gamma_2(t \wedge S_m, T, y)$ satisfy the conditions of Proposition 7.2.5 and therefore the stochastic integrals

$$\int_0^t \int_U (\varphi_s^n, \Gamma_2^m(s,\cdot,y))\tilde{\pi}(ds, dy), \quad t \in [0, T^*],$$

are well defined. They can be identified with the integrals

$$\int_0^{t\wedge T_n\wedge S_m} (\varphi_s, dP(s,\cdot)),$$

for which the limit, as $n, m \to +\infty$, is the required integral.

(ii) It is enough to show that the process $|\varphi(t)|_{G^*}$ has bounded trajectories. This follows, however, from the following proposition of independent interest. □

Proposition 7.2.6 *The G^*-norm of an arbitrary finite measure μ on $[0, +\infty)$ viewed as a functional on $G = W^{1,\gamma}$ is given by the formula,*

$$|\mu|_{G^*} = \left((\mu(\mathbb{R}_+))^2 + \int_0^{+\infty} e^{-\gamma u}(\mu([u, +\infty))^2 du \right)^{1/2},$$

so

$$|\mu|_{G^*} \leq |\mu|(1 + 1/\gamma)^{1/2},$$

where $|\mu|$ denotes the total variation of μ, i.e., $|\mu| := (\mu_+ + \mu_-)(\mathbb{R}_+)$.

Proof Let h be an arbitrary element of G. Then

$$\begin{aligned}
(\mu, h) &= \int_0^{+\infty} h(s)\mu(ds) = \int_0^{+\infty} (h(0) + \int_0^s h'(u)du)\mu(ds) \\
&= h(0)\mu(\mathbb{R}_+) + \int_0^{+\infty}\int_0^{+\infty} 1_{[0,s]}(u)h'(u)du \ \mu(ds) \\
&= h(0)\mu(\mathbb{R}_+) + \int_0^{+\infty} h'(u)\mu([u, +\infty))du \\
&= h(0)\mu(\mathbb{R}_+) + \int_0^{+\infty} e^{\gamma u}(h'(u))(e^{-\gamma u}\mu([u, +\infty))du \\
&= < h, g >_G,
\end{aligned}$$

where

$$g(u) = \mu(\mathbb{R}_+) + \int_0^u e^{-\gamma s}\mu([s, +\infty))ds, \ \ u \geq 0.$$

Thus $|\mu|_{G^*} = |g|_G$ and the result follows. □

Remark 7.2.7 The following formula can be useful if one wants to approximate strategies by simpler ones, for instance based on a finite number of bonds, without losing much from the gains. Assume that $\mu_1 - \mu_2$ are two finite measures on $\mathbb{R}_+$ and let F_{μ_1}, F_{μ_2}, be their cumulative distribution functions:

$$F_{\mu_i}(u) = \mu_i([0, u]), \ \ u \geq 0, \ \ i = 1, 2.$$

It follows from the proposition that

$$|\mu_1 - \mu_2|_{G^*} = \left((\mu_1(\mathbb{R}_+) - \mu_2(\mathbb{R}_+))^2 + \int_0^{+\infty} e^{-\gamma u}(F_{\mu_1}(u)) - F_{\mu_2}(u)))^2 du \right)^{1/2}.$$

Thus, in particular, an arbitrary measure can be approximated, as close as one wishes, by atomic measures with a finite number of atoms.

7.3 Non-arbitrage, Claims and Their Prices

A self-financing strategy φ is called an *arbitrage opportunity* if for the corresponding wealth process starting from zero:

$$X(t) = \int_0^t (\varphi_s, dP(s,\cdot)), \; t \in [0, T^*],$$

one has

$$\mathbb{P}(X(s) \geq 0, \; s \in [0, T^*]) = 1, \quad \text{and, for some } t \in [0, T^*], \quad \mathbb{P}(X(t) > 0) > 0.$$

We say that the bond market is *arbitrage free*, if there is no arbitrage opportunity in the class of self-financing strategies based on a finite number of bonds. This means that for self-financing strategies of the form

$$\varphi_t = \sum_{k=0}^{M} a_k(t)\delta_{T_k}, \quad t \in [0, T^*], \tag{7.3.1}$$

the condition

$$X(t) = \int_0^t (\varphi_s, dP(s,\cdot)) \geq 0, \qquad t \in [0, T^*] \tag{7.3.2}$$

implies that

$$X(t) = \int_0^t (\varphi_s, dP(s,\cdot)) = 0, \qquad t \in [0, T^*]. \tag{7.3.3}$$

In view of Proposition 7.2.2 and Corollary 7.2.3 one can reformulate conditions (7.3.2) and (7.3.3) with the use of discounted quantities and forget the self-financing requirement. So, the market is arbitrage free if, for arbitrary strategy (7.3.1) such that

$$\int_0^t (\varphi_s, d\hat{P}(s,\cdot)) \geq 0, \quad t \in [0, T^*], \tag{7.3.4}$$

one has

$$\int_0^t (\varphi_s, d\hat{P}(s,\cdot)) = 0, \quad t \in [0, T^*]. \tag{7.3.5}$$

Important criteria for non-arbitrage uses the idea of local martingale. It is based on the fact that the stochastic integral, with respect to a local martingale, is a local martingale, under very general conditions on the integrand. In particular, this is the case when the integrand is locally bounded.

Proposition 7.3.1 *Let, for each $T > 0$, the discounted bond price process $\hat{P}(t,T)$, $t \in [0,T^*]$ be a local martingale on a filtered probability space $(\Omega, \mathcal{F}, (\mathcal{F}_t), \mathbb{P})$. Then there are no arbitrage opportunities in the class of locally bounded strategies of the form* (7.3.1).

Proof For φ given by (7.3.1), the corresponding discounted wealth process

$$\hat{X}(t) = \int_0^t (\varphi_s, d\hat{P}(s,\cdot)) = \sum_{k=1}^{M} \int_0^t a_k(s) d\hat{P}(s,T_k), \quad t \in [0,T^*]$$

is a local martingale. For the localizing sequence τ_n we have thus

$$0 = \hat{X}(0) = \mathbb{E}(\hat{X}(t \wedge \tau_n)), \quad t \in [0,T^*], \quad n = 1,2,\ldots.$$

If (7.3.4) holds then we obtain $\hat{X}(t \wedge \tau_n) = 0$ for arbitrary $n, t \in [0,T^*]$ and thus $\hat{X}$ disappears. So, (7.3.5) is satisfied and the market is arbitrage free. □

An important sufficient condition for non-arbitrage is based on the concept of martingale measure. A probability measure $\mathbb{Q}$ is a *martingale measure* for the bond market defined on $(\Omega, \mathcal{F}, (\mathcal{F}_t), \mathbb{P})$ if it is equivalent to $\mathbb{P}$ and for arbitrary $T > 0$ the process

$$\hat{P}(t,T), \quad t \in [0,T^*]$$

is a local martingale on $(\Omega, \mathcal{F}, (\mathcal{F}_t), \mathbb{Q})$. It is clear that each step in the proof of Proposition 7.3.1 remains true if $\mathbb{P}$ is replaced by a martingale measure $\mathbb{Q}$. Thus if there exists a martingale measure for a bond model then the model is arbitrage free. For markets with a finite number of trading assets this sufficient condition is, under rather weak assumptions, also a necessary condition and constitutes the First Fundamental Theorem of Asset Pricing. An extensive study of this issue can be found in Delbaen and Schachermayer [36] and Delbaen and Schachermayer [37].

A *contingent claim* executed at time $T^* > 0$ is an arbitrary $\mathcal{F}_{T^*}$-measurable random variable X. It is interpreted as an amount of money that the *writer of the contract*, or agent, agrees to pay to the *buyer of the contract* at time T^*. A self-financing strategy φ, taking values in $\mathcal{M}$, with the initial capital x is a *replicating strategy*, or *hedging strategy* for X if

$$X = x + \int_0^{T^*} (\varphi_s, dP(s,\cdot)). \tag{7.3.6}$$

Then X is called *attainable*. Equivalently, using the discounted bond prices we see that the discounted attainable claim,

$$\hat{X} = e^{-\int_0^{T^*} R(s)ds} X,$$

admits the representation

$$\hat{X} = x + \int_0^{T^*} (\varphi_s, d\hat{P}(s,\cdot)). \tag{7.3.7}$$

Representation (7.3.6) means that the writer of the contract, can recover the claim starting with the initial capital x and cleverly investing in the bond market. The capital x is then, rightly called, the *price of the claim* X. If $\mathbb{Q}$ is a martingale measure then, under proper integrability requirement, we have by (7.3.7) the following formula for the price $p(X)$ of X

$$p(X) = \mathbb{E}^{\mathbb{Q}}[\hat{X}], \tag{7.3.8}$$

where $\mathbb{E}^{\mathbb{Q}}[\cdot]$ stands for the expectation under $\mathbb{Q}$.

7.4 HJM Modelling

One of the most important ways of describing bond markets starts from models where the dynamics of forward rates consists of two parts, the drift term, reflecting basic market trends and the noise term, corresponding to chaotic fluctuations. In particular, in the Heath–Jarrow–Morton (HJM) model the field $f(t,T)$ is described as a family of stochastic processes $f(\cdot,T)$ parametrized by $T>0$ and assumed to be of the following form. For fixed T, one requires that

$$f(t,T) = f(0,T) + \int_0^t \alpha(s,T)ds + \int_0^t \langle \sigma(s,T), dZ(s)\rangle, \quad t \le T, \tag{7.4.1}$$

where Z is a Lévy process taking values in $U = \mathbb{R}^d$, $\alpha(\cdot,T)$ a real-valued process and $\sigma(\cdot,T)$ a U-valued predictable process. In the preceding $\langle\cdot,\cdot\rangle$ stands for the scalar product in U. This modelling was introduced by Heath, Jarrow and Morton in [67] with Z being a one-dimensional Wiener process and afterwards developed by many authors by considering more general processes including discontinuous ones. The HJM model with forward rates (7.4.1) will be denoted by (α,σ,Z).

We introduce conditions on the model (α,σ,Z) that guarantee that the bond prices

$$P(t,T) = e^{-\int_t^T f(t,s)ds}, \quad 0 \le t \le T, \tag{7.4.2}$$

as well as the bank account process

$$B(t) = e^{\int_0^t R(s)ds} = e^{\int_0^t f(s,s)ds}, \quad t \ge 0, \tag{7.4.3}$$

are well defined on some finite time interval $[0, T^*]$. They should be regarded as instructive examples rather than a universal set of sufficient conditions. Recall that the definition of $P(t, T)$ can be extended also for $t > T$ by cashing the bond at time T and putting 1 on the savings account. This can be achieved in the model (7.4.1) by assuming that

$$\alpha(t, T) = 0, \quad \sigma(t, T) = 0 \quad \text{for} \quad t > T. \tag{7.4.4}$$

Then

$$f(t, T) = f(0, T) + \int_0^T \alpha(s, T)ds + \int_0^T \langle \sigma(s, T), dZ(s) \rangle = f(T, T) = R(T), \quad t \geq T, \tag{7.4.5}$$

and consequently (see (7.1.12) and (7.1.13)) one obtains

$$P(t, T) = e^{\int_T^t R(s)ds} = e^{-\int_t^T f(t,s)ds}, \quad t > T.$$

Proposition 7.4.1 *Assume that in the HJM model (α, σ, Z), for arbitrary $S > 0$,*

$$\int_0^S | f(0, T) | \, dT < +\infty, \tag{7.4.6}$$

$$\int_0^{T^*} \int_0^S | \alpha(t, T) | \, dt \, dT < +\infty, \quad \int_0^{T^*} \int_0^S | \sigma(t, T) |^2 \, dt \, dT < +\infty, \tag{7.4.7}$$

and (7.4.4) *holds. Then the family of bond prices*

$$P(t, T) = e^{-\int_t^T f(t,u)du}, \qquad t \in [0, T^*], \ T \in (0, +\infty), \tag{7.4.8}$$

and the bank account process

$$B(t) = e^{\int_0^t R(s)ds}, \qquad t \in [0, T^*],$$

are well defined. Moreover, the discounted bond prices are given by

$$\hat{P}(t, T) := e^{-\int_0^T f(t,s)ds}, \qquad t \in [0, T^*], \ T \in (0, +\infty). \tag{7.4.9}$$

Proof We have to show that the integrals

$$\int_t^T f(t, u)du, \tag{7.4.10}$$

which define the bond prices by (7.4.8), are well defined for any $t \in [0, T^*]$, $T \in (0, +\infty)$. It follows from (7.4.1) that

$$\int_t^T f(t, u)du = \int_t^T f(0, u)du + \int_t^T \left(\int_0^t \alpha(s, u)ds \right) du + \int_t^T \left(\int_0^t \langle \sigma(s, u), dZ(s) \rangle \right) du.$$

Taking S so large that $T < S$ we see that (7.4.6) and (7.4.7) are sufficient conditions for applicability of the Fubini and stochastic Fubini theorem (see Theorem 5.4.5). In particular, the preceding double integrals are then well defined.

In view of (7.4.1), for almost all $t \in [0, T^*]$, the short-rate process is given by

$$R(t) = f(t,t) = f(0,t) + \int_0^t \alpha(s,t)ds + \int_0^t \langle \sigma(s,t), dZ(s)\rangle.$$

The integrals

$$\int_0^t R(u)du = \int_0^t f(0,u)du + \int_0^t \left(\int_0^u \alpha(s,u)ds\right) du + \int_0^t \left(\int_0^u \langle \sigma(s,u), dZ(s)\rangle\right) du, \tag{7.4.11}$$

are, however, well defined for each $t \in [0, T^*]$. To see this we write the double integrals in the form

$$\int_0^t \left(\int_0^u \alpha(s,u)ds\right) du = \int_0^t \left(\int_0^t \alpha(s,u)\mathbf{1}_{[0,u]}(s)ds\right) du,$$

$$\int_0^t \left(\int_0^u \langle \sigma(s,u), dZ(s)\rangle\right) du = \int_0^t \left(\int_0^t \langle \sigma(s,u)\mathbf{1}_{[0,u]}(s), dZ(s)\rangle\right) du,$$

and use again (7.4.6)–(7.4.7). Consequently, the bank account process (7.4.3) is well defined. In view of (7.4.11) one can show that the discounted bond prices

$$\hat{P}(t,T) := e^{-\int_0^t R(s)ds} P(t,T),$$

admit the representation (7.4.9). □

Under additional conditions forward curves can be regular functions of maturities.

Proposition 7.4.2 *Let us assume that in the HJM model* (α, σ, Z) *the conditions* (7.4.6), (7.4.7) *and* (7.4.4) *are satisfied. If* $f(0,\cdot)$ *is differentiable on* $[0, +\infty)$, $\alpha(t,\cdot)$, $\sigma(t,\cdot)$ *are differentiable on* $[0,+\infty)$ *for any* $t \in [0, T^*]$, *and for arbitrary* $S > 0$,

$$\int_0^{T^*} \sup_{T\in(0,S)} \left| \frac{\partial \alpha}{\partial T}(s,T) \right| ds < +\infty, \quad \mathbb{E}\left(\int_0^{T^*} \sup_{T\in(0,S)} \left| \frac{\partial \sigma}{\partial T}(s,T) \right|^2 ds\right) < +\infty, \tag{7.4.12}$$

then $f(t,T)$ *is differentiable over* T *and the following formulas hold*

$$\frac{\partial f}{\partial T}(t,T) = \frac{\partial f}{\partial T}(0,T) + \int_0^t \frac{\partial \alpha}{\partial T}(s,T)ds + \int_0^t \left\langle \frac{\partial \sigma}{\partial T}(s,T), dZ(s)\right\rangle, \quad t \in [0,T^*], T>0, \tag{7.4.13}$$

$$R(t) = R(0) + \int_0^t \frac{\partial f}{\partial T}(s,s)ds, \quad t \in [0,T^*]. \tag{7.4.14}$$

In particular, the short-rate process has differentiable paths.

Proof We choose S greater than T. By (7.4.12), it follows from the dominated convergence theorem that

$$\frac{\partial}{\partial T}\int_0^t \alpha(s,T)ds = \int_0^t \frac{\partial \alpha}{\partial T}(s,T)ds, \quad t \in [0,T^*].$$

Similarly, since for any $s \in [0,T^*]$ and small $\varepsilon > 0$ we have

$$\left|\frac{\sigma(s,T+\varepsilon)-\sigma(s,T)}{\varepsilon} - \frac{\partial \sigma}{\partial T}(s,T)\right|^2 \leq 2 \sup_{T\in(0,S)} \left| \frac{\partial \sigma}{\partial T}(s,T) \right|^2,$$

it follows that (7.4.12) implies that

$$\mathbb{E}\left(\int_0^{T^*} \left|\frac{\sigma(s,T+\varepsilon)-\sigma(s,T)}{\varepsilon} - \frac{\partial \sigma}{\partial T}(s,T)\right|^2 ds\right) \underset{\varepsilon\to 0}{\longrightarrow} 0.$$

The significance of the preceding convergence is that in the Lévy–Itô decomposition of Z we can apply the square integrability criteria and show that

$$\frac{\partial}{\partial T}\int_0^t \langle \sigma(s,T), dZ(s)\rangle = \int_0^t \left\langle \frac{\partial \sigma}{\partial T}(s,T), dZ(s)\right\rangle, \qquad t \in [0,T^*].$$

Hence, if $f(0,\cdot)$ is differentiable and (7.4.12) holds then $f(t,T)$ is also differentiable over T and (7.4.13) holds.

Since $R(t) = f(t,t)$, the short-rate process admits the following representation

$$\begin{aligned} R(t) &= f(0,t) + \int_0^t \alpha(s,t)ds + \int_0^t \langle \sigma(s,t), dZ(s)\rangle \\ &= f(0,0) + \int_0^t \frac{\partial f}{\partial T}(0,u)du + \int_0^t \left(\alpha(s,s) + \int_s^t \frac{\partial \alpha}{\partial T}(s,u)du\right) ds \\ &\quad + \int_0^t \left\langle \sigma(s,s) + \int_s^t \frac{\partial \sigma}{\partial T}(s,u)du, dZ(s)\right\rangle, t \in [0,T^*]. \end{aligned}$$

By (7.4.12) we can apply the Fubini and stochastic Fubini theorem and change the order of integration. This yields

$$\begin{aligned} R(t) &= f(0,0) + \int_0^t \left(\frac{\partial f}{\partial T}(0,u)du + \int_0^u \frac{\partial \alpha}{\partial T}(s,u)ds + \int_0^u \left\langle \frac{\partial}{\partial T}\sigma(s,u), dZ(s)\right\rangle\right) du \\ &\quad + \int_0^t \alpha(s,s)ds + \int_0^t \langle \sigma(s,s), dZ(s)\rangle, \quad t \in [0,T^*], \end{aligned}$$

which, in view of (7.4.13), (7.4.4), and the continuity of $\alpha(t,T)$ and $\sigma(t,T)$ over T, yields (7.4.14). □

7.4.1 Bond Prices Formula

In this section we establish the formula for bond prices in terms of the representation of forward rates (7.4.1) in the HJM model. In fact, we consider a generalized form of (7.4.1) where the volatility is splitted into three parts and integrated separately over components of the Lévy process Z. So, (7.4.1) is replaced by

$$df(t,T) = \alpha(t,T)dt + \langle \sigma^0(t,T), dW(t)\rangle + \int_{|y|\le 1} \sigma^1(t,T,y)\tilde{\pi}(dt,dy) + \int_{|y|>1} \sigma^2(t,T,y)\pi(dt,dy), \quad t\in[0,T^*],\ T>0, \tag{7.4.15}$$

where $W, \pi, \tilde{\pi}$ come from the Lévy–Itô decomposition of the Lévy process Z and $f(0,T) = f_0(T)$ is a given function. The HJM model with forward rates (7.4.15) will be denoted $(\alpha, \sigma^0, \sigma^1, \sigma^2, Z)$. Let us notice that for a given process $\sigma(t,T)$ and

$$\sigma^0(t,T) := \sigma(t,T), \quad \sigma^1(t,T,y) := \sigma(t,T)y, \quad \sigma^2(t,T,y) := \sigma(t,T)y,$$

the dynamics (7.4.15) reduces to (7.4.1) with slightly modified $\alpha(t,T)$. Similarly, as in the model (α, σ, Z), we assume that

$$\alpha(t,T) = 0, \quad \sigma^0(t,T) = 0, \quad \sigma^1(t,T,y) = 0, \quad \sigma^2(t,T,y) = 0 \quad \text{for } t > T, \tag{7.4.16}$$

and to enable application of the Fubini type theorems we need that for arbitrary $S>0$,

$$\int_0^S |\, f(0,T)\,|\, dT < +\infty, \qquad \int_0^{T^*}\int_0^S |\,\alpha(t,T)\,|\, dt\, dT < +\infty, \tag{7.4.17}$$

$$\int_0^{T^*}\int_0^S |\,\sigma^0(t,T)\,|^2\, dt\, dT < +\infty, \qquad \int_0^{T^*}\int_0^S\int_{|y|\le 1} |\,\sigma^1(t,T)\,|^2\, dt\, dT\, \nu(dy) < +\infty, \tag{7.4.18}$$

$$\int_0^{T^*}\int_0^S\int_{|y|>1} |\,\sigma^2(t,T)\,|\, dt\, dT\, \nu(dy) < +\infty. \tag{7.4.19}$$

Proposition 7.4.3 *In the HJM model $(\alpha, \sigma^0, \sigma^1, \sigma^2, Z)$ satisfying* (7.4.16), (7.4.17)–(7.4.19) *the discounted bond prices are given by*

$$\frac{d\hat{P}(t,T)}{\hat{P}(t-,T)} = D(t,T)dt - \langle \Sigma^0(t,T), dW(t)\rangle + \int_{|y|\le 1} \Gamma^1(t,T,y)\tilde{\pi}(dt,dy) + \int_{|y|>1} \Gamma^2(t,T,y)\pi(dt,dy), \quad t\in[0,T^*],\ T>0, \tag{7.4.20}$$

and the bond prices by

$$\frac{dP(t,T)}{P(t-,T)} = f(t,t)dt + \frac{d\hat{P}(t,T)}{\hat{P}(t-,T)}, \quad t \in [0,T^*], T > 0, \tag{7.4.21}$$

where

$$\begin{aligned} D(t,T) := & -\int_0^T \alpha(t,s)ds + \frac{1}{2} \mid Q^{\frac{1}{2}} \int_0^T \sigma^0(t,s)ds \mid^2 \\ & + \int_{|y|\le 1} \left(e^{-\int_0^T \sigma^1(t,s,y)ds} - 1 + \int_0^T \sigma^1(t,s,y)ds \right) \nu(dy), \end{aligned}$$

$$\Sigma^0(t,T) := \int_0^T \sigma^0(t,s)ds, \quad \Gamma^1(t,T,y) := \left(e^{-\int_0^T \sigma^1(t,s,y)ds} - 1 \right),$$

$$\Gamma^2(t,T,y) := \left(e^{-\int_0^T \sigma^2(t,s,y)ds} - 1 \right).$$

Proof Let us write $\hat{P}$ in the form

$$\hat{P}(t,T) = e^{-X(t,T)}, \quad \text{where} \quad X(t,T) := \int_0^T f(t,s)ds, \quad t \in [0,T^*],\ T > 0.$$

Application of the stochastic Fubini theorem yields

$$\begin{aligned} X(t,T) = & \int_0^T f(0,s)ds + \int_0^T \left(\int_0^t \alpha(v,s)dv \right) ds + \int_0^T \left(\int_0^t \langle \sigma^0(v,s), dW(v) \rangle \right) ds \\ & + \int_0^T \left(\int_0^t \int_{|y|\le 1} \sigma^1(v,s,y) \tilde{\pi}(dv,dy) \right) ds + \int_0^T \left(\int_0^t \int_{|y|>1} \sigma^2(v,s,y) \pi(dv,dy) \right) ds \\ = & \int_0^T f(0,s)ds + \int_0^t \left(\int_0^T \alpha(v,s)ds \right) dv + \int_0^t \left\langle \int_0^T \sigma^0(v,s)ds, dW(v) \right\rangle \\ & + \int_0^t \int_{|y|\le 1} \left(\int_0^T \sigma^1(v,s,y)ds \right) \tilde{\pi}(dv,dy) + \int_0^t \int_{|y|>1} \left(\int_0^T \sigma^2(v,s,y)ds \right) \pi(dv,dy). \end{aligned}$$

Hence $X(t,T), T > 0$ satisfies

$$\begin{aligned} dX(t,T) = & A(t,T)dt + \langle \Sigma^0(t,T), dW(t) \rangle + \int_{|y|\le 1} \Sigma^1(t,T,y) \tilde{\pi}(dt,dy) \\ & + \int_{|y|>1} \Sigma^2(t,T,y) \pi(dt,dy), \end{aligned}$$

with

$$A(t,T) := \int_0^T \alpha(t,s)ds, \quad \Sigma^0(t,T) = \int_0^T \sigma^0(t,s)ds,$$

$$\Sigma^1(t,T,y) := \int_0^T \sigma^1(t,y,s)ds, \quad \Sigma^2(t,T,y) = \int_0^T \sigma^2(t,y,s)ds.$$

The Itô formula yields

$$\begin{aligned}\hat{P}(t,T) = e^{-X(t,T)} &= e^{-X(0,T)} - \int_0^t e^{-X(s-,T)} dX(s,T) \\ &+ \frac{1}{2}\int_0^t e^{-X(s-,T)} d[X(s,T),X(s,T)]^c \\ &\sum_{s\in[0,t]} \left(e^{-X(s,T)} - e^{-X(s-,T)} + e^{X(s-,T)} \triangle X(s,T)\right), \quad t\in[0,T^*].\end{aligned} \tag{7.4.22}$$

Since

$$\triangle X(s,T) = \left(\Sigma^1(s,T,\triangle Z(s)) + \Sigma^2(s,T,\triangle Z(s))\right)\mathbf{1}_{\{\triangle Z(s)\neq 0\}},$$

and

$$d[X(s,T),X(s,T)]^c = \mid Q^{\frac{1}{2}}\Sigma^0(s,T)\mid^2 ds,$$

the jump part in (7.4.22) equals

$$\begin{aligned} I(t,T) &:= \sum_{s\in[0,t]} \left(e^{-X(s,T)} - e^{-X(s-,T)} + e^{X(s-,T)} \triangle X(s,T)\right) \\ &= \sum_{s\in[0,t]} e^{-X(s-,T)}(e^{-\triangle X(s,T)} - 1 + \triangle X(s,T)) \\ &= \sum_{s\in[0,t]} e^{-X(s-,T)}\left(e^{-\Sigma^1(s,T,\triangle Z(s))-\Sigma^2(s,T,\triangle Z(s))} - 1 + (\Sigma^1(s,T,\triangle Z(s)) + \Sigma^2(s,T,\triangle Z(s))\right) \\ &= \int_0^t\int_U e^{-X(s-,T)}\left(e^{-\Sigma^1(s,T,y)-\Sigma^2(s,T,y)} - 1 + (\Sigma^1(s,T,y)+\Sigma^2(s,T,y)\right)\pi(ds,dy) \\ &= \int_0^t\int_{|y|\le 1}\int_U e^{-X(s-,T)}\left(e^{-\Sigma^1(s,T,y)} - 1 + \Sigma^1(s,T,y)\right)\tilde{\pi}(ds,dy) \\ &\quad + \int_0^t\int_{|y|\le 1}\int_U e^{-X(s-,T)}\left(e^{-\Sigma^1(s,T,y)} - 1 + \Sigma^1(s,T,y)\right) ds\nu(dy \\ &\quad + \int_0^t\int_{|y|>1} e^{-X(s-,T)}\left(e^{-\Sigma^2(s,T,y)} - 1 + \Sigma^2(s,T,y))\right)\pi(ds,dy). \end{aligned}$$

Thus

$$\begin{aligned} dI(t,T) &= e^{-X(t-,T)}\int_{|y|\le 1}\left(e^{-\Sigma^1(t,T,y)} - 1 + \Sigma^1(t,T,y)\right)\tilde{\pi}(dt,dy) \\ &\quad + e^{-X(t-,T)}\int_{|y|>1}\left(e^{-\Sigma^2(t,T,y)} - 1 + \Sigma^2(t,T,y)\right)\tilde{\pi}(dt,dy) \\ &\quad + e^{-X(t-,T)}\left(\int_{|y|\le 1}\left(e^{-\Sigma^1(t,T,y)} - 1 + \Sigma^1(t,T,y)\right)\nu(dy)\right) dt. \end{aligned}$$

Consequently, by (7.4.22) we obtain

$$d\hat{P}(t,T) = -P(t-,T)\Big(A(t,T)dt + \langle \Sigma^0(t,T), dW(t)\rangle + \int_{|y|\leq 1} \Sigma^1(t,T,y)\tilde{\pi}(dt,dy)$$

$$+ \int_{|y|>1} \Sigma^2(t,T,y)\pi(dt,dy)\Big) + \frac{1}{2}P(t-,T) \mid Q^{\frac{1}{2}}\Sigma^0(t,T) \mid^2 dt + dI(t,T),$$

which yields

$$\frac{d\hat{P}(t,T)}{\hat{P}(t-,T)} = \left(-A(t,T) + \frac{1}{2} \mid Q^{\frac{1}{2}}\Sigma^0(t,T)\mid^2 + \int_{|y|\leq 1} (e^{-\Sigma^1(t,T,y)} - 1 + \Sigma^1(t,T,y))\nu(dy)\right) dt$$

$$- \langle \Sigma^0(t,T), dW(t)\rangle + \int_{|y|\leq 1} (e^{-\Sigma^1(t,T,y)} - 1)\tilde{\pi}(dt,dy) + \int_{|y|>1} (e^{-\Sigma^2(t,T,y)} - 1)\pi(dt,dy),$$

and (7.4.20) follows. Formula (7.4.21) is a consequence of (7.1.23). □

7.4.2 Forward Curves in Function Spaces

The HJM modelling with forward rates

$$f(t,T) = f(0,T) + \int_0^t \alpha(s,T)ds + \int_0^t \langle \sigma(s,T), dZ(s)\rangle, \quad t \in [0,T^*], T > 0, \tag{7.4.23}$$

can be regarded as a model where forward curves $f_t := \{f(t,T), T > 0\}$ live in a Hilbert space H, for instance $L^{2,\gamma}$ defined by (7.1.28). Let us define function valued processes

$$\alpha_t(T) := \alpha(t,T), \quad \sigma_t^i(T) := \sigma^j(t,T), \quad j = 1,2,\ldots,d,$$

and assume that they take values in $L^{2,\gamma}$, are adapted and predictable and such that with probability 1

$$\int_0^{T^*} |\alpha_s|_{L^{2,\gamma}} ds < +\infty, \quad \sum_{j=1}^d \int_0^{T^*} |\sigma_s^j|_{L^{2,\gamma}} ds < +\infty. \tag{7.4.24}$$

Then the integrals

$$\int_0^t \alpha_s ds, \quad \sum_{j=1}^d \int_0^t \sigma_s^j dZ_j(s), \quad t \in [0,T^*], \tag{7.4.25}$$

are well-defined $L^{2,\gamma}$- valued stochastic processes.

Proposition 7.4.4 *Setting $f_0(T) := f(0,T), T \geq 0$, the process f_t given by*

$$f_t = f_0 + \int_0^t \alpha_s ds + \sum_{j=1}^d \int_0^t \sigma_s^j dZ_j(s), \quad t \in (0,T^*] \tag{7.4.26}$$

is such that for each $t \in [0, T^*]$ the function $T \to f(t, T)$ is a representation of the element f_t in $L^{2,\gamma}$.

Proof Assume that $\mathbb{P}$-a.s. (7.4.23) holds for arbitrary $T > 0$ and $t \in [0, T^*]$. Let h be a continuous function on $[0, +\infty)$ with bounded support. Then for each $t \in [0, T^*]$, by changing the order of integration, we obtain

$$\int_0^{+\infty} f(t,T)h(t)e^{\gamma T}dT = \int_0^{+\infty} f(0,T)h(T)e^{\gamma T}dT + \int_0^{+\infty}\Big(\int_0^t \alpha(s,T)ds\Big)h(T)e^{\gamma T}dT$$

$$+\sum_{j=1}^{d}\int_0^{+\infty}\Big(\int_0^t \sigma^j(s,T)dZ_j(s)\Big)h(T)e^{\gamma T}dT$$

$$= \int_0^{+\infty} f(0,T)h(T)e^{\gamma T}dT + \int_0^t\Big(\int_0^{+\infty}\alpha(s,T)h(T)e^{\gamma T}dT\Big)ds$$

$$+\sum_{j=1}^{d}\int_0^t\Big(\int_0^{+\infty}\sigma^j(s,T)h(T)e^{\gamma T}dT\Big)dZ_j(s), \quad t \in [0, T^*].$$

The preceding relation can be written in the Hilbert space setting as

$$\langle f(t,\cdot), h(\cdot)\rangle_{L^{2,\gamma}} = \langle f(0,\cdot), h(\cdot)\rangle_{L^{2,\gamma}} + \int_0^t \langle \alpha(s,\cdot), h(\cdot)\rangle_{L^{2,\gamma}}ds$$

$$+\sum_{j=1}^{d}\int_0^t \langle \sigma^j(s,\cdot), h(\cdot)\rangle_{L^{2,\gamma}}dZ_j(s), \quad t \in [0, T^*].$$

It follows from (7.4.26) that

$$\langle f_t, h\rangle_{L^{2,\gamma}} = \langle f_0, h\rangle_{L^{2,\gamma}} + \int_0^t \langle \alpha_s, h\rangle_{L^{2,\gamma}}ds + \sum_{j=1}^{d}\int_0^t \langle \sigma_s^j, h\rangle_{L^{2,\gamma}}dZ_i(s), \quad t \in [0, T^*].$$

Since

$$\langle f_0, h\rangle_{L^{2,\gamma}} = \langle f(0,\cdot), h(\cdot)\rangle_{L^{2,\gamma}}, \quad \langle \alpha_s, h\rangle_{L^{2,\gamma}} = \langle \alpha(s,\cdot), h(\cdot)\rangle_{L^{2,\gamma}}, \quad s \in [0, T^*],$$

$$\langle \sigma_s^j, h\rangle_{L^{2,\gamma}} = \langle \sigma^j(s,\cdot), h(\cdot)\rangle_{L^{2,\gamma}}, \quad s \in [0, T^*], \quad j = 1, 2, \ldots, d,$$

we see that $\mathbb{P}$-a.s.

$$\langle f_t, h\rangle_{L^{2,\gamma}} = \langle f(t,\cdot), h(\cdot)\rangle_{L^{2,\gamma}}, \quad t \in [0, T^*],$$

and therefore $f_t = f(t,\cdot)$, as required. □

Note also that, $f(t,T) = f(T,T)$ for $T \leq t$ and therefore $f_t(T) = f(T,T)$ for almost all $T \leq t$. Consequently,

$$P(t,T) = e^{-\int_t^T f_t(s)ds}, \quad \hat{P}(t,T) = e^{-\int_0^T f_t(s)ds} \quad t \in [0, T^*], \ T \geq 0. \tag{7.4.27}$$

In general, if one starts from (7.4.26) and

$$\alpha_t(s) = 0, \ \sigma_t(s) = 0, \quad \text{for almost all } s \leq t,$$

then for arbitrary $0 \leq t_1 \leq t_2$

$$f_{t_2}(s) = f_{t_1}(s), \quad \text{for } s \leq t_1.$$

One therefore defines $R(t) = f_T(t), t \leq T$.

Instead of the spaces $L^{2,\gamma}$ one can arrive at similar results for other Hilbert spaces such as $\hat{L}^{2,\gamma}$ (see Rusinek [113]).

7.5 Factor Models and the Musiela Parametrization

In many situations it is convenient to regard forward rates and bond prices as curves from sets of functions given in advance. In particular, one can assume that the *forward curves*

$$T \mapsto f(t,T), \quad T \in [t, +\infty)$$

are of the form

$$f(t,T) = G(T-t, X(t)), \quad 0 \leq t \leq T, \tag{7.5.1}$$

where G is a real-valued function defined on $[0,+\infty) \times E$ and $X(t), t \geq 0$ is a stochastic process. It is called a *factor process*, takes values in a closed subset E of $\mathbb{R}^m$ and could be interpreted as a description of economical environment. So, for fixed value of the factor process $X(t)$, the function

$$u \mapsto G(u, X(t)), \quad u \geq 0$$

yields the forward rate as a function of *time to maturity* $u = T - t$. Therefore the short-rate process is given by

$$R(t) = G(0, X(t)), \quad t \geq 0.$$

This very convenient way of parametrizing forward rates in terms of time to maturity was introduced by Musiela and is called the *Musiela parametrization*.

From the definition of the bond prices we have that

$$P(t,T) = e^{-\int_t^T f(t,s)ds} = e^{-\int_t^T G(u-t,X(t))du}, \quad 0 \leq t \leq T,$$

and therefore the *bond curves* have the form

$$P(t,T) = F(T-t, X(t)), \quad 0 \leq t \leq T,$$

where

$$F(T-t, x) := e^{-\int_t^T G(u-t,x)du} = e^{-\int_0^{T-t} G(s,x)ds}, \quad 0 \leq t \leq T.$$

For F and G we have the relations

$$G(u,x) = -\frac{\frac{\partial F}{\partial u}(u,x)}{F(u,x)}, \qquad F(u,x) = e^{-\int_0^u G(v,x)dv}, \quad u \geq 0,\ x \in E. \tag{7.5.2}$$

If the function F has the form

$$F(u,x) = e^{-C(u)-\langle D(u),x\rangle},$$

and the factor $X(t) = R(t)$ is the short rate then the resulting model is called *affine term structure* or briefly *affine model*. If, additionally, the short rate $R(t)$ is given by the stochastic equation

$$dR(t) = A(R(t))dt + \langle B(R(t-)), dZ(t)\rangle, \quad R(0) = R_0, \tag{7.5.3}$$

with some functions A, B, then the affine model can be viewed as a particular HJM model.

Proposition 7.5.1 *Let C and D be twice differentiable functions on $[0, +\infty)$. Then the affine model with the bond prices*

$$P(t,T) = e^{-C(T-t)-\langle D(T-t),R(t)\rangle}, \quad T \geq t \geq 0,$$

and the short rate (7.5.3) is the HJM model (α, σ, Z) with forward rate

$$df(t,T) = \alpha(t,T)dt + \langle \sigma(t,T), dZ(t)\rangle,$$

where

$$\alpha(t,T) := A(R(t))D'(T-t) - C''(T-t) - D''(T-t)R(t), \quad T \geq t \geq 0, \tag{7.5.4}$$

$$\sigma(t,T) := D'(T-t)B(R(t-)), \quad T \geq t \geq 0. \tag{7.5.5}$$

Proof Writing the bond prices in the affine model with the use of forward rates yields

$$e^{-C(T-t)-\langle D(T-t),R(t)\rangle} = e^{-\int_t^T f(t,s)ds}, \quad T \geq t.$$

Consequently,

$$f(t,T) = C'(T-t) + D'(T-t)R(t).$$

Using Itô's formula and (7.5.3) we obtain that, for each $T > 0$,

$$\begin{aligned} df(t,T) &= -C''(T-t)dt + D'(T-t)dR(t) - R(t)D''(T-t)dt \\ &= \alpha(t,T)dt + \langle \sigma(t,T), dZ(t)\rangle \end{aligned}$$

where α and σ are given by (7.5.4) and (7.5.5). □

8

Arbitrage-Free HJM Markets

In this chapter we characterize arbitrage-free HJM models by describing the set of all martingale measures. In deriving the corresponding Heath–Jarrow–Morton drift conditions we use in an essential way the martingale representation theorem and Girsanov's formula for equivalent measures.

8.1 Heath–Jarrow–Morton Conditions

In this section we work with the HJM model (α, σ, Z) given, for each $T > 0$, by the forward rate dynamics

$$f(t,T) = f(0,T) + \int_0^t \alpha(s,T)ds + \int_0^t \langle \sigma(s,T), dZ(s) \rangle, \quad t \in [0,T^*], \tag{8.1.1}$$

driven by a Lévy process Z in $U = \mathbb{R}^d$ with characteristic triplet (a, Q, ν). For any $T > 0$, $\alpha(\cdot, T)$ is real-valued and $\sigma(\cdot, T)$ is a U-valued predictable process. In view of Theorem 7.4.1, the assumptions

$$\int_0^S \mid f(0,T) \mid dT < +\infty, \quad S > 0, \tag{8.1.2}$$

$$\int_0^{T^*} \int_0^S \mid \alpha(t,T) \mid dt\, dT < +\infty, \qquad \int_0^{T^*} \int_0^S \mid \sigma(t,T) \mid_U^2 dt\, dT + \infty, \quad S > 0, \tag{8.1.3}$$

$$\alpha(t,T) = 0, \quad \sigma(t,T) = 0 \quad \text{for} \quad t > T \tag{8.1.4}$$

guarantee that the bond prices

$$P(t,T) = e^{-\int_t^T f(t,u)du}, \qquad t \in [0,T^*], \quad T > 0$$

and the bank account process

$$B(t) = e^{\int_0^t R(s)ds} = e^{\int_0^t f(s,s)ds}, \qquad t \in [0,T^*]$$

are well defined. Moreover, the discounted bond prices are given then by

$$\hat{P}(t,T) := e^{-\int_0^T f(t,s)ds}, \qquad t \in [0,T^*], \quad T > 0.$$

Our aim now is to characterize models that satisfy the following (MM) condition.

There exists a measure $\mathbb{Q} \sim \mathbb{P}$ such that for each $T > 0$ the discounted bond price process

$$\hat{P}(t,T) := e^{-\int_0^T f(t,s)ds}, \qquad t \in [0,T^*], \tag{MM}$$

is a $\mathbb{Q}$-local martingale.

Recall that the measure $\mathbb{Q}$ appearing in (MM) is called a martingale measure and its existence is sufficient for the model to be arbitrage free (see Section 7.3 for details). Our analysis leads to a generalized version of the famous HJM drift condition from Heath, Jarrow and Morton [67] that yields a special type of dependence between α and σ in (8.1.1). They are formulated in terms of the processes

$$A(t,T) := \int_{t\wedge T}^T \alpha(t,v)dv, \qquad \Sigma(t,T) := \int_{t\wedge T}^T \sigma(t,v)dv, \quad t \in [0,T^*], \quad T > 0. \tag{8.1.5}$$

Let us recall that a predictable process $\phi(t) = (\phi_1(t),\dots,\phi_d(t))$, resp. $\psi(t,y)$ belongs to $\Phi(U)$ resp. $\psi \in \Psi_{1,2}$ if and only if

$$\int_0^{T^*} \sum_{i=1}^d \phi_i^2(t)dt < +\infty, \quad \mathbb{P}-a.s.,$$

resp.

$$\int_0^{T^*}\int_U \Big(\mid \psi(s,y) \mid^2 \wedge \mid \psi(s,y) \mid \Big) ds\nu(dy) < +\infty, \quad \mathbb{P}-a.s.$$

(see Section 5.4.1 and Section 5.4.2 for details). The Laplace exponent J of Z is given by

$$J(u) = -\langle a,u\rangle + \frac{1}{2}\langle Qu,u\rangle + \int_U \left(e^{-\langle u,y\rangle} - 1 + \mathbf{1}_{\{|y|\le 1\}}\langle u,y\rangle\right)\nu(dy) \tag{8.1.6}$$

(see Section 5.2).

Theorem 8.1.1 *Let (α,σ,Z) be an HJM model satisfying* (8.1.2)–(8.1.4), *where Z is a Lévy process with characteristic triplet (a,Q,ν) and the processes $A(t,T)$, $\Sigma(t,T)$ are given by* (8.1.5). *Let us assume that*

$$\sup_{t\in[0,T^*]} \int_0^S \mid \sigma(t,v) \mid dv < +\infty, \quad S > 0. \tag{8.1.7}$$

(a) *If the model satisfies (MM), then there exist processes* (ϕ, ψ) *such that* $\phi \in \Phi(U)$, $e^{\psi} - 1 \in \Psi_{1,2}$, *which satisfy in addition*

$$\int_0^{T^*} \int_{\{|y| \le 1\}} | e^{\psi(s,y)} - 1 | \, ds \, \nu(dy) < +\infty, \quad \mathbb{P} - a.s., \tag{8.1.8}$$

and, for each $T \in (0, +\infty)$,

$$\int_0^{T^*} \int_{\{|y| > 1\}} e^{-\langle \Sigma(s,T), y \rangle} \cdot e^{\psi(s,y)} ds \, \nu(dy) < +\infty, \quad \mathbb{P} - a.s. \tag{8.1.9}$$

(b) *If* (8.1.8) *and* (8.1.9) *are satisfied for some processes* (ϕ, ψ) *with* $\phi \in \Phi(U)$, $e^{\psi} - 1 \in \Psi_{1,2}$ *then the model satisfies (MM) if and only if for each* $T \in (0, +\infty)$ *almost all* $s \in [0, T^*]$, $\mathbb{P}$*-a.s.:*

$$A(s,T) = -\langle \Sigma(s,T), a \rangle + \frac{1}{2} \langle Q\Sigma(s,T), \Sigma(s,T) \rangle - \langle Q\phi(s), \Sigma(s,T) \rangle$$

$$+ \int_U \left(e^{\psi(s,y)} (e^{-\langle \Sigma(s,T), y \rangle} - 1) + \mathbf{1}_{\{|y| \le 1\}} \langle \Sigma(s,T), y \rangle \right) \nu(dy). \tag{8.1.10}$$

(c) *If* (8.1.8) *and* (8.1.9) *are satisfied for some processes* (ϕ, ψ) *with* $\phi \in \Phi(U)$, $e^{\psi} - 1 \in \Psi_{1,2}$ *and for each* $T > 0$,

$$\int_0^{T^*} \int_{\{|y| > 1\}} e^{-\langle \Sigma(s,T), y \rangle} ds \nu(dy) < +\infty, \quad \mathbb{P} - a.s., \tag{8.1.11}$$

then the model satisfies (MM) if and only if for each $T > 0$ *almost all* $s \in [0, T^*]$, $\mathbb{P}$*-a.s.:*

$$A(s,T) = J(\Sigma(s,T)) - \langle Q\phi(s), \Sigma(s,T) \rangle + \int_U (e^{\psi(s,y)} - 1)(e^{-\langle \Sigma(s,T), y \rangle} - 1) \nu(dy). \tag{8.1.12}$$

Theorem 8.1.1 constitutes a fundamental tool for the study of arbitrage-free HJM models. Conditions (8.1.10) and (8.1.12), which are called *HJM conditions*, enable, at least theoretically, to check if a given HJM model (α, σ, Z) is arbitrage free and to construct models that are arbitrage free. We comment now on the direct consequences of Theorem 8.1.1.

Remark 8.1.2 For a given HJM model (α, σ, Z) the processes (ϕ, ψ) appearing in the formulation of Theorem 8.1.1 are the generating pair of a martingale measure $\mathbb{Q}$. The problem of the existence of $\mathbb{Q}$ is thus equivalent to the solvability of the equation (8.1.10) or (8.1.12) over (ϕ, ψ). The existence of such (ϕ, ψ) in the general HJM model is, however, an open problem. Even particular examples of models for which it is known if (ϕ, ψ) exist are hardly found in the literature. In Section 8.2.2,

Section 8.2.3 and Section 8.2.4 we partially fill this gap by presenting a couple of models for which the required pair (ϕ, ψ) exists and some for which it does not exist.

Remark 8.1.3 If the pair $\phi \equiv 0$, $\psi \equiv 0$ satisfies conditions of Theorem 8.1.1 then the original measure $\mathbb{P}$ is a martingale measure. In this case, the necessary conditions (8.1.8), (8.1.9) amount to

$$\int_0^t \int_{\{|y|>1\}} e^{-\langle \Sigma(s,T),y\rangle} ds\nu(dy) < +\infty. \tag{8.1.13}$$

This implies that, for each $T > 0$, $\Sigma(s,T)$ belongs to the domain of the Laplace exponent J of Z almost ω-surely for almost all s. In particular, this means that Z has finite some exponential moments. Moreover, (8.1.10) and (8.1.12) reduce to

$$A(t,T) = J(\Sigma(t,T)),$$

which leads to the condition

$$\alpha(t,T) = J'(\Sigma(t,T))\sigma(t,T). \tag{8.1.14}$$

The last relation opens the door to construction of arbitrage-free models. Starting from any Lévy process Z and volatility process $\sigma(t,T)$, such that (8.1.13) is satisfied, one can define the drift $\alpha(t,T)$ by (8.1.14). Then the resulting model is clearly arbitrage free and $\mathbb{P}$ is a martingale measure.

Remark 8.1.4 If Z is a Q-Wiener process then the HJM condition has the form

$$A(t,T) = \frac{1}{2}\langle Q\Sigma(t,T), \Sigma(t,T)\rangle - \langle Q\phi(t), \Sigma(t,T)\rangle,$$

which was originally obtained by Heath, Jarrow and Morton in [67] in the one-dimensional case.

Remark 8.1.5 It follows from the inequality (8.1.22), which will be obtained in the proof of Theorem 8.1.1, that condition (8.1.8) together with the fact that $e^{\psi} - 1 \in \Psi_{1,2}$ is equivalent to

$$\int_0^{T^*} \int_U \mid e^{\psi(t,y)} - 1 \mid dt\nu(dy) < +\infty, \quad \mathbb{P} - a.s.$$

This means that $e^{\psi} - 1 \in \Psi_1$, where

$$\Psi_1 := \{g = g(s,y) - \text{predictable} : \int_0^{T^*} \int_U \mid g(s,y) \mid ds\nu(dy) < +\infty, \qquad \mathbb{P} - a.s.\}.$$

Remark 8.1.6 If the volatility σ of the model (α, σ, Z) is such that $\langle \Sigma(t,T), y\rangle \geq 0$ for $\mid y \mid > 1$ then both conditions (8.1.9) and (8.1.11) are automatically satisfied.

8.1.1 Proof of Theorem 8.1.1

For the proof of Theorem 8.1.1 we need the representation of the discounted bond price process in the HJM model. It can be deduced from Proposition 7.4.3.

Proposition 8.1.7 *In the HJM model* (α, σ, Z) *satisfying* (8.1.2)–(8.1.4) *the discounted bond price process*

$$\hat{P}(t,T) = e^{-\int_0^T f(t,s)ds}, \quad t \in [0,T^*],\ T > 0$$

admits the following representation

$$\frac{d\hat{P}(t,T)}{\hat{P}(t-,T)} = \left\{ \frac{1}{2} \langle Q\Sigma(t,T), \Sigma(t,T)\rangle - A(t,T) - \langle \Sigma(t,T), a\rangle \right\} dt$$

$$- \langle \Sigma(t,T), dW(t)\rangle - \int_U \langle \Sigma(t,T), y\rangle \tilde{\pi}(dt,dy)$$

$$+ \int_U \left[e^{-\langle \Sigma(t,T), y\rangle} - 1 + \mathbf{1}_{\{|y|\le 1\}} \langle \Sigma(t,T), y\rangle \right] \pi(dt,dy). \qquad (8.1.15)$$

Proof of Theorem 8.1.1 (*a*) In view of Lemma 6.2.2 we examine the process $\rho(\cdot)\hat{P}(\cdot,T)$ under the measure $\mathbb{P}$ and use the integral representation of $\hat{P}(\cdot,T)$ given by Proposition 8.1.7. Recall that

$$\hat{P}(t,T) = e^{-X(t,T)}, \quad \text{with} \quad X(t,T) := \int_0^T f(t,s)ds, \quad t \in [0,T^*],\ T > 0,$$

$$\rho(t)\hat{P}(t,T) = e^{V_t^T}, \quad \text{with} \quad V_t^T := Y(t) - X(t,T), \quad t \in [0,T^*],\ T > 0,$$

where $Y(t)$, defined by $\rho(t) = e^{Y(t)}$, is equal to

$$Y(t) = \int_0^t \langle \phi(s), dW(s)\rangle - \frac{1}{2}\int_0^t |\, Q^{\frac{1}{2}}\phi(s)\,|_U^2\, ds + \int_0^t \int_U (e^{\psi(s,y)} - 1)\tilde{\pi}(ds,dy)$$

$$- \int_0^t \int_U (e^{\psi(s,y)} - 1 - \psi(s,y))\pi(ds,dy), t \in [0,T^*]$$

(see formula (6.2.6) in Section 6.2). Moreover,

$$X(t,T) = \int_0^T f_0(s)ds + \int_0^T \left(\int_0^t \alpha(v,s)dv \right) ds + \int_0^T \left(\int_0^t \langle \sigma(v,s), dZ(v)\rangle \right) ds$$

$$= \int_0^T f_0(s)ds + \int_0^t A(v,T)dv + \int_0^t \langle \Sigma(v,T), dZ(v)\rangle, \quad t \in [0,T^*],\ T>0,$$

or, equivalently,

$$dX(t,T) = A(t,T)dt + \langle \Sigma(t,T), dZ(t)\rangle, \qquad t \in [0,T^*],\ T > 0. \qquad (8.1.16)$$

Since

$$\Delta V_t^T = \Delta Y(t) - \Delta X(t,T) = \psi(t, \Delta Z(t)) - \langle \Sigma(t,T), \Delta Z(t) \rangle,$$

it follows from Itô's formula that

$$e^{V_t^T} = e^{V_0^T} + \int_0^t e^{V_{s-}^T} dV_s^T + \frac{1}{2}\int_0^t e^{V_{s-}^T} d[V_s^T, V_s^T]^c + \sum_{s\le t} e^{V_{s-}^T}(e^{\Delta V_s^T} - 1 - \Delta V_s^T)$$

$$= e^{V_0^T} + \int_0^t e^{V_{s-}^T}\Big\{\langle \phi(s), dW(s)\rangle - \frac{1}{2} \mid Q^{\frac{1}{2}}\phi(s) \mid^2 ds + \int_U (e^{\psi(s,y)} - 1)\tilde{\pi}(ds,dy)$$

$$- \int_U (e^{\psi(s,y)} - 1 - \psi(s,y))\pi(ds,dy) - (A(s,T) + \langle \Sigma(s,T), a\rangle)ds - \langle \Sigma(s,T), dW(s)\rangle$$

$$- \int_{\{|y|\le 1\}} \langle \Sigma(s,T), y\rangle \tilde{\pi}(ds,dy) - \int_{\{|y|>1\}} \langle \Sigma(s,T), y\rangle \pi(ds,dy)\Big\}$$

$$+ \frac{1}{2}\int_0^t e^{V_{s-}^T} \mid Q^{\frac{1}{2}}(\phi(s) - \Sigma(s,T)) \mid^2 ds$$

$$+ \int_0^t \int_U e^{V_{s-}^T}\Big(e^{\psi(s,y) - \langle \Sigma(s,T), y\rangle} - 1 - \big(\psi(s,y) - \langle \Sigma(s,T), y\rangle\big)\Big)\pi(ds,dy).$$

Finally, we have

$$\rho(t)\hat{P}(t,T) = e^{V_t^T} = e^{V_0^T} + M(t) + A(t) + B(t), \tag{8.1.17}$$

where

$$M(t) := \int_0^t e^{V_{s-}^T}\langle \phi(s), dW(s)\rangle + \int_0^t \int_U e^{V_{s-}^T}(e^{\psi(s,y)} - 1)\tilde{\pi}(ds,dy)$$

$$- \int_0^t \int_B e^{V_{s-}^T}\langle \Sigma(s,T), y\rangle \tilde{\pi}(ds,dy),$$

$$A(t) := \int_0^t e^{V_{s-}^T}\Big(- \frac{1}{2} \mid Q^{\frac{1}{2}}\phi(s) \mid^2 - A(s,T) - \langle \Sigma(s,T), a\rangle + \frac{1}{2} \mid Q^{\frac{1}{2}}(\phi(s) - \Sigma(s,T)) \mid^2 \Big)ds,$$

$$B(t) := \int_0^t \int_U e^{V_{s-}^T}\Big(e^{\psi(s,y) - \langle \Sigma(s,T), y\rangle} - e^{\psi(s,y)} + \mathbf{1}_B\langle \Sigma(s,T), y\rangle\Big)\pi(ds,dy),$$

and $B := \{y : \mid y \mid \le 1\}$. Let us notice that $M(t)$ is a local martingale, $A(t)$ and $B(t)$ have finite variation and $A(t)$ is predictable. Since $\rho(t)\hat{P}(t,T)$ is also a local martingale, it follows from (8.1.17) that $A(t) + B(t)$ is a local martingale as well and thus, in view of Proposition 4.2.9, that it is a process of locally integrable variation. In view of Proposition 4.2.8 the process $A(t)$ has also locally integrable variation, so does $B(t)$. Hence, it follows from Theorem 4.4.3 that the compensator of $B(t)$ exists and is of the form

$$B^p(t) = \int_0^t \int_U e^{V^T_{s-}} \Big(e^{\psi(s,y) - \langle \Sigma(s,T), y \rangle} - e^{\psi(s,y)} + \mathbf{1}_B \langle \Sigma(s,T), y \rangle \Big) ds \nu(dy),$$
$$t \in [0, T^*]. \tag{8.1.18}$$

Since the function $s \mapsto V_s^T$ is càdlàg and thus bounded on finite intervals, the preceding integral exists if and only if

$$\int_0^{T^*} \int_U \left| e^{\psi(s,y) - \langle \Sigma(s,T), y \rangle} - e^{\psi(s,y)} + \mathbf{1}_B \langle \Sigma(s,T), y \rangle \right| ds \nu(dy) < +\infty, \tag{8.1.19}$$

for each T almost ω-surely. In view of (8.1.7) the function $s \to \Sigma(s,T)$ is bounded and it follows that

$$\int_0^{T^*} \int_U \left| e^{-\langle \Sigma(s,T), y \rangle} - 1 + \langle \Sigma(s,T), y \rangle \right| \mathbf{1}_B(y) ds \nu(dy) < +\infty, \tag{8.1.20}$$

and further from (8.1.20) and (8.1.19) that

$$\int_0^{T^*} \int_U \left| (e^{-\langle \Sigma(s,T), y \rangle} - 1)(e^{\psi(s,y)} - \mathbf{1}_B(y)) \right| ds \nu(dy) < +\infty.$$

Let us split the preceding integral into two integrals over B and $B^c := U \setminus B$. Since the function

$$(s, y) \mapsto (e^{-\langle \Sigma(s,T), y \rangle} - 1)$$

is bounded on $[0, T^*] \times B$, (8.1.8) follows. To show that (8.1.9) is satisfied, we show that

$$\int_0^{T^*} \int_{B^c} e^{\psi(s,y)} ds \nu(dy) < +\infty. \tag{8.1.21}$$

Let us denote

$$\underline{A} := \{(s, y) \in [0, T^*] \times U : | e^{\psi(s,y)} - 1 | \leq 1\},$$
$$\bar{A} := \{(s, y) \in [0, T^*] \times U : | e^{\psi(s,y)} - 1 | > 1\}.$$

It follows from the fact that $e^{\psi} - 1 \in \Psi_{1,2}$ that

$$\int_0^{T^*} \int_{B^c} | e^{\psi(s,y)} - 1 | ds \nu(dy)$$
$$= \int_{[0,T^*] \times B^c \cap \underline{A}} | e^{\psi(s,y)} - 1 | \nu(dy) + \int_{[0,T^*] \times B^c \cap \bar{A}} | e^{\psi(s,y)} - 1 | ds \nu(dy)$$
$$\leq \left(\int_{[0,T^*] \times B^c \cap \underline{A}} 1 ds \nu(dy) \right)^{\frac{1}{2}} \left(\int_{[0,T^*] \times B^c \cap \underline{A}} | e^{\psi(s,y)} - 1 |^2 ds \nu(dy) \right)^{\frac{1}{2}}$$

$$+\int_{\bar{A}} | e^{\psi(s,y)} - 1 | dsv(dy) \leq \left(T^* v(B^c)\right)^{\frac{1}{2}} \left(\int_{\underline{A}} | e^{\psi(s,y)} - 1 |^2 dsv(dy)\right)^{\frac{1}{2}}$$

$$+\int_{\bar{A}} | e^{\psi(s,y)} - 1 | dsv(dy) < +\infty. \tag{8.1.22}$$

Condition (8.1.21) follows from the preceding estimation and the fact that $v(B^c) < +\infty$.

(b) If (8.1.8) and (8.1.9) are satisfied then the process $B(t)$ admits a compensator $B^p(t)$ of the form (8.1.18) and

$$\rho(t)\hat{P}(t,T) - e^{V_0^T} - M(t) - (B(t) - B^p(t)) = A(t) + B^p(t), \qquad t \in [0, T^*].$$

The process $\rho(t)\hat{P}(t,T)$ is a local martingale if and only if $A(t) + B^p(t)$ is a local martingale. But $A(t) + B^p(t)$ is a predictable process of finite variation, so, in view of Proposition 4.2.10, it is a local martingale if and only if

$$A(t) + B^p(t) = 0, \quad t \in [0, T^*].$$

Using formulas for $A(t)$ and $B^p(t)$ and rearranging terms yields

$$\begin{aligned} A(s,T) = &- \langle\Sigma(s,T), a\rangle + \frac{1}{2}\langle Q\Sigma(s,T), \Sigma(s,T)\rangle - \langle Q\phi(s), \Sigma(s,T)\rangle \\ &+ \int_U \left(e^{\psi(s,y)}(e^{-\langle\Sigma(s,T),y\rangle} - 1) + \mathbf{1}_B\langle\Sigma(s,T), y\rangle\right) v(dy), \end{aligned} \tag{8.1.23}$$

for each $T > 0$ almost all s almost ω-surely.

(c) If (8.1.11) is satisfied then $\Sigma(s,T)$ belongs to the domain of J for almost all s and we can rearrange terms in (8.1.23) in the following way

$$\begin{aligned} A(s,T) = &- \langle\Sigma(s,T), a\rangle + \frac{1}{2}\langle Q\Sigma(s,T), \Sigma(s,T)\rangle \\ &+ \int_U \left(e^{-\langle\Sigma(s,T),y\rangle} - 1 + \mathbf{1}_B\langle\Sigma(s,T), y\rangle\right) v(dy) - \langle Q\phi(s), \Sigma(s,T)\rangle \\ &+ \int_U \left(e^{\psi(s,y)-\langle\Sigma(s,T),y\rangle} - e^{\psi(s,y)} - e^{-\langle\Sigma(s,T),y\rangle} + 1\right) v(dy). \end{aligned}$$

Taking into account (8.1.6) and simplifying the last term yields

$$A(s,T) = J(\Sigma(s,T)) - \langle Q\phi(s), \Sigma(s,T)\rangle + \int_U (e^{\psi(s,y)} - 1)(e^{-\langle\Sigma(s,T),y\rangle} - 1) v(dy).$$

□

8.2 Martingale Measures

In view of Theorem 8.1.1 martingale measures in the HJM model (α, σ, Z) are described in terms of their generating pairs (ϕ, ψ), which are supposed to satisfy the relation:

$$A(t,T) = -\langle \Sigma(t,T), a\rangle + \frac{1}{2}\langle Q\Sigma(t,T), \Sigma(t,T)\rangle - \langle Q\phi(t), \Sigma(t,T)\rangle$$
$$+ \int_U \big(e^{\psi(t,y)}(e^{-\langle \Sigma(t,T),y\rangle} - 1) + \mathbf{1}_{\{|y|\le 1\}}\langle \Sigma(t,T), y\rangle\big)\nu(dy). \tag{8.2.1}$$

The relation (8.2.1) simplifies considerably in the case when the process Z is real valued.

Proposition 8.2.1 *Let (α, σ, Z) be an HJM model where Z is a one-dimensional Lévy process with characteristics $(a, q = 1, \nu)$, i.e. the Gaussian part is a standard Wiener process. Assume that* (8.1.2)–(8.1.4), (8.1.7) *are satisfied and for each $t \in [0, T^*]$ the function*

$$T \mapsto \frac{\alpha(t,T)}{\sigma(t,T)},$$

is right continuous at point t.

(a) Assume that (MM) holds. If (ϕ, ψ) is the generating pair of some martingale measure, then for almost all $t \in [0, T^*]$, $\mathbb{P} - a.s.$:

$$\phi(t) = -\frac{\alpha(t,t)}{\sigma(t,t)} - a - \int_{\mathbb{R}} \left(e^{\psi(t,y)} - \mathbf{1}_{\{|y|\le 1\}}\right) y\ \nu(dy). \tag{8.2.2}$$

(b) Assume that there exists ψ with $e^{\psi} - 1 \in \Psi_1$ such that ϕ given by (8.2.2) *belongs to Φ and $\mathbb{E}[\rho_{T^*}] = 1$ with ρ given by* (6.2.1). *Then ψ determines a martingale measure if and only if for each $T > 0$, almost all $t \in [0, T^*]$, $\mathbb{P}$-a.s.:*

$$A(t,T) - \frac{1}{2}(\Sigma(t,T))^2 - \Sigma(t,T)\frac{\alpha(t,t)}{\sigma(t,t)} = \int_{\mathbb{R}} e^{\psi(t,y)}\big(e^{-\Sigma(t,T)y} - 1 + \Sigma(t,T)y\big)\nu(dy). \tag{8.2.3}$$

Proof (*a*) Dividing both sides of (8.2.1) by $\Sigma(t,T)$ yields

$$\frac{A(t,T)}{\Sigma(t,T)} = -a + \frac{1}{2}\Sigma(t,T) - \phi(t) + \int_{\mathbb{R}} \left(e^{\psi(t,y)}\frac{e^{-\Sigma(t,T)y} - 1}{\Sigma(t,T)} + \mathbf{1}_{\{|y|\le 1\}} y\right)\nu(dy). \tag{8.2.4}$$

Since

$$\lim_{T\downarrow t} \frac{A(t,T)}{\Sigma(t,T)} = \lim_{T\downarrow t} \frac{\alpha(t,T)}{\sigma(t,T)} = \frac{\alpha(t,t)}{\sigma(t,t)},$$

letting $T \downarrow t$ in (8.2.4) yields

$$\frac{\alpha(t,t)}{\sigma(t,t)} = -a - \phi(t) + \int_{\mathbb{R}} \left(-e^{\psi(t,y)} + \mathbf{1}_{\{|y|\le 1\}}\right) y\nu(dy),$$

which is (8.2.2).

(*b*) Inserting (8.2.2) into (8.2.1) yields

$$\begin{aligned}A(t,T) = & -a\Sigma(t,T) + \frac{1}{2}(\Sigma(t,T))^2 \\ & + \Sigma(t,T)\left(\frac{\alpha(t,t)}{\sigma(t,t)} + a + \int_{\mathbb{R}}(e^{\psi(t,y)} - \mathbf{1}_{\{|y|\le 1\}})y\nu(dy)\right) \\ & + \int_{\mathbb{R}}\left(e^{\psi(t,u)}(e^{-y\Sigma(t,T)} - 1) + \mathbf{1}_{\{|y|\le 1\}}\Sigma(t,T)y\right)\nu(dy).\end{aligned}$$

Rearranging terms provides (8.2.3). □

It follows from Proposition 8.2.1 that if a given HJM model (α, σ, Z) admits a martingale measure $\mathbb{Q}$ then the process ϕ of its generating pair (ϕ, ψ) is determined by ψ, drift and volatility in a unique way. In particular, if ψ satisfying (8.2.3) is unique, so is ϕ, and the related martingale measure is unique as well. With the use of Proposition 8.2.1 we can characterize martingale measures in the case in which Z is continuous.

Corollary 8.2.2 *Let us assume that the HJM model $(\alpha, \sigma, at + W(t))$ with $a \in \mathbb{R}$ and a Wiener process W satisfies the assumptions of Proposition 8.2.1. Then the following statements are true.*

(a) (necessity) If there exists a martingale measure then its generating process ϕ is given by

$$\phi(t) = -\frac{\alpha(t,t)}{\sigma(t,t)} - a, \quad t \in [0, T^*]. \tag{8.2.5}$$

This follows from (8.2.2) *with $\nu \equiv 0$.*

(b) (sufficiency) Assume that $\phi(t)$ given by (8.2.5) *belongs to Φ and that $\mathbb{E}[\rho_{T^*}] = 1$ with ρ defined by* (6.2.1). *If for each $T > 0$, almost all $t \in [0, T^*]$, $\mathbb{P}$-a.s.:*

$$\alpha(t,T) = \sigma(t,T)\left\{\int_t^T \sigma(t,u)du + \frac{\alpha(t,t)}{\sigma(t,t)}\right\}, \qquad t \in [0,T^*], \quad T > 0, \tag{8.2.6}$$

then there exists a martingale measure and it is generated by ϕ. This follows from (8.2.3) *by putting $\nu = 0$ and differentiation over T.*

(c) (uniqueness) If there exists a martingale measure then it is unique. This follows from the fact that (8.2.5) *uniquely determines ϕ.*

8.2.1 Specification of Drift

An interesting question concerned with an HJM model (α, σ, Z) is if the existence of a martingale measure implies that the volatility σ determines the drift α in a unique way. It is easily seen that the answer is positive in the case in which the original measure $\mathbb{P}$ is a martingale measure. Indeed, condition (8.1.10) in Theorem 8.1.1 with $\phi \equiv 0, \psi \equiv 0$ yields

$$A(t,T) = J(\Sigma(t,T)),$$

for each $T > 0$, almost all $t \in [0, T^*]$, $\mathbb{P}$-a.s. Differentiation over T yields the formula for drift, i.e.

$$\alpha(t,T) = \langle DJ(\Sigma(t,T)), \sigma(t,T)\rangle.$$

In the particular case, when Z is a real-valued standard Wiener process we obtain the well-known classical HJM drift condition

$$\alpha(t,T) = \Sigma(t,T)\sigma(t,T).$$

In the case when the martingale measure is different from $\mathbb{P}$, the problem becomes more involved and the answer is, in general, negative. We explain this in the one-dimensional case with the use of condition (8.2.3) in Proposition 8.2.1. Recall that (8.2.3) yields

$$A(t,T) = \frac{1}{2}(\Sigma(t,T))^2 + \Sigma(t,T)\frac{\alpha(t,t)}{\sigma(t,t)} + \int_{\mathbb{R}} e^{\psi(t,y)}\big(e^{-\Sigma(t,T)y} - 1 + \Sigma(t,T)y\big)\nu(dy),$$

which, by differentiation, yields

$$\alpha(t,T) = \sigma(t,T)\Big[\int_t^T \sigma(t,u)du + \frac{\alpha(t,t)}{\sigma(t,t)} + \int_U e^{\psi(t,y)}(e^{-y\int_t^T \sigma(t,u)du} - 1)y\nu(dy)\Big]. \tag{8.2.7}$$

Now we see that even if Z has no jumps then $\alpha(t,T)$ is determined by the volatility and the boundary value of the drift/volatility ratio, that is, by the transformations

$$(t,T) \to \sigma(t,T), \qquad t \to \frac{\alpha(t,t)}{\sigma(t,t)}.$$

So, even in this degenerated case volatility does not determine the drift. It follows from (8.2.7) that in the general case we need additionally a function ψ and then $\alpha(t,T)$ is determined by the triplet

$$(t,T) \to \sigma(t,T), \qquad t \to \frac{\alpha(t,t)}{\sigma(t,t)}, \qquad \psi \quad \text{such that} \quad e^{\psi} - 1 \in \Psi_1.$$

8.2.2 Models with No Martingale Measures

It is instructive to analyze some examples of HJM models that do not admit martingale measures. Since the conditions for drift and volatility of the model (α, σ, Z) in the following examples are quite general, they throw some light on the necessary conditions for the existence of martingale measures.

Proposition 8.2.3 *Let the coefficients of the model (α, σ, Z), where Z is an arbitrary Lévy process, satisfy with positive $d\mathbb{P} \times dt$-measure (on $\Omega \times [0, T^*]$) the following two conditions*

$$\lim_{T\to+\infty} \sigma(t,T) > 0, \tag{8.2.8}$$

and for each $T > 0$,

$$\alpha(t,T) \leq H(\sigma(t,T)), \tag{8.2.9}$$

where H is a concave function of sublinear growth, i.e.

$$H(x) \leq cx, \quad x \in \mathbb{R},$$

with some $c \in \mathbb{R}$. Then the model does not satisfy (MM).

Proposition 8.2.4 *Let the coefficients in the model (α,σ,Z) satisfy with positive $d\mathbb{P} \times dt$-measure the following conditions*

$$\lim_{T\to+\infty} \alpha(t,T) > 0, \tag{8.2.10}$$

and

$$\int_t^{+\infty} |\, \sigma(t,u) \,|\, du < +\infty, \tag{8.2.11}$$

and let Z be a Lévy process with bounded jumps. Then the model does not satisfy (MM).

The next example is concerned with the affine model

$$P(t,T) = e^{-C(T-t)-D(T-t)R(t)}, \quad t \in [0,T^*], \quad T > 0, \tag{8.2.12}$$

with positive short rate of the form

$$R(t) = F(R(t))dt + G(R(t-))dZ(t), \quad t \in [0,T^*], \tag{8.2.13}$$

where Z is a Lévy process. This model was introduced in Section 7.5 where we also showed that it can be viewed as an HJM model (α,σ,Z) with relevant coefficients.

Proposition 8.2.5 *Let in the affine model* (8.2.12)–(8.2.13) *Z be an arbitrary Lévy process.*

(a) If $G(x) > 0, x > 0$, $F(x) \leq G(x), x \geq 0$, $C(\cdot), D(\cdot)$ are convex functions and

$$D'(u) \geq 0, \; u \geq 0, \qquad \lim_{u\to+\infty} D'(u) > 0, \tag{8.2.14}$$

then the model does not satisfy (MM).

(b) If $C(\cdot), D(\cdot)$ are twice differentiable and

$$\int_0^{+\infty} |\, D'(u) \,|\, du < +\infty, \tag{8.2.15}$$

$$\lim_{u\to+\infty} C''(u) < 0, \qquad \lim_{u\to+\infty} D''(u) < 0 \tag{8.2.16}$$

then the model, with any function F, does not satisfy (MM).

The proofs of Proposition 8.2.3 and Proposition 8.2.4 are based on Proposition 8.2.1. Proposition 8.2.5 can be deduced from Proposition 8.2.3 and Proposition 8.2.4.

Proof of Proposition 8.2.3 We prove that there are no processes ψ solving the equation (8.2.3) in Proposition 8.2.1. With the use of (8.2.9), the concavity of H and Jensen's inequality we obtain, for $t < T$,

$$\begin{aligned} A(t,T) &= \int_t^T \alpha(t,u)du \leq \int_t^T H(\sigma(t,u))du \\ &\leq (T-t)H\left(\frac{1}{T-t}\Sigma(t,T)\right), \quad t \in [0,T^*]. \end{aligned}$$

Consequently, since H is sublinear, we can estimate the left side of (8.2.3) as follows

$$\begin{aligned} A(t,T) &- \frac{1}{2}(\Sigma(t,T))^2 - \Sigma(t,T)\frac{\alpha(t,t)}{\sigma(t,t)} \qquad (8.2.17) \\ &\leq (T-t)H\left(\frac{1}{T-t}\Sigma(t,T)\right) - \frac{1}{2}(\Sigma(t,T))^2 - \Sigma(t,T)\frac{\alpha(t,t)}{\sigma(t,t)} \\ &\leq c\Sigma(t,T) - \frac{1}{2}(\Sigma(t,T))^2 - \Sigma(t,T)\frac{\alpha(t,t)}{\sigma(t,t)} \\ &\leq -\frac{1}{2}\Sigma(t,T)\left[\Sigma(t,T) - 2\left(c - \frac{\alpha(t,t)}{\sigma(t,t)}\right)\right], \quad t \in [0,T^*],\ T>t. \end{aligned}$$

For (ω, t) satisfying (8.2.8) the function $T \to \Sigma(t,T)$ is unbounded and consequently, for large T the expression (8.2.17) becomes strictly negative. Since

$$e^{-z} + z - 1 \geq 0, \quad z \in \mathbb{R},$$

the right side of (8.2.3) is nonnegative. So, (8.2.3) is not satisfied for any Lévy measure ν. □

Proof of Proposition 8.2.4 Let Z be a Lévy process with bounded jumps, i.e. $| \triangle Z(t) | \leq \bar{y}, t \in [0,T^*]$, for some $\bar{y} > 0$. Using the inequality $e^x - 1 - x \leq \frac{1}{2}e^{|x|}x^2, x \in \mathbb{R}$ we can estimate the right side of (8.2.3), for (ω, t) such that (8.2.10) and (8.2.11) are satisfied, as follows

$$\int_U e^{\psi(t,y)}\left(e^{-\Sigma(t,T)y} - 1 + \Sigma(t,T)y\right)\nu(dy) \leq \frac{1}{2}e^{K(t)\bar{y}}K^2(t)\bar{y}^2 \int_{\mathbb{R}} e^{\psi(t,y)}\nu(dy), \tag{8.2.18}$$

where $K(t) := \int_t^{+\infty} | \sigma(t,u) | \, du$. Hence for any ψ solving (8.2.3),

$$A(t,T) - \frac{1}{2}(\Sigma(t,T))^2 - \Sigma(t,T)\frac{\alpha(t,t)}{\sigma(t,t)} \leq \frac{1}{2}e^{K(t)\bar{y}}K^2(t)\bar{y}^2 \int_{\mathbb{R}} e^{\psi(t,y)}\nu(dy). \tag{8.2.19}$$

By (8.2.10) and (8.2.11) the left side of (8.2.19) tends to $+\infty$ for $T \to +\infty$ while the right side is finite, so we arrive at a contradiction. $\square$

Proof of Proposition 8.2.5 In view of Proposition 7.5.1 the affine model can be written in the HJM parametrization

$$df(t,T) = \alpha(t,T)dt + \sigma(t,T)dZ(t), \quad t \in [0,T^*], T > 0,$$

with

$$\alpha(t,T) := F(R(t))D'(T-t) - C''(T-t) - D''(T-t)R(t),$$

$$\sigma(t,T) := D'(T-t)G(R(t-)).$$

(*a*) By the positivity of G and (8.2.14) we have

$$\lim_{T\to+\infty} \sigma(t,T) = G(R(t-)) \lim_{T\to+\infty} D'(T-t) > 0, \quad d\mathbb{P}\times dt - a.s.$$

Since $F \le G$, the convexity of C and D yields

$$\begin{aligned}\alpha(t,T) &\le G(R(t))D'(T-t) - C''(T-t) - D''(T-t)R(t)\\ &\le \sigma(t,T), \quad T > 0,\ d\mathbb{P}\times dt - a.s.\end{aligned}$$

The assertion follows from Proposition 8.2.3.

(*b*) In view of (8.2.15) we have

$$\int_t^{+\infty} \mid \sigma(t,u) \mid du = \mid G(R(t-)) \mid \cdot \int_0^{+\infty} \mid D'(u) \mid du < +\infty,$$

and consequently

$$\begin{aligned}\lim_{T\to+\infty} \alpha(t,T) &= F(R(t)) \lim_{T\to+\infty} D'(T-t) - \lim_{T\to+\infty} C''(T-t) - \lim_{T\to+\infty} D''(T-t)\\ &= -\lim_{u\to+\infty} C''(u) - \lim_{u\to+\infty} D''(u)R(t).\end{aligned}$$

By (8.2.16) the last term in the preceding is strictly positive and the assertion follows from Proposition 8.2.4. $\square$

8.2.3 Invariance of Lévy Noise

Our aim now is to construct an HJM model that admits a martingale measure $\mathbb{Q}$ different from the original measure $\mathbb{P}$. Additionally, we require that Z remains a Lévy process under $\mathbb{Q}$. Although this narrows the class of examined models, it also offers a possibility of treating the model after measure change with the use of the HJM conditions. First we prove that for an HJM model (α, σ, Z) for which Z is still a Lévy process under $\mathbb{Q}$, the drift α is determined by volatility σ in a deterministic manner.

Using this property we construct a class of possible drifts in the HJM model driven by a compound Poisson process such that the model admits a martingale measure.

Theorem 8.2.6 *Let the HJM model (α, σ, Z) satisfy the assumptions* (8.1.2)–(8.1.4) *and* (8.1.7) *and let Z be a Lévy process with characteristic triplet (a, Q, ν) under the measure $\mathbb{P}$. If (MM) is satisfied and Z is a Lévy process under some martingale measure $\mathbb{Q}$, then*

$$\alpha(t,T) = H\left(\sigma(t,T), \int_t^T \sigma(t,u)du\right), \qquad t \in [0,T^*], \quad T > 0,$$

where H is a deterministic function:

$$H(x,z) = \langle x, Q(z-\phi) - a\rangle - \int_U \left(e^{\psi(y)-\langle z,y\rangle}\langle x,y\rangle + \mathbf{1}_{\{|y|\le 1\}}\langle x,y\rangle\right)\nu(dy), \quad x, z \in U,$$

where $\phi \in U$ and $\psi(y)$ is a deterministic function such that

$$\int_U |\, e^{\psi(y)} - 1 \,|\, \nu(dy) < +\infty.$$

Proof Let $\mathbb{Q}$ be a martingale measure under which Z remains a Lévy process. In view of Theorem 6.2.1 (d) its generating pair (ϕ, ψ) is such that $\phi(\omega, t) = \phi$ and $\psi(\omega, t, y) = \psi(y)$, where ϕ is a deterministic vector in U and $\psi(\cdot)$ a deterministic function. Then (see (8.1.10)) we have

$$\begin{aligned} A(t,T) = &- \langle \Sigma(t,T), a\rangle + \frac{1}{2}\langle Q\Sigma(t,T), \Sigma(t,T)\rangle - \langle Q\phi, \Sigma(t,T)\rangle \\ &+ \int_U \left(e^{\psi(y)}(e^{-\langle \Sigma(t,T),y\rangle} - 1) + \mathbf{1}_{\{|y|\le 1\}}\langle \Sigma(t,T), y\rangle\right)\nu(dy). \end{aligned}$$

Differentiation over T yields

$$\begin{aligned} \alpha(t,T) &= -\langle \sigma(t,T), a\rangle + \frac{1}{2}\langle Q\sigma(t,T), \Sigma(t,T)\rangle + \frac{1}{2}\langle Q\Sigma(t,T), \sigma(t,T)\rangle - \langle Q\phi, \sigma(t,T)\rangle \\ &\quad - \int_U \left(e^{\psi(y)}e^{-\langle \Sigma(t,T),y\rangle}\langle \sigma(t,T), y\rangle + \mathbf{1}_{\{|y|\le 1\}}\langle \sigma(t,T), y\rangle\right)\nu(dy) \\ &= -\langle \sigma(t,T), a\rangle + \langle Q\sigma(t,T), \Sigma(t,T)\rangle - \langle Q\phi, \sigma(t,T)\rangle \\ &\quad - \int_U \left(e^{\psi(y)}e^{-\langle \Sigma(t,T),y\rangle}\langle \sigma(t,T), y\rangle + \mathbf{1}_{\{|y|\le 1\}}\langle \sigma(t,T), y\rangle\right)\nu(dy) \\ &= H\Big(\sigma(t,T), \Sigma(t,T)\Big). \end{aligned}$$

□

Proposition 8.2.7 *Let Z be a compound Poisson process with Lévy measure ν. Let us consider the class of functions G_ν satisfying*

$$g(y) > 0, \qquad \int_{\mathbb{R}} g(y)\nu(dy) < +\infty,$$

and define

$$F_g(z) := -\int_{\mathbb{R}} ye^{-zy} g(y)\nu(dy),$$

providing that the right side is well defined. If in the model (α, σ, Z) the drift is determined by

$$\alpha(t,T) = F_g\left(\int_t^T \sigma(t,u)du\right)\sigma(t,T), \qquad t \in [0,T^*],\ T > t, \tag{8.2.20}$$

for some function $g \in G_\nu$, then the model satisfies (MM) and Z is a compound Poisson process under some martingale measure.

Proof The characteristic triplet of the process Z equals

$$\left(\int_0^1 y\nu(dy), 0, \nu(dy)\right).$$

For $g \in G_\nu$ let us consider the measure $\mathbb{Q}_g \sim \mathbb{P}$ with generating pair $(\phi \equiv 0, \psi(t,y) \equiv \ln g(y))$. Since ν is finite and $g \in G_\nu$, it is clear that

$$\int_0^{T^*}\int_{\mathbb{R}} | e^{\psi(t,y)} - 1 | \, dt\nu(dy) = T^* \int_{\mathbb{R}} | g(y) - 1 | \, \nu(dy) < +\infty,$$

so the measure $\mathbb{Q}_g$ satisfies the necessary condition to be a martingale measure (see (8.1.8) and Remark 8.1.5). Moreover, the process Z remains under $\mathbb{Q}_g$ a Lévy process and its characteristic triplet under $\mathbb{Q}_g$ equals

$$\left(\int_0^1 yg(y)\nu(dy), 0, g(y)\nu(dy)\right)$$

(see Theorem 6.2.1). Hence its Laplace exponent under Q_g equals

$$J^{\mathbb{Q}_g}(z) = \int_{\mathbb{R}} (e^{-zy} - 1)g(y)\nu(dy)$$

and $\frac{\partial}{\partial z} J^{\mathbb{Q}_g}(z) = F_g(z)$. In view of Theorem 8.1.1, condition (8.2.20) means that $\mathbb{Q}_g$ is a martingale measure for the model (α, σ, Z). □

8.2.4 Volatility-Based Models

In this section we continue our discussion from Section 8.2.3 on the construction of HJM models that admit a martingale measure $\mathbb{Q}$ different from $\mathbb{P}$, the original one. We focus, however, on a more special dependence of the drift on volatility. To motivate our idea, let us recall that the existence of a martingale measure for the model (α, σ, Z) is equivalent to the following drift condition

$$\alpha(t,T) = \sigma(t,T)\Big(-a + \Sigma(t,T) - \phi(t) - \int_{\mathbb{R}} (e^{\psi(t,y)} e^{-\Sigma(t,T)y} - \mathbf{1}_{\{|y|\le 1\}}) y\nu(dy)\Big)$$

for $t \in [0, T^*], T > 0$ (see formula (8.1.10) for a one-dimensional noise in Theorem 8.1.1). This means that any volatility $\sigma(t,T)$ together with corresponding $\Sigma(t,T)$ determine $\alpha(t,T)$:

$$\alpha(t,T) = H\Big(\sigma(t,T), \Sigma(t,T)\Big), \quad t \in [0,T^*], T > 0, \tag{8.2.21}$$

where $H = H(u,v), u, v \in \mathbb{R}$ is a random field.

The randomness of H is hidden in the generating pair (ϕ, ψ) of a martingale measure. In the sequel we focus on the particular class of HJM models where drift is some deterministic function of volatility only, that is,

$$\alpha(t,T) = g(\sigma(t,T)), \quad t \in [0,T^*], T > 0. \tag{8.2.22}$$

We formulate the problem as follows. For a given Lévy process, find a volatility $\sigma(t,T)$ and a function g such that the HJM model

$$df(t,T) = g(\sigma(t,T))dt + \sigma(t,T)dZ(t) \tag{8.2.23}$$

admits a martingale measure. Moreover, we focus on time homogeneous volatilities $\sigma(t,T) = \sigma(T-t)$ that are given by the ordinary differential equation

$$\sigma'(v) = h(\sigma(v)), \quad \sigma(0) = \sigma_0, \quad v \ge 0, \tag{8.2.24}$$

where $h\colon \mathbb{R} \to \mathbb{R}$ is a given function. The function g is assumed to be differentiable and such that $g(\sigma(v))/\sigma(v)$ is also differentiable on $[0, +\infty)$. This setting comprises a reasonably wide subclass of arbitrage-free HJM models. We determine g such that at least one martingale measure exists.

Theorem 8.2.8 *Let the volatility $\sigma(t,T) = \sigma(T-t)$ be given by* (8.2.24). *Let $\psi(y)$ be a deterministic function satisfying*

$$\int_{\mathbb{R}} | e^{\psi(y)} - 1 | \nu(dy) < +\infty. \tag{8.2.25}$$

If the function g in the HJM model (8.2.23) *is given by*

$$g(x) := c_1 x + x\int_{c_2}^{x} \frac{K(z)}{z^2} dz, \quad x \in \mathbb{R}, \tag{8.2.26}$$

where c_1, c_2 are constants,

$$K(z) = K(h, \psi)(z) := \frac{z^3}{h(z)}\Big(1 + \int_{\mathbb{R}} e^{\psi(y)} e^{\{w(\sigma_0) - w(z)\}y} y^2 \, \nu(dy)\Big), \quad z \in \mathbb{R}, \tag{8.2.27}$$

and

$$w(z) := \int_0^z \frac{u}{h(u)} du, \quad z \in \mathbb{R}, \tag{8.2.28}$$

then the model admits a martingale measure.

Proof In the first part of the proof we show that the model admits a martingale measure if and only if, for $t \in [0, T^*], u \geq 0$,

$$\sigma'(u)\Big(g'(\sigma(u))\sigma(u) - g(\sigma(u))\Big) = \sigma(u)^3\Big(1 + \int_{\mathbb{R}} e^{\psi(t,y)} e^{-\Sigma(u)y} y^2 \nu(dy)\Big), \tag{8.2.29}$$

with some process $\psi(t, y)$ such that $e^{\psi(t,y)} - 1 \in \Psi_1$. Since $\sigma(t, T) = \sigma(T - t)$ is time homogeneous, so $\alpha(t, T) = g(\sigma(T - t))$ and

$$A(t, T) = \int_t^T \alpha(u - t)du = \int_0^{T-t} \alpha(u)du = A(T - t), \quad 0 \leq t \leq T,$$

$$\Sigma(t, T) = \int_t^T \sigma(u - t)du = \int_0^{T-t} \sigma(u)du = \Sigma(T - t), \quad 0 \leq t \leq T$$

for some functions $A(\cdot)$ and $\Sigma(\cdot)$. Since

$$\lim_{T \downarrow t} \frac{\alpha(T - t)}{\sigma(T - t)} = \lim_{T \downarrow t} \frac{g(\sigma(T - t))}{\sigma(T - t)} = \frac{g(\sigma_0)}{\sigma_0} = \frac{\alpha(0)}{\sigma(0)}$$

is finite, we can use Proposition 8.2.1. Setting $v := T - t$ we see from (8.2.3) that the model admits a martingale measure if and only if

$$A(v) - \frac{1}{2}(\Sigma(v))^2 - \Sigma(v)\frac{\alpha(0)}{\sigma(0)} = \int_{\mathbb{R}} e^{\psi(t,y)}\big(e^{-\Sigma(v)y} - 1 + \Sigma(v)y\big)\nu(dy), \tag{8.2.30}$$
$$t \in [0, T^*], v \geq 0,$$

with some function $\psi(t, y)$ such that $e^{\psi(t,y)} - 1 \in \Psi_1$. Differentiation of (8.2.30) yields

$$\alpha(v) = \sigma(v)\Big[\Sigma(v) + \frac{\alpha(0)}{\sigma(0)} - \int_{\mathbb{R}} e^{\psi(t,y)}(e^{-\Sigma(v)y} - 1)y\nu(dy)\Big], \quad t \in [0, T^*], v \geq 0,$$

and further

$$\frac{\alpha(v)}{\sigma(v)} - \frac{\alpha(0)}{\sigma(0)} = \Sigma(v) - \int_{\mathbb{R}} e^{\psi(t,y)}(e^{-\Sigma(v)y} - 1)y\nu(dy), \quad t \in [0, T^*], v \geq 0.$$

By (8.2.22), we obtain

$$\frac{g(\sigma(v))}{\sigma(v)} - \frac{g(\sigma_0)}{\sigma_0} = \Sigma(v) - \int_{\mathbb{R}} e^{\psi(t,y)}(e^{-\Sigma(v)y} - 1)y\nu(dy), \quad t \in [0,T^*], v \geq 0.$$

The preceding left side can be written as the integral of the derivative of the function $v \to \frac{g(\sigma(v))}{\sigma(v)}$. In the same way we represent the right side. This yields

$$\int_0^v \frac{\sigma'(u)(g'(\sigma(u))\sigma(u) - g(\sigma(u)))}{\sigma^2(u)} du = \int_0^v \sigma(u)du$$
$$+ \int_0^v \int_{\mathbb{R}} e^{\psi(t,y)} e^{-\Sigma(u)y} \sigma(u) y^2 du \, \nu(dy), \quad t \in [0,T^*], v \geq 0.$$

Comparison of the integrands yields (8.2.29).

In the second part of the proof, with the use of (8.2.29), we determine g. It can be checked that (8.2.24) is equivalent to the equation

$$\sigma'(v) = \frac{\sigma(v)}{w'(\sigma(v))}, \quad v \geq 0,$$

where w is given by (8.2.28). It follows that

$$\sigma(v) = w'(\sigma(v))\sigma'(v), \quad v \geq 0,$$

and

$$\Sigma(v) = \int_0^v \sigma(u)du = \int_0^v w'(\sigma(u))\sigma'(u)du = w(\sigma(v)) - w(\sigma_0), \quad v \geq 0.$$

Using this and putting any $\psi(t,y)$ which does not depend on t, i.e. $\psi(t,y) = \psi(y)$ satisfying (8.2.25) into (8.2.29), yields

$$\frac{\sigma(v)}{w'(\sigma(v))}\Big(g'(\sigma(v))\sigma(v) - g(\sigma(v))\Big) = \sigma^3(v)\Big(1 + \int_{\mathbb{R}} e^{\psi(y)} e^{\{w(\sigma_0) - w(\sigma(v))\}y} y^2 \nu(dy)\Big),$$
$$v \geq 0.$$

Setting $x := \sigma(v)$ leads to

$$\frac{x}{w'(x)}\Big(g'(x)x - g(x)\Big) = x^3\Big(1 + \int_{\mathbb{R}} e^{\psi(y)} e^{\{w(\sigma_0) - w(x)\}y} y^2 \nu(dy)\Big), \quad x \in \mathbb{R},$$

which yields

$$g'(x)x - g(x) = x^2 w'(x)\Big(1 + \int_{\mathbb{R}} e^{\psi(y)} e^{\{w(\sigma_0) - w(x)\}y} y^2 \nu(dy)\Big), \quad x \in \mathbb{R}.$$

Finally we obtain (see (8.2.27)) that

$$g'(x)x - g(x) = K(x), \quad x \in \mathbb{R}. \tag{8.2.31}$$

Since $g(0) = 0$ (see (8.1.4)) the solution of the equation above is given by (8.2.26).

□

Remark 8.2.9 In the HJM model in Proposition 8.2.8 there exist many functions g that guarantee that the model (8.2.23) admits a martingale measure. Indeed, the constants c_1 and c_2 in (8.2.26) are arbitrary. Moreover, the function K in (8.2.26) depends on the function ψ, which can be chosen freely.

Remark 8.2.10 If Z is a Wiener process with drift then the integral in (8.2.29) disappears and it follows from the proof of Theorem 8.2.8 that (8.2.29) is equivalent to (8.2.31). Consequently, the model (8.2.23) admits a martingale measure if and only if

$$g(x) = c_1 x + x \int_{c_2}^{x} \frac{z}{h(z)} dz, \qquad x \in \mathbb{R}.$$

8.2.5 Uniqueness of the Martingale Measure

Our first result on the uniqueness of the martingale measure in the HJM model is concerned with a finite activity Lévy process Z in $\mathbb{R}^d$, which additionally has a finite number of jump sizes. This means that the support of the Lévy measure ν is concentrated on a finite set, i.e.

$$\text{supp}\{\nu\} = \{y_1, y_2, \ldots, y_n\}, \quad y_i \in U = \mathbb{R}^d,\ i = 1, 2, \ldots, n.$$

We will assume that the original measure $\mathbb{P}$ is a martingale measure. In the analysis we use the following functions:

$$\Sigma(t,T) := (\Sigma^1(t,T), \ldots, \Sigma^d(t,T)),\ e^{-\langle \Sigma(t,T), y_1 \rangle} - 1,\ \ldots\ , e^{-\langle \Sigma(t,T), y_n \rangle} - 1. \tag{8.2.32}$$

To simplify notation their dependence on ω is not indicated. For fixed (ω, t) they will be viewed as functions of $T \in [0, +\infty)$.

Theorem 8.2.11 *Let $\mathbb{P}$ be a martingale measure for an HJM model, where Z is a finite activity Lévy process in $\mathbb{R}^d$ with a finite number of jump sizes. Then $\mathbb{P}$ is a unique martingale measure if and only if functions* (8.2.32) *are linearly independent $d\mathbb{P} \times dt$-a.s.*

Proof Since $\mathbb{P}$ is a martingale measure, the drift is determined by the volatility in the following way

$$A(t,T) = J(\Sigma(t,T)), \tag{8.2.33}$$

where J stands for the Laplace exponent of Z. Let us now formulate conditions for the generating pair (ϕ, ψ) of an equivalent measure $\mathbb{Q}$ that ensure that $\mathbb{Q}$ is also a martingale measure. Recall that by (8.1.12) we know that ϕ, ψ satisfy

$$\begin{aligned} A(t,T) &= J(\Sigma(t,T)) - \langle Q\psi(t), \Sigma(t,T) \rangle \\ &\quad + \int_U (e^{\psi(t,y)} - 1)(e^{-\langle \Sigma(t,T), y \rangle} - 1)\nu(dy), \qquad t \in [0, T^*]. \end{aligned} \tag{8.2.34}$$

By (8.2.33) and (8.2.34) we have

$$-\langle Q\phi(t), \Sigma(t,T)\rangle + \int_U (e^{\psi(t,y)} - 1)(e^{-\langle\Sigma(t,T),y\rangle} - 1)\nu(dy) = 0,$$

which, in the finite activity setting, can be written as

$$\langle Q\phi(t), \Sigma(t,T)\rangle = \sum_{i=1}^{n} (e^{\psi(t,y_i)} - 1)(e^{-\langle\Sigma(t,T),y_i\rangle} - 1)\delta_i, \tag{8.2.35}$$

where $\delta_i := \nu(\{y_i\}), i = 1, 2, \ldots, n$. So, $\mathbb{Q}$ is a martingale measure if and only if (8.2.35) is satisfied.

If (8.2.32) are linearly independent then one can find maturities $T_1, \ldots, T_K$ such that the vectors

$$w_i := \begin{pmatrix} \Sigma^i(t,T_1) \\ \vdots \\ \Sigma^i(t,T_K) \end{pmatrix}, \quad v_j := \begin{pmatrix} e^{-\Sigma(t,T_1)y_j} - 1 \\ \vdots \\ e^{-\Sigma(t,T_K)y_j} - 1 \end{pmatrix}, \quad i = 1, 2, \ldots, d,\ j = 1, 2, \ldots, n \tag{8.2.36}$$

are linearly independent. For these maturities (8.2.35) has the form

$$\begin{bmatrix} \Sigma^1(t,T_1) & \ldots & \Sigma^d(t,T_1) \\ \vdots & & \vdots \\ \Sigma^1(t,T_K) & \ldots & \Sigma^d(t,T_K) \end{bmatrix} \begin{bmatrix} (Q\phi(t))^1 \\ \vdots \\ (Q\phi(t))^d \end{bmatrix}$$
$$= \begin{bmatrix} e^{-\langle\Sigma(t,T_1),y_1\rangle} - 1 & \ldots & e^{-\langle\Sigma(t,T_1),y_n\rangle} - 1 \\ \vdots & & \vdots \\ e^{-\langle\Sigma(t,T_K),y_1\rangle} - 1 & \ldots & e^{-\langle\Sigma(t,T_K),y_n\rangle} - 1 \end{bmatrix} \begin{bmatrix} \delta_1(e^{\psi(t,y_1)} - 1) \\ \vdots \\ \delta_n(e^{\psi(t,y_n)-1}) \end{bmatrix}, \tag{8.2.37}$$

where $(Q\phi(t))^i, i = 1, 2, \ldots, d$ are the coordinates of the vector $Q\phi(t)$. It follows from the linear independence of $\{w_i, v_j\}, i = 1, 2, \ldots, d; j = 1, 2, \ldots, n$ that (8.2.37) may hold if and only if $\phi(t) \equiv 0$ and $\psi(t, y_i) \equiv 0, i = 1, 2, \ldots, n$. So, $\mathbb{Q} = \mathbb{P}$.

Conversely, if (8.2.32) are linearly dependent then there are constants $\alpha_i = \alpha_i(\omega, t), i = 1, 2, \ldots, d$ and $\beta_j = \beta_j(\omega, t), j = 1, 2, \ldots, n$ such that

$$\alpha_1\Sigma^1(t,T) + \cdots + \alpha_d\Sigma^d(t,T)$$
$$= \beta_1(e^{-\langle\Sigma(t,T),y_1\rangle} - 1) + \cdots + \beta_n(e^{-\langle\Sigma(t,T),y_n\rangle} - 1), \qquad T \in [t, +\infty).$$

Moreover, one can choose these constants such that $\alpha_i, \beta_j \in (-1, 1)$, $i = 1, 2, \ldots, d; j = 1, 2, \ldots, n$, and define $(\phi(t), \psi(t, y_i))$ by

$$(Q\phi(t))_i := \alpha_i, \quad \delta_j(e^{\psi(t,y_j)} - 1) := \beta_j, \quad i = 1, 2, \ldots, d; j = 1, 2, \ldots, n.$$

Then (8.2.35) is satisfied for each $T > 0$, and $\phi \in \Phi(U)$, $e^{\psi} - 1 \in \Psi_{1,2}$. This means that (ϕ, ψ) is a generating pair of some martingale measure. □

The problem of uniqueness of the martingale measure was studied in Eberlein, Jacod and Raible [47] in a slightly different framework. The bond market model in [47] consists of bonds with maturities forming a dense subset J of $[0, T^*]$, that is, only bonds

$$P(t, T),\ t \in [0, T^*], \quad T \in J, \quad \text{are tradeable.} \tag{8.2.38}$$

In the forward-rate dynamics the process Z is an $\mathbb{R}^d$-valued process with independent but not necessarily stationary increments and the coefficients are assumed to satisfy the condition

$$\sup_{t,T\in[0,T^*]} \Big(\mid \alpha(t, T) \mid + \mid \sigma(t, T) \mid_{\mathbb{R}^d} \Big) < +\infty, \qquad P - a.s. \tag{8.2.39}$$

Now we present the main results of [47] adopting them to processes Z with stationary increments.

The first result is concerned with the case $d = 1$.

Theorem 8.2.12 *Assume that* (8.2.38) *holds and Z is a one-dimensional Lévy process with a characteristic triplet (a, q, ν).*

(a) If $q + \nu(\mathbb{R}) > 0$ then the set of all martingale measures contains at most one element if and only if for almost all $t \in [0, T^*]$:

$$\int_t^{T^*} \mid \sigma(t, T) \mid dT > 0. \tag{8.2.40}$$

(b) If $q + \nu(\mathbb{R}) = 0$ then the set of all martingale measures contains at most one element.

The condition $q + \nu(\mathbb{R}) > 0$ means that the Lévy process is not trivial, i.e. it contains a Wiener part or has nonzero jumps. Condition (8.2.40) means that σ is not degenerated. In concrete examples it can easily be fulfilled. The next result shows that condition (8.2.40) can be omitted if the model coefficients are deterministic.

Theorem 8.2.13 *Assume that* (8.2.38) *holds and Z is a one-dimensional Lévy process. If the processes*

$$\alpha(t, T), \quad \sigma(t, T), \quad t \in [0, T^*], \quad T \in J$$

are deterministic then the set of all martingale measures is either empty or a singleton.

The preceding result can be generalized to the case $d \geq 2$.

Theorem 8.2.14 *Assume that J is a dense subset in $[0, T^*]$ and Z is a Lévy process in $\mathbb{R}^d$. If the processes*

$$\alpha(t,T), \quad \sigma(t,T), \quad t \in [0,T^*], \quad T \in J$$

are deterministic and the spaces

$$E_t := span\{\Sigma(t,T), T \in J\}$$

are such that

$$\dim E_t \leq 1, \quad t \in [0,T^*], \tag{8.2.41}$$

then the set of all martingale measures is either empty or a singleton.

Condition (8.2.41) is automatically satisfied if $d = 1$ and then Theorem 8.2.14 reduces to Theorem 8.2.13. If $d \geq 2$ then (8.2.41) is very restrictive. It tells that the vectors

$$\Sigma(t,T), \qquad T \in J$$

in $\mathbb{R}^d$ for each $t \in [0, T^*]$ satisfy

$$\Sigma(t,T) = c\Sigma(t,S), \qquad T, S \in J$$

for some $c = c(t)$, which means that it has an extremely degenerated structure.

9

Arbitrage-Free Forward Curves Models

We investigate the non-arbitrage problem for forward curves models moved by a Markov process. The main result of the chapter is the term structure equation. Some applications to special factor processes like multiplicative or Ornstein–Uhlenbeck processes are presented.

9.1 Term Structure Equation

We now study the problem of absence of arbitrage in the model given by *forward curves* of the form

$$f(t,T) = G(T-t, X(t)), \quad 0 \le t \le T, \tag{9.1.1}$$

where G is a real-valued function defined on $[0,+\infty) \times E$ and $X(t), t \ge 0$ is a factor process taking values in a closed subset E of $\mathbb{R}^n$. Recall that the short rate is given by

$$R(t) = G(0, X(t)), \quad t \ge 0,$$

and consequently

$$B(t) = e^{\int_0^t G(0,X(s))ds}, \quad t \ge 0$$

defines the bank account process. The corresponding *bond curves*

$$P(t,T) = F(T-t, X(t)), \quad 0 \le t \le T \tag{9.1.2}$$

are described by the function F, defined on $[0,+\infty) \times E$, by

$$F(T-t,x) := e^{-\int_t^T G(u-t,x)du} = e^{-\int_0^{T-t} G(s,x)ds}, \quad 0 \le t \le T.$$

Thus for F and G, we have the important relations

$$G(u,x) = -\frac{\frac{\partial F}{\partial u}(u,x)}{F(u,x)}, \qquad F(u,x) = e^{-\int_0^u G(v,x)dv}, \quad u \ge 0,\ x \in E \tag{9.1.3}$$

(see also Section 7.5).

We start with a given process X on E and our aim is to determine forward curves $G(\cdot,x), x \in E$ and bond curves $F(\cdot,x), x \in E$, for which the bond prices satisfy the following condition:

For each $T > 0$ the discounted bond price process

$$\hat{P}(t,T) = \frac{P(t,T)}{B(t)}, \qquad t \in [0,T] \tag{MP}$$

is a martingale.

As a first approximation of the financial reality one regards the case of X being a Markov process say, with transition semigroup $(\mathcal{P}_t)$ acting on the space $C_b(E)$ of continuous bounded functions on E equipped with the supremum norm. If E is unbounded the elements of $C_b(E)$ are required to have finite limits at infinity, and then $C_b(E)$ is separable. The infinitesimal generator of $(\mathcal{P}_t)$ and its domain will be denoted by $\mathcal{A}$ and $D(\mathcal{A})$. The (MP) condition should hold for processes X^x starting from an arbitrary $x \in E$.

Let $(\mathcal{Q}_u, u \geq 0)$ be the *discounted semigroup*

$$\mathcal{Q}_u\varphi(x) := \mathbb{E}\Big(\varphi(X^x(u))e^{-\int_0^u G(0,X^x(s))ds}\Big), \quad u \geq 0, \quad x \in E \quad \varphi \in C_b(E),$$

also called the *pricing semigroup*.

Theorem 9.1.1 *(a) Condition (MP) is satisfied if and only if the function F satisfies one of the following two relations:*

$$\mathcal{Q}_u(F(r,\cdot))(x) = F(r+u,x), \quad r \geq 0,\ u \geq 0,\ x \in E, \tag{9.1.4}$$

or

$$\mathcal{Q}_u\mathbf{1}(x) = F(u,x), \quad u \geq 0,\ x \in E. \tag{9.1.5}$$

(b) If $\mathbf{1}$ is in the domain of the infinitesimal generator of $\mathcal{Q}_u$ and $G(0,\cdot) \in C_b(E)$, then the bond curves $F(\cdot,x)$, $x \in E$ are solutions of the following evolution equation

$$\frac{\partial}{\partial u}F(u,x) = \mathcal{A}F(u,\cdot)(x) - G(0,x)F(u,x), \quad F(0,x) = \mathbf{1}(x), \quad u > 0,\ x \in E. \tag{9.1.6}$$

Moreover, they have the following stochastic representation:

$$F(u,x) = \mathbb{E}[e^{-\int_0^u G(0,X^x(s))ds}]. \tag{9.1.7}$$

Proof (*a*) Property (MP) means that, for any $T > 0$, the processes

$$\begin{aligned}\hat{P}(t,T) &= e^{-\int_t^T G(u-t,X^x(t))du} \cdot e^{-\int_0^t G(0,X^x(u))du}\\ &= F(T-t,X^x(t)) \cdot e^{-\int_0^t G(0,X^x(u))du}, \quad 0 \leq t \leq T\end{aligned}$$

are martingales. So, for $s < t$,

$$\mathbb{E}\left(F(T-t, X^x(t))e^{-\int_0^t G(0,X^x(u))du} \mid \mathcal{F}_s\right) = F(T-s, X^x(s))e^{-\int_0^s G(0,X^x(u))du}.$$

Equivalently, setting $r = T - t$,

$$\mathbb{E}\left(F(r, X^x(t))e^{-\int_s^t G(0,X^x(u))du} \mid \mathcal{F}_s\right) = F(T-s, X^x(s)). \tag{9.1.8}$$

Applying to the left side of (9.1.8) the Markov property of X, we get that

$$\mathcal{Q}_{t-s}(F(r,\cdot))(X^x(s)) = F(T-s, X^x(s)), \tag{9.1.9}$$

or

$$\mathcal{Q}_{t-s}(F(r,\cdot))(X^x(s)) = F(r+t-s, X^x(s)),$$

which implies (9.1.4).

Conversely, if (9.1.4) holds, then (9.1.9) is true as well. But (9.1.9) is equivalent to (9.1.8), so the martingale property takes place.

Since $F(0,x) = 1, x \in E$, it follows from (9.1.4) that

$$\mathcal{Q}_u\mathbf{1}(x) = F(u,x), \quad u \geq 0,\ x \in E.$$

Conversely, if (9.1.5) holds, then by the semigroup property of $\mathcal{Q}_t, t \geq 0$ (9.1.4) holds as well.

(*b*) Since $\varphi = \mathcal{Q}_t\mathbf{1}$ is in the domain of the infinitesimal operator of the pricing semigroup, the limit

$$\lim_{h\to 0} \frac{\mathcal{Q}_h\varphi(x) - \varphi(x)}{h} \tag{9.1.10}$$

exists uniformly in x. However, uniformly in x,

$$\frac{\mathbb{E}[\varphi(X_h^x)(e^{-\int_0^h G(0,X_s^x)ds} - 1)]}{h} \underset{h\to 0}{\longrightarrow} -\varphi(x)G(0,x),$$

and therefore

$$\begin{aligned}\frac{\mathcal{Q}_h\varphi(x) - \varphi(x)}{h} &= \frac{\mathbb{E}[\varphi(X_h^x)e^{-\int_0^h G(0,X_s^x)ds}] - \varphi(x)}{h} \\ &= \frac{\mathbb{E}[\varphi(X_h^x)(e^{-\int_0^h G(0,X_s^x)ds} - 1)]}{h} + \frac{\mathbb{E}[\varphi(X_h^x)] - \varphi(x)}{h} \\ &\underset{h\to 0}{\longrightarrow} -\varphi(x)G(0,x) + \mathcal{A}\varphi(x).\end{aligned}$$

Since the semigroup $(\mathcal{Q}_u)$ defines a solution to the Cauchy problem (9.1.6), the assertion follows.

To show the final part of the theorem, recall that the short-rate process $R^x(t)$, $t \geq 0$, is exactly $G(0, X_t^x), t \geq 0$. Since $\hat{P}(\cdot, u)$ is a martingale for any $u > 0$, we have the following relation

$$F(u,x) = e^{-\int_0^u G(v,x)dv} = \hat{P}(0,u) = \mathbb{E}[\hat{P}(u,u)]$$
$$= \mathbb{E}\left(e^{-\int_0^u R^x(s)ds}\right) = \mathbb{E}\left(e^{-\int_0^u G(0,X^x(s))ds}\right), \quad u \geq 0, x \in E. \qquad \square$$

Remark 9.1.2 The equation (9.1.6) is called a *term structure equation* and can serve as a starting point for the construction of models satisfying (MP). Here it was established under stringent assumptions using the semigroup concepts. In specific cases, when, for instance, the process X is given by a stochastic equation, it can be derived directly under less restrictive conditions on $G(0, \cdot)$ and with operators $\mathcal{A}$ extended to much greater sets.

Theorem 9.1.1 is of great heuristical value as, in many cases, it allows to obtain formulae for F and G. We illustrate the strength of the term structure equation by deriving, formally, several explicit formulae for forward and bond curves.

9.1.1 Markov Chain and CIR as Factor Processes

We start by characterizing bond curves in the case in which the factor process is a Markov chain or Cox–Ingersoll–Ross (CIR) process.

Example 9.1.3 Let X be a Markov process with a finite state space $E = \{1, 2, \ldots, M\}$ and the generator matrix

$$\mathcal{A} = (a_{ij})_{i,j \in E},$$

where $a_{ij} \geq 0$ for $i \neq j$ and $a_{ii} = -\sum_{j \neq i} a_{i,j}$. Denoting $F_j(u) := F(u,j), j = 1, 2, \ldots, M$, we see that (9.1.6) amounts to the system

$$\begin{cases} F_j'(u) = \sum_{k=1}^M a_{jk} F_k(u) + b_j F_j(u), \\ F_j(0) = 1, \end{cases}$$

with $b_j := \frac{F_j'(0)}{F_j(0)} = F_j'(0), j = 1, 2, \ldots, M$. For $M = 2$ the preceding system boils down to

$$\begin{cases} F_1'(u) = (a_{11} + b_1)F_1(u) - a_{11}F_2(u), \\ F_2'(u) = -a_{22}F_1(u) + (a_{22} + b_2)F_2(u), \\ F_1(0) = F_2(0) = 1. \end{cases}$$

The form of the solution (F_1, F_2) depends on the roots of the characteristic polynomial

$$det\begin{bmatrix} a_{11}+b_1-\lambda & -a_{11} \\ -a_{22} & a_{22}+b_2-\lambda \end{bmatrix} = \lambda^2 - \lambda(a_{11}+a_{22}+b_1+b_2)+a_{11}b_2 + a_{22}b_1+b_1b_2.$$

If

$$\gamma := (a_{11}+a_{22}+b_1+b_2) - 4(a_{11}b_2+a_{22}b_1+b_1b_2)^2 > 0,$$

then the roots are given by

$$\lambda_1 = \frac{1}{2}\Big(a_{11}+a_{22}+b_1+b_2 - \sqrt{(a_{11}+a_{22}+b_1+b_2)^2 - 4(a_{11}b_2+a_{22}b_1+b_1b_2)}\Big),$$

$$\lambda_2 = \frac{1}{2}\Big(a_{11}+a_{22}+b_1+b_2 + \sqrt{(a_{11}+a_{22}+b_1+b_2)^2 - 4(a_{11}b_2+a_{22}b_1+b_1b_2)}\Big),$$

and corresponding eigenvectors of the system are

$$v_1 = \left(-\frac{a_{11}-a_{22}+b_1-b_2-\sqrt{(a_{11}+a_{22}+b_1+b_2)^2-4(a_{11}b_2+a_{22}b_1+b_1b_2)}}{2a_{22}}, 1\right)^*,$$

$$v_2 = \left(-\frac{a_{11}-a_{22}+b_1-b_2+\sqrt{(a_{11}+a_{22}+b_1+b_2)^2-4(a_{11}b_2+a_{22}b_1+b_1b_2)}}{2a_{22}}, 1\right)^*.$$

Consequently,

$$\begin{pmatrix} F_1(u) \\ F_2(u) \end{pmatrix} = c_1 v_1 e^{\lambda_1 u} + c_2 v_2 e^{\lambda_2 u},$$

where the constants c_1, c_2 are such that $F_1(0) = F_2(0) = 1$. If $\gamma = 0$ then there is only one eigenvalue λ_0 of multiplicity 2. Then either there are two linearly independent eigenvectors w_1, w_2 or there is only one eigenvector w_0. In the first case

$$\begin{pmatrix} F_1(u) \\ F_2(u) \end{pmatrix} = c_1 w_1 e^{\lambda_0 u} + c_2 w_2 e^{\lambda_0 u},$$

while in the second

$$\begin{pmatrix} F_1(u) \\ F_2(u) \end{pmatrix} = c_1 w_0 u e^{\lambda_0 u} + c_2 \hat{w}_0 e^{\lambda_0 u},$$

where $\hat{w}_0$ is a certain vector. If $\gamma < 0$ then there are two complex eigenvalues $\lambda_1 = \alpha+\beta i$ and $\lambda_2 = \alpha-\beta i$ with corresponding complex vectors w_1 and $\bar{w_1}$ - the conjugate of w_1. Then $w_2 := \frac{1}{2}(w_1+\bar{w}_1)$ and $w_3 := \frac{i}{2}(w_1-\bar{w}_1)$ are real vectors and

$$\begin{pmatrix} F_1(u) \\ F_2(u) \end{pmatrix} = c_1\Big(w_2\cos(\beta u)+w_3\sin(\beta u)\Big)e^{\alpha u} + c_2\Big(w_3\cos(\beta u)-w_2\sin(\beta u)\Big)e^{\alpha u}.$$

Example 9.1.4 Let $G(0,x) = x$ and the Markov process $X(t) = R(t)$ be the short rate with CIR dynamics, i.e.

$$dX(t) = (aX(t) + b)dt + \sqrt{X(t)}dW(t), \quad t > 0.$$

Then the term structure equation is of the form:

$$\begin{cases} \frac{\partial}{\partial t}F(t,x) & = (ax+b)\frac{\partial F}{\partial x}(t,x) + \frac{1}{2}x\frac{\partial^2 F}{\partial x^2}(t,x) - xF(t,x), \\ F(0,x) & = 1, \quad x \geq 0. \end{cases}$$

One easily checks that its solution is

$$F(t,x) = e^{-C(t)-D(t)x}, \quad t,x \geq 0,$$

where the functions C and D are solutions of the equations

$$C'(t) = bD(t),\ D'(t) = aD(t) - 1/2D^2(t) + 1,\ C(0) = D(0) = 0, \quad t \geq 0.$$

The preceding explicit formulae will be rederived in the context of the general affine term structure model (see Section 10.2.2).

9.1.2 Multiplicative Factor Process

Here we extend Proposition 2.4.15 to the continuous time setting. Namely, we determine bond curves in the model with random factor of the form

$$dX(t) = aX(t)dt + bX(t)dZ(t), \quad X(0) = x > 0, \quad t > 0, \tag{9.1.11}$$

where Z is a real Lévy process, $a, b \in \mathbb{R}$. By Theorem 4.4.6 we have that

$$X(t) = xe^{(a-\frac{1}{2}b^2)t+bZ(t)} \cdot S(t), \quad t > 0,$$

where

$$S(t) := \prod_{s\in[0,t]} (1 + \Delta Z(s))e^{-\Delta Z(s)}.$$

Proposition 9.1.5 *Let us assume that the forward curve model with factor X given by (9.1.11) and $G(0,x) = -\gamma \ln x$, $x > 0$, $\gamma \neq 0$, satisfies (MP). Then the bond curves are given by*

$$F(u,x) = x^{\gamma u}\ e^{\frac{\gamma}{2}(a-\frac{1}{2}b^2)u^2}\mathbb{E}[e^{\gamma(b\int_0^u Z(s)ds+\int_0^u \ln(S(s))ds)}]. \tag{9.1.12}$$

In particular, if Z is a Wiener process then

$$F(u,x) = x^{\gamma u}e^{\frac{\gamma}{2}(a-\frac{1}{2}b^2)u^2+\frac{\gamma^2}{6}b^2u^3}, \tag{9.1.13}$$

and if Z is a compound Poisson process with jumps greater than -1 then

$$F(u,x) = x^{\gamma u} e^{-u\nu(\mathbb{R}) + \frac{\gamma}{2}(a-\frac{1}{2})u^2 - \int_{-\infty}^{+\infty} \frac{1}{\gamma \ln(1+y)}(e^{-\gamma u \ln(1+y)}-1)\nu(dy)}, \tag{9.1.14}$$

providing that $b = 1$ and $\gamma < 0$.

Proof By (9.1.7),

$$F(u,x) = \mathbb{E}[e^{\gamma \int_0^u \ln(X(s))ds}]$$

$$= \mathbb{E}[e^{\gamma \int_0^u \{\ln x + (a-\frac{1}{2}b^2)s + bZ(s) + \ln(S(s))\}ds}],$$

which yields (9.1.12). If Z is a standard Wiener process, then the integral

$$\int_0^u W(s)ds$$

has a zero mean Gaussian distribution. Its second moment equals

$$\mathbb{E}\left[\left(\int_0^u W(s)ds\right)^2\right] = \mathbb{E}\left[\int_0^u W(s)ds \int_0^u W(v)dv\right] = \mathbb{E}\left[\int_0^u \int_0^u W(s)W(v)dvds\right]$$

$$= \int_0^u \int_0^u \mathbb{E}[W(s)W(v)]dvds = \int_0^u \int_0^u (s \wedge v)dvds$$

$$= \int_0^u \left(\int_0^s vdv + \int_s^u sdv\right) ds = \frac{u^3}{3}.$$

Hence

$$\mathbb{E}\left[e^{\gamma b \int_0^u W(s)ds}\right] = e^{\frac{\gamma^2 b^2 u^3}{6}}$$

and (9.1.13) follows.

If Z is a compound Poisson process

$$Z(t) = \sum_{Y=1}^{N(t)} Y_i, \quad t \geq 0,$$

with $Y_i > -1$ and $b = 1$, then

$$X(t) = xe^{(a-\frac{1}{2})t} \prod_{s\in[0,t]} (1 + \triangle Z(s)) = xe^{(a-\frac{1}{2})t} e^{\tilde{Z}(s)},$$

where

$$\tilde{Z}(t) := \sum_{i=1}^{N(t)} \ln(1 + Y_i), \quad t \geq 0$$

is also a compound Poisson process. Then

$$F(u,x) = \mathbb{E}[e^{\gamma \int_0^u \ln(X(s))ds}] = x^{\gamma u} e^{\frac{\gamma}{2}(a-\frac{1}{2})u^2} \mathbb{E}\left[e^{\gamma \int_0^u \tilde{Z}(s)ds}\right].$$

To determine the latter expectation, we write the integral in exponent in the form

$$\begin{aligned}\int_0^u \tilde{Z}(s)ds &= \int_0^u \left(\int_0^s \int_{-1}^{+\infty} \ln(1+y)\pi(dr,dy)\right) ds \\ &= \int_0^{+\infty} \left(\int_0^{+\infty} \int_{-\infty}^{+\infty} \ln(1+y)\mathbf{1}_{[0,u]}(s)\mathbf{1}_{[0,s]}(r)\pi(dr,dy)\right) ds \\ &= \int_0^u \int_{-\infty}^{+\infty} \ln(1+y)(u-r)\pi(dr,dy),\end{aligned}$$

where π stands for the jump measure of Z. Now with the use of Theorem 6.6 in Peszat and Zabczyk [100], for $\gamma < 0$, we obtain

$$\begin{aligned}\mathbb{E}\left[e^{\gamma \int_0^u \tilde{Z}(s)ds}\right] &= \mathbb{E}\left[e^{\gamma \int_0^u \int_{-\infty}^{+\infty} \ln(1+y)(u-r)\pi(dr,dy)}\right] \\ &= e^{-\int_0^u \int_{-\infty}^{+\infty}(1-e^{\gamma(\ln(1+y)(u-r))})dr\nu(dy)} \\ &= e^{-u\nu(\mathbb{R})-\int_{-\infty}^{+\infty}\frac{1}{\gamma\ln(1+y)}(e^{-\gamma u\ln(1+y)}-1)\nu(dy)}.\end{aligned}$$

Finally, we obtain (9.1.14). □

One can combine the results from Proposition 9.1.5 to characterize bond curves in the case in which Z is a finite activity process.

9.1.3 Affine Term Structure Model

In this section we are concerned with the affine model

$$P(t,T) = e^{-C(T-t)-D(T-t)R(t)}, \quad 0 \le t \le T,$$

with the jump diffusion short rate

$$dR(t) = F(R(t))dt + \langle G(R(t-)), dZ(t)\rangle, \quad t \ge 0, \tag{9.1.15}$$

where Z stands for a d-dimensional Lévy process with characteristic triplet (a, Q, ν). We derive conditions on C, D, F, G implied by the term structure equation. These conditions will be derived again and studied in detail in Chapter 10.

Proposition 9.1.6 *Let the bond curve $F(\cdot,\cdot)$ be of the form*

$$F(u,x) = e^{-C(u)-D(u)x}, \quad u, x \ge 0,$$

and the short rate be given by (9.1.15). *If (MP) is satisfied then*

$$J\Big(G(x)D(u)\Big) = -C'(u) - [D'(u) - 1]x + D(u)F(x), \quad u, x \ge 0, \tag{9.1.16}$$

where J stands for the Laplace exponent of Z, providing that all terms in (9.1.16) *are well defined.*

Proof We stress that in the proof the functions $F(x), G(x)$ should be distinguished from the functions $F(u,x), G(u,x)$.
The bond curves satisfy

$$F(u,x) = e^{-C(u)-D(u)x}, \quad \frac{\partial F}{\partial u}(u,x) = e^{-C(u)-D(u)x}\big(-C'(u) - D'(u)x\big). \quad (9.1.17)$$

Recall that the generator of R given by (9.1.15) is

$$\mathcal{A}f(x) = f'(x)F(x) + f'(x)\langle G(x), a\rangle + \frac{1}{2}f''(x)\langle QG(x), G(x)\rangle$$
$$+ \int_U \Big\{f(x + \langle G(x), y\rangle) - f(x) - \mathbf{1}_{\{|y|\le 1\}}(y)f'(x)\langle G(x), y\rangle\Big\}\nu(dy)$$

(see Section B.1.1). Let us now determine $\mathcal{A}f_\gamma(x)$ for $f_\gamma(x) := e^{-\gamma x}$. Since

$$f_\gamma'(x) = -\gamma e^{-\gamma x}, \quad f_\gamma''(x) = \gamma^2 e^{-\gamma x},$$

we obtain

$$\mathcal{A}f_\gamma(x) = -\gamma e^{-\gamma x}\Big(F(x) + \langle G(x), a\rangle\Big) + \frac{1}{2}\gamma^2 e^{-\gamma x}\langle QG(x), G(x)\rangle$$
$$+ \int_U \Big\{e^{-\gamma(x+\langle G(x),y\rangle)} - e^{-\gamma x} + \mathbf{1}_{\{|y|\le 1\}}(y)\gamma e^{-\gamma x}\langle G(x), y\rangle\Big\}\nu(dy).$$

Using the formula for J (see (5.2.9)), we have

$$\mathcal{A}f_\gamma(x) = e^{-\gamma x}\Big[-\gamma F(x) - \langle \gamma G(x), a\rangle\Big] + \frac{1}{2}e^{-\gamma x}\langle Q\gamma G(x), \gamma G(x)\rangle$$
$$+ \int_U \Big\{e^{-\langle \gamma G(x),y\rangle} - 1 + \mathbf{1}_{\{|y|\le 1\}}(y)\langle \gamma G(x), y\rangle\Big\}\nu(dy)$$
$$= e^{-\gamma x}\Big[-\gamma F(x) + J(\gamma G(x))\Big]. \quad (9.1.18)$$

Since $G(0,x) = x$, the term structure equation (9.1.6) has the form

$$\mathcal{A}F(u,\cdot)(x) - xF(u,x) = \frac{\partial F}{\partial u}(u,x). \quad (9.1.19)$$

By (9.1.18),

$$\mathcal{A}F(u,\cdot)(x) = e^{-C(u)-D(u)x}\Big[-D(u)F(x) + J(D(u)G(x))\Big]. \quad (9.1.20)$$

Taking into account (9.1.17) and (9.1.20), we can write (9.1.19) in the form

$$e^{-C(u)-D(u)x}\Big[-D(u)F(x) + J(D(u)G(x)) - x\Big] = e^{-C(u)-D(u)x}\big(-C'(u) - D'(u)x\big),$$

which yields (9.1.16). □

Notice that (9.1.16) is identical with (10.2.4) in Theorem 10.2.1.

9.1.4 Ornstein–Uhlenbeck Factors

In this section we characterize bond curves for factor models where X is the Ornstein–Uhlenbeck process driven by a Lévy process Z. Explicit formulae can be obtained when Z is a general real Lévy process or a multidimensional Wiener process.

Proposition 9.1.7 *Let $G(0,x) = x$ and the short-rate process, identical with the factor process, be an Ornstein–Uhlenbeck process,*

$$dR^x(t) = (a + bR^x(t))dt + dZ(t), \quad R^x(0) = x, \quad t \geq 0,$$

where $a \in \mathbb{R}_+, b \in \mathbb{R}$ and Z is a real-valued Lévy process with Laplace exponent J. If (MP) is satisfied then

$$F(u,x) = e^{-\left(x(e^{bu}-1)/b + a(e^{bu}-1)/b^2 - au/b\right)} e^{J\left((e^{bu}-1)/b^2 - u/b\right)}. \tag{9.1.21}$$

Proof Since R is given by

$$dR^x(t) = xe^{bt} + \int_0^t e^{b(t-s)} a ds + \int_0^t e^{b(t-s)} dZ(s),$$

we have

$$\int_0^t R^x(s)ds = x\int_0^t e^{bs}ds + a\int_0^t \left(\int_0^s e^{b(s-u)}du\right) ds + \int_0^t \left(\int_0^s e^{b(s-u)}dZ(u)\right) ds,$$

and, by the stochastic Fubini theorem,

$$\int_0^t R^x(s)ds = x\int_0^t e^{bs}ds + a\int_0^t \left(\int_0^s e^{b(s-u)}du\right) ds + \int_0^t \left(\int_u^t e^{b(s-u)}ds\right) dZ(u).$$

It follows from (9.1.7) that

$$\begin{aligned} F(u,x) &= \mathbb{E}\left[e^{-\int_0^u R^x(s)ds}\right] \\ &= e^{-\left(x\int_0^u e^{bs}ds + a\int_0^u (\int_0^s e^{b(s-v)}dv)ds\right)} \mathbb{E}\left[e^{-\int_0^u (\int_v^u e^{b(s-v)}ds)dZ(v)}\right] \\ &= e^{-\left(x\int_0^u e^{bs}ds + a\int_0^u (\int_0^s e^{b(s-v)}dv)ds\right)} e^{J\left(\int_0^u (\int_v^u e^{b(s-v)}ds)dv\right)}. \end{aligned}$$

Above we used the following elementary formula valid for an arbitrary continuous function h

$$\mathbb{E}[e^{-\int_0^t h(s)dZ(s)}] = e^{\int_0^t J(h(s))ds}, \quad t > 0.$$

This implies the result. □

Proposition 9.1.8 *Let the factor process X^x be an Ornstein–Uhlenbeck process*

$$dX^x(t) = (a + AX^x(t))dt + dW(t), \quad X^x(0) = x \in \mathbb{R}^d, \tag{9.1.22}$$

where, $a \in \mathbb{R}^d$, A is a $d \times d$-matrix and W is a Wiener process in $\mathbb{R}^d$ with identity covariance matrix. Let us assume that $G(0,x) := \langle Kx, x\rangle$, where K is a positive $d \times d$-matrix. The corresponding bond curve is of the form

$$F(u,x) = e^{-\langle M(u)x,x\rangle - \langle b(u),x\rangle - c(u)}, \quad u \geq 0, x \in \mathbb{R}^d, \tag{9.1.23}$$

where $M(u)$ is a matrix solution of the matrix Riccati equation

$$M'(u) = K + 2A^*M(u) - 2M^2(u), \quad u \geq 0, \tag{9.1.24}$$

with $M(0) = 0$, $b(u)$ is a solution of the equation

$$b'(u) = (A^* - 2M(u))b(u) + 2M(u)a, \quad u \geq 0, \tag{9.1.25}$$

with $b(0) = 0$ and $c(u)$ satisfies

$$c'(u) = \langle b(u), a\rangle - \frac{1}{2} \mid b(u) \mid^2 + TraceM(u), \quad u \geq 0, \quad c(0) = 0. \tag{9.1.26}$$

The corresponding forward curve has the form

$$G(u,x) = \langle M'(u)x, x\rangle + \langle b'(u), x\rangle + c'(u), \quad u \geq 0, x \in \mathbb{R}^d.$$

In particular, if $a = 0$ then

$$F(u,x) = e^{-\langle M(u)x,x\rangle - \int_0^u TraceM(s)ds}, \quad u \geq 0, x \in \mathbb{R}^d,$$

and, for $u \geq 0, x \in \mathbb{R}^d$,

$$G(u,x) = TraceM(u) + \langle (K + 2A^*M(u) - 2M^2(u))x, x\rangle).$$

Proof The generator of the process $X^x(t)$ acts on regular functions φ as follows:

$$\mathcal{A}\varphi(x) = \frac{1}{2}\Delta\varphi(x) + \langle a + Ax, D\varphi(x)\rangle \tag{9.1.27}$$

(see Section B.1.1). Here Δ, D denote the Laplacian and the gradient operator, respectively. We show that the term structure equation

$$\frac{\partial}{\partial u}F(u,x) = \mathcal{A}F(u,x) - G(0,x)F(u,x), \quad F(0,x) = \mathbf{1}(x), \quad u > 0, \ x \in \mathbb{R}^d \tag{9.1.28}$$

is solved by the function (9.1.23) with functions $M(u), b(u), c(u)$ satisfying $M(0) = 0, b(0) = 0, c(0) = 0$. It follows from (9.1.23) that

$$\frac{\partial F}{\partial u}(u,x) = -F(u,x)[\langle M'(u)x, x\rangle + \langle b'(u), x\rangle + c'(u)] \tag{9.1.29}$$

and

$$\frac{\partial F}{\partial x_i}(u,x) = -F(u,x)\left[\frac{\partial}{\partial x_i}\langle M(u)x, x\rangle + \frac{\partial}{\partial x_i}\langle b(u), x\rangle\right]. \tag{9.1.30}$$

By (9.1.30) we obtain, for $i = 1, 2, \ldots, d$,

$$\begin{aligned}\frac{\partial^2}{\partial x_i^2}F(u,x) &= -\frac{\partial F(u,x)}{\partial x_i}\left[\frac{\partial}{\partial x_i}\langle M(u)x,x\rangle + \frac{\partial}{\partial x_i}\langle b(u),x\rangle\right]\\ &\quad - F(u,x)\left[\frac{\partial^2}{\partial x_i^2}\langle M(u)x,x\rangle + \frac{\partial^2}{\partial x_i^2}\langle b(u),x\rangle\right]\\ &= F(u,x)\left[\left(\frac{\partial}{\partial x_i}\langle M(u)x,x\rangle + \frac{\partial}{\partial x_i}\langle b(u),x\rangle\right)^2\right.\\ &\qquad \left. -\frac{\partial^2}{\partial x_i^2}\langle M(u)x,x\rangle - \frac{\partial^2}{\partial x_i^2}\langle b(u),x\rangle\right]. \end{aligned} \tag{9.1.31}$$

It follows from (9.1.30) that

$$\begin{aligned} DF(u,x) &= -F(u,x)[D\langle M(u)x,x\rangle + b(u)]\\ &= -F(u,x)[M(u)x + M^*(u)x + b(u)], \end{aligned} \tag{9.1.32}$$

and from (9.1.31) that

$$\begin{aligned}\Delta F(u,x) &= F(u,x)\left(\sum_{i=1}^{d}\left\{\left(\frac{\partial}{\partial x_i}\langle M(u)x,x\rangle\right)^2 + 2b_i(u)\frac{\partial}{\partial x_i}\langle M(u)x,x\rangle + b_i^2(u)\right.\right.\\ &\qquad \left.\left. -\frac{\partial^2}{\partial x_i^2}\langle M(u)x,x\rangle\right\}\right)\\ &= F(u,x)\Big(\mid M(u)x + M^*(u)x\mid^2 + \mid b(u)\mid^2 + 2\langle b(u), M(u)x + M^*(u)x\rangle\\ &\qquad - 2\mathrm{Trace}M(u)\Big). \end{aligned} \tag{9.1.33}$$

Taking into account (9.1.27), (9.1.29), (9.1.32) and (9.1.33), we insert (9.1.23) into (9.1.28). This yields

$$\begin{aligned} &-F(u,x)[\langle M'(u)x,x\rangle + \langle b'(u),x\rangle + c'(u)]\\ &\quad = -F(u,x)[\langle (M(u)+M^*(u))x + b(u), a + Ax\rangle] + \frac{1}{2}F(u,x)[\mid (M(u)+M^*(u))x\mid^2\\ &\qquad + \mid b(u)\mid^2 + 2\langle b(u),(M(u)+M^*(u))x\rangle - 2\mathrm{Trace}M(u)] - F(u,x)\langle Kx,x\rangle. \end{aligned} \tag{9.1.34}$$

Comparing quadratic terms yields

$$\begin{aligned} -\langle M'(u)x,x\rangle &= -\langle (M(u)+M^*(u))x, Ax\rangle + \frac{1}{2}\langle (M(u)+M^*(u))x, (M(u)+M^*(u))x\rangle\\ &\quad - \langle Kx,x\rangle. \end{aligned}$$

Since $M^*(u) = M(u)$:

$$\langle M'(u)x, x\rangle = \langle Kx, x\rangle + 2\langle A^*M(u), x\rangle - 2\langle M^2(u)x, x\rangle.$$

This implies that $M(u)$ satisfies (9.1.24). Similarly, comparing linear terms and free terms in (9.1.34) yields (9.1.25) and (9.1.26). The formula for $G(u, x)$ follows from (9.1.3).

For $a = 0$ we see from (9.1.25) that $b(u) \equiv 0$ and, consequently, by (9.1.26), $c(u) = \int_0^u TraceM(s)ds$. □

Remark 9.1.9 As a byproduct we derived the so-called Cameron–Martin formula (see Lipster and Shiryaev [89]), which in the case $a = 0$ takes the form

$$\mathbb{E}[e^{-\int_0^u \langle KX^x(s), X^x(s)\rangle ds}] = e^{-<M(u)x,x>-\int_0^u TraceM(s)ds}.$$

The matrix Riccati equation (9.1.24) is of great importance in the control theory (see Zabczyk [123]).

Example 9.1.10 For $d = 1$ and $K = k$, we obtain

$$M(u) = \sqrt{\frac{k}{2}}\tanh(\sqrt{2k}u) = \sqrt{\frac{k}{2}}\,\frac{e^{2\sqrt{2k}u} - 1}{e^{2\sqrt{2k}u} + 1}. \tag{9.1.35}$$

10

Arbitrage-Free Affine Term Structure

This chapter is concerned with affine models of bond prices. Short rates are given as solutions of stochastic equations driven by a Lévy process or they can be general Markov processes. Conditions are found under which the discounted bond prices are local martingales. Generalizations of the Cox–Ingersoll–Ross and the Vasiček short-rate models to the framework with Lévy factors are presented. The discrete time case was solved by Filipović and Zabczyk [59], [58] and is discussed in Section 2.4.

10.1 Preliminary Model Requirements

Bond prices in the affine term structure model are required to be of the form

$$P(t,T) = e^{-C(T-t)-D(T-t)R(t)}, \quad 0 \le t \le T, \tag{10.1.1}$$

where C, D are deterministic continuous functions and R stands for the short rate. To make the model regular (see Section 7.1), we assume that the short rate $R(t)$ is a nonnegative process and that

$$C(u), D(u) \ge 0, \quad C(u) \le C(v), D(u) \le D(v), \qquad 0 \le u \le v. \tag{10.1.2}$$

As in the previous chapter, our goal is to figure out which affine models have the following martingale property:

For each $T > 0$ the discounted bond price process

$$\hat{P}(t,T) = \frac{P(t,T)}{B(t)}, \qquad t \in [0,T] \tag{MP}$$

is a local martingale, which implies that the model does not allow arbitrage.

We want to find short-rate processes and functions C and D such that (MP) is satisfied. Since $P(T,T) = 1$, necessarily

$$C(0) = 0, \quad D(0) = 0. \tag{10.1.3}$$

Writing the model in terms of forward rates we obtain from (10.1.1) that

$$C(T-t)+D(T-t)R(t)=\int_t^T f(t,u)du, \quad 0 \le t \le T.$$

Thus if C, D are absolutely continuous then

$$f(t,T)=C'(T-t)+D'(T-t)R(t), \quad 0 \le t \le T. \tag{10.1.4}$$

In fact, we will require that C and D are continuously differentiable on $[0,+\infty)$. Since $R(t)=f(t,t)$, we have

$$C'(0)=0, \quad D'(0)=1. \tag{10.1.5}$$

We divide our considerations into two parts. First we assume that the short-rate process R solves a stochastic differential equation of the form

$$dR(t)=F(R(t))dt+\langle G(R(t-)), dZ(t)\rangle, \quad t \ge 0, \quad R(0)=x, \tag{10.1.6}$$

where Z is a Lévy process and $F(\cdot), G(\cdot)$ are deterministic functions. If there exist required functions C, D such that the resulting affine model satisfies (MP), we say that (10.1.6) *generates* an affine model with (MP) property. We try to characterize all equations generating affine models with (MP) property. A complete answer is presented in the case in which Z is a one-dimensional martingale. For subordinators Z we consider special cases. For multidimensional noise we provide examples only. In the second part we will merely require that R is a general Markov process and present theory developed by Filipović in [53]. This setting covers, in particular, short rates of the form (10.1.6), but characterization of models satisfying (MP) is not so explicit.

10.2 Jump Diffusion Short Rate

Our basic requirement is that (10.1.6) is *positive invariant*, i.e. has a unique nonnegative strong solution for each initial condition $R(0)=x \ge 0$. We assume also that

$$\text{supp}\{\nu\} \subseteq \mathbb{R}^d_+, \qquad G(x) \in \mathbb{R}^d_+, \quad \text{for } x \ge 0. \tag{10.2.1}$$

The first condition in (10.2.1) implies that

$$\int_{|y|>1} e^{-\langle \lambda, y\rangle}\nu(dy) < +\infty, \quad \lambda \in \mathbb{R}^d_+, \tag{10.2.2}$$

and consequently the Laplace exponent J of Z (see (5.2.9)) given by

$$J(\lambda)=-\langle a,\lambda\rangle+\frac{1}{2}\langle Q\lambda,\lambda\rangle+\int_U \left(e^{-\langle \lambda, y\rangle}-1+\mathbf{1}_{\{|y|\le 1\}}\langle \lambda, y\rangle\right)\nu(dy) \tag{10.2.3}$$

is well defined on $\mathbb{R}^d_+$. On the other hand, for $\lambda \notin \mathbb{R}^d_+$ condition (10.2.2) enforces exponential moments of ν, which is rather restrictive. So, $J(\lambda)$ does not exist in this case, in general.

Our aim now is to find conditions for F, G and Z in (10.1.6) that guarantee that the affine model (10.1.1), with some functions C, D, satisfies (MP).

10.2.1 Analytical HJM Condition

Our starting point in examining the (MP) condition for affine models is to show that under (MP) the coefficients F, G in (10.1.6) and the functions C, D in (10.1.1) satisfy a certain analytical equation that involves also the Laplace exponent of the noise process Z. This equation (see (10.2.4)) appears by treating the affine model as a particular form of the HJM model.

Let us recall that (10.2.4) has already been derived, in a bit informal way, in the previous chapter as a consequence of the term structure equation (see (9.1.16)). We stress that although the condition (10.2.4) implies that the discounted bond prices are local martingales it does not imply the positivity of the short rate and does not exclude explosions of the rates. A related example will be discussed in Section 10.2.3.

The *right directional derivative* $D^+_w J(\lambda)$ of J at a point λ along a vector $w \in \mathbb{R}^d$ is defined by the formula

$$D^+_w J(\lambda) := \lim_{h \downarrow 0} \frac{J(\lambda + hw) - J(\lambda)}{h}.$$

Theorem 10.2.1 *Assume that* (10.1.6) *is positive invariant,* (10.2.1) *is satisfied and that F, G are continuous. Let C, D be twice differentiable functions satisfying* (10.1.3) *and* (10.1.5)*. Then the affine model* (10.1.1) *satisfies (MP) if and only if*

$$J\Big(G(x)D(v)\Big) = -C'(v) - [D'(v) - 1]x + D(v)F(x), \quad v \geq 0,\ x \geq 0. \tag{10.2.4}$$

If (10.2.4) *holds and J is differentiable in the interior of $\mathbb{R}^d_+$, then*

$$\Big\langle DJ\Big(G(x)D(v)\Big), D'(v)G(x)\Big\rangle = F(x)D'(v) - C''(v) - D''(v)x \tag{10.2.5}$$

for $v > 0$ and x such that $G(x)$ is in the interior of $\mathbb{R}^d_+$.
If (10.2.4) *holds and $D^+_w J(0)$ exists for any vector $w \in \mathbb{R}^d_+$, then*

$$F(x) = D^+_{G(x)} J(0) + C''(0) + D''(0)x, \quad x \geq 0. \tag{10.2.6}$$

Remark 10.2.2 Formula (10.2.5) is obtained by differentiating (10.2.4) over v outside of zero. The case $v = 0$ must be handled separately, because $DJ(0)$ does not exist for a general Lévy process and therefore the right directional derivative of J in (10.2.6) is introduced.

Proof of Theorem 10.2.1 We convert the model to the HJM framework. In view of Proposition 7.5.1 the forward rate satisfies

$$\begin{aligned} df(t,T) &= -C''(T-t)dt + D'(T-t)dR(t) - R(t)D''(T-t)dt \\ &= \alpha(t,T)dt + \langle \sigma(t,T), dZ(t) \rangle, \end{aligned}$$

where

$$\alpha(t,T) := F(R(t))D'(T-t) - C''(T-t) - D''(T-t)R(t), \tag{10.2.7}$$

$$\sigma(t,T) := D'(T-t)G(R(t-)). \tag{10.2.8}$$

It follows from Theorem 8.1.1 and Remark 8.1.3 that (MP) is satisfied if and only if

$$A(t,T) = J\big(\Sigma(t,T)\big). \tag{10.2.9}$$

By (10.2.7) and (10.2.8), for $t < T$,

$$\begin{aligned} A(t,T) &= \int_t^T \alpha(t,s)ds = \int_t^T \Big(F(R(t))D'(s-t) - C''(s-t) - D''(s-t)R(t)\Big)ds \\ &= F(R(t))[D(T-t)-D(0)] - [C'(T-t)-C'(0)] - [D'(T-t) - D'(0)]R(t), \end{aligned}$$

$$\Sigma(t,T) = \int_t^T \sigma(t,s)ds = \int_t^T D'(s-t)G(R(t-))ds = G(R(t-))[D(T-t) - D(0)].$$

Taking into account (10.1.3) and (10.1.5) we can write (10.2.9) in the form

$$J\Big(G(R(t-))D(T-t)\Big) = -C'(T-t) - [D'(T-t) - 1]R(t) + D(T-t)F(R(t)) \tag{10.2.10}$$

for each $T > 0$, $\mathbb{P}$-almost surely, for almost all $t \in [0,T]$. In (10.2.10) we can replace $R(t-)$ by $R(t)$ because $R(t) = R(t-)$ for almost all $t \geq 0$. Thus it is clear that (10.2.4) is sufficient for (10.2.10) to hold. To see that (10.2.4) is also necessary for (10.2.10) let us assume that for some $\bar{x} \geq 0$ and $\bar{v} > 0$,

$$J(G(\bar{x})D(\bar{v})) > -C'(\bar{v}) - [D'(\bar{v}) - 1]\bar{x} + D(\bar{v})F(\bar{x}).$$

Then, by the continuity of J, F, G and D , there exists $\delta > 0$ such that

$$J(G(x)D(v)) > -C'(v) - [D'(v) - 1]x + D(v)F(x)$$

for $x \in ((\bar{x} - \delta) \wedge 0, \bar{x} + \delta)$ and $v \in (\bar{v} - \delta, \bar{v} + \delta)$. Let us consider the solution R of (10.1.6) starting from $\bar{x}$ and define

$$\tau := \inf\{t \geq 0 : | R(t) - \bar{x} | > \delta\}.$$

For $t \in (0, \tau)$ and T such that $T - t \in (\bar{v} - \delta, \bar{v} + \delta)$, we have

$$J(G(R(t-))D(T-t)) > -C'(T-t) - [D'(T-t) - 1]R(t) + D(T-t)F(R(t)),$$

which is a contradiction.

Now we prove (10.2.6). From (10.2.4) we obtain

$$\frac{J\Big(G(x)D(v)\Big) - J(0)}{v} = \frac{-C'(v) - [D'(v) - 1]x + D(v)F(x)}{v}, \qquad v > 0,\ x \geq 0. \tag{10.2.11}$$

But

$$\frac{J\Big(G(x)D(v)\Big) - J(0)}{v} = \frac{J\Big(G(x)D(v)\Big) - J(0)}{D(v)} \cdot \frac{D(v)\Big)}{v} \underset{v\downarrow 0}{\longrightarrow} D^+_{G(x)}J(0),$$

so taking into account (10.1.3) and (10.1.5) and letting $v \downarrow 0$ in (10.2.11), we arrive at (10.2.6). □

Remark 10.2.3 Let us consider the short-rate equation

$$dR(t) = F(R(t))dt + \langle G(R(t-)), dZ(t)\rangle, \quad t \geq 0, \quad R(0) = x,$$

where Z is a martingale and assume that this short rate generates an affine model satisfying (MP). Then

(a) $DJ(\lambda)$ exists for any λ in the interior of $\mathbb{R}^d_+$ and (10.2.5) holds,
(b) the drift is a linear function, i.e.

$$F(x) = C''(0) + D''(0)x, \quad x \geq 0. \tag{10.2.12}$$

Point (*b*) follows from (10.2.6) in Theorem 10.2.1. In fact, if Z is a martingale then $D^+_{G(x)}J(0) = 0$. This and the differentiability of J in the interior of $\mathbb{R}^d_+$ are shown in the following proposition.

Proposition 10.2.4 *Let Z be a $\mathbb{R}^d$-valued Lévy martingale with characteristic triplet (a, Q, ν). If the Lévy measure ν is concentrated on $\mathbb{R}^d_+$ then*

(a) the right directional derivative along a vector w, at point zero, exists and

$$D^+_w J(0) = 0,$$

for any $w \in \mathbb{R}^d_+$,

(b) $DJ(\lambda)$ exists for any λ in the interior $Int(\mathbb{R}^d_+)$ of $\mathbb{R}^d_+$.

Proof The Laplace exponent of Z equals

$$J(\lambda) = \frac{1}{2}\langle Q\lambda, \lambda\rangle + \int_{y\geq 0}\Big(e^{-\langle\lambda,y\rangle} - 1 + \langle\lambda, y\rangle\Big)\,\nu(dy), \quad \lambda \in \mathbb{R}^d_+ \tag{10.2.13}$$

(see Proposition 5.3.4). Since $\frac{1}{2}\langle Q\lambda, \lambda\rangle$ is differentiable on $\mathbb{R}^d_+$ and disappears at zero, we examine the differentiability and existence of directional derivatives along $w \in \mathbb{R}^d_+$ at zero for the functions

$$J^0(\lambda) := \int_{y\geq 0, |y|\leq 1} \left(e^{-\langle\lambda,y\rangle} - 1 + \langle\lambda, y\rangle\right) \nu(dy), \quad \lambda \in \mathbb{R}^d_+,$$

$$J^1(\lambda) := \int_{y\geq 0, |y|>1} \left(e^{-\langle\lambda,y\rangle} - 1 + \langle\lambda, y\rangle\right) \nu(dy), \quad \lambda \in \mathbb{R}^d_+.$$

(*a*) For $h > 0$ we have

$$\begin{aligned} J^0(hw) &= \int_{y\geq 0, |y|\leq 1} \left(e^{-h\langle w,y\rangle} - 1 + h\langle w, y\rangle\right) \nu(dy) \\ &= h^2 \int_{y\geq 0, |y|\leq 1} \frac{\left(e^{-h\langle w,y\rangle} - 1 + h\langle w, y\rangle\right)}{h^2(\langle w, y\rangle)^2}(\langle w, y\rangle)^2 \nu(dy) \end{aligned} \tag{10.2.14}$$

and, by dominated convergence,

$$\int_{y\geq 0, |y|\leq 1} \frac{\left(e^{-h\langle w,y\rangle} - 1 + h\langle w, y\rangle\right)}{h^2(\langle w, y\rangle)^2}(\langle w, y\rangle)^2 \nu(dy) \underset{h\downarrow 0}{\longrightarrow} \frac{1}{2}\int_{y\geq 0, |y|\leq 1} (\langle w, y\rangle)^2 \nu(dy).$$

So, it follows from (10.2.14) that

$$D^+_w J^0(0) = \lim_{h\downarrow 0} \frac{J^0(hw)}{h} = 0.$$

Since Z is integrable, $\int_{|y|>1} \mid y \mid \nu(dy) < +\infty$, and therefore we can split $J^1(hw)$:

$$J^1(hw) = \int_{y\geq 0, |y|>1} \left(e^{-h\langle w,y\rangle} - 1\right) \nu(dy) + h\langle w, \int_{y\geq 0, |y|>1} y\nu(dy)\rangle, \quad h > 0.$$

It follows that

$$\begin{aligned} \frac{J^1(hw)}{h} &= \int_{y\geq 0, |y|>1} \frac{e^{-h\langle w,y\rangle} - 1}{h\langle w, y\rangle}\langle w, y\rangle \nu(dy) + \langle w, \int_{y\geq 0, |y|>1} y\nu(dy)\rangle \\ &\underset{h\downarrow 0}{\longrightarrow} -\int_{y\geq 0, |y|>1} \langle w, y\rangle \nu(dy) + \langle w, \int_{y\geq 0, |y|>1} y\nu(dy)\rangle = 0. \end{aligned}$$

So, $D^+_w J^1(0) = 0$ and the assertion (*a*) follows.

(*b*) The following formulae are true

$$DJ^0(\lambda) = \int_{y\geq 0, |y|\leq 1} y(1 - e^{-\langle\lambda,y\rangle})\nu(dy), \quad \lambda \in \mathrm{Int}(\mathbb{R}^d_+),$$

$$DJ^1(\lambda) = \int_{y\geq 0, |y|>1} y(1 - e^{-\langle\lambda,y\rangle})\nu(dy), \quad \lambda \in \mathrm{Int}(\mathbb{R}^d_+).$$

In fact

$$\int_{y\geq 0,|y|\leq 1} | y(1-e^{-\langle\lambda,y\rangle}) | \nu(dy) \leq \int_{y\geq 0,|y|\leq 1} | y\langle\lambda,y\rangle | \nu(dy)$$

$$\leq |\lambda| \int_{y\geq 0,|y|\leq 1} |y|^2 \nu(dy) < +\infty, \quad \lambda\in \mathrm{Int}(\mathbb{R}^d_+),$$

and

$$\int_{y\geq 0,|y|>1} | y(1-e^{-\langle\lambda,y\rangle}) | \nu(dy) \leq \int_{y\geq 0,|y|>1} | y | \nu(dy) < +\infty, \quad \lambda \in \mathrm{Int}(\mathbb{R}^d_+).$$

□

10.2.2 Generalized CIR Equations

In this section we characterize affine models satisfying (MP) with short rates of the form

$$dR(t) = F(R(t))dt + G(R(t-))dZ(t), \quad t \geq 0, \tag{10.2.15}$$

driven by a one-dimensional martingale Z. We present the results by formulating separately necessary conditions for equation (10.2.15) in Theorem 10.2.5 and sufficient conditions in Theorem 10.2.7, where also functions C, D from (10.1.1) are described. We consider two cases concerning the diffusion coefficient G, that is, either

$$\exists\, \bar{x} > 0 \quad \text{such that} \quad G(\bar{x}) > 0, \quad G'(\bar{x}) \neq 0, \tag{10.2.16}$$

or

$$G(x) \equiv \sigma \tag{10.2.17}$$

is a positive constant. In the first case, we obtain generalization of the well-known CIR equation while the second extends the Vasiček model on nonnegative short rates. The requirement (10.2.16) is very natural; however, to simplify the presentation we will assume in addition that

$$G(0) = 0. \tag{10.2.18}$$

The full proof can be found in Barski and Zabczyk [9], see also an earlier version [8]. Let us also notice that condition (10.2.17) is, in a sense, a complement of (10.2.16). Indeed, if there exists a number $\bar{x} > 0$ such that $G(\bar{x}) > 0$, and for all such numbers $G'(\bar{x}) = 0$, then the function G should be a positive constant on $[0, +\infty)$.

A particular role in the sequel will be played by the α-stable martingale Z^α, where $\alpha \in (1, 2]$, with positive jumps only. Recall, see Example 5.3.6, that Z^α with $\alpha \in (1, 2)$ has the form

$$Z^\alpha(t) = \int_0^t \int_0^{+\infty} y\, \tilde{\pi}(ds, dy), \quad t \geq 0,$$

and its Lévy measure is given by

$$\nu(dy) = \frac{1}{y^{1+\alpha}} \mathbf{1}_{[0,+\infty)}(y)dy, \quad \alpha \in (1,2).$$

Its Laplace exponent equals

$$J(z) = c_\alpha z^\alpha, \quad \alpha \in (1,2),$$

where $c_\alpha := \frac{1}{\alpha(\alpha-1)}\Gamma(2-\alpha)$. By Z^α with $\alpha = 2$ we denote the Wiener process.

Theorem 10.2.5 *[Necessity] Let us assume that the equation (10.2.15), with functions F, G which are continuous on $[0,+\infty)$, is positive invariant. Let C, D be twice differentiable functions satisfying* (10.1.3) *and* (10.1.5) *such that the corresponding affine model* (10.1.1) *has the (MP) property.*

(I) *If G is differentiable on $(0,+\infty)$ and $G(\bar{x}) > 0$, $G'(\bar{x}) \neq 0$ for some $\bar{x} > 0$, then*
  (a) *$Z = Z^\alpha$ is a stable Lévy process with index $\alpha \in (1,2]$ with positive jumps only,*
  (b) *$F(x) = ax + b$ with $a \in \mathbb{R}$, $b \geq 0$, $x \geq 0$,*
  (c) *$G(x) = c^{\frac{1}{\alpha}} x^{\frac{1}{\alpha}}$, $c > 0$, $x \geq 0$.*

(II) *If G is a positive constant σ, then*
  (d) *Z has no Wiener part, i.e. Q in* (10.2.3) *disappears,*
  (e) *the martingale Z has positive jumps only and $\int_0^{+\infty} y\nu(dy) < +\infty$,*
  (f) *$F(x) = ax + b$, $x \geq 0$, with $a \in \mathbb{R}$, $b \geq \sigma \int_0^{+\infty} y\nu(dy)$.*

Remark 10.2.6 Part (II) of Theorem 10.2.5 indicates that in the short-rate equation with constant volatility the process Z must be a martingale of finite variation.

Proof of Part (I) of Theorem 10.2.5 We present the proof under additional assumptions that Z has positive jumps only, $G(x) \geq 0, x \geq 0$ and $G(0) = 0$. Let us recall that positive jumps of Z imply that $J(z)$ is well defined for $z \geq 0$ and $J'(z)$ is well defined for $z > 0$. The proof is divided into four steps.

Let us assume that (MP) is satisfied.

Step 1: We prove the linear form of F.

Since Z is a martingale, we have that $J'(0+) = 0$ and, by Remark 10.2.3, we conclude that

$$F(x) = C''(0) + D''(0)x := b + ax, \quad x \geq 0. \tag{10.2.19}$$

To show that $b \geq 0$, assume, by contradiction, that $b < 0$ and consider a solution R of (10.2.15) starting from 0. Since R is nonnegative, we have:

$$R(t) = b\int_0^t e^{a(t-s)}ds + \int_0^t e^{a(t-s)}G(R(s-))dZ(s) \geq 0, \; t \geq 0.$$

Hence

$$-|b|\int_0^t e^{-as}ds+\int_0^t e^{-as}G(R(s-))dZ(s)\geq 0,\quad t\geq 0,$$

or equivalently

$$\int_0^t e^{-as}G(R(s-))dZ(s)\geq |b|\int_0^t e^{-as}ds\quad t\geq 0.$$

Since the preceding stochastic integral is a local nonnegative martingale starting from 0, it must be identically 0. Thus the process

$$R(t)=b\int_0^t e^{a(t-s)}ds,\ t\in[0,T]$$

is strictly negative and we have a contradiction.

Step 2: We prove that $G(x)=\rho x^{\frac{1}{\alpha}}, x\geq 0$ with some constants α,ρ.
Formula (10.2.19) and (10.2.4) yield

$$\begin{aligned}J(G(x)D(v))&=-C'(v)-[D'(v)-1]x+D(v)[b+ax]\\&=bD(v)-C'(v)+[aD(v)-D'(v)+1]x,\quad x,v\geq 0.\end{aligned}\tag{10.2.20}$$

Since $G(0)=0$ and $J(0)=0$, $bD(v)$ and (10.2.20) can be written in the form

$$J(D(v)G(x))=[aD(v)-D'(v)+1]x,\quad x,v\geq 0.\tag{10.2.21}$$

Since Z is a martingale, $J'(\lambda)$ exists for $\lambda>0$ (see Proposition 10.2.4). Differentiation of (10.2.21) over x and v yields

$$J'\Big(D(v)G(x)\Big)G'(x)D(v)=aD(v)-D'(v)+1,\quad x>0,\ v>0,\tag{10.2.22}$$

$$J'\Big(D(v)G(x)\Big)G(x)D'(v)=[aD'(v)-D''(v)]x,\quad x>0,\ v>0.\tag{10.2.23}$$

Since D is continuously differentiable and $D'(0)=1$, we can find $\varepsilon>0$ such that

$$D(v)>0,\quad D'(v)>0,\quad v\in(0,\varepsilon).$$

Let us assume that the right side of (10.2.22) is zero for $v\in(0,\varepsilon)$. Then D solves

$$D'(v)=aD(v)+1,\quad D(0)=0,\quad v\in(0,\varepsilon),$$

and, on the interval $(0,\varepsilon)$, $D(v)=\frac{1}{a}(e^{av}-1)$ if $a\neq 0$ or $D(v)=v$ if $a=0$. Since the left side of (10.2.22) equals zero and $D(v)>0$ for $v\in(0,\varepsilon)$ and $G'(\bar{x})\neq 0$, we obtain

$$J'(G(\bar{x})D(v))=0,\quad v\in(0,\varepsilon).$$

Hence J' disappears on some interval and consequently must disappear on $[0,+\infty)$ as an analytic function. Since $J(0)=0$, this implies that $J(\lambda)=0$ for $\lambda\in[0,+\infty)$, which is impossible. It follows thus that the right side of (10.2.22) is different from zero for some $\bar{v} \in (0, \varepsilon)$. This implies that

$$D(\bar{v}) \neq 0, \quad G'(x) \neq 0, \quad J'(G(x)D(\bar{v})) \neq 0, \quad x > 0.$$

Hence, by (10.2.22),

$$J'(G(x)D(\bar{v})) = \frac{D(\bar{v})a + 1 - D'(\bar{v})}{G'(x)D(\bar{v})}, \quad x > 0.$$

Putting this into (10.2.23) with $v = \bar{v}$ yields

$$\frac{D(\bar{v})a + 1 - D'(\bar{v})}{G'(x)D(\bar{v})} \cdot G(x)D'(\bar{v}) = x[D'(\bar{v})a - D''(\bar{v})], \quad x > 0,$$

and, consequently,

$$\frac{G(x)}{G'(x)} = \frac{x[D'(\bar{v})a - D''(\bar{v})]}{D(\bar{v})a + 1 - D'(\bar{v})} \cdot \frac{D(\bar{v})}{D'(\bar{v})}, \quad x > 0. \tag{10.2.24}$$

If $D'(\bar{v})a - D''(\bar{v}) = 0$ then $G(x) = 0, x > 0$, which is impossible. Hence it follows that

$$\frac{G'(x)}{G(x)} = \frac{\bar{A}}{x}, \quad x > 0,$$

with $\bar{A} := [D(\bar{v})a + 1 - D'(\bar{v})]D'(\bar{v})/[D'(\bar{v})a - D''(\bar{v})]D(\bar{v})$. The last identity can be written in the form

$$\frac{d}{dx}\ln(G(x)) = \bar{A}\frac{d}{dx}\ln x, \quad x > 0$$

and yields

$$\ln G(x) - \ln x^{\bar{A}} = k, \quad x > 0$$

with some constant k. Using this and the continuity of G at zero we obtain

$$G(x) = e^k x^{\bar{A}}, \quad x \geq 0.$$

Relabelling the constants yields the assertion.

Step 3: We prove that the Laplace exponent of Z is of the form $J(z) = \beta z^\alpha, z \geq 0$ with some constant β and α from the previous step.

Putting $v = \bar{v}$ from the previous step into (10.2.21) yields

$$J(D(\bar{v})G(x)) = [aD(\bar{v}) - D'(\bar{v}) + 1]x, \quad x \geq 0.$$

Setting $z := D(\bar{v})G(x)$ yields $x = G^{-1}(\frac{z}{D(\bar{v})})$. To find $G^{-1}(u)$, we have to solve the equation

$$G(x) = u,$$

which is equivalent to

$$\rho x^{\frac{1}{\alpha}} = u.$$

Thus we obtain

$$G^{-1}(u) = x = \left(\frac{u}{\rho}\right)^{\alpha},$$

and consequently

$$\begin{aligned} J(z) &= [aD(\bar{v}) - D'(\bar{v}) + 1] \cdot G^{-1}\left(\frac{z}{D(\bar{v})}\right) \\ &= \frac{[aD(\bar{v}) - D'(\bar{v}) + 1]}{(D(\bar{v})\rho)^{\alpha}} z^{\alpha}. \end{aligned}$$

Finally we obtain $J(z) = \beta z^{\alpha}$, $z \geq 0$ with $\beta := \frac{[aD(\bar{v}) - D'(\bar{v}) + 1]}{(D(\bar{v})\rho)^{\alpha}}$.

Step 4: We show that the constant α from the previous steps belongs to $(1, 2]$. This means that Z is a stable Lévy process.

Since $J'(0+) = 0$, we have that $\gamma > 1$. We show that the case $\gamma > 2$ can be excluded. Indeed, by Proposition 5.3.4, the Laplace exponent of a Lévy martingale has the form

$$\begin{aligned} J(z) &= \frac{1}{2}qz^2 + \int_0^{+\infty} \left(e^{-zy} - 1 + zy\right) \nu(dy) \\ &= \frac{1}{2}qz^2 + \int_0^{1} \left(e^{-zy} - 1 + zy\right) \nu(dy) + \int_1^{+\infty} \left(e^{-zy} - 1 + zy\right) \nu(dy) \\ &= \frac{1}{2}qz^2 + z^2 \int_0^{1} \frac{e^{-zy} - 1 + zy}{(zy)^2} y^2 \nu(dy) + \int_1^{+\infty} \left(e^{-zy} - 1 + zy\right) \nu(dy), \quad z \geq 0. \end{aligned}$$

Since the function

$$x \to \frac{e^{-x} - 1 + x}{x^2}, \quad x \geq 0$$

is bounded, the measure $y^2 \nu(dy)$ is finite on $[0, 1]$ and

$$\int_1^{+\infty} \left(e^{-zy} - 1 + zy\right) \nu(dy) \leq z \int_1^{+\infty} y \nu(dy)$$

we see that $J(z) \leq cz^2 + d$ for some positive constants c, d. ❑

Proof of Part (II) of Theorem 10.2.5 By elementary arguments, positivity of solutions to the equation (10.2.15) with $G(x) \equiv \sigma$ implies that Z has no Wiener part and can have only positive jumps. Repeating the arguments from the proof of Part (I) one can show that that $F(x) = ax + b, x \geq 0$, and $b \geq 0$. We will establish now that

$$\int_0^{+\infty} y\nu(dy) < +\infty, \quad b \geq \sigma \int_0^{+\infty} y\nu(dy). \tag{10.2.25}$$

Let $\tilde{\pi}$ be the compensated jump measure corresponding to the martingale Z. Then for $\epsilon > 0$,

$$Z(t) = \int_0^t \int_0^{+\infty} y\tilde{\pi}(ds, dy) = \int_0^t \int_0^{\epsilon} y\tilde{\pi}(ds, dy) + \int_0^t \int_{\epsilon}^{+\infty} y\tilde{\pi}(ds, dy) \tag{10.2.26}$$

$$= Z_\epsilon(t) + P_\epsilon(t) - t\int_\epsilon^{+\infty} y\nu(dy). \tag{10.2.27}$$

Here Z_ϵ is a Lévy martingale with positive jumps bounded by ϵ, P_ϵ is a compound Poisson process with the Lévy measure ν restricted to the interval $[\epsilon, +\infty)$. For the solution R of the stochastic equation, starting from 0, we have for all $t \in [0, T]$:

$$\begin{aligned} e^{-at}R(t) &= b\int_0^t e^{-as}ds + \sigma\int_0^t e^{-as}dZ(s) \\ &= (b - \sigma\int_\epsilon^{+\infty} y\nu(dy))\int_0^t e^{-as}ds + \sigma\int_0^t e^{-as}dZ_\epsilon(s) + \int_0^t e^{-as}dP_\epsilon(s) \geq 0. \end{aligned} \tag{10.2.28}$$

If $\int_0^{+\infty} y\nu(dy) = +\infty$, then by taking ϵ close to 0, the number $(b - \sigma\int_\epsilon^{+\infty} y\nu(dy))$ can be made arbitrary small negative. The stochastic integrals with respect to Z_ϵ, P_ϵ are independent processes. The integral with respect to Z_ϵ is less on $[0, T]$ than a given number with positive probability. The integral with respect to P_ϵ is 0 on $[0, T]$ with positive probability. Thus $e^{-at}R(t)$ is negative for some $t \in [0, T]$, with positive probability, which is a contradiction. Now we show that $b \geq \sigma\int_0^{+\infty} y\nu(dy)$. In the opposite case we have that the difference

$$b - \sigma\int_\epsilon^{+\infty} y\nu(dy)$$

is negative for sufficiently small $\epsilon > 0$ and decreases as $\epsilon \downarrow 0$. It follows from the Chebyshev inequality that for any $\gamma > 0$ and $t > 0$

$$\mathbb{P}\Big(\sigma\int_0^t e^{-as}dZ_\epsilon(s) > \gamma\Big) \leq \frac{\sigma^2\mathbb{E}(\int_0^t e^{-as}dZ_\epsilon(s))^2}{\gamma^2} = \frac{\sigma^2\int_0^t\int_0^\epsilon e^{-2as}y^2 ds\nu(dy)}{\gamma^2} \underset{\epsilon\to 0}{\to} 0,$$

and consequently

$$\mathbb{P}(\sigma\int_0^t e^{-as}dZ_\epsilon(s) \leq \gamma) \underset{\epsilon\to 0}{\to} 1.$$

Since the integral over P_ε disappears with positive probability, we have by (10.2.28) that $R(t) < 0$, which is a contradiction. □

Theorem 10.2.7 [*Sufficiency*]

(I) *The equation*

$$dR(t) = (aR(t) + b)dt + c^{\frac{1}{\alpha}} R(t-)^{\frac{1}{\alpha}} dZ^{\alpha}(t), \quad R(0) = x \geq 0, \quad a \in \mathbb{R}, b \geq 0, c > 0 \tag{10.2.29}$$

with $\alpha \in (1,2]$ *is positive invariant. There exist functions* C, D *which together with the solution of* (10.2.29) *form an affine model satisfying (MP). Moreover,* D *solves the equation*

$$D'(v) = -cc_{\alpha} D^{\alpha}(v) + aD(v) + 1, \quad v \geq 0, \quad D(0) = 0 \tag{10.2.30}$$

and C *is given by* $C'(v) = bD(v), v \geq 0,\ C(0) = 0.$

(II) *If* G *is a positive constant* σ *and (d), (e), (f) in Theorem 10.2.5, hold, then the equation*

$$dR(t) = (aR(t) + b) + \sigma dZ(t), \;\; R(0) = x \geq 0, \;\; t > 0 \tag{10.2.31}$$

is positive invariant. The solution of (10.2.31) *and functions* C, D *given by*

$$D'(v) = D(v)a + 1, \;\; D(0) = 0, \tag{10.2.32}$$

$$C'(v) = D(v)\Big(b - \sigma \int_0^{+\infty} y\nu(dy)\Big) + \int_0^{+\infty} (1 - e^{-\sigma D(v) y})\nu(dy), \;\; C(0) = 0 \tag{10.2.33}$$

form an affine model satisfying (MP).

Remark 10.2.8 Equation (10.2.29) defines the generalized CIR model of the short rate. For $\alpha = 2$ one obtains the original CIR model introduced in Cox, Ingersoll and Ross [30]. In this case equation (10.2.30) becomes the Riccati equation, which can be solved explicitly. Properties of solutions of equation (10.2.29) have been an object of many investigations (see Kyprianou and Pardo [87], Kawazu and Watanabe [81], Ikeda and Watanabe [72], Li, Mytnik [88] and Jeanblanc, Yor and Chesney [78, p. 357–365] for a discussion of various topics and derivation of formulas of financial interests). Equation (10.2.31) generalizes the well-known Vasiček model based on the Wiener process (see Vasiček [121]). Passing from the Wiener process to the general Lévy martingale of finite variation in (10.2.31) enables preserving the positivity of the short-rate process that fails in the original Vasiček model.

Example 10.2.9 Requiring $C''(0) = 0$ and $D''(0) = 0$ one arrives at the simplified CIR model

$$dR(t) = \sqrt{2c}\sqrt{R(t)}dW(t), \quad t \geq 0.$$

Then

$$\begin{cases} C'(v) = 0, & v \geq 0, \\ D'(v) + cD^2(v) - 1 = 0, & v \geq 0, \end{cases}$$

with initial conditions $C(0) = 0, C'(0) = 0, D(0) = 0, D'(0) = 1$ and some $c > 0$. The solutions C, D are given by $C(v) \equiv 0$ and

$$D(v) = \frac{1}{\sqrt{c}} \frac{e^{2\sqrt{cv}} - 1}{e^{2\sqrt{cv}} + 1}, \quad v \geq 0.$$

Proof of Theorem 10.2.7 (I) It was shown in Fu and Li [60] that equation (10.2.29) has a unique nonnegative strong solution. Now we use Theorem 10.2.1 with $J(\lambda) = c_\alpha \lambda^\alpha$, $F(x) = ax + b$ and $G(x) = c^{\frac{1}{\alpha}} x^{\frac{1}{\alpha}}$. Then (10.2.4) becomes

$$c_\alpha \left(c^{\frac{1}{\alpha}} x^{\frac{1}{\alpha}} D(v) \right)^\alpha = -C'(v) - [D'(v) - 1]x + D(v)[ax + b], \quad x \geq 0, \quad v \geq 0.$$

Consequently,

$$c_\alpha cxD^\alpha(v) = (aD(v) - D'(v) + 1)x + bD(v) - C'(v), \quad x \geq 0, \quad v \geq 0. \quad (10.2.34)$$

Putting $x = 0$ yields

$$bD(v) - C'(v) = 0, \quad v \geq 0,$$

which is the required formula for C. It follows from (10.2.34) that

$$c_\alpha cD^\alpha(v) = aD(v) - D'(v) + 1, \quad v \geq 0,$$

which yields the equation for D.

(II) Note that functions C, D should satisfy, for all $x \geq 0$, $v \geq 0$, the equation

$$\begin{aligned} J(\sigma D(v)) &= -C'(v) - (D'(v) - 1)x + D(v)(ax + b) \\ &= x(D(v)a - D'(v) + 1) + D(v)b - C'(v). \end{aligned}$$

Consequently,

$$\begin{aligned} D'(v) &= D(v)a + 1, \quad D(0) = 0, \\ C'(v) &= D(v)b - J(\sigma D(v)), \quad C(0) = 0. \end{aligned}$$

However,

$$\begin{aligned} &D(v)b - J(\sigma D(v)) \\ &= D(v)\Big(b - \sigma \int_0^{+\infty} y\nu(dy)\Big) + \int_0^{+\infty} (1 - e^{-\sigma D(v)y})\nu(dy), \end{aligned}$$

and the proof is complete. □

Remark 10.2.10 Both models described in Parts (I) and (II) of Theorem 10.2.7 are regular because the short rate R is positive and the functions C and D turn out to be positive and increasing. Moreover, D can be shown to be bounded. This follows by an elementary analysis of the differential equations for C and D.

10.2.3 Exploding Short Rates

In this section we present an affine model satisfying (MP) with exploding short rates. In particular, after some time, bonds might be of value 0. If, for some $t_0 > 0$,

$$R(t) \underset{t \to t_0}{\longrightarrow} +\infty,$$

then the discounted bond prices with maturities $T > t_0$ satisfy

$$\hat{P}(t,T) = e^{-C(T-t) - D(T-t)R(t) - \int_0^t R(s)ds} \underset{t \to t_0}{\longrightarrow} 0.$$

As nonnegative local martingales reaching zero, $\hat{P}(t,T)$ are zero for $t \geq t_0$. Consequently, $P(t,T) = 0, t \geq t_0$ for $T > t_0$.

We focus on the model of the form

$$dR(t) = (aR(t) + b)dt + (cR(t-) + d)^{\frac{1}{\beta}} dZ(t), \quad R(0) = x \geq 0, \quad t \geq 0, \tag{10.2.35}$$

where Z is a stable subordinator with index $\beta \in (0,1)$ and a,b,c,d are nonnegative constants. Recall that the Lévy measure of Z is given by $\nu(dy) = \frac{1}{y^{1+\beta}} \mathbf{1}_{(0,+\infty)}(y)$ and its Laplace exponent equals

$$J(z) = -c_\beta z^\beta, \quad z > 0 \tag{10.2.36}$$

with $c_\beta := \frac{\Gamma(1-\beta)}{\beta}$, see Example 5.3.3. Since $J'(0+)$ is infinite in this case, the analysis differs from the case in which Z was a martingale (see Remark 10.2.3) and this is why we assume a specific form of drift and volatility in (10.2.35). Since both drift and volatility in (10.2.35) are locally Lipschitz functions, local solutions exist. As we will see, in this case, the affine models satisfying (MP) are regular but the short-rate process explodes in finite time.

Proposition 10.2.11 *Let the short rate be given by* (10.2.35).

(a) Then R generates an affine model satisfying (MP) and the corresponding functions C, D are given by

$$\begin{cases} C'(v) = D(v)b + c_\beta d D(v)^\beta, & C(0) = 0, \quad v \geq 0, \\ D'(v) = c_\beta c D(v)^\beta + aD(v) + 1, & D(0) = 0, \quad v \geq 0. \end{cases}$$

(b) The short rate R explodes in finite time with positive probability.

Remark 10.2.12 By elementary analysis of the differential equations for C and D one gets that the resulting model is regular.

Proof of Proposition 10.2.11 (*a*) Condition (10.2.4) takes the form

$$J((cx+d)^{\frac{1}{\beta}}D(v)) = -C'(v) - [D'(v)-1]x + D(v)(ax+b), \quad x, v \geq 0.$$

By (10.2.36):

$$-c_\beta(cx+d)D(v)^\beta = -C'(v) - (D'(v)-1)x + D(v)(ax+b), \quad x, v \geq 0, \tag{10.2.37}$$

or equivalently,

$$(c_\beta c D(v)^\beta + aD(v) - D'(v) + 1)x - C'(v) + D(v)b + c_\beta d D(v)^\beta = 0, \quad x, v \geq 0.$$

As a consequence we obtain differential equations for C and D.

(*b*) By a comparison argument for equation (10.2.35) driven by Z without small jumps (see e.g. the proof of Theorem 13.1.1) one can show that the solution of the equation (10.2.35), starting from $x \geq 0$, dominates the solution of the equation

$$dR^x(t) = (cR^x(t-))^{\frac{1}{\beta}}dZ(t), \quad t \geq 0, \; R^x(0) = x. \tag{10.2.38}$$

It is therefore sufficient to show explosion for the latter equation. In the proof of the following Lemma 10.2.13 we follow Ikeda and Watanabe [72].

□

Lemma 10.2.13 *Assume that R^x is a solution of the equation (10.2.38) where Z is a stable subordinator with index $\beta \in (0,1)$ and $c > 0$. Then, for $f_\lambda(z) := e^{-\lambda z}$,*

$$\mathcal{P}_t f_\lambda(x) := \mathbb{E}\Big(e^{-\lambda R^x(t)}; R^x(t) < +\infty\Big) = e^{-x(\lambda^{1-\beta}+tcc_\beta(1-\beta))^{\frac{1}{1-\beta}}}, \tag{10.2.39}$$

where $c_\beta = \frac{1}{\beta}\Gamma(1-\beta)$. In particular,

$$\mathbb{P}(R^x(t) < +\infty) = \lim_{\lambda \downarrow 0} \mathcal{P}_t f_\lambda(x) = e^{-x(tcc_\beta(1-\beta))^{\frac{1}{1-\beta}}} < 1.$$

Proof The generator $\mathcal{L}$ of Z is of the form

$$\mathcal{L}f(x) = \int_0^{+\infty} \Big(f(x+y) - f(x)\Big)\frac{1}{y^{1+\beta}}dy,$$

and therefore the generator $\mathcal{A}$ of the process R^x is

$$\begin{aligned}
\mathcal{A}f(x) &= \int_0^{+\infty} \Big(f(x+(cx)^{\frac{1}{\beta}}y) - f(x)\Big)\frac{1}{y^{1+\beta}}dy \\
&= \int_0^{+\infty} \Big(f(x+z) - f(x)\Big)\frac{1}{(\frac{z}{c^{1/\beta}x^{1/\beta}})^{1+\beta}}\frac{1}{c^{\frac{1}{\beta}}x^{\frac{1}{\beta}}}dz \\
&= cx\int_0^{+\infty} \Big(f(x+z) - f(x)\Big)\frac{1}{z^{1+\beta}}dz
\end{aligned}$$

(see Section B.1.1). Fix $x \geq 0$ and let us denote

$$h(t, \lambda) := \mathcal{P}_t f_\lambda(x) = \mathbb{E}(e^{-\lambda R^x(t)}) = \mathbb{E}(e^{-\lambda R^x(t)}; R^x(t) < +\infty).$$

Then

$$\frac{\partial h}{\partial t}(t, \lambda) = \mathcal{A}\mathcal{P}_t f_\lambda(x) = \mathcal{P}_t \mathcal{A} f_\lambda(x). \tag{10.2.40}$$

Since

$$\begin{aligned} \mathcal{A} f_\lambda(x) &= x \int_0^{+\infty} \left(e^{-\lambda(x+z)} - e^{-\lambda x}\right) \frac{1}{z^{1+\beta}} dz = c x e^{-\lambda x} \int_0^{+\infty} \left(e^{-\lambda z} - 1\right) \frac{1}{z^{1+\beta}} dz \\ &= -c_\beta \lambda^\beta c x e^{-\lambda x}, \end{aligned}$$

so

$$\mathcal{P}_t \mathcal{A} f_\lambda(x) = -c c_\beta \lambda^\beta \, \mathbb{E}(R^x(t) e^{-\lambda R^x(t)}) = c c_\beta \lambda^\beta \, \frac{\partial}{\partial \lambda} \mathbb{E}(e^{-\lambda R^x(t)}). \tag{10.2.41}$$

Therefore, by (10.2.40) and (10.2.41),

$$\frac{\partial}{\partial t} h(t, \lambda) = c c_\beta \lambda^\beta \, \frac{\partial}{\partial \lambda} h(t, \lambda) \tag{10.2.42}$$

$$h(0, \lambda) = e^{-\lambda x}. \tag{10.2.43}$$

The solution of the preceding equation is

$$h(t, \lambda) = e^{-x(\lambda^{1-\beta} + t c c_\beta (1-\beta))^{\frac{1}{1-\beta}}}. \tag{10.2.44}$$

In fact, for h given by (10.2.44) we have

$$\begin{aligned} \frac{\partial}{\partial t} h(t, \lambda) &= h(t, \lambda) \Big[- x \Big(\lambda^{1-\beta} + t c c_\beta (1-\beta)\Big)^{\frac{1}{1-\beta} - 1} c c_\beta \Big], \\ \frac{\partial}{\partial \lambda} h(t, \lambda) &= h(t, \lambda) \Big[- x \Big(\lambda^{1-\beta} + t c c_\beta (1-\beta)\Big)^{\frac{1}{1-\beta} - 1} \lambda^{-\beta} \Big], \end{aligned}$$

which shows that h solves (10.2.42). This way, formula (10.2.39) was established. The final part follows from (10.2.39). □

10.2.4 Multidimensional Noise

We pass now to affine models with multidimensional noise. We restrict our attention to a four-dimensional Lévy processes $Z := (W, Z_1, Z_2, L)$ where W is a Wiener process, Z_1 is a stable martingale with index $\gamma_1 \in (1, 2)$, Z_2 is a stable subordinator with index $\gamma_2 \in (0, 1)$ and L is some subordinator. The noise processes are assumed to be independent. Arbitrary dimensions can be treated similarly. We formulate conditions for the short-rate dynamics

$$dR(t)=F(R(t))dt+G_0(R(t))dW(t)+G_1(R(t-))dZ_1(t)+G_2(R(t-))dZ_2(t)+dL(t), \tag{10.2.45}$$

to generate an affine model satisfying (MP). The independence of (W, Z_1, Z_2, L) implies that the Laplace exponent of Z has the form

$$J(z_0, z_1, z_2, z_3) = J_0(z_0) + J_1(z_1) + J_2(z_2) + J_3(z_3),$$

where

$$J_0(z_0) = \frac{1}{2}z_0^2, \quad J_1(z_1) = \int_0^{+\infty}(e^{-z_1 y} - 1 + z_1 y)\nu_1(dy) = c_1 z_1^{\gamma_1},$$

$$J_2(z_2) = \int_0^{+\infty}(e^{-z_2 y} - 1)\nu_2(dy) = c_2 z_2^{\gamma_2}, \quad J_3(z_3) = \int_0^{+\infty}(e^{-z_3 y} - 1)m(dy).$$

where c_1, c_2 are some constants. Moreover,

$$\nu_1(dy)=\frac{1}{y^{1+\gamma_1}}\mathbf{1}_{[0,+\infty)}(y)dy,\ \nu_2(dy)=\frac{1}{y^{1+\gamma_2}}\mathbf{1}_{[0,+\infty)}(y)dy,\ \int_0^{+\infty}(y\wedge 1)m(dy)<+\infty.$$

Our aim is to discuss the solvability of the equation (10.2.4), which is equivalent to (MP), i.e.

$$J\Big(G(x)D(v)\Big) = -C'(v) - [D'(v) - 1]x + D(v)F(x), \quad v \geq 0, \quad x \geq 0. \tag{10.2.46}$$

To simplify considerations we assume additionally that the drift F is linear.

Theorem 10.2.14 *Assume that $F(x) = ax + b$, $x \geq 0$ with $a \in \mathbb{R}, b \geq 0$, $G(x) := (G_0(x), G_1(x), G_2(x), 1)$ has nonnegative coordinates and that the equation* (10.2.45) *is positive invariant. Let $C(\cdot), D(\cdot)$ be twice differentiable functions on $[0, +\infty)$. If R generates an affine model satisfying (MP) then*

$$G_0(x) = \sqrt{2}\sqrt{e_0 x + f_0}, \quad G_1(x) = (e_1 x + f_1)^{\frac{1}{\gamma_1}}, \quad G_2(x) = (e_2 x + f_2)^{\frac{1}{\gamma_2}} \quad x \geq 0, \tag{10.2.47}$$

with some nonnegative constants $e_0, e_1, e_2, f_0, f_1, f_2$. Moreover,

$$D'(v) = 1 + D(v)a - D^2(v)e_0 - c_1 e_1 D^{\gamma_1}(v) - c_2 e_2 D^{\gamma_2}(v), \quad D(0) = 0, \tag{10.2.48}$$

$$C'(v) = -J_3(D(v)) + D(v)b - D^2(v)f_0 - c_1 f_1 D^{\gamma_1}(v) - c_2 f_2 D^{\gamma_2}(v), \quad C(0) = 0. \tag{10.2.49}$$

Conversely, if G is of the form (10.2.47) *and $f_0 = f_1 = f_2 = 0$ then equation* (10.2.45) *generates an affine model satisfying (MP).*

Proof It follows from (10.2.46) that the function

$$J_0\Big(D(v)G_0(x)\Big) + J_1\Big(D(v)G_1(x)\Big) + J_2\Big(D(v)G_2(x)\Big) + J_3\Big(D(v)\Big), \quad x, v \geq 0,$$

is linear over x, hence its second derivative disappears. Thus we obtain

$$D^2(v)\Big(\frac{1}{2}G_0^2(x)\Big)'' + c_1 D^{\gamma_1}(v)\Big(G_1^{\gamma_1}(x)\Big)'' + c_2 D^{\gamma_2}(v)\Big(G_2^{\gamma_2}(x)\Big)'' = 0 \quad x, v \geq 0.$$

Denoting $z := D(v)$ we write the preceding relation in the form

$$z^2\Big(\frac{1}{2}G_0^2(x)\Big)'' + c_1 z^{\gamma_1}\Big(G_1^{\gamma_1}(x)\Big)'' + c_2 z^{\gamma_2}\Big(G_2^{\gamma_2}(x)\Big)'' = 0, \quad x \geq 0, \qquad (10.2.50)$$

true for all z from an open interval.

Since functions $z^2, z^{\gamma_1}, z^{\gamma_2}$ are linearly independent, the coefficients in (10.2.50), should be functions of x identically equal zero. Hence

$$\frac{1}{2}G_0^2(x) = e_0 x + f_0, \quad G_1^{\gamma_1}(x) = e_1 x + f_1, \quad G_2^{\gamma_2}(x) = e_2 x + f_2,$$

which yields (10.2.47).

Plugging the obtained formulas for coordinates of G into (10.2.46) and comparing coefficients of both sides we obtain (10.2.48) and (10.2.49). □

Remark 10.2.15 The equation for D has always a positive increasing solution. For C to be positive and increasing it is enough that $f_0 = f_1 = f_2 = 0$. Then the resulting bond market is regular. It is also clear that if $f_i,\ i = 0, 1, 2$ are large then C might be decreasing for large times. We leave aside the specification of those $f_i,\ i = 0, 1, 2$ for which the model is regular.

10.3 General Markovian Short Rate

It is of interest to characterize general Markov processes R and functions C, D such that the affine model (10.1.1) satisfies (MP) condition for all initial values $R(0) = x \geq 0$. This problem was solved by Filipović [53] in the class of nonnegative Markov Feller processes on $[0, +\infty)$ with càdlàg trajectories. We refer also to a seminal paper by Duffie, Filipović and Schachermeyer [42] where a more general problem is treated.

The short-rate process starting from $x \geq 0$ will be often denoted by $R^x(t)$, $t \geq 0$ and $(\mathcal{P}_t)$ stands for its transition semigroup, i.e.

$$\mathcal{P}_t\varphi(x) := \mathbb{E}(\varphi(R^x(t))), \ x \geq 0, \ t \geq 0.$$

10.3.1 Filipović's Theorems

The following theorems are due to Filipović [53], who derived them using the results of Kawazu and Watanabe [81] on branching processes.

Characterization of positive short-rate processes R^x, which generate affine models satisfying (MP), is formulated in terms of the generator of its transition semigroup $(\mathcal{P}_t)$ (see Section B.1). The semigroup is regarded on the space $C_0([0,+\infty))$ of continuous functions on $[0,+\infty)$, vanishing at $+\infty$. In the following theorem Λ is the linear hull of the set of all functions f_λ, $\lambda > 0$, where

$$f_\lambda(x) := e^{-\lambda x}, \ x \geq 0.$$

Theorem 10.3.1 *Assume that $\mathcal{A}$ is the infinitesimal operator of the transition semigroup $(\mathcal{P}_t)$ on $C_0([0,+\infty))$ corresponding to a Feller process R that generates an affine term structure satisfying (MP). Then Λ is a core of the operator $\mathcal{A}$ and for $f \in \Lambda$*

$$\mathcal{A}f(x) = \alpha x f''(x) + (\beta x + b)f'(x) + \int_0^{+\infty} (f(x+y) - f(x))m(dy)$$
$$+ x\int_0^{+\infty} \big(f(x+y) - f(x) - \mathbf{1}_{[0,1]}(y)f'(x)y\big)\,\mu(dy), \qquad (10.3.1)$$

where μ and m are measures on $(0,+\infty)$ such that

$$\int_0^{+\infty} (1\wedge y)\,m(dy) + \int_0^{+\infty} (1\wedge y^2)\,\mu(dy) < +\infty \qquad (10.3.2)$$

and

$$\alpha, b \in \mathbb{R}_+,\ \beta \in \mathbb{R}. \qquad (10.3.3)$$

In addition, either

$$\int_1^{+\infty} y\mu(dy) < +\infty, \qquad (10.3.4)$$

or

$$\int_1^{+\infty} y\mu(dy) = +\infty \quad \text{and} \quad \int_0^{\epsilon} \frac{1}{\mathcal{R}_0(y)}dy = +\infty \qquad (10.3.5)$$

for small $\epsilon > 0$, where

$$\mathcal{R}_0(\lambda) = -\alpha\lambda^2 + \beta\lambda + \int_0^{+\infty} (1 - e^{-\lambda y} - \lambda y \mathbf{1}_{[0,1]}(y))\,\mu(dy), \qquad \lambda \geq 0. \quad (10.3.6)$$

Condition (10.3.5) implies that the Markov process R does not explode.

The preceding theorem has the following converse, which in addition describes the form of the functions C, D.

Theorem 10.3.2 *If the infinitesimal operator of the transition semigroup $(\mathcal{P}_t)$ of the process R is of the form* (10.3.1) *with parameters satisfying* (10.3.2), (10.3.3) *and* (10.3.4) *or* (10.3.5) *then the process R generates an affine term structure satisfying (MP) with functions C, D, determined by the following formulas:*

$$D'(t) = \mathcal{R}(D(t)), \quad D(0) = 0, \quad C(t) = \int_0^t \mathcal{F}(D(s))ds, \quad t \geq 0, \qquad (10.3.7)$$

where

$$\mathcal{R}(\lambda) := -\alpha\lambda^2 + \beta\lambda + 1 + \int_0^{+\infty} (1 - e^{-\lambda y} - \lambda y \mathbf{1}_{[0,1]}(y))\, \mu(dy) \qquad (10.3.8)$$

$$= \mathcal{R}_0(\lambda) + 1, \quad \lambda \geq 0,$$

$$\mathcal{F}(\lambda) := b\lambda + \int_0^{+\infty} (1 - e^{-\lambda y})\, m(dy), \qquad \lambda \geq 0. \qquad (10.3.9)$$

We will not prove the theorems but provide some explanatory arguments in the following subsection.

10.3.2 Comments on Filipović's Theorems

We give now some hints to the proofs of Theorem 10.3.1 and Theorem 10.3.2. Complete proofs can be found in [53]. Denote the discounted transition semigroup of the process R by $(\mathcal{Q}_t)$

$$\mathcal{Q}_t\varphi(x) = \mathbb{E}\left(\varphi(R^x(t))e^{-\int_0^t R^x(s)\,ds}\right), \qquad \varphi \in C_0([0, +\infty)), \quad t \geq 0,\ x > 0.$$

The first part of the following result shows that an affine model satisfies (MP) if and only if the discounted semigroup transforms exponential functions on positive multiples of exponential functions.

Proposition 10.3.3 *For a Feller process R the requirement (MP) is satisfied with some continuous and increasing functions C, D, $C(0) = D(0) = 0$, if and only if there exist functions $\varphi(\cdot,\cdot)$, $\psi(\cdot,\cdot)$ such that*

$$\mathcal{Q}_t f_\lambda(x) = e^{-\varphi(t,\lambda)-\psi(t,\lambda)x}, \quad t, \lambda, x \geq 0. \qquad (10.3.10)$$

If (10.3.10) *is satisfied then for* $0 \leq t \leq T$,

$$C(T) = C(T - t) + \varphi(t, D(T - t)), \qquad (10.3.11)$$

$$D(T) = \psi(t, D(T - t)). \qquad (10.3.12)$$

Conversely, if (10.3.11) *and* (10.3.12) *hold for some functions* φ, ψ *then (MP) is satisfied.*

Proposition 10.3.4 *1) Functions $\mathcal{R}, \mathcal{F}$, given by* (10.3.8), (10.3.9) *are related to* ψ *and* ϕ, *defined by* (10.3.10), *by the formulae:*

$$\frac{\partial\psi}{\partial t}(0, \lambda) = \mathcal{R}(\lambda), \quad \frac{\partial\phi}{\partial t}(0, \lambda) = \mathcal{F}(\lambda), \quad \lambda \geq 0. \qquad (10.3.13)$$

2) For each $\lambda \geq 0$ the functions $\psi(\cdot,\lambda)$ and $\varphi(\cdot,\lambda)$ satisfy the following ordinary differential equation:

$$\frac{d\psi}{dt}(t,\lambda) = \mathcal{R}(\psi(t,\lambda)), \qquad t \geq 0, \quad \psi(0,\lambda) = \lambda, \geq 0, \tag{10.3.14}$$

$$\frac{d\varphi}{dt}(t,\lambda) = \mathcal{F}(\psi(t,\lambda), \qquad t \geq 0, \quad \varphi(0,\lambda) = 0,\ \lambda \geq 0. \tag{10.3.15}$$

Proof of Proposition 10.3.3 For each $T > 0$ the discounted bond price process

$$\hat{P}(t,T) := e^{-\int_0^t R^x(s)\,ds} P(t,T), \quad 0 \leq t \leq T, \quad x \in [0,+\infty)$$

is, by (MP) condition, a martingale with respect to the underlying filtration. In particular, for $t \in [0,T]$,

$$\hat{P}(0,T) = \mathbb{E}(\hat{P}(t,T) \mid \mathcal{F}_0).$$

Thus, taking into account that R^x is Markovian and $R^x(0) = x$, we obtain

$$\begin{aligned} e^{-C(T)-D(T)x} &= \mathbb{E}\left(P(t,T)e^{-\int_0^t R^x(s)\,ds} \mid \mathcal{F}_0\right) \\ &= \mathbb{E}\left(e^{-C(T-t)-D(T-t)R^x(t)} e^{-\int_0^t R^x(s)\,ds} \mid \mathcal{F}_0\right) \\ &= e^{-C(T-t)} \mathcal{Q}_t f_{D(T-t)}(x). \end{aligned} \tag{10.3.16}$$

Therefore, for $\lambda = D(T-t)$, and $x \geq 0$,

$$\mathcal{Q}_t f_\lambda(x) = e^{-\varphi(t,\lambda)-\psi(t,\lambda)x}$$

for some $\varphi(t,\lambda)$ and $\psi(t,\lambda)$. In fact let $[0,\lambda_0] := D([0,u_0])$, and $D^{-1}\colon [0,\lambda_0] \to [0,u_0]$ is such that

$$D(D^{-1}(\lambda)) = \lambda, \qquad \lambda \in [0,\lambda_0],$$

and thus

$$\begin{aligned} \psi(t,\lambda) &= D(t+T-t) = D(t+D^{-1}(\lambda)), \\ \varphi(t,\lambda) &= C(t+D^{-1}(\lambda)) - C(D^{-1}(\lambda)), \qquad \lambda \in [0,\lambda_0]. \end{aligned}$$

However, for fixed $t > 0$ and x, the left side of (10.3.10) is the Laplace transform of a finite measure and therefore is an analytic function of $\lambda > 0$. Since

$$\ln \mathcal{Q}_t f_\lambda(0) = -\varphi(t,\lambda), \tag{10.3.17}$$

$$\ln \mathcal{Q}_t f_\lambda(1) = -\varphi(t,\lambda) - \psi(t,\lambda), \qquad \lambda \in [0,\lambda_0], \tag{10.3.18}$$

functions $\varphi(t,\cdot)$ and $\psi(t,\cdot)$ have unique, analytic extensions to the whole interval $[0,+\infty)$. It is also clear that (10.3.10) must hold for all $t \geq 0$, $\lambda \geq 0$ and $x \geq 0$. By (10.3.16), the functions C and D necessarily should be such that

$$e^{-C(T)-D(T)x} = e^{-C(T-t)}e^{-\varphi(t,D(T-t))-\psi(t,D(T-t))x}$$

and therefore (10.3.11) and (10.3.12) are satisfied.

Conversely, assume that (10.3.11) and (10.3.12) hold. We show that (MP) condition holds as well. In fact, if $0 \leq s \leq t \leq T$ then

$$\begin{aligned}
&\mathbb{E}\left(e^{-C(T-t)-D(T-t)R(t)}e^{-\int_0^t R(u)\,du} \mid \mathcal{F}_s\right)\\
&= e^{-C(T-t)}e^{-\int_0^s R(u)\,du}\mathbb{E}\left(e^{-\int_s^t R(u)\,du}e^{-D(T-t)R(t)} \mid \mathcal{F}_s\right)\\
&= e^{-C(T-t)}e^{-\int_0^s R(u)\,du}Q_{t-s}f_{D(T-t)}(R(s))\\
&= e^{-C(T-t)}e^{-\int_0^s R(u)\,du}e^{-\varphi(t-s,D(T-t))}e^{-\psi(t-s,D(T-t))R(s)},
\end{aligned}$$

and we need to show that this expression is equal to

$$e^{-C(T-s)-D(T-s)R(s)}e^{-\int_0^s R(u)\,du}.$$

Thus we should have that

$$C(T-s) = C(T-t) + \varphi(t-s, D(T-t)), \tag{10.3.19}$$

$$D(T-s) = \psi(t-s, D(T-t)). \tag{10.3.20}$$

Changing, in (10.3.11) and (10.3.12), T to $T-s$ and t to $t-s$ we arrive at (10.3.19) and (10.3.20). □

Proof of Proposition 10.3.4 1) Let $\hat{\mathcal{A}}$ be the generator of (Q_t). Then, at least for f in the domain of $\mathcal{A}$, and thus for all f_λ, $\lambda > 0$,

$$\hat{\mathcal{A}}f(x) = \mathcal{A}f(x) - xf(x), \quad x \geq 0. \tag{10.3.21}$$

By (10.3.10)

$$\frac{d}{dt}Q_tf_\lambda(x) = -Q_tf_\lambda(x)\left(\frac{\partial\varphi}{\partial t}(t,\lambda) + \frac{\partial\psi}{\partial t}(t,\lambda)\,x\right). \tag{10.3.22}$$

On the other hand

$$\frac{d}{dt}Q_tf_\lambda(x) = \hat{\mathcal{A}}(Q_tf_\lambda)(x), \ t \geq 0, x \geq 0, \tag{10.3.23}$$

and, in particular, for the derivative at $t = 0$:

$$\frac{d}{dt}Q_tf_\lambda(x)|_{t=0} = \hat{\mathcal{A}}(Q_0f_\lambda)(x) = \hat{\mathcal{A}}(f_\lambda)(x), x \geq 0. \tag{10.3.24}$$

Combining (10.3.22) and (10.3.24),

$$\hat{\mathcal{A}}(f_\lambda)(x) = -f_\lambda(x)(\frac{\partial\varphi}{\partial t}(0,\lambda) + \frac{\partial\psi}{\partial t}(0,\lambda)\,x). \tag{10.3.25}$$

However, from the formula for $\mathcal{A}$ and (10.3.21)

$$\begin{aligned}\hat{\mathcal{A}}(f_\lambda)(x) = -f_\lambda(x)[\alpha x\lambda^2 - \lambda(\beta x + b) - x + \int_0^{+\infty}(e^{-\lambda y} - 1)m(dy)\\ + x\int_0^{+\infty}(e^{-\lambda y} - 1 + \lambda y\mathbf{1}_{[0,1]}(y))(\mu(dy))].\end{aligned} \tag{10.3.26}$$

Comparing (10.3.25) and (10.3.26) and coefficients by x one gets the first part of the proposition.

2) The proof is based only on the semigroup property of the discounted semigroup and its specific representation (10.3.10). Indeed, it follows that for all $\lambda \geq 0$ and $t, s \geq 0$,

$$\begin{aligned}\mathcal{Q}_t(\mathcal{Q}_s f_\lambda)(x) &= e^{-\varphi(s,\lambda)}\mathcal{Q}_t f_{\psi(s,\lambda)}(x)\\ &= e^{-(\varphi(s,\lambda)+\varphi(t,\psi(s,\lambda)))-\psi(t,\psi(s,\lambda))x}\\ &= \mathcal{Q}_{t+s} f_\lambda(x) = e^{-\varphi(t+s,\lambda)-\psi(t+s,\lambda)x},\end{aligned}$$

and therefore

$$\varphi(t+s,\lambda) = \varphi(s,\lambda) + \varphi(t,\psi(s,\lambda)), \tag{10.3.27}$$

$$\psi(t+s,\lambda) = \psi(t,\psi(s,\lambda)). \tag{10.3.28}$$

Thus the family of transformation $\psi(t,\cdot)$, $t \geq 0$ is a flow, so it defines a solution of an ordinary differential equation. In fact, for $h > 0$,

$$\frac{1}{h}\Big(\psi(h+s,\lambda) - \psi(s,\lambda)\Big) = \frac{1}{h}\Big(\psi(h,\psi(s,\lambda)) - \psi(s,\lambda)\Big),$$

so, since $\psi(0,\lambda) = \lambda$, we have

$$\frac{\partial\psi}{\partial t}(s,\lambda) = \frac{\partial\psi}{\partial t}(0,\psi(s,\lambda)), \quad s \geq 0,\ \lambda \geq 0.$$

Similarly, since $\varphi(0,\lambda) = 0$,

$$\frac{d\varphi}{ds}(s,\lambda) = \frac{\partial\varphi}{\partial s}(0,\psi(s,\lambda)), \quad s \geq 0,\ \lambda \geq 0,$$

and thus (10.3.13) follows. □

10.3.3 Examples

Theorem 10.3.2 allows to find, at least numerically, functions C and D in the formula for the bond prices. Since $\mathcal{R}(0) = 1$,

$$\bar{\lambda} := \inf\{\lambda : \mathcal{R}(\lambda) \le 0\} \tag{10.3.29}$$

is a positive number or $+\infty$. Define

$$\mathcal{G}(x) := \int_0^x \frac{1}{\mathcal{R}(y)} dy, \quad x \in [0, \bar{\lambda}), \quad \bar{t} := \lim_{\lambda \to \bar{\lambda}.} \mathcal{G}(\lambda). \tag{10.3.30}$$

By (10.3.7) and (10.3.30) function D is the inverse of $\mathcal{G}$:

$$D(t) = \mathcal{G}^{-1}(t), \quad t < \bar{t}.$$

Example 10.3.5 Let $\alpha > 0$, $\beta > 0$ and

$$\mathcal{R}(\lambda) = 1 - \alpha\lambda^2 - \beta\lambda, \quad \lambda \ge 0.$$

Then $\bar{\lambda} = \frac{-\beta+\sqrt{\beta^2+4\alpha}}{2\alpha}$ and $\bar{t} = +\infty$. For $x \in \left(0, \frac{-\beta+\sqrt{\beta^2+4\alpha}}{2\alpha}\right)$,

$$\mathcal{G}(x) = \frac{\alpha}{\sqrt{\beta^2+4\alpha}} \ln \left[\left(\frac{\frac{-\beta+\sqrt{\beta^2+4\alpha}}{2\alpha} - x}{x - \frac{-\beta-\sqrt{\beta^2+4\alpha}}{2\alpha}} \right) \left(\frac{\beta + \sqrt{\beta^2+4\alpha}}{-\beta + \sqrt{\beta^2+4\alpha}} \right) \right]$$

and

$$D(t) = \mathcal{G}^{-1}(t) = \frac{2(e^{t\frac{\sqrt{\beta^2+4\alpha}}{\alpha}} - 1)}{(e^{t\frac{\sqrt{\beta^2+4\alpha}}{\alpha}} - 1)\beta + \sqrt{\beta^2+4\alpha} - 2\sqrt{\beta^2+4\alpha}}, \quad 0 < t < +\infty.$$

If, in addition,

$$\mathcal{F}(\lambda) = b\lambda + \int_0^{+\infty} (1 - e^{-\lambda y}) m(dy),$$

then the short-rate process $R(t)$ has generator

$$\mathcal{A}f(x) = \alpha x f''(x) + (\beta x + b) f'(x) + \int_0^{+\infty} (f(x+y) - f(x)) m(dy), \tag{10.3.31}$$

which corresponds to the solution of the following stochastic equation

$$dR(t) = (\beta R(t) + b)dt + \sqrt{2\alpha R(t)} dW(t) + dL(t), \quad R(0) = x \ge 0$$

(see Section B.1.1). The process L can be an arbitrary subordinator.

Example 10.3.6 Consider the case in which $\mu(dy) = \frac{1}{y^{1+\gamma}}\mathbf{1}_{[0,+\infty)}(y)dy$ and $\gamma \in (1,2)$. Since

$$\int_0^{+\infty} (1 - e^{-\lambda y} - \lambda y)\frac{1}{y^{1+\gamma}}dy = -\frac{1}{\gamma(\gamma-1)}\Gamma(2-\gamma)\lambda^{\gamma}$$

(see Example 5.3.6),

$$\begin{aligned}
\mathcal{R}(\lambda) &= 1 - \alpha\lambda^2 + \beta\lambda + \int_0^{+\infty} (1 - e^{-\lambda y} - \lambda y\mathbf{1}_{[0,1]}(y))\frac{1}{y^{1+\gamma}}dy \\
&= 1 - \alpha\lambda^2 + \beta\lambda - \frac{1}{\gamma(\gamma-1)}\Gamma(2-\gamma)\lambda^{\gamma} + \lambda\int_1^{+\infty}\frac{1}{y^{\gamma}}dy \\
&= 1 - \alpha\lambda^2 + \Big(\beta + \frac{1}{\gamma-1}\Big)\lambda - \frac{1}{\gamma(\gamma-1)}\Gamma(2-\gamma)\lambda^{\gamma}.
\end{aligned} \tag{10.3.32}$$

To calculate function $\mathcal{G}$ and its inverse $\mathcal{G}^{-1}$ is not a straightforward task. For instance, if $\alpha = 0,\ \beta = -\frac{1}{\gamma-1}$, then

$$\mathcal{R}(\lambda) = 1 - c_{\gamma}\lambda^{\gamma}, \quad \lambda < \bar{\lambda},$$

$$c_{\gamma} = \frac{1}{\gamma(\gamma-1)}\Gamma(2-\gamma), \ \ \bar{\lambda} = (c_{\gamma})^{-\frac{1}{1+\gamma}}.$$

Then $\bar{t}$ defined in (10.3.30) equals $+\infty$ and $\mathcal{G}^{-1}$, and thus also D, can be found only numerically.

10.3.4 Back to Short-Rate Equations

We are returning now to the problem of describing those stochastic equations that generate affine models satisfying (MP) condition, but using Theorem 10.3.1 and Theorem 10.3.2. We rederive some earlier results. We restrict our attention to equations of the form

$$dR(t) = F(R(t)dt + G(R(t-))dZ(t), \quad R(0) = x \geq 0, \tag{10.3.33}$$

where Z is a real-valued Lévy process with characteristics (a, q, ν). Since processes R should be nonnegative we will require that G is nonnegative and that jumps of Z are nonnegative as well.

As before we denote by $C_0([0, +\infty))$ the space of all continuous functions defined on $[0, +\infty)$ and vanishing at infinity, equipped with the supremum norm. Let Λ be the linear hull of the set of all exponential functions f_{λ}, $\lambda > 0$, where

$$f_{\lambda}(x) = e^{-\lambda x}, \ x \geq 0.$$

If ν is a measure concentrated on $(0, +\infty)$ and p a positive number then, as before, by ν_p we denote the image of ν by the linear transformation $z \to pz,\ z \geq 0$. We set also $\nu_0 = 0$. The generators corresponding to stochastic equations are discussed in Appendix B (see Proposition B.1.2).

The conditions that appear in the next proposition play a similar role in the theory of the affine term structure as the analytical HJM conditions of the Proposition 10.2.1.

Proposition 10.3.7 *Assume that the transition semigroup* $(\hat{\mathcal{P}}_t)$ *of the solution of the equation* (10.3.33) *is strongly continuous and that the domain of its generator* $\hat{\mathcal{A}}$ *contains* Λ. *It coincides with the transition semigroup* $(\mathcal{P}_t)$ *having the generator specified in Theorem* (10.3.1) *if and only if*

$$x\mu + m = \nu_{G(x)}, \qquad \frac{1}{2}qG^2(x) = \alpha x,\ x \geq 0, \tag{10.3.34}$$

$$\beta x + b + \int_0^1 ym(dy) = F(x) + \bar{F}(x) + G(x)a, \tag{10.3.35}$$

where

$$\bar{F}(x) = \int_{(0,+\infty)} \Big\{\mathbf{1}_{[0,1]}(G(x)y) - \mathbf{1}_{[0,1]}(y)\Big\}G(x)y\nu(dy). \tag{10.3.36}$$

Remark 10.3.8 It should be noticed that the function $\bar{F}$ is well defined. In fact

(a) $\bar{F}(x) = 0$ if $G(x) = 0$,
(b) $\bar{F}(x) = \int_1^{\frac{1}{G(x)}} y\nu(dy)$ if $G(x) \in (0, 1]$,
(c) $\bar{F}(x) = -\int_{\frac{1}{G(x)}}^1 y\nu(dy)$ if $G(x) > 1$.

Proof of Proposition 10.3.7 The generator of the solution of the stochastic equation (10.3.33) (see Section B.1.1) has the form

$$\begin{aligned}\hat{\mathcal{A}}f(x) =& f'(x)(F(x) + G(x)a) + \frac{1}{2}qf''(x)G^2(x)\\ &+ \int_{(0,+\infty)} \Big\{f(x + G(x)y) - f(x) - \mathbf{1}_{[0,1]}(y)G(x)y\Big\}\nu(dy).\end{aligned} \tag{10.3.37}$$

Equivalently,

$$\begin{aligned}\hat{\mathcal{A}}f(x) =& f'(x)(F(x) + G(x)a) + \frac{1}{2}qf''(x)G^2(x)\\ &+ \int_{(0,+\infty)} \Big\{f(x + G(x)y) - f(x) - \mathbf{1}_{[0,1]}(G(x)y)f'(x)G(x)y\Big\}\nu(dy)\end{aligned} \tag{10.3.38}$$

$$+f'(x)\int_{(0,+\infty)} \Big\{\mathbf{1}_{[0,1]}(G(x)y) - \mathbf{1}_{[0,1]}(y)\Big\}G(x)y\nu(dy), \tag{10.3.39}$$

and thus

$$\hat{\mathcal{A}}f(x) = f'(x)(F(x) + G(x)a) + \frac{1}{2}qf''(x)G^2(x)$$
$$+ \int_{(0,+\infty)} \Big\{ f(x+y) - f(x) - f'(x)\mathbf{1}_{[0,1]}(y)y \Big\} \nu_{G(x)}(dy) + f'(x)\bar{F}(x). \tag{10.3.40}$$

Note that if $G(x) = 0$ then $\nu_{G(x)} = 0$, the preceding integral makes sense and $\bar{F}(x) = 0$.
Rewriting this formula we arrive at the following formula for $\hat{\mathcal{A}}$:

$$\hat{\mathcal{A}}f(x) = f'(x)(F(x) + \bar{F}(x) + G(x)a) + \frac{1}{2}qf''(x)G^2(x)$$
$$+ \int_{(0,+\infty)} \Big\{ f(x+y) - f(x) - f'(x)(\mathbf{1}_{[0,1]}(y)y) \Big\} \nu_{G(x)}(dy). \tag{10.3.41}$$

On the other hand the operator $\mathcal{A}$ is of the form

$$\mathcal{A}f(x) = \alpha x f''(x) + (\beta x + b)f'(x) + \int_0^{+\infty} (f(x+y) - f(x))m(dy)$$
$$+ x \int_C^{+\infty} \big(f(x+y) - f(x) - \mathbf{1}_{[0,1]}(y)f'(x)y \big)\, \mu(dy), \tag{10.3.42}$$

and can be written as:

$$\mathcal{A}f(x) = \alpha x f''(x) + (\beta x + b + \int_0^1 y m(dy))f'(x)$$
$$+ \int_0^{+\infty} \big(f(x+y) - f(x) - \mathbf{1}_{[0,1]}(y)f'(x)y \big)\, (x\mu(dy) + m(dy)). \tag{10.3.43}$$

The transition semigroups $(\hat{\mathcal{P}}_t)$ and $(\mathcal{P}_t)$ coincide if their generators coincide on the set Λ, that is, if and only if:

$$\mathcal{A}f_\lambda(x) = \hat{\mathcal{A}}f_\lambda(x), \;\; \lambda > 0, \;\; x \geq 0. \tag{10.3.44}$$

By direct calculations one gets:

$$\mathcal{A}f_\lambda(x) = f_\lambda(x)\Big\{ -\lambda[\beta x + b + \int_0^1 y m(dy)] + \lambda^2 \alpha x$$
$$+ \int_0^{+\infty} [e^{-\lambda y} - 1 + \lambda \mathbf{1}_{[0,1]}(y)y](x\mu + m)(dy) \Big\}, \tag{10.3.45}$$

$$\hat{\mathcal{A}}f_\lambda(x) = f_\lambda(x)\Big\{ -\lambda[F(x) + \bar{F}(x) + G(x)a] + \frac{1}{2}\lambda^2 q G^2(x)$$
$$+ \int_0^{+\infty} [e^{-\lambda y} - 1 + \lambda \mathbf{1}_{[0,1]}(y)y]\nu_{G(x)}(dy) \Big\}. \tag{10.3.46}$$

Thus the identity(10.3.44) holds if and only if for each $x \geq 0$, the following Laplace exponent:

$$- \lambda[\beta x + b + \int_0^1 ym(dy)] + \lambda^2 \alpha x$$
$$+ \int_0^{+\infty} [e^{-\lambda y} - 1 + \lambda \mathbf{1}_{[0,1]}(y)y](x\mu + m)(dy), \quad \lambda > 0, \tag{10.3.47}$$

coincides, as a function of λ, with the following one:

$$- \lambda[F(x) + \bar{F}(x) + G(x)a] + \frac{1}{2}\lambda^2 q G^2(x)$$
$$+ \int_0^{+\infty} [e^{-\lambda y} - 1 + \lambda \mathbf{1}_{[0,1]}(y)y]\nu_{G(x)}(dy), \quad \lambda > 0. \tag{10.3.48}$$

However, the Laplace exponents, say,

$$J(\lambda) = a\lambda + b\lambda^2 + \int_{-\infty}^{+\infty} [e^{-\lambda} y - 1 + \lambda \mathbf{1}_{[0,1]}(y)y]\zeta(dy), \quad \lambda \geq 0,$$

uniquely determine the parameters a, b, ζ and consequently the specified in the proposition identities hold. □

Remark 10.3.9 Proposition 10.3.7 implies that if the stochastic equation generates a process for which the (MP) property holds, the functions F, G and the characteristics (a, Q, ν) of the process Z should solve the functional identities (10.3.34), (10.3.35) and (10.3.36).

Remark 10.3.10 In a similar way, one can derive analogical conditions in the multidimensional case, when

$$dR(t) = F(R(t)dt + \langle G(R(t-)), dZ(t)\rangle, \quad R(0) = x \geq 0 \tag{10.3.49}$$

and the process Z has characteristics (a, Q, ν). Namely, the conditions (10.3.34), (10.3.35) and (10.3.36), become:

$$x\mu + m = \nu_{G(x)}, \qquad \frac{1}{2}|Q^{1/2}G(x)|^2 = \alpha x, \; x \geq 0, \tag{10.3.50}$$

$$\beta x + b + \int_0^1 ym(dy) = F(x) + \bar{F}(x) + \langle G(x), a\rangle, \tag{10.3.51}$$

where

$$\bar{F}(x) = \int_{(0,+\infty)} \Big\{\mathbf{1}_{[0,1]}(|\langle G(x), y\rangle|) - \mathbf{1}_{[0,1]}(|y|)\Big\} G(x)y\nu(dy), \tag{10.3.52}$$

and ν_p, where $p \in \mathbb{R}^d$ is the image of the measure ν by the map that transforms $z \in \mathbb{R}^d$ onto $\langle p, z\rangle$. A precise description of all solutions of the equations is an open problem.

To illustrate the applicability of Proposition 10.3.7 we consider a very special case.

Proposition 10.3.11 *Assume that ν and μ are locally finite measures on $(0, +\infty)$, μ has a continuous density g and $G(\cdot)$ is a continuous, invertible transformation of $(0, +\infty)$ onto $(0, +\infty)$, such that for all φ bounded with compact support in $(0, +\infty)$ and all $x > 0$*

$$x \int_0^{+\infty} \varphi(y)\mu(dy) = \int_0^{+\infty} \varphi(G(x)y)\nu(dy). \tag{10.3.53}$$

Then ν has also density, say h, and for some $\gamma \in \mathbb{R}, \gamma \neq 0$, and positive constants c_1, c_2

$$g(x) = c_1 \frac{1}{x^{1+\gamma}}, \quad h(x) = c_2 \frac{1}{x^{1+\gamma}}, \quad G(x) = (x/c_2)^{\frac{1}{\gamma}}, \quad x \geq 0.$$

Proof The assumption (10.3.53) is an explicit formulation of the fact that the measure $x\mu$, for any x, is the image of the measure ν by the linear transformation $y \to G(x)y, y > 0$. Thus if one of the measures μ or ν has density, the same follows for the other one. Note that

$$\begin{aligned}\int_0^{+\infty} \varphi(G(x)y)\nu(dy) &= \int_0^{+\infty} \varphi(G(x)y)h(y)dy \\ &= \int_0^{+\infty} \varphi(z)h\left(\frac{z}{G(x)}\right)\frac{1}{G(x)}dz = x\int_0^{+\infty} \varphi(z)g(z)dz,\end{aligned}$$

and we have that for all $x > 0$,

$$h\left(\frac{z}{G(x)}\right)\frac{1}{G(x)} = xg(z), \quad z > 0. \tag{10.3.54}$$

In particular, for $x = 1$ we obtain

$$h\left(\frac{z}{G(1)}\right)\frac{1}{G(1)} = g(z), \quad z > 0.$$

To simplify calculations we assume that $G(1) = 1$. Then $h(z) = g(z), z > 0$ and

$$g\left(\frac{z}{G(x)}\right)\frac{1}{G(x)} = xg(z), \quad x, z > 0. \tag{10.3.55}$$

For given $z > 0$, let x be such that $G(x) = z$. Then

$$g(z) = g(1)\frac{1}{zG^{-1}(z)}, \quad z > 0, \tag{10.3.56}$$

which, by (10.3.55), for any $x, z > 0$ gives

$$g(1)\frac{1}{\frac{z}{G(x)}G^{-1}(\frac{z}{G(x)})} \cdot \frac{1}{G(x)} = xg(1)\frac{1}{zG^{-1}(z)}.$$

Hence, for all $x, z > 0$,

$$G^{-1}\left(\frac{z}{G(x)}\right) = \frac{1}{x}G^{-1}(z),$$

and, consequently,

$$G^{-1}\Big(\frac{z}{y}\Big) = \frac{G^{-1}(z)}{G^{-1}(y)}, \quad y, z > 0.$$

Rearranging terms yields

$$G^{-1}\Big(\frac{z}{y}\Big)G^{-1}(y) = G^{-1}(z) = G^{-1}\Big(\frac{z}{y}\cdot y\Big), \quad y, z > 0.$$

Finally, we have that for any $a, b > 0$,

$$G^{-1}(a)G^{-1}(b) = G^{-1}(a \cdot b).$$

It is well known that if ψ is a continuous positive function on $(0, +\infty)$ such that

$$\psi(a \cdot b) = \psi(a)\psi(b), \quad a, b > 0,$$

then $\psi(a) = a^\gamma, a > 0$ with some $\gamma \in \mathbb{R}$. It follows that $G^{-1}(a) = a^\gamma, a > 0, \gamma \in \mathbb{R}$ and, by (10.3.56), $g(x) = c_1 \frac{1}{x^{1+\gamma}}, x > 0$. □

Assume that $m = 0$, the measures μ, ν have densities and G is a one-to-one transformation of $\mathbb{R}_+$ and

$$Z(t) = \int_0^t \int_0^1 y\tilde{\pi}(ds, dy) + \int_0^t \int_1^{+\infty} y\pi(ds, dy).$$

Let us see to what stochastic equations we arrive by applying Proposition 10.3.7 and Proposition 10.3.11.

We know that, up to positive multiplicative constants,

$$\mu(dx) = g(x)dx = \frac{1}{x^{1+\gamma}}dx, \quad \nu(dx) = h(x)dx = \frac{1}{x^{1+\gamma}}dx, \quad G(x) = \frac{1}{x^{1+\gamma}}.$$

Example 10.3.12 If $\gamma \in (0, 1)$, by direct calculations

$$\bar{F}(x) = x\left(\frac{1}{\gamma(1-\gamma)}\right) - x^{\frac{1}{\gamma}}\frac{1}{1-\gamma},$$

$$F(x) = x\left(\beta - \frac{1}{\gamma(1-\gamma)}\right) + b + x^{\frac{1}{\gamma}}\frac{1}{1-\gamma}.$$

This leads to the equation

$$dR(t) = \left(R(t)\left(\beta - \frac{1}{\gamma(1-\gamma)}\right) + b\right)dt + (R(t))^{\frac{1}{\gamma}}\frac{1}{1-\gamma}dt + (R(t-))^{\frac{1}{\gamma}}dZ(t).$$

However,

$$Z(t) = \tilde{Z}(t) - \frac{t}{1-\gamma}, \tag{10.3.57}$$

where $\tilde{Z}$ is the α-stable subordinator, so we arrive at the equation

$$dR(t) = \left(R(t) \left(\beta - \frac{1}{\gamma(1-\gamma)} \right) + b \right) dt + (R(t-))^{\frac{1}{\gamma}} d\tilde{Z}(t). \tag{10.3.58}$$

Example 10.3.13 Similarly, if $\gamma \in (1, 2)$,

$$\bar{F}(x) = -x \left(\frac{1}{\gamma(\gamma-1)} \right) + x^{\frac{1}{\gamma}} \frac{1}{\gamma - 1}. \tag{10.3.59}$$

Moreover,

$$Z(t) = \tilde{Z}(t) + \frac{t}{\gamma - 1}, \tag{10.3.60}$$

where $\tilde{Z}(t)$ is the γ-stable martingale. We therefore arrive at the equation

$$dR(t) = \left(R(t) \left(\beta + \frac{1}{\gamma(\gamma-1)} \right) + b \right) dt + (R(t-))^{\frac{1}{\gamma}} d\tilde{Z}(t). \tag{10.3.61}$$

11

Completeness

In this chapter we consider the completeness problem for models where the discounted bond prices are local martingales. Most models turn out not to be complete because completeness forces very strong assumptions on jumps of the underlying Lévy process. Therefore we also describe conditions for approximate completeness, which are less demanding. Our considerations cover the models discussed in the previous sections, i.e. HJM models and factor models including also affine models.

11.1 Problem of Completeness

As we already know (see Section 7.3), the fair price of an $\mathcal{F}_{T^*}$-measurable contingent claim X is a number $x \in \mathbb{R}$ such that

$$X = x + \int_0^{T^*} (\varphi_s, dP(s, \cdot)), \qquad \mathbb{P} - a.s. \tag{11.1.1}$$

holds for some self-financing strategy (φ_t). Recall that (φ_t) is a predictable process taking values in the set $\mathcal{M} = \mathcal{M}([0, +\infty))$ of finite signed measures on $[0, +\infty)$ equipped with the weak topology. Since (φ) is assumed to be self-financing, (11.1.1) can be written in the following equivalent form:

$$\hat{X} = x + \int_0^{T^*} (\varphi_t, d\hat{P}(t, \cdot)), \qquad \mathbb{P} - a.s. \tag{11.1.2}$$

(see Section 7.2 for details), where $\hat{X}$ and $\hat{P}(t, \cdot)$ stand for the discounted claim and the discounted bond curve, respectively. In fact, the representation (11.1.2) is more convenient for analysis as it allows one to ignore the self-financing condition. If the self-financing condition is violated, one can modify the measures $\varphi(t, \cdot), t \in [0, T^*]$, at zero so that the modification is self-financing and (11.1.2) still holds for it (see Corollary 7.2.3). So, as we prefer to work with (11.1.2) rather than with (11.1.1), conditions for the claim X to be replicated will be formulated in terms of its discounted value $\hat{X}$.

Recall, if (11.1.2) is satisfied for some $x \in \mathbb{R}$ and an admissible $\mathcal{M}$-valued strategy (φ_t), then X, or $\hat{X}$, is called *attainable* (see Section 7.3). *Completeness* of the bond market with a given filtration $(\mathcal{F}_t, t \in [0, T^*])$ means that each

$$\hat{X} \in L^\infty(\Omega, \mathcal{F}_{T^*}, \mathbb{P})$$

is attainable, where $L^\infty(\Omega, \mathcal{F}_{T^*}, \mathbb{P})$ stands for the set of all bounded $\mathcal{F}_{T^*}$-measurable random variables. If this is the case, we will also say that the *market is complete in the class* $\hat{X} \in L^\infty(\Omega, \mathcal{F}_{T^*}, \mathbb{P})$. Complete markets are important because in such markets the price of each bounded claim is defined in a natural way and given by the formula (7.3.8).

As we shall see completeness takes place for a rather narrow class of models. One is therefore using a more general concept of *approximate completeness* where $\hat{X}$ is approximated in the mean-square sense. One is looking for a sequence (x_n, φ_n) such that

$$\lim_{n \to +\infty} \mathbb{E}\Big[\Big(\Big(x_n + \int_0^{T^*} (\varphi_t^n, d\hat{P}(t, \cdot))\Big) - \hat{X}\Big)^2\Big] = 0. \tag{11.1.3}$$

Since bounded random variables are dense in $L^2(\Omega, \mathcal{F}_{T^*}, \mathbb{P})$, we can examine (11.1.3) for $\hat{X} \in L^2(\Omega, \mathcal{F}_{T^*}, \mathbb{P})$. The limit of $\{x_n\}$, if exists, can be regarded as a generalized price of X.

The completeness problem will be examined for models defined on a probability space $(\Omega, \mathcal{F}, \mathbb{P})$ equipped with filtration generated by a Lévy process Z in specific classes of *admissible strategies* (see Section 11.3). We assume that the original measure $\mathbb{P}$ is a martingale measure and thus the following assumption holds.

For each $T > 0$ the discounted bond price process

$$\hat{P}(t, T) = \frac{P(t, T)}{B(t)}, \qquad t \in [0, T] \tag{MP}$$

is a local martingale.

Our aim is to describe conditions for completeness and approximate completeness for HJM models, affine models and factor models.

11.2 Representation of Discounted Bond Prices

With the use of (MP) we describe first discounted bond prices and introduce the class of admissible strategies in the following sections.

Let (a, Q, ν) be the characteristic triplet of the underlying $U = \mathbb{R}^d$-valued Lévy process Z with the Lévy–Itô decomposition:

$$Z(t) = at + W(t) + \int_0^t \int_{|y| \le 1} y\, \tilde{\pi}(ds, dy) + \int_0^t \int_{|y| > 1} y\, \tilde{\pi}(ds, dy), \qquad t \ge 0,$$

where W is a Wiener process and $\tilde{\pi}$ the compensated jump measure of Z.

Since, by (MP), the discounted price of any T-bond is a local martingale, it can be, by Theorem 6.1.1, represented in the form

$$d\hat{P}(t,T) = \langle \Gamma_1(t,T), dW(t)\rangle + \int_U \Gamma_2(t,T,y)\tilde{\pi}(dt,dy), \quad t \in [0,T^*],\ T > 0 \tag{11.2.1}$$

for some processes $(\Gamma_1(t,T)) \in \Phi(U)$ and $(\Gamma_2(t,T,y)) \in \Psi_{1,2}$, thus satisfying, for each $T > 0$,

$$\int_0^{T^*} \mid \Gamma_1(s,T) \mid^2 ds < +\infty,$$

$$\int_0^{T^*} \int_U \Big(\mid \Gamma_2(s,T,y) \mid^2 \wedge \mid \Gamma_2(s,T,y) \mid \Big) ds\, \nu(dy) < +\infty.$$

Since $\hat{P}(t,T)$ is constant for $t > T$, it follows that

$$\Gamma_1(t,T) = 0, \quad \Gamma_2(t,T,y) = 0, \quad t > T.$$

Now we derive explicit forms of $\Gamma_1(t,T)$ and $\Gamma_2(t,T,y)$ in the decomposition (11.2.1) for models studied in the sequel.

Proposition 11.2.1 *Let us assume that in the following models (a), (b) and (c) the condition (MP) is satisfied. Let $\Gamma_1(t,T)$ and $\Gamma_2(t,T,y)$ be given by* (11.2.1).

(a) For HJM models $df(t,T) = \alpha(t,T)dt + \langle \sigma(t,T), dZ(t)\rangle$,

$$\Gamma_1(t,T) = -\hat{P}(t-,T)\Sigma(t,T), \quad \Gamma_2(t,T,y) = \hat{P}(t-,T)[e^{-\langle \Sigma(t,T),y\rangle} - 1], \tag{11.2.2}$$

where $\Sigma(t,T) := \int_{t\wedge T}^T \sigma(t,u)du$.

(b) For affine models $P(t,T) = e^{-C(T-t)-D(T-t)R(t)}$, with short rate

$$dR(t) = F(R(t))dt + \langle G(R(t-)), dZ(t)\rangle:$$

$$\Gamma_1(t,T) = -\hat{P}(t-,T)D(T-t)G(R(t-)),$$
$$\Gamma_2(t,T,y) = \hat{P}(t-,T)[e^{-\langle D(T-t)G(R(t-)),y\rangle} - 1]. \tag{11.2.3}$$

(c) For factor models $f(t,T) = G(T-t, X(t))$, with factor

$$dX(t) = a(X(t))dt + \langle b(X(t-)), dZ(t)\rangle:$$

$$\Gamma_1(t,T) = -\hat{P}(t,T)b(X(t-)) \int_0^{T-t} G'_x(u, X(t-))du, \tag{11.2.4}$$

$$\Gamma_2(t,T,y) = \hat{P}(t,T)\big(e^{\int_0^{T-t}\{G(s,X(t-))-G(s,X(t)+y)\}ds} - 1\big). \tag{11.2.5}$$

Proof (*a*) It follows from Proposition 8.1.7 that

$$\hat{P}(t,T) = \hat{P}(0,T) + \int_0^t \hat{P}(s-,T)\left\{\frac{1}{2}\langle Q\Sigma(s,T),\Sigma(s,T)\rangle - A(s,T) - \langle\Sigma(s,T),a\rangle\right\}ds$$

$$- \int_0^t \hat{P}(s-,T)\langle\Sigma(s,T),dW(s)\rangle - \int_0^t \hat{P}(s-,T)\langle\Sigma(s,T),dZ_0(s)\rangle$$

$$+ \int_0^t\int_U \hat{P}(s-,T)\left[e^{-\langle\Sigma(s,T),y\rangle} - 1 + \mathbf{1}_{\{|y|\le 1\}}\langle\Sigma(s,T),y\rangle\right]\pi(ds,dy).$$

Since we know that $\hat{P}(t,T)$ is a local martingale, we can compensate the last integral and then cancel all dt-integrals. This yields directly the required representation.

(*b*) The affine model can be written as an HJM model with coefficients

$$\alpha(t,T) := F(R(t))D'(T-t) - C''(T-t) - D''(T-t)R(t),$$
$$\sigma(t,T) := D'(T-t)G(t,R(t-))$$

(see Proposition 7.5.1). It follows that

$$\Sigma(t,T) = \int_{t\wedge T}^T \sigma(t,u)du = G(R(t-))\int_{t\wedge T}^T D'(u-t)du = G(R(t-))D(T-t\wedge T).$$

Now we can use 11.2.2.

(*c*) The required representation can be obtained by repetitive application of the Itô formula. Recall that the short rate is given by $R(t) = f(t,t) = G(0,X(t))$, so

$$\hat{P}(t,T) = e^{-\int_0^t R(u)du}P(t,T) = A(t)C(t),$$

where, for fixed T, A and C are given by

$$A(t) := e^{-\int_0^t R(u)du} = e^{-\int_0^t G(0,X(u))du}, \quad C(t) := e^{-\int_t^T G(u-t,X(t))du}.$$

By the Itô formula

$$d\hat{P}(t,T) = A(t)dC(t) + C(t)dA(t), \tag{11.2.6}$$

so we need to determine the dynamics of A and C. Clearly,

$$dA(t) = -A(t)G(0,X(t))dt.$$

Since $C(t) = e^{-Y(t)}$ with

$$Y(t) := \int_0^{T-t} G(s,X(t))ds = h(t,X(t)), \quad \text{where } h(t,x) := \int_0^{T-t} G(s,x)ds,$$

by the Itô formula

$$dY(t) = a_Y(t)dt + \langle b_Y(t), dZ(t)\rangle + \{h(t,X(t)) - h(t,X(t-)) - \langle b_Y(t), \triangle Z(t)\rangle\},$$

with $b_Y(t) := h'_x(t,X(t-))b(t)$ and some coefficient $a_Y(t)$. Here we write $b(t) := b(X(t-))$ for short. It follows, in particular, that

$$\triangle Y(t) = h(t,X(t)) - h(t,X(t-)). \tag{11.2.7}$$

Consequently, again by the Itô formula,

$$\begin{aligned} dC(t) &= -C(t-)dY(t) + \frac{1}{2}C(t-)b_Y^2(t)dt + C(t-)\{e^{-\triangle Y(t)} - 1 + \triangle Y(t)\} \\ &= -C(t-)\Big[\big(aY(t) - \frac{1}{2}b_Y^2(t)\big)dt + \langle b_Y(t), dZ(t)\rangle \\ &\qquad + \{-\langle b_Y(t), \triangle Z(t)\rangle - e^{-\triangle Y(t)} + 1\}\Big]. \end{aligned}$$

Coming back to (11.2.6) we obtain

$$\begin{aligned} d\hat{P}(t,T) = &- C(t)B(t-)\Big[\big(aY(t) - \frac{1}{2}b_Y^2(t)\big)dt + \langle b_Y(t), dZ(t)\rangle \\ &+ \{-\langle b_Y(t), \triangle Z(t)\rangle - e^{-\triangle Y(t)} + 1\}\Big] - C(t)A(t)G(0,X(t))dt. \end{aligned}$$

The use of the Lévy–Itó decomposition of Z and (11.2.7) leads to

$$\begin{aligned} d\hat{P}(t,T) = &- \hat{P}(t-,T)\Big[\big(aY(t) - \frac{1}{2}b_Y^2(t)\big)dt + \langle b_Y(t), a\rangle dt + \langle b_Y(t), dW(t)\rangle \\ &+ \int_{|y|\le 1} \langle b_Y(t), y\rangle \tilde{\pi}(dt,dy) + \int_{|y|>1} \langle b_Y(t), y\rangle \pi(dt,dy) \\ &+ \int_U \{1 - e^{\{-h(t,X(t-)+y) - h(t,X(t-))\}} - \langle b_Y(t), y\rangle\}\pi(dt,dy)\Big] \\ &- \hat{P}(t,T)G(0,X(t))dt. \end{aligned}$$

Since $\hat{P}(t,T)$ is a local martingale, we can compensate the integrals over $\pi(dt,dy)$ on the right side and, after that, cancel all dt-integrals. This gives

$$d\hat{P}(t,T) = -\hat{P}(t-,T)\Big[\langle b_Y(t), dW(t)\rangle + \int_U \big(1 - e^{h(t,X(t-)) - h(t,X(t-)+y)}\big)\tilde{\pi}(dt,dy)\Big].$$

The required representation follows by putting

$$h(t,x) = \int_0^{T-t} G(s,x)ds, b_Y(t) := h'_x(t,X(t-))b(t)$$

and $h'_x(t,x) = \int_0^{T-t} G'_x(s,x)ds$. □

11.3 Admissible Strategies

Under (MP) we can deduce more specific sufficient conditions defining trading strategies than those described in Section 7.2.3. Since our aim is to give a meaning to the integral

$$\int_0^t (\varphi_s, d\hat{P}(s,\cdot)), \quad t \in [0,T^*],$$

we can make use of the decomposition (11.2.1) and define, for $t \in [0,T^*]$,

$$\int_0^t (\varphi_s, d\hat{P}(s,\cdot)) := \int_0^t \langle(\varphi_s, \Gamma_1(s,\cdot)), dW(s)\rangle + \int_0^t \int_U (\varphi_s, \Gamma_2(s,\cdot,y))\tilde{\pi}(ds,dy), \tag{11.3.1}$$

providing that the right side is well defined. Therefore we define admissible strategies as follows.

Definition 11.3.1 Let (MP) be satisfied and the discounted bond prices admit, for any $T > 0$, the decomposition

$$d\hat{P}(t,T) = \langle \Gamma_1(t,T), dW(t)\rangle + \int_U \Gamma_2(t,T,y)\tilde{\pi}(dt,dy), \quad t \in [0,T^*], \quad T > 0, \tag{11.3.2}$$

where $(\Gamma_1(t,T)) \in \Phi(U)$ and $(\Gamma_2(t,T,y)) \in \Psi_{1,2}$. An *admissible strategy* is an $\mathcal{M}([0,+\infty))$-valued predictable process (φ_t) such that the processes

$$(\varphi_t, \Gamma_1(t,\cdot)) := \int_0^{+\infty} \Gamma_1(t,T)\varphi(t,dT), \quad t \in [0,T^*],$$

$$(\varphi_t, \Gamma_2(t,\cdot,y)) := \int_0^{+\infty} \Gamma_2(t,T,y)\varphi(t,dT), \quad t \in [0,T^*],$$

belong to $\Phi(U)$, $\Psi_{1,2}$ (see Section 6), respectively, and such that the corresponding discounted wealth process

$$\begin{aligned}\hat{X}(t) &= X(0) + \int_0^t (\varphi_s, d\hat{P}(s,\cdot)) \\ &:= X(0) + \int_0^t \langle(\varphi_s, \Gamma_1(s,\cdot)), dW(s)\rangle + \int_0^t \int_U (\varphi_s, \Gamma_2(s,\cdot,y))\tilde{\pi}(ds,dy), \quad t \in [0,T^*]\end{aligned} \tag{11.3.3}$$

is a martingale.

Let us notice that if, for instance, for each $s \in [0,T^*]$,

$$T \longrightarrow \Gamma_1(s,T), \quad T \longrightarrow \Gamma_2(s,T,y)$$

are continuous and bounded on $[0, +\infty)$, then

$$(\varphi_s, \Gamma_1(s,\cdot)) = \int_0^{+\infty} \Gamma_1(s,T)\varphi_s(dT), \quad (\varphi_s, \Gamma_2(s,\cdot,y)) = \int_0^{+\infty} \Gamma_2(s,T,y)\varphi_s(dT)$$

are finite. So, in this case the integrands on the right side of (11.3.1) are well defined.

We consider sufficient conditions for φ to be an admissible strategy. Recall that if φ is admissible then

$$\int_0^{T^*} |\, (\varphi_s, \Gamma_1(s,\cdot)) \,|^2 \, ds < +\infty, \tag{11.3.4}$$

$$\int_0^{T^*} \int_U \Big(|\, (\varphi_s, \Gamma_2(s,\cdot,y)) \,|^2 \wedge |\, (\varphi_s, \Gamma_2(s,\cdot,y)) \,| \Big) ds \, \nu(dy) < +\infty. \tag{11.3.5}$$

It is clear that if there exists a constant $K > 0$ such that

$$\left| \int_0^t (\varphi_s, d\hat{P}(s,\cdot)) \right| < K, \qquad t \in [0, T^*], \tag{11.3.6}$$

then φ is admissible because a bounded local martingale is a martingale. However, (11.3.6) is very rarely satisfied. More checkable conditioins are formulated below.

Proposition 11.3.2 *Let the strategy φ satisfy* (11.3.4), (11.3.5) *and*

$$\mathbb{E}\left(\int_0^{T^*} | Q^{\frac{1}{2}} (\varphi_s, \Gamma_1(s,\cdot)) |^2 ds \right)^{\frac{1}{2}} + \mathbb{E}\left(\int_0^{T^*} \int_U |\, (\varphi_s, \Gamma_2(s,\cdot,y)) \,|^2 \, ds\nu(dy) \right)^{\frac{1}{2}} < +\infty.$$

Then φ is admissible.

Proof The quadratic variation of the discounted wealth, for $t \in [0, T^*]$, equals

$$\left[\int_0^{\cdot} (\varphi_s, d\hat{P}(s,\cdot) \right]_t = \int_0^t |\, Q^{\frac{1}{2}} (\varphi_s, \Gamma_1(s,\cdot)) |^2 \, ds + \int_0^t \int_U |(\varphi_s, \Gamma_2(s,\cdot,y))|^2 \, \pi(ds,dy),$$

and in this case satisfies

$$\mathbb{E}\left(\left[\int_0^{\cdot} (\varphi_s, d\hat{P}(s,\cdot)) \right]_{T^*} \right)^{\frac{1}{2}} < +\infty.$$

Hence, it follows from the Burkholder–Davies–Gundy inequality (4.3.4) that the integral $\int_0^{\cdot} (\varphi_s, d\hat{P}(s,\cdot))$ is a martingale. □

Remark 11.3.3 The requirement that discounted wealth processes are supposed to be martingales, although the integrator is assumed to be only a local martingale, is necessary for further analysis of the completeness problem. It implies that the integrands in (11.3.3) are determined in a unique way. This property fails if we require instead, for instance, that $\hat{X}$ is a positive local martingale. We clarify this issue in Proposition 11.3.4 by presenting an example of a positive local martingale adapted to the filtration generated by a Wiener process, which admits two different integral representations. The proof is based on Example 8, page 237 in Liptser and Shiryaev [89].

Proposition 11.3.4 *There exists a positive local martingale $M(t), t \in [0, 1]$ adapted to the filtration generated by a one-dimensional Wiener process W such that $M(1)$ is a bounded random variable admitting two different representations*

$$M(1) = \mathbb{E}[M(1)] + \int_0^1 \gamma(s)dW(s), \tag{11.3.7}$$

$$M(1) = 1 + \int_0^1 \psi(s)dW(s), \tag{11.3.8}$$

where $\mathbb{E}[M(1)] \neq 1$, and $\gamma \neq \psi$.

Proof The following stopping time

$$\tau := \inf\{t \in [0, 1] : W^2(t) + t = 1\}$$

satisfies $\mathbb{P}(0 < \tau < 1) = 1$. Let us define

$$X(t) := -\frac{2W(t)}{(1-t)^2}\mathbf{1}_{\{t\leq\tau\}}$$

and the Doléans-Dade exponent $M := \mathcal{E}(X)$ of X. The process X is stochastically integrable with respect to W because

$$\int_0^1 X(s)^2 ds = 4\int_0^\tau \frac{W(s)^2}{(1-s)^4}ds < +\infty.$$

It follows from the Itô formula applied to the process $W(t)^2/(1-t)^2$ that

$$\begin{aligned}
&\int_0^1 X(s)dW(s) - \frac{1}{2}\int_0^t X(s)^2 ds \\
&\quad = -\frac{W^2(\tau)}{(1-\tau)^2} + \int_0^\tau \left\{2W(s)^2\left(\frac{1}{(1-s)^3} - \frac{1}{(1-s)^4}\right) + \frac{1}{(1-s)^2}\right\} ds \\
&\quad \leq -\frac{1}{(1-\tau)} + \int_0^\tau \frac{1}{(1-s)^2}ds \leq -1.
\end{aligned}$$

As a consequence, the Doléans-Dade exponent $M = \mathcal{E}(X)$, which is a local martingale, is not a martingale because

$$\mathbb{E}(M(1)) = \mathbb{E}\left(e^{\int_0^1 X(s)dW(s) - \frac{1}{2}X(s)^2 ds}\right) \leq e^{-1} < M(0) = 1.$$

The random variable $M(1)$ satisfies $0 < M(1) \leq e^{-1}$ and thus application of the martingale representation theorem to the square integrable martingale $\mathbb{E}[M(1) \mid \mathcal{F}_t]$ yields (11.3.7) where $\mathbb{E}\int_0^1 \gamma^2(s)ds < +\infty$. On the other hand, application of the martingale representation theorem to the local martingale M provides (11.3.8), where $\mathbb{P}(\int_0^1 \psi^2(s)ds < +\infty) = 1$. Since M is not a martingale, it follows that $\mathbb{E}\int_0^1 \psi^2(s)ds = +\infty$. Therefore $\psi \neq \gamma$. □

11.4 Hedging Equation

In this section we introduce the so-called *hedging equation* – a basic tool for examining the completeness problem. It allows us to construct the required representation

$$\hat{X} = x + \int_0^{T^*} (\varphi_t, d\hat{P}(t,\cdot)), \qquad \mathbb{P}-a.s. \tag{11.4.1}$$

for any $\hat{X} \in L^1(\Omega, \mathcal{F}_{T^*}, \mathbb{P})$ from the decomposition of its conditional expectation

$$\mathbb{E}[\hat{X} \mid \mathcal{F}_t] = \mathbb{E}\hat{X} + \int_0^t \langle f_{\hat{X}}(s), dW(s)\rangle + \int_0^t \int_U g_{\hat{X}}(s,y)\tilde{\pi}(ds,dy), \qquad t \in [0,T^*]. \tag{11.4.2}$$

Recall that the existence of a unique pair $f_{\hat{X}} \in \Phi(U)$, $g_{\hat{X}} \in \Psi_{1,2}$ in (11.4.2) follows from Theorem 6.1.1.

Theorem 11.4.1 *Let $\hat{X} \in L^1(\Omega, \mathcal{F}_{T^*}, \mathbb{P})$ and $f_{\hat{X}}$, $g_{\hat{X}}$ be given by* (11.4.2). *Then an admissible strategy (φ_t) with initial capital $x \in \mathbb{R}$ replicates $\hat{X}$, i.e.* (11.4.1) *holds, if and only if $x = \mathbb{E}[\hat{X}]$ and φ solves the hedging equation*

$$\begin{cases} f_{\hat{X}}(s) = (\varphi_s, \Gamma_1(s,\cdot)), & d\mathbb{P}\times ds - a.s., \\ g_{\hat{X}}(s,y) = (\varphi_s, \Gamma_2(s,\cdot,y)), & d\mathbb{P}\times ds \times dv - a.s. \end{cases} \tag{11.4.3}$$

Proof Writing (11.4.1) in the form

$$x + \int_0^{T^*} (\varphi_s, d\hat{P}(s,\cdot)) = \mathbb{E}(\hat{X}) + \int_0^{T^*} \langle f_{\hat{X}}(s), dW(s)\rangle + \int_0^{T^*}\int_U g_{\hat{X}}(s,y)\tilde{\pi}(ds,dy), \tag{11.4.4}$$

and taking expectations of both sides, we obtain $x = \mathbb{E}(\hat{X})$. The process

$$\begin{aligned} M_t := &\int_0^t (\varphi_s, d\hat{P}(s,\cdot)) - \int_0^t \langle f_{\hat{X}}(s), dW(s)\rangle - \int_0^t\int_U g_{\hat{X}}(s,y)\tilde{\pi}(ds,dy) \\ = &\int_0^t \langle((\varphi_s, \Gamma_1(s,\cdot)) - f_{\hat{X}}(s)), dW(s)\rangle \\ &+ \int_0^t\int_U \big((\varphi_s, \Gamma_2(s,\cdot,y)) - g_{\hat{X}}(s,y)\big)\,\tilde{\pi}(ds,dy), \qquad t \in [0,T^*] \end{aligned}$$

is thus a martingale equal to zero. With the use of Theorem 6.1.1 we obtain (11.4.3). □

In view of (11.3.3) and (11.4.2) it is easy to see that (11.4.3) is sufficient for φ to replicate X. Necessity is, however, not so obvious and requires the assumption that admissible strategies generate discounted wealth processes $\hat{X}(t)$, which are martingales. If we require instead that $\hat{X}(t)$ is a positive local martingale then (11.4.3) does not have to hold, see Proposition 11.3.4.

11.5 Completeness for the HJM Model

This section is concerned with the completeness problem of the HJM model

$$df(t,T) = \alpha(t,T)dt + \langle\sigma(t,T), dZ(t)\rangle, \quad t \in [0,T^*], \quad T > 0, \tag{11.5.1}$$

with a $U = \mathbb{R}^d$-valued Lévy process Z. Recall that the model satisfies the (MP) condition. We consider two cases when the support of the Lévy measure ν of Z is finite (finite activity noise) and when the support of ν has a concentration point. The results involve the following random field

$$\Sigma(t,T) := \int_{t\wedge T}^{T} \sigma(t,s)ds, \quad t \in [0,T^*], T > 0,$$

which appeared in the formulation of Theorem 8.1.1 on the HJM drift conditions. To be precise, we should write $\Sigma(t,T,\omega)$, but the dependence on ω will be omitted to keep the notation simple. It is clear that $\Sigma(t,T)$ is a continuous function of $T \geq 0$.

11.5.1 Lévy Measure with Finite Support

First, we treat the completeness problem in the case in which support of the Lévy measure of Z is finite:

$$\text{supp}\{\nu\} = \{y_1, y_2, \ldots, y_n\}, \quad y_i \in U, \; i = 1, 2, \ldots, n. \tag{11.5.2}$$

Then Z is a finite activity process with a finite number of jump sizes. We will often replicate contingent claims with the use of a finite number of bonds with maturities, say $T_1, T_2, \ldots, T_K$, with some K. Then

$$\varphi_t = \sum_{k=1}^{K} \varphi(t,T_k)\delta_{\{T_k\}}, \quad t \in [0,T^*],$$

with some predictable processes $\varphi(t,T_1), \ldots, \varphi(t,T_K)$, $t \in [0,T^*]$. Now we formulate conditions for φ to replicate a given contingent claim X. Solution of the completeness problem in the one-dimensional case is given by the following result.

Theorem 11.5.1 *Let Z be a real-valued Lévy process with jumps satisfying (11.5.2). Let us assume that there exist maturities $T_1, T_2, \ldots, T_{n+2}$ in $[T^*, +\infty)$ such that $d\mathbb{P} \times dt$-a.s.:*

$$\Sigma(t,T_i) \neq \Sigma(t,T_j), \quad i \neq j, \quad t \in [0,T^*].$$

Then each $\hat{X} \in L^1(\Omega, \mathcal{F}_{T^*}, \mathbb{P})$ *can be replicated by strategies involving* T_i*-bonds only with* $i = 1, 2, \ldots, n+2$. *In particular, the market is complete in the class* $\hat{X} \in L^\infty(\Omega, \mathcal{F}_{T^*}, \mathbb{P})$.

If $d \geq 1$ then the conditions for completeness involve the following functions

$$\Sigma(t,T) := (\Sigma^1(t,T), \ldots, \Sigma^d(t,T)), \; e^{-\langle \Sigma(t,T), y_1 \rangle} - 1, \; \ldots \;, e^{-\langle \Sigma(t,T), y_n \rangle} - 1. \tag{11.5.3}$$

For fixed (ω, t), the functions in (11.5.3) depend continuously on the argument $T > 0$.

Theorem 11.5.2 *(a) If the functions* (11.5.3) *of argument T are linearly dependent with positive* $d\mathbb{P} \times dt$*-measure, then the market is not complete in the class* $\hat{X} \in L^\infty(\Omega, \mathcal{F}_{T^*}, \mathbb{P})$ *and* $\mathcal{M}$*-valued admissible strategies.*

(b) If the functions (11.5.3) *are linearly independent* $d\mathbb{P} \times dt$*-a.s., then each* $\hat{X} \in L^\infty(\Omega, \mathcal{F}_{T^*}, \mathbb{P})$ *can be replicated by a strategy consisting, at any time, of* $d+n$ *bonds with different maturities. In particular, the market is complete in the class* $\hat{X} \in L^\infty(\Omega, \mathcal{F}_{T^*}, \mathbb{P})$.

Replicating strategies in Theorem 11.5.2 (b) consist of $n+d$ bonds, but the choice of bonds is time and ω-dependent. The result is thus weaker than in Theorem 11.5.1 where maturities of the involved bonds are fixed. In the multidimensional case we can, however, impose stronger assumptions that also guarantee that maturities of replicating bonds can be fixed in advance.

Theorem 11.5.3 *Let us assume functions* (11.5.3) *of argument* $T \geq T^*$ *are linearly independent* $d\mathbb{P} \times dt$*- a.s. Moreover, assume that the functions*

$$t \longrightarrow \Sigma(t,T), \quad t \longrightarrow e^{-\langle \Sigma(t,T), y_i \rangle} - 1, \qquad i = 1, 2, \ldots, n; \; T \in [T^*, +\infty)$$

are analytic on the interval $[0, T^*]$. *Then there exists a set of dates* $T_1, T_2, \ldots, T_{d+n} \in [T^*, +\infty)$*, such that each* $\hat{X} \in L^1(\Omega, \mathcal{F}_{T^*}, \mathbb{P})$ *can be replicated with the use of bonds with maturities* $T_1, T_2, \ldots, T_{d+n}$. *Specifically, the market is complete.*

Before we present proofs of the preceding results, we present first some of their implications.

In view of Theorem 11.5.2 and Theorem 8.2.11 we obtain the following result connecting completeness and the uniqueness of the martingale measure.

Theorem 11.5.4 *Let Z be a finite activity Lévy process in* $\mathbb{R}^d$ *with a finite number of jump sizes. Then the HJM model satisfying (MP) is complete in the class* $\hat{X} \in L^\infty(\Omega, \mathcal{F}_{T^*}, \mathbb{P})$ *if and only if the martingale measure is unique.*

Theorem 11.5.2 is ilustrated by a model with constant volatility.

Proposition 11.5.5 *Let Z be a Lévy process in $\mathbb{R}^d$ without Wiener part and jumps satisfying* (11.5.2). *If the volatility is constant: $\sigma(t,T) \equiv \sigma$ and*

$$y_1 + y_2 + \cdots + y_n \neq 0,$$

then each $\hat{X} \in L^1(\Omega, \mathcal{F}_{T^*}, \mathbb{P})$ is attainable. Hence, the market is complete in the class $\hat{X} \in L^\infty(\Omega, \mathcal{F}_{T^*}, \mathbb{P})$.

Proof In the proof we use Theorem 11.5.2. Since $\Sigma(t,T) = (T-t)\sigma$ the condition for completeness amounts to the linear independence of the functions

$$h_1(T) := e^{-(T-t)\langle\sigma, y_1\rangle} - 1, \dots, h_n(T) := e^{-(T-t)\langle\sigma, y_n\rangle} - 1.$$

It is clear that the linear independence of $h_1, \dots, h_n$ is implied by the linear independence of $\tilde{h}_1, \dots, \tilde{h}_n, 1$ where $\tilde{h}_i(T) = h_i(T) + 1, i = 1, 2, \dots, n$. Now we examine linear independence of the latter set. Its Wrońskian $W(T)$ (see Pontriagin [104], Section 3.18) equals

$$W(T) = DET \begin{bmatrix} \tilde{h}_1(T) & \tilde{h}_2(T) & \dots & \tilde{h}_n(T) & 1 \\ \tilde{h}'_1(T) & \tilde{h}'_2(T) & \dots & \tilde{h}'_n(T) & 0 \\ \vdots & \vdots & \vdots & \vdots & \vdots \\ \tilde{h}^{(n)}_1(T) & \tilde{h}^{(n)}_2(T) & \dots & \tilde{h}^{(n)}_n(T) & 0 \end{bmatrix}$$

$$= (-1)^{n+2} DET \begin{bmatrix} \tilde{h}'_1(T) & \tilde{h}'_2(T) & \dots & \tilde{h}'_n(T) \\ \vdots & \vdots & \vdots & \vdots \\ \tilde{h}^{(n)}_1(T) & \tilde{h}^{(n)}_2(T) & \dots & \tilde{h}^{(n)}_n(T) \end{bmatrix}.$$

Since

$$\tilde{h}^{(j)}_i(T) = (-\langle y_i, \sigma\rangle)^j \cdot \tilde{h}_i(T), \quad i = 1, 2, \dots, n, \quad j = 0, 1, \dots, n,$$

we have

$$W(T)$$

$$= (-1)^{n+2} DET \begin{bmatrix} (-\langle y_1, \sigma\rangle)\tilde{h}_1(T) & (-\langle y_2, \sigma\rangle)\tilde{h}_2(T) & \dots & (-\langle y_n, \sigma\rangle)\tilde{h}_n(T) \\ (-\langle y_1, \sigma\rangle)^2\tilde{h}_1(T) & (-\langle y_2, \sigma\rangle)^2\tilde{h}_2(T) & \dots & (-\langle y_n, \sigma\rangle)^2\tilde{h}_n(T) \\ \vdots & \vdots & \vdots & \vdots \\ (-\langle y_1, \sigma\rangle)^n\tilde{h}_1(T) & (-\langle y_2, \sigma\rangle)^n\tilde{h}_2(T) & \dots & (-\langle y_n, \sigma\rangle)^n\tilde{h}_n(T) \end{bmatrix}.$$

This yields

$$W(T) = (-1)^{n+1}\langle \sigma, y_1 + y_2 + \cdots + y_n\rangle e^{-(T-t)\langle \sigma, y_1+y_2+\cdots+y_n\rangle}$$

$$\cdot\, DET \begin{bmatrix} 1 & 1 & \dots & 1 \\ (-\langle y_1,\sigma\rangle) & (-\langle y_2,\sigma\rangle) & \dots & (-\langle y_n,\sigma\rangle) \\ \vdots & \vdots & \vdots & \vdots \\ (-\langle y_1,\sigma\rangle)^{n-1} & (-\langle y_2,\sigma\rangle)^{n-1} & \dots & (-\langle y_n,\sigma\rangle)^{n-1} \end{bmatrix}.$$

However, the matrix is of the Vandermonde type (see Knapp [83, p. 215]) and thus

$$W(T) = (-1)^{n+1}\langle \sigma, y_1+y_2+\cdots+y_n\rangle e^{-(T-t)\langle \sigma, y_1+y_2+\cdots+y_n\rangle} \prod_{1\le i<j\le n} \sigma\cdot(y_i-y_j) \neq 0.$$

Consequently, functions $h_1(T), \dots, h_n(T)$ are linearly independent. □

11.5.2 Proofs of Theorems 11.5.1–11.5.3

The proofs involve the hedging equation (11.4.3), which, in the present setting, can be transformed to the system of linear equations. We first describe this system and after that prove the results.

Proposition 11.5.6 *Let $\hat{X} \in L^1(\Omega, \mathcal{F}_{T^*}, \mathbb{P})$ and its conditional expectation has the form*

$$\mathbb{E}[\hat{X} \mid \mathcal{F}_t] = \mathbb{E}[\hat{X}] + \int_0^t \langle f_{\hat{X}}(s), dW(s)\rangle + \int_0^t \sum_{i=1}^n g_{\hat{X}}(s, y_i)\tilde{\pi}(ds, \{y_i\}), \quad t \in [0, T^*]. \tag{11.5.4}$$

Then a strategy φ involving only bonds with maturities $T_1, T_2, \dots, T_K$ replicates $\hat{X}$ if and only if, for each $t \in [0, T^*]$,

$$\begin{bmatrix} -\Sigma^1(t,T_1) & \dots & -\Sigma^1(t,T_K) \\ \vdots & & \vdots \\ -\Sigma^d(t,T_1) & \dots & -\Sigma^d(t,T_K) \\ e^{-\langle\Sigma(t,T_1),y_1\rangle}-1 & \dots & e^{-\langle\Sigma(t,T_K),y_1\rangle}-1 \\ \vdots & & \vdots \\ e^{-\langle\Sigma(t,T_1),y_n\rangle}-1 & \dots & e^{-\langle\Sigma(t,T_K),y_n\rangle}-1 \end{bmatrix} \begin{bmatrix} \hat{P}(t-,T_1)\cdot\varphi(t,T_1) \\ \hat{P}(t-,T_2)\cdot\varphi(t,T_2) \\ \vdots \\ \hat{P}(t-,T_K)\cdot\varphi(t,T_K) \end{bmatrix} = \begin{bmatrix} f^1_{\hat{X}}(t) \\ \vdots \\ f^d_{\hat{X}}(t) \\ g_{\hat{X}}(t,y_1) \\ \vdots \\ g_{\hat{X}}(t,y_n) \end{bmatrix}. \tag{11.5.5}$$

In the preceding, $\Sigma^1(t,T), \Sigma^2(t,T), \dots, \Sigma^d(t,T)$ stand for the coordinates of $\Sigma(t,T)$.

Proof Formula (11.5.5) is a special form of (11.4.2) in our setting with finite activity noise and a finite number of bonds in the portfolio. The corresponding to φ discounted wealth process equals

$$\hat{X}(t) = X(0) + \int_0^t \Big\langle \sum_{j=1}^{K} \varphi(s, T_j)\Gamma_1(s), dW(s)\Big\rangle + \sum_{i=1}^{n} \int_0^t \Big\{\sum_{j=1}^{K} \varphi(s, T_j)\Gamma_2(s, T_j, y)\Big\}\tilde{\pi}(ds, \{y_i\}).$$

Consequently, the hedging equation (11.4.3) amounts to

$$f_{\hat{X}}(t) = \sum_{j=1}^{K} \varphi(t, T_j)\Gamma_1(t, T_j), \qquad g_{\hat{X}}(t, y_i) = \sum_{j=1}^{K} \varphi(t, T_j)\Gamma_2(t, T_j, y_i),\ i = 1, 2, \ldots, n,$$
$$d\mathbb{P} \times dt - a.s. \tag{11.5.6}$$

By Proposition 11.2.1, we have,

$$\Gamma_1(t, T) = -\hat{P}(t-, T)\Sigma(t, T), \quad \Gamma_2(t, T, y) = \hat{P}(t-, T)[e^{-\langle\Sigma(t,T),y\rangle} - 1],$$

and, consequently, (11.5.2) can be written as the system of linear equations (11.5.5). □

Proof of Theorem 11.5.1 We show that for $d = 1$ and $K = n + 2$ the rank of the matrix in (11.5.5) is $n + 1$. This will imply that X can be replicated by a strategy involving bonds with maturities $T_1, T_2, \ldots, T_{n+2}$. We prove that the vectors

$$v_0 := \begin{pmatrix} \Sigma(t, T_1) \\ \vdots \\ \Sigma(t, T_{n+2}) \end{pmatrix}, v_1 := \begin{pmatrix} e^{-\Sigma(t,T_1)y_1} - 1 \\ \vdots \\ e^{-\Sigma(t,T_{n+2})y_1} - 1 \end{pmatrix}, \ldots, v_n := \begin{pmatrix} e^{-\Sigma(t,T_1)y_n} - 1 \\ \vdots \\ e^{-\Sigma(t,T_{n+2})y_n} - 1 \end{pmatrix} \tag{11.5.7}$$

are linearly independent in $\mathbb{R}^{n+2}$, i.e.

$$\sum_{j=0}^{n} c_j v_j = 0 \quad \Longrightarrow \quad c_0 = c_1 = \cdots = c_n = 0 \tag{11.5.8}$$

is satisfied. Set $x(T) := e^{-\Sigma(t,T)} > 0$ and consider the function

$$f(x) := c_0 \ln x + \sum_{j=1}^{n} c_j(x^{y_j} - 1), \quad x > 0$$

for a non-trivial sequence $c_0, \ldots, c_n$, and its derivative

$$f'(x) = -\frac{c_0}{x} + \sum_{j=1}^{n} c_j y_j x^{y_j - 1} = \frac{1}{x}\left(-c_0 + \sum_{j=1}^{n} c_j y_j x^{y_j}\right), \qquad x > 0.$$

By Lemma 2.2.5 the number of positive roots of f' is no greater than n. Since between each two consecutive roots of f there is at least one root of f', the function f has at most $n+1$ positive roots. Since $x(T_1), \ldots, x(T_{n+2})$ are $n+2$ different numbers, it follows that $f(x(T_i)) \neq 0$ for some T_i. This implies (11.5.8). □

For the proof of Theorem 11.5.2 we need the following auxiliary lemma on the linear independence of infinite sequences. The proof is left to the reader.

Lemma 11.5.7 *Let M be an infinite matrix of the form*

$$M = \begin{pmatrix} z^1 \\ z^2 \\ \vdots \\ z^r \end{pmatrix} = \begin{bmatrix} z_1^1 & z_2^1 & z_3^1 & \cdots \\ z_1^2 & z_2^2 & z_3^2 & \cdots \\ \vdots & \vdots & \vdots & \vdots \\ z_1^r & z_2^r & z_3^r & \cdots \end{bmatrix},$$

with linearly independent rows $z^1, z^2, \ldots, z^r$. Then there exists a set of r linearly independent columns of the matrix M.

Proof of Theorem 11.5.2 (*a*) Recall that $\mathcal{M}([0, +\infty))$ stands for the space of finite signed measures on $[0, +\infty)$. For fixed $t \in [0, T^*]$ and given vectors (11.5.3) let us define the linear transformation $\mathcal{K}_t : \mathcal{M}([0, +\infty)) \longrightarrow \mathbb{R}^{d+n}$ by

$$\mathcal{K}_t(\mu) := \begin{pmatrix} (\mu, \Gamma_1(t, \cdot)) \\ (\mu, \Gamma_2(t, \cdot, y_i)) \end{pmatrix}_{i=1,2,\ldots,n} = \begin{pmatrix} -(\mu, \hat{P}(t-, \cdot)\Sigma^1(t, \cdot)) \\ \vdots \\ -(\mu, \hat{P}(t-, \cdot)\Sigma^d(t, \cdot)) \\ \left(\mu, \hat{P}(t-, \cdot)(e^{-\langle \Sigma(t, \cdot), y_1\rangle} - 1)\right) \\ \vdots \\ \left(\mu, \hat{P}(t-, \cdot)(e^{-\langle \Sigma(t, \cdot), y_n\rangle} - 1)\right) \end{pmatrix}.$$

Let $A \subseteq \Omega \times [0, T^*]$ be the set of positive $d\mathbb{P} \times dt$-measure where (11.5.3) are linearly dependent. On the set A the dimension of the image of $\mathcal{K}_t$ is strictly less than $d+n$, so the orthogonal complement of $Im(\mathcal{K}_t)$ is not trivial. Let $e_1(t), e_2(t), \ldots, e_k(t)$, with some $k > 0$, be an orthonormal basis of $Im(\mathcal{K}_t)^{\perp}$.

This basis can be chosen in a predictable way which follows from Proposition 11.5.8 and Remark 11.5.9. With the use of $e_1(t) = (e_1^1(t), \dots, e_1^{d+n}(t))$ let us define the following processes

$$v_1(t) := e_1^1(t), \ \dots, \ v_d(t) := e_1^d(t), \quad v_{d+1}(t, y_i) := e_1^{d+i}(t), \quad i = 1, 2, \dots, n,$$

on A and on A^c we set $v_1(t) = \dots = v_d(t) = v_{d+1}(t, y_i) = 0$ for $i = 1, 2, \dots, n$. Let us define the process

$$V(t) := \int_0^t \langle (v_1(s), \dots, v_d(s)), dW(s) \rangle + \int_0^t \int_U v_{d+1}(s, y) \tilde{\pi}(ds, dy), \quad t \in [0, T^*]$$

and stopping times

$$\tau_k := \inf\{t \in [0, T^*] : V(t) \geq k\} \wedge T^*.$$

With the use of the sequence

$$A_k := \{(\omega, t) \in A, t \in [0, \tau_k)\},$$

we fix an index k_0 such that $(d\mathbb{P} \times dt)(A_{k_0}) > \varepsilon$ for some $\varepsilon > 0$. For the bounded random variable $\hat{X} := V(T^* \wedge \tau_{k_0})$ and the discounted wealth process related to a pair (x, φ)

$$\hat{X}(T^*) = x + \int_0^{T^*} \langle (\varphi_s, \Gamma_1(s, \cdot)), dW(s) \rangle + \int_0^{T^*} \int_U (\varphi_s, \Gamma_2(s, \cdot, y)) \tilde{\pi}(ds, dy),$$

we have the following estimation

$$\begin{aligned} \mathbb{E}(\hat{X}(T^*) - \hat{X})^2 &= x^2 + \mathbb{E}\Big(\int_0^{T^*} \langle \mathbf{1}_{[0, \tau_{k_0})}(v_1(s), \dots, v_d(s)) - (\varphi_s, \Gamma_1(s)), dW(s) \rangle \Big)^2 \\ &\quad + \mathbb{E}\Big(\int_0^t \int_U \Big(\mathbf{1}_{[0, \tau_{k_0})} v_{d+1}(s, y) - (\varphi_s, \Gamma_2(s, \cdot, y)) \Big) \tilde{\pi}(ds, dy) \Big)^2 \\ &\geq \mathbb{E} \int_0^{\tau_{k_0}} \Big(v_1^2(s) + \dots + v_d^2(s) \Big) ds + \mathbb{E} \int_0^{\tau_{k_0}} \sum_{i=1}^n v_{d+1}^2(s, y_i) ds \\ &= \mathbb{E} \int_0^{\tau_{k_0}} | \, e_1(s) \, |^2 \, ds = (d\mathbb{P} \times dt)(A_{k_0}) > \varepsilon. \end{aligned}$$

In the preceding we used the fact that $e_1(t)$ is orthogonal to $Im(\mathcal{K}_t)$. So, we showed that $\hat{X}$ cannot be replicated by any measure-valued strategy φ.

(*b*) In virtue of Lemma 11.5.7 one can find maturities $T_1, T_2, \ldots, T_{d+n} \in (t, +\infty)$ such that

$$\begin{pmatrix} \Sigma^1(t, T_j) \\ \vdots \\ \Sigma^d(t, T_j) \\ e^{-\langle \Sigma(t,T_j), y_1 \rangle} - 1 \\ \vdots \\ e^{-\langle \Sigma(t,T_j), y_n \rangle} - 1 \end{pmatrix}, j = 1, 2, \ldots, d+n; \tag{11.5.9}$$

are linearly independent vectors in $\mathbb{R}^{d+n}$. The hedging equation when one is trading only with bonds with maturities $T_1, T_2, \ldots, T_{d+n}$ is equivalent to (11.5.5) with $K = n+d$. The solution $\varphi(t, T_1), \ldots, \varphi(t, T_{d+n})$ of (11.5.5) exists because the matrix in (11.5.5) is nonsingular. The solution is clearly a replicating strategy for $\hat{X}$. □

Proposition 11.5.8 *Assume that E is a Banach space and $K_1, \ldots, K_N$ are E-valued random variables on a probability space $(\Omega, \mathcal{G}, \mathbb{P})$. There exists a sequence $e_1, \ldots, e_N$, where $e_j = (e_j^1, \ldots, e_j^N), j = 1, \ldots, N$, of orthonormal vectors in $\mathbb{R}^N$ which are $\mathcal{G}$ measurable and such that:*

$$\inf\left\{ \|\sum_{j=1}^N K_j(\omega) z^j\| \, : |z| = 1 \right\} = \|\sum_{j=1}^N K_j(\omega) e_1^j(\omega)\|,$$

and for $k = 1, \ldots, N-1$,

$$\inf\left\{ \|\sum_{j=1}^N K_j(\omega) z^j\| \, : |z| = 1, < z, e_j >= 0, j = 1, \ldots, k \right\} = \|\sum_{j=1}^N K_j(\omega) e_{k+1}^j(\omega)\|.$$

Proof Let S_N be the unit sphere in $\mathbb{R}^N$. The function

$$M(\omega, z) = \|\sum_{j=1}^N K_j(\omega) z^j\|, \ \omega \in \Omega, \ z = (z^1, \ldots, z^N) \in S_N$$

is continuous in z and $\mathcal{G}$-measurable in ω. By the Kuratowski–Ryll–Nardzewski theorem on measurable selectors (see e.g. Peszat and Zabczyk [100], p. 115) for arbitrary compact subset of Z of S_N, there exists $\mathcal{G}$-measurable Z-valued function e such that

$$\inf\left\{ \|\sum_{j=1}^N K_j(\omega) z^j\| \, : z \in Z \right\} = \|\sum_{j=1}^N K_j(\omega) e^j\|. \qquad \square$$

Remark 11.5.9 Note that if $K_j(\omega)$, $j = 1, \ldots, N$ are linearly dependent then

$$\inf\left\{||\sum_{j=1}^{N} K_j(\omega)z^j|| \; : z \in S_N\right\} = 0.$$

Moreover, $||\sum_{j=1}^{N} K_j(\omega)z^j|| = 0$ if and only if the vector e is orthogonal to the image of the following transformation $K(\omega)$ from E^* into $\mathbb{R}^N$:

$$K(\omega)x^* = (x^*(K_1(\omega)), \ldots, x^*(K_N(\omega))).$$

Before we prove Theorem 11.5.3 let us explain why we need to use bonds with maturities from $[T^*, +\infty)$. If $T_1, T_2, \ldots, T_{d+n} \in [0, T^*]$ then the columns of the matrix in (11.5.5) become zero vectors if $T_i < t$ and thus the system does not have a solution, in general.

Proof of Theorem 11.5.3 Fix any $t \in [0, T^*]$ such that the functions (11.5.3) restricted to $[T^*, +\infty)$ are linearly independent. In virtue of Lemma 11.5.7 we can find maturities $T_1, T_2, \ldots, T_{d+n} \in [T^*, +\infty)$ such that the matrix

$$A(t) := \begin{bmatrix} -\Sigma^1(t, T_1) & \ldots & -\Sigma^1(t, T_{d+n}) \\ \vdots & & \vdots \\ -\Sigma^d(t, T_1) & \ldots & -\Sigma^d(t, T_{d+n}) \\ e^{-\langle\Sigma(t,T_1),y_1\rangle} - 1 & \ldots & e^{-\langle\Sigma(t,T_{d+n}),y_1\rangle} - 1 \\ \vdots & & \vdots \\ e^{-\langle\Sigma(t,T_1),y_n\rangle} - 1 & \ldots & e^{-\langle\Sigma(t,T_{d+n}),y_n\rangle} - 1 \end{bmatrix}$$

is invertible. Moreover, the function $t \to detA(t), t \in [0, T^*]$ is analytic and thus can be equal to zero at a finite number of points only. As a consequence, the matrix $A(t)$ is invertible for almost all $t \in [0, T^*]$ and thus the system (11.5.5) has a solution on the interval $[0, T^*]$. □

11.5.3 Incomplete Markets

In this section we show that for rather general HJM models with the process Z, having Lévy measure with infinite support completeness fails. Our analysis covers the case in which the measure ν has a so-called *concentration point* or it is supported by an infinite sequence that diverges to infinity.

Definition 11.5.10 The point $y_0 \in U = \mathbb{R}^d$ is a concentration point of the Lévy measure ν if there exists a sequence of positive numbers $\{\varepsilon_n\}_{n=1}^{\infty}$ s.t. $\varepsilon_n \downarrow 0$ satisfying

$$\nu(D_n) > 0, \quad \forall\, n = 1, 2, \ldots, \tag{11.5.10}$$

where $D_n := \{y \in U : \varepsilon_{n+1} <| y - y_0 |\leq \varepsilon_n\}$.

Let us notice that the condition formulated in Definition 11.5.10 is very often satisfied. For example every Lévy measure with density has a concentration point. In fact, in this case each point is a concentration point. Thus the following theorem comprises a large class of models.

Theorem 11.5.11 *If the Lévy measure ν has a concentration point $y_0 \neq 0$ and*

$$\int_0^{+\infty} \mid \sigma(t,T) \mid dT < +\infty, \quad t \in [0,T^*], \quad \mathbb{P}-a.s., \tag{11.5.11}$$

then the market is not complete in the class $\hat{X} \in L^\infty(\Omega, \mathcal{F}_{T^*}, \mathbb{P})$. In particular, if ν has a density, then the market is not complete in the class $\hat{X} \in L^\infty(\Omega, \mathcal{F}_{T^*}, \mathbb{P})$.

For the proof of Theorem 11.5.11, we use two auxiliary results that proofs will be given at the end of the section.

Lemma 11.5.12 *Let E be a normed linear space and A an arbitrary set. Let $g\colon A \longrightarrow \mathbb{R}$ and $h\colon A \longrightarrow E$. Then there exists $e^* \in E^*$ such that*

$$g(a) = < e^*, h(a) >_E, \quad \forall a \in A, \tag{11.5.12}$$

if and only if

$$\exists\, \gamma > 0 \quad \forall\, n \in \mathbb{N} \quad \forall\, \{\beta_i\}_{i=1}^n,\ \beta_i \in \mathbb{R} \quad \forall\, \{a_i\}_{i=1}^n,\ a_i \in A \quad \textit{holds}$$

$$\left| \sum_{i=1}^n \beta_i g(a_i) \right| \leq \gamma \left| \sum_{i=1}^n \beta_i h(a_i) \right|_E. \tag{11.5.13}$$

The proof of Lemma 11.5.12 is postponed until the final part of this section.

Lemma 11.5.13 *Let $(E_1, \mathcal{E}_1, \mu_1)$, $(E_2, \mathcal{E}_2, \mu_2)$ be measurable spaces with sigma-finite measures μ_1, μ_2 and $(E_1 \times E_2, \mathcal{E}_1 \times \mathcal{E}_2, \mu_1 \times \mu_2)$ be their product space. If two measurable functions $f_1 : E_1 \times E_2 \longrightarrow \mathbb{R}$, $f_2 \colon E_1 \times E_2 \longrightarrow \mathbb{R}$ satisfy the condition*

$$f_1 = f_2, \qquad d\mu_1 \times d\mu_2 - a.s., \tag{11.5.14}$$

then there exists a set $\hat{E}_1 \in \mathcal{E}_1$ such that

$$\hat{E}_1 \quad \textit{is of full } \mu_1 - \textit{measure} \tag{11.5.15}$$

$$\forall x \in \hat{E}_1 \quad \textit{the set} \quad \{y : f_1(x,y) = f_2(x,y)\} \quad \textit{is of full } \mu_2 - \textit{measure}. \tag{11.5.16}$$

Proof of Theorem 11.5.11 In the proof we construct a bounded random variable $\hat{X}$ of the form

$$\hat{X} = \int_0^{T^*} g_{\hat{X}}(s,y)\tilde{\pi}(ds,dy),$$

such that $\int_0^{\cdot}\int_U g_{\hat{X}}(s,y)\tilde{\pi}(ds,dy)$ is a martingale and the hedging equation

$$g_{\hat{X}}(s,y) = (\varphi_s, \Gamma_2(s,\cdot,y)), \quad d\mathbb{P}\times ds\times d\nu - a.s. \tag{11.5.17}$$

does not have a solution. This implies that X is not attainable.

Let $\{\varepsilon_n\}_{n=1}^{\infty}$ be a sequence satisfying (11.5.10) and define an auxiliary deterministic function g by the formula

$$g(y) = (|y|\wedge 1)\mathbf{1}_{\{y\notin D\}} - (|y|\wedge 1)\mathbf{1}_{\{y\in D\}}, \text{ where } D := \bigcup_{k=1}^{+\infty} D_{2k} \quad y\in U.$$

In particular, $g(y) = (|y|\wedge 1)$ for $y\in D_{2k+1}, k = 0,1,\dots$ and $g(y_0) = (|y_0|\wedge 1)$. First we show that there is no admissible strategy φ such that

$$(\varphi_t, \Gamma_2(t,\cdot,y)) = g(y) \tag{11.5.18}$$

holds $d\mathbb{P}\times dt\times d\nu$-a.s. The left side of (11.5.18) is well defined because, by (11.5.11), the function $\Gamma_2(t,\cdot,y)$ is continuous and bounded. Let us fix any pair $(\omega,t)\in\Omega\times[0,T^*]$ and assume that (11.5.18) holds $d\nu$ - a.s. In view of Lemma 11.5.13 there exists a set $A_\nu(\omega,t)$ of a full ν-measure such that (11.5.18) is satisfied for each $y\in A_\nu(\omega,t)$. Due to Lemma 11.5.12 there exists $\gamma = \gamma(\omega,t) > 0$ such that

$$\forall\, n\in\mathbb{N} \quad \forall\,\{\beta_i\}_{i=1}^n,\ \beta_i\in\mathbb{R} \quad \forall\,\{y_i\}_{i=1}^n,\ y_i\in A_\nu(\omega,t)$$

$$\Big|\sum_{i=1}^n \beta_i g(y_i)\Big| \le \gamma\Big|\sum_{i=1}^n \beta_i\Gamma_2(t,y_i)\Big|_{C_b(\mathbb{R}_+)}. \tag{11.5.19}$$

Here $C_b(\mathbb{R}_+)$ stands for the Banach space of continuous bounded functions on $\mathbb{R}_+$ equipped with the supremum norm. Clearly, (11.5.10) implies that

$$\nu\Big\{A_\nu(\omega,t)\cap D_k\Big\} > 0, \quad k = 1,2,\dots,$$

so we can choose a sequence $\{a_k\}_{k=1}^{\infty}$ s.t.

$$a_k\in A_\nu(\omega,t)\cap D_k \quad \forall\, k = 1,2,\dots.$$

Let us examine now (11.5.19) with $n = 2$, $\beta_1 = 1, \beta_2 = -1$ and $y_1 = a_{2k+1}$, $y_2 = a_{2k+2}$ for $k = 0,1,\dots$. After dividing by γ, the left side of (11.5.19) has the form

$$\frac{1}{\gamma}\Big|\beta_1 g(a_{2k+1}) + \beta_2 g(a_{2k+2})\Big| = \frac{1}{\gamma}\Big((|a_{2k+1}|\wedge 1) + (|a_{2k+2}|\,|\wedge 1)\Big)$$

and thus satisfies

$$\lim_{k\longrightarrow\infty}\frac{1}{\gamma}\Big|\beta_1 g(a_{2k+1}) + \beta_2 g(a_{2k+2})\Big| = \frac{2(|y_0|\wedge 1)}{\gamma} \ne 0. \tag{11.5.20}$$

For estimating the right side of (11.5.19) we use the following

$$\begin{aligned}
&\left|\Gamma_2(t,a_{2k+1})-\Gamma_2(t,a_{2k+1})\right|_{C_b(\mathbb{R}_+)}\\
&\quad=\sup_{T\in[0,+\infty)}\left|\hat{P}(t-,T)(e^{-\langle\Sigma(t,T),a_{2k+1}\rangle}-1)-\hat{P}(t-,T)(e^{-\langle\Sigma(t,T),a_{2k+1}\rangle}-1)\right|\\
&\quad\le\sup_{T\in[0,+\infty)}|\hat{P}(t-,T)|\cdot\sup_{T\in[0,+\infty)}\left|e^{-\langle\Sigma(t,T),a_{2k+1}\rangle}-e^{-\langle\Sigma(t,T),a_{2k+2}\rangle}\right|.
\end{aligned}$$

The first supremum is clearly finite. To deal with the second supremum let us fix $\delta>0$ such that $\mid a_{2k+1}-a_{2k+2}\mid<\delta$ for $k=0,1,2,\ldots$. Then

$$\begin{aligned}
&\sup_{T\in[0,+\infty)}\left|e^{-\langle\Sigma(t,T),a_{2k+1}\rangle}-e^{-\langle\Sigma(t,T),a_{2k+2}\rangle}\right|\\
&\qquad\le\sup_{T\in[0,+\infty)}\sup_{|y-y_0|<\delta}\left|De^{-\langle\Sigma(t,T),y\rangle}\right|_{L(U,\mathbb{R})}\cdot|a_{2k+1}-a_{2k+2}|\\
&\qquad\le\sup_{T\in[0,+\infty)}\sup_{|y-y_0|<\delta}\left\{e^{|\Sigma(t,T)|\cdot|y|}\cdot|\Sigma(t,T)|\right\}\cdot|a_{2k+1}-a_{2k+2}|.
\end{aligned}$$

Since $|a_{2k+1}-a_{2k+2}|\to 0$ and (11.5.11) implies that $\sup_{T\in[0,+\infty)}|\Sigma(t,T)|<+\infty$, we see that

$$\left|\Gamma_2(t,a_{2k+1})-\Gamma_2(t,a_{2k+1})\right|_{C_b(\mathbb{R}_+)}\underset{k\to+\infty}{\longrightarrow}0. \tag{11.5.21}$$

It follows from (11.5.20) and (11.5.21) that (11.5.19) is not satisfied for any $(\omega,t)\in\Omega\times[0,T^*]$. Consequently (11.5.18) does not hold $\nu-a.s.$ for any $(\omega,t)\in\Omega\times[0,T^*]$.

Now, with the use of function g, we construct a bounded random variable $\hat{X}$ which cannot be replicated. Since

$$\int_0^{T^*}\int_U(\mid g(y)\mid^2\wedge\mid g(y)\mid)\,ds\nu(dy)\le\int_0^{T^*}\int_U(\mid y\mid^2\wedge 1)\,ds\nu(dy)<+\infty,$$

we see that $g\in\Psi_{1,2}$, so g is integrable over $\tilde{\pi}(ds,sy)$. Let τ_k be the stopping time defined by

$$\tau_k=\inf\{t:\left|\int_0^t\int_U g(y)\tilde{\pi}(ds,dy)\right|\ge k\}\wedge T^*,$$

and choose a number k_0 s.t. the set $\{(\omega,\tau_{k_0}(\omega));\omega\in\Omega\}\subseteq\Omega\times[0,T^*]$ is of positive $d\mathbb{P}\times dt$- measure. Then the process

$$g_{\hat{X}}(s,y):=g(y)\mathbf{1}_{(0,\tau_{k_0}]}(s),\quad s\in[0,T^*],\ y\in U$$

is predictable, bounded and also belongs to $\Psi_{1,2}$. The process $\int_0^\cdot\int_U g_{\hat{X}}(s,y)\tilde{\pi}(ds,dy)$ is bounded because $|\Delta\int_0^\cdot\int_U g_{\hat{X}}(x,y)\tilde{\pi}(ds,dy)|\le 1$ and thus is a martingale as a bounded local martingale. Consequently,

$$\hat{X} := \int_0^{T^*} \int_U g_{\hat{X}}(s, y)\tilde{\pi}(ds, dy) \tag{11.5.22}$$

is a bounded random variable that is not attainable. Indeed, for any $(\omega, t) \in \{(\omega, \tau_{k_0}(\omega)); \omega \in \Omega\}$ we have that $g_{\hat{X}}(t) = g(t)$ and, since (11.5.18) is not satisfied ν-a.s., also (11.5.17) is not satisfied ν-a.s. By Proposition 11.5.13 we conclude that (11.5.17) does not hold $d\mathbb{P} \times dt \times d\nu$-a.s. for any admissible strategy φ. □

Now we proceed to the case when the Lévy measure is such that

$$supp\{\nu\} = \{y_1, y_2, \ldots\}, y_i \in U.$$

To exclude the existence of the concentration point for the support we reqiure that

$$\lim_{i\to\infty} |y_i| = +\infty. \tag{11.5.23}$$

Moreover, since $\int_U (|y|^2 \wedge 1)\nu(dy) < +\infty$, we have

$$\nu(U) = \sum_{i=1}^{\infty} \nu(\{y_i\}) < +\infty. \tag{11.5.24}$$

Theorem 11.5.14 *Assume* (11.5.11) *and that the following set*

$$A = \Big\{(\omega, t) \in \Omega \times [0, T^*] \quad s.t. \quad \langle \Sigma(t, T), y_i\rangle \geq 0 \qquad T > 0, \quad i = 1, 2, \ldots\Big\}$$

is of positive $d\mathbb{P} \times dt$- measure. Then there exists $\hat{X} \in L^2(\Omega, \mathcal{F}_{T^*}, \mathbb{P})$, which cannot be replicated.

Theorem 11.5.15 *Assume* (11.5.11) *and that there exists $\tilde{\Sigma} > 0$ such that the set*

$$A = \Big\{(\omega, t) \in \Omega \times [0, T^*] : \sup_{T>0} \mid \Sigma(t, T) \mid \leq \tilde{\Sigma}\Big\}$$

is of positive $d\mathbb{P} \times dt$ measure and

$$\mathbb{E}\left(\int_0^{T^*} \sum_{i=1}^{+\infty} e^{2(\tilde{\Sigma}+\varepsilon)|y_i|}\nu(\{y_i\})ds\right) < +\infty$$

for some $\varepsilon > 0$. Then there exists $\hat{X} \in L^2(\Omega, \mathcal{F}_{T^*}, \mathbb{P})$ which cannot be replicated.

Proof of Theorem 11.5.14 Let us define

$$g_{\hat{X}}(t, y_i) := \begin{cases} \sqrt{k} & \text{for} \quad i = i_k, \\ 0 & \text{for} \quad i \neq i_k, \end{cases} \tag{11.5.25}$$

where $i_k := \inf\Big\{i : \nu(\{y_i\}) \leq \frac{1}{k^3}\Big\}$. Due to (11.5.24) the process $g_{\hat{X}}$ is well defined and, by (11.5.23),

$$\limsup_{i\to\infty} \mid g_{\hat{X}}(t, y_i) \mid = +\infty. \tag{11.5.26}$$

It follows from

$$\int_0^{T^*} \sum_{i=1}^{\infty} \mid g_{\hat{X}}(t, y_i) \mid^2 \nu(\{y_i\})dt \leq T^* \sum_{k=1}^{\infty} \frac{1}{k^2} < +\infty \tag{11.5.27}$$

that $g_{\hat{X}} \in \Psi_{1,2}$ and that $\int_0^{\cdot} \int_U g_{\hat{X}}(s, y)\tilde{\pi}(ds, dy)$ is a square integrable martingale. We will show that the square integrable random variable,

$$\hat{X} := \int_0^{T^*} \int_U g_{\hat{X}}(s, y)\tilde{\pi}(ds, dy), \tag{11.5.28}$$

cannot be replicated. Let us fix $(\omega, t) \in A$ and assume that (11.5.17) is satisfied by some φ. Then, by Lemma 11.5.12, there exists $\gamma = \gamma(\omega, t) > 0$ such that

$$\forall\, n \in \mathbb{N} \quad \forall\, \{\beta_k\}_{k=1}^n,\ \beta_k \in \mathbb{R} \quad \forall\, \{y_{i_k}\}_{k=1}^n$$

$$\left| \sum_{k=1}^{n} \beta_k g_{\hat{X}}(t, y_{i_k}) \right| \leq \gamma \left| \sum_{k=1}^{n} \beta_k \Gamma_2(t, y_i) \right|_{C_b(\mathbb{R}_+)}. \tag{11.5.29}$$

Let us check (11.5.29) with $n = 1$, $\beta_1 = 1$ and for $i_1 = 1, 2, \ldots$ successively, that is,

$$|g_{\hat{X}}(t, y_i)| \leq \gamma \sup_{T \in [0,+\infty)} \left| \hat{P}(t-, T)(e^{-\langle \Sigma(t,T), y_i \rangle} - 1) \right| \qquad \forall i = 1, 2, \ldots. \tag{11.5.30}$$

By definition of the set A, for any $i = 1, 2, \ldots$, we have

$$\limsup_{i \to \infty} \sup_{T \in [0,+\infty)} \left| \hat{P}(t-, T)(e^{-\langle \Sigma(t,T), y_i \rangle} - 1) \right| \leq \sup_{T \in [0,T^*]} \left| \hat{P}(t-, T) \right|$$

$$\cdot \limsup_{i \to \infty} \sup_{T \in [0,+\infty)} \mid e^{-\langle \Sigma(t,T), y_i \rangle} - 1 \mid < +\infty.$$

However, recall that the left side of (11.5.30) satisfies (11.5.26), so the required constant γ does not exist. Hence (11.5.17) is no satisfied for any $(\omega, t) \in A$. But the set A is of positive $d\mathbb{P} \times dt$-measure, so in view of Lemma 11.5.13 (11.5.17) does not hold $d\mathbb{P} \times ds \times d\nu$-a.s. for any admissible strategy φ. □

Proof of Theorem 11.5.15: The arguments are similar as in the proof of Theorem 11.5.14. Let us define

$$g_{\hat{X}}(t, y_i) := e^{(\tilde{\Sigma}+\varepsilon)|y_i|}, \qquad i = 1, 2, \ldots$$

Then $g_{\hat{X}} \in \Psi_{1,2}$ and $\int_0^t \int_U g_{\hat{X}}(s, y)\tilde{\pi}(ds, dy)$ is a square integrable martingale. The square integrable random variable

$$\hat{X} := \int_0^{T^*} \int_U g_{\hat{X}}(y)\tilde{\pi}(ds, dy),$$

cannot be replicated because for any $(\omega, t) \in A$ we have

$$\limsup_{i\to\infty} \frac{| g_{\hat{X}}(t, y_i) |}{| e^{-\langle \Sigma(t,T), y_i \rangle} - 1 |_{C_b(\mathbb{R}_+)}} \geq \limsup_{i\to\infty} \frac{e^{(\tilde{\Sigma}+\varepsilon)|y_i|}}{e^{\tilde{\Sigma}|y_i|} + 1} = +\infty,$$

and, consequently, (11.5.17) does not have a solution. □

Proof of Lemma 11.5.12 The result is an extension of the moment problem solution (see Yosida [122]). Necessity is obvious, (11.5.13) holds with $\gamma = |e^*|_{E^*}$. To prove sufficiency let us define a linear subspace M of E by

$$M = \Big\{ e \in E : e = \sum_{i=1}^{n} \beta_i h(a_i); \quad n \in \mathbb{N},\ \beta_i \in \mathbb{R},\ a_i \in A \Big\}$$

and a linear transformation $\tilde{e}^* : M \longrightarrow \mathbb{R}$ by the formula

$$\tilde{e}^*\Big(\sum_{i=1}^{n} \beta_i h(a_i) \Big) = \sum_{i=1}^{n} \beta_i g(a_i).$$

Notice, that for $e_1, e_2 \in M$ with $e_1 = \sum_{i=1}^{n} \beta_i h(a_i)$ and $e_2 = \sum_{j=1}^{m} \beta_j' h(u_j)$, by (11.5.13), we obtain

$$\begin{aligned} \Big| \tilde{e}^*(e_1) - \tilde{e}^*(e_2) \Big| &= \Big| \sum_{i=1}^{n} \beta_i g(a_i) - \sum_{j=1}^{m} \beta_j' g(a_j) \Big| \\ &\leq \gamma \Big| \sum_{i=1}^{n} \beta_i h(a_i) - \sum_{j=1}^{m} \beta_j' h(a_j) \Big|_E = \gamma |e_1 - e_2|_E. \end{aligned}$$

If $e_1 = e_2$ then $\tilde{e}^*(e_1) = \tilde{e}^*(e_2)$, so this transformation is well defined because its value does not depend on the representation. It is also continuous and thus by the Hahn–Banach theorem it can be extended to the functional $e^* \in E^*$, which clearly satisfies (11.5.12). □

Proof of Lemma 11.5.13 The assertion follows from the Fubini theorem applied to the function $h = \mathbf{1}_A$ where $A := \{(x, y) \in E_1 \times E_2 : f_1(x, y) \neq f_2(x, y)\}$. □

11.6 Completeness for Affine Models

Let us consider an affine model

$$P(t, T) = e^{-C(T-t)-D(T-t)R(t)}, \quad t \in [0, T^*], \quad T > 0,$$

with short rate satisfying

$$dR(t) = F(R(t))dt + G(R(t-))dZ(t), \quad t \geq 0. \tag{11.6.1}$$

Here, Z is a one-dimensional Lévy process and the functions $C(\cdot), D(\cdot), F(\cdot), G(\cdot)$ are deterministic. The model is assumed to satisfy the (MP) condition.

Since

$$R(t) = -\frac{1}{D(T-t)}\Big(\ln P(t,T) + C(T-t)\Big),$$

it follows that, for each $T > 0$, the filtration $(\mathcal{F}_t^R)$ generated by R is identical with $(\mathcal{F}_t^{P^T})$ generated by $P(\cdot, T)$, i.e.

$$\mathcal{F}_t^R = \mathcal{F}_t^{P^T}, \quad t \in [0, T].$$

From (11.6.1) we have that

$$dZ(t) = \frac{dR(t)}{G(R(t-))} - \frac{F(R(t))}{G(R(t-))}dt,$$

which implies that $(\mathcal{F}_t^R)$ is identical with the filtration $(\mathcal{F}_t)$ generated by Z. Hence, for each $T > 0$,

$$\mathcal{F}_t^{P^T} = \mathcal{F}_t^R = \mathcal{F}_t, \quad t \in [0, T]. \tag{11.6.2}$$

Recall that by Proposition 7.5.1 the model can be written as the HJM model

$$df(t,T) = \alpha(t,T)dt + \sigma(t,T)dZ(t), \tag{11.6.3}$$

with

$$\alpha(t,T) := F(R(t))D'(T-t) - C''(T-t) - D''(T-t)R(t), \tag{11.6.4}$$

$$\sigma(t,T) := D'(T-t)G(R(t-)). \tag{11.6.5}$$

This enables us to examine the completeness problem for affine models by using the results proven in previous sections concerned with the HJM model.

Theorem 11.6.1 *Let the affine model satisfying (MP) have short rate of the form*

(a)

$$dR(t) = (\beta R(t) + c)dt + \sqrt{R(t)}dW(t), \quad R(0) > 0, \quad \beta \in \mathbb{R},\ c \geq 0, \quad t \in [0, T^*], \tag{11.6.6}$$

where W is a one-dimensional Wiener process. Then any claim $\hat{X} \in L^1(\Omega, \mathcal{F}_{T^*}, \mathbb{P}) = L^1(\Omega, \mathcal{F}_{T^*}^R, \mathbb{P})$ is attainable and can be replicated by a strategy involving one T-bond only, where T is any number such that $T \geq T^*$. In particular, the model is complete.

(b)

$$dR(t) = (\beta R(t) + c)dt + (R(t-))^{\frac{1}{\alpha}}dZ^{\alpha}(t), \quad \beta \in \mathbb{R},\ c \geq 0, \quad t \in [0, T^*], \tag{11.6.7}$$

where Z^{α} is an α-stable martingale with index $\alpha \in (1,2)$ and positive jumps only. Then the market is not complete in the class $\hat{X} \in L^{\infty}(\Omega, \mathcal{F}_{T^*}, \mathbb{P})$. Moreover, *for any* $T \geq T^*$*, there exists a bounded claim of the form* $\hat{X} = h(P(t,T), t \in [0,T^*])$*, where* h *is a deterministic function, which is not attainable in the class of admissible* $\mathcal{M}$*-valued strategies.*

Proof It follows from Theorem 10.2.7 that both equations generate affine models which satisfy (MP) with some functions $C(\cdot), D(\cdot)$. Writing the model in the HJM parametrization, we obtain, by (11.6.3)–(11.6.5), that the volatility equals

$$\sigma(t,T) = D'(T-t)\sqrt{R(t)},$$

for (11.6.6) and

$$\sigma(t,T) = D'(T-t)(R(t))^{\frac{1}{\alpha}},$$

for (11.6.7).

(*a*) The assertion follows from Theorem 11.5.2 and the relation

$$\mathcal{F}_t^{P^T} = \mathcal{F}_t, \quad t \in [0,T^*], \quad T \geq T^*, \tag{11.6.8}$$

which follows from (11.6.2).

(*b*) We can apply Theorem 11.5.11. Since D is bounded (see Remark 10.2.10) we have

$$\int_0^{+\infty} \mid \sigma(t,T) \mid dT = \mid (R(t-))^{\frac{1}{\alpha}} \mid \cdot \int_0^{+\infty} \mid D'(T-t) \mid dT < +\infty,$$

and thus (11.5.11) is satisfied. Since the Lévy measure of Z^α has a concentration point, the market is not complete in the class $\hat{X} \in L^\infty(\Omega, \mathcal{F}_{T^*}, \mathbb{P})$. The rest of the assertion follows from (11.6.8). □

11.7 Completeness for Factor Models

We present now a partial solution of the completeness problem for a factor model

$$f(t,T) = G(T-t, X(t)), \tag{11.7.1}$$

with $X(t)$ solving

$$dX(t) = a(t)dt + \langle b(t), dZ(t)\rangle, \quad t \geq 0. \tag{11.7.2}$$

Here Z is a d-dimensional Lévy process, $(a(t), t \geq 0)$- an adapted process and $(b(t), t \geq 0)$- a predictable process.

Theorem 11.7.1 *Let us assume that for the model given by* (11.7.1)–(11.7.2) *the (MP) condition is satisfied.*

(*a*) *Let* Z *be a Wiener process with a drift. If* $d = 1$ *then the market is complete and any* $\hat{X} \in L^1(\Omega, \mathcal{F}_{T^*}, \mathbb{P})$ *is attainable. If* $d > 1$ *then the market is not complete.*

(b) *Let the Lévy measure of Z have a concentration point $y_0 \neq 0$. If G is linear in x, i.e. $G(t,x) = G_1(t)x + G_2(t)$ and*

$$\int_0^{+\infty} \mid G_1(s) \mid ds < +\infty,$$

then the market is not complete.

Proof By the Itô formula

$$\begin{aligned} df(t,T) &= dG(T-t, X(t)) \\ &= -G'_t(T-t, X(t))dt + G'_x(T-t, X(t-))dX(t) + \frac{1}{2}G''_{xx}(T-t, X(t))b^2(t)dt \\ &\quad + \Big\{G(T-t, X(t)) - G(T-t, X(t-)) - G'_x(T-t, X(t-))\triangle X(t)\Big\}. \end{aligned} \quad (11.7.3)$$

(a) If Z is continuous then the last line in (11.7.3) disappears and, taking into account (11.7.2), we obtain

$$\begin{aligned} df(t,T) = \Big[&- G'_t(T-t, X(t)) + G'_x(T-t, X(t))a(t) + \frac{1}{2}G''_{xx}(T-t, X(t))b^2(t)\Big]dt \\ &+ \langle G'_x(T-t, X(t))b(t), dZ(t)\rangle. \end{aligned} \quad (11.7.4)$$

Coming back to the HJM notation, it follows that $\Sigma(t,T) = (\Sigma^1(t,T), \ldots, \Sigma^d(t,T))$ is given by

$$\Sigma^i(t,T) = \int_t^T \sigma^i(t,u)du = \int_t^T b^i(t)G'_x(u-t, X(t))du = b^i(t)\int_0^{T-t} G'_x(s, X(t))ds,$$
$$i = 1, 2, \ldots, d,$$

so, for fixed t, vectors $\Sigma^1(t,T), \ldots, \Sigma^d(t,T)$ are linearly dependent functions of T in the case $d > 1$. The assertion follows from Theorem 11.5.2.

(b) If $G(t,x) = G_1(t)x + G_2(t)$ then $G'_x = G_1(t)$ and consequently

$$\begin{aligned} &G(T-t, X(t)) - G(T-t, X(t-)) - G'_x(T-t, X(t-))\triangle X(t) \\ &\qquad = G_1(T-t)\big(X(t) - X(t-) - \triangle X(t)\big) = 0. \end{aligned}$$

From (11.7.3) we obtain

$$\begin{aligned} df(t,T) = &\big[G_1(T-t)a(t) - G'_1(T-t)X(t) - G'_2(T-t)\big]dt \\ &+ \langle G_1(T-t)b(t), dZ(t)\rangle \end{aligned}$$

and can apply Theorem 11.5.11 with $\sigma(t,T) = G_1(T-t)b(t)$. Since

$$\int_0^{+\infty} |\,\sigma(t,T)\,| dT = |\,b(t)\,| \int_t^{+\infty} |\,G_1(u-t)\,| du \leq |\,b(t)\,| \int_0^{+\infty} |\,G_1(s)\,| ds < +\infty,$$

so (11.5.11) is satisfied, and the Lévy measure has a concentration point, the assertion follows. □

As an application of Theorem 11.7.1 we solve the completeness problem for models with Ornstein–Uhlenbeck factors (see Chapter 9).

Proposition 11.7.2 *Let the factor process be given by*

$$dX^x(t) = (a + AX^x(t))dt + dW(t), \quad X^x(0) = x \in \mathbb{R}^d,$$

where $a \in \mathbb{R}^d$, A is a $d \times d$-matrix and W the Wiener process in $\mathbb{R}^d$ with identity covariance matrix. Let the factor model with factor X^x and initial curve $G(0,x) := \langle Kx, x\rangle$, where K is a positive $d \times d$-matrix, satisfy (MP). Then the following statements are true.

(a) If $d = 1$ then the market is complete. Moreover, any claim $\hat{X} \in L^1(\Omega, \mathcal{F}_{T^*}, \mathbb{P})$ of the form $\hat{X} = \hat{X}(P(t,T), t \in [0,T^*])$, for some $T \geq T^*$, can be replicated by a strategy involving the T-bond only.

(b) For $d > 1$ the model is not complete. Moreover, for any $T \geq T^*$, there exists a bounded claim of the form $\hat{X} = \hat{X}(P(t,T), t \in [0,T^*])$, which is not attainable.

Proof Note that since $P(t,T) = F(T-t, X(t))$, where the function F is described in Proposition 9.1.8, we have for each T

$$\mathcal{F}_t^{P^T} \subseteq \mathcal{F}_t, \quad t \in [0,T].$$

Therefore the assertion for claims of the form $\hat{X} = \hat{X}(P(t,T), t \in [0,T^*])$ follows from Theorem 11.7.1. □

Proposition 11.7.3 *Let us assume that the factor is the short rate given by*

$$dR^x(t) = (a + bR^x(t))dt + dZ(t), \quad R^x(0) = x, \quad t \geq 0,$$

where $a \geq 0, b < 0$ and (MP) is satisfied. If the Lévy measure of Z has a non-zero concentration point then the model is not complete. Moreover, for any $T \geq T^*$ there exists a bounded claim of the form $\hat{X} = \hat{X}(P(t,T), t \in [0,T^*])$, which is not attainable.

Proof We have shown in Proposition 9.1.7 that if the model satisfies (MP) then the bond curve is given by

$$F(t,x) = e^{-\left(x(e^{bt}-1)/b + a(e^{bt}-1)/b^2 - at/b\right)} e^{J\left((e^{bt}-1)/b^2 - t/b\right)}. \tag{11.7.5}$$

Consequently,

$$G(t,x) = G_1(t)x + G_2(t),$$

where

$$G_1(t) := e^{bt}, \quad G_2(t) := \frac{d}{dt}\Big(-a(e^{bt}-1)/b^2 + at/b - J\big((e^{bt}-1)/b^2 - t/b\big)\Big).$$

Since $b < 0$, we have

$$\int_0^{+\infty} G_1(s)ds = \int_0^{+\infty} e^{bs}ds < +\infty,$$

and, by Theorem 11.7.1 (b), the market is not complete providing that the Lévy measure has a nonzero concentration point.

Since, for any $T > 0$, $P(t,T) = F(T-t, R(t)), t \in [0,T]$ we obtain, by (11.7.5), that

$$R(t) = \frac{b}{1-e^{bt}}\left(\ln P(t,T) + \frac{a}{b^2}(e^{bt}-1) - \frac{at}{b} - J\left(\frac{e^{bt}-1}{b^2} - \frac{t}{b}\right)\right), \quad t \in [0,T].$$

This implies that $\mathcal{F}_t^{P^T} = \mathcal{F}_t^R$ for $t \in [0,T]$. Since the filtrations generated by Z and R are also equal, we obtain

$$\mathcal{F}_t^{P^T} = \mathcal{F}_t^R = \mathcal{F}_t, \quad t \in [0,T^*],$$

for $T \geq T^*$ and the assertion follows. □

11.8 Approximate Completeness

Here we work, as before, with a bond market defined on a probability space $(\Omega, \mathcal{F}, \mathbb{P})$ with filtration $(\mathcal{F}_t)$ generated by a Lévy process Z and assume that (MP) is satisfied. Recall that the discounted bond prices have the form

$$d\hat{P}(t,T) = \langle \Gamma_1(t,T), dW(t)\rangle + \int_U \Gamma_2(t,T,y)\tilde{\pi}(dt,dy), \quad t \in [0,T^*], \quad T > 0, \tag{11.8.1}$$

where $(\Gamma_1(t,T)) \in \Phi(U)$ and $(\Gamma_2(t,T,y)) \in \Psi_{1,2}$ and the discounted wealth process of an $\mathcal{M}([0,+\infty))$-valued strategy (φ_t) is given by

$$\hat{X}(t) = X(0) + \int_0^t (\varphi_s, d\hat{P}(s,\cdot))$$

$$:= X(0) + \int_0^t \langle(\varphi_s, \Gamma_1(s,\cdot)), dW(s)\rangle + \int_0^t\int_U (\varphi_s, \Gamma_2(s,\cdot,y))\tilde{\pi}(ds,dy), \quad t \in [0,T^*].$$

We use a subclass $\mathcal{A}$ of admissible strategies defined by

$$\mathcal{A} := \left\{\varphi_s, s\in[0,T^*] : \mathbb{E}\int_0^{T^*} (\varphi_s, \Gamma_1(s,\cdot))^2 ds + \mathbb{E}\int_0^{T^*}\int_U (\varphi_s, \Gamma_2(s,\cdot,y))^2 ds\nu(dy) < +\infty\right\}. \tag{11.8.2}$$

For $(\varphi_t) \in \mathcal{A}$ the process $(\hat{X}(t))$ is a square integrable martingale.

If a claim $\hat{X} \in L^2(\Omega, \mathcal{F}_{T^*}, \mathbb{P})$ is not replicable at time T^* by strategies from $\mathcal{A}$, one may try to find a sequence $(x_n, \varphi^n), n = 1, 2, \dots$ where $x_n \in \mathbb{R}$ and $\varphi^n \in \mathcal{A}$ such that

$$\lim_{n\to+\infty} \mathbb{E}\Big[\Big(\big(x_n + \int_0^{T^*} (\varphi_t^n, d\hat{P}(t,\cdot))\big) - \hat{X}\Big)^2\Big] = 0. \tag{11.8.3}$$

Such a sequence will be called an *approximating sequence* for $\hat{X}$. If each $\hat{X} \in L^2(\Omega, \mathcal{F}_{T^*}, \mathbb{P})$ admits an approximating sequence, the market is called *approximately complete* in the class $\mathcal{A}$. First we characterize approximating sequences for $\hat{X} \in L^2(\Omega, \mathcal{F}_{T^*}, \mathbb{P})$ with the use of the decomposition

$$\mathbb{E}[\hat{X} \mid \mathcal{F}_t] = \mathbb{E}\hat{X} + \int_0^t \langle f_{\hat{X}}(s), dW(s)\rangle + \int_0^t \int_U g_{\hat{X}}(s,y)\tilde{\pi}(ds,dy), \qquad t \in [0,T^*], \tag{11.8.4}$$

where $f_{\hat{X}}, g_{\hat{X}}$ are uniquely determined and satisfy

$$\mathbb{E}\int_0^{T^*} | f_{\hat{X}}(t) |^2 \, dt < +\infty, \quad \mathbb{E}\int_0^{T^*}\int_U | g_{\hat{X}}(t,y) |^2 \, dt\nu(dy) < +\infty.$$

Proposition 11.8.1 *(x_n, φ^n) is an approximating sequence for $\hat{X} \in L^2(\Omega, \mathcal{F}_{T^*}, \mathbb{P})$ if and only if*

$$x_n - \mathbb{E}\hat{X} \longrightarrow 0, \tag{11.8.5}$$

$$\mathbb{E}\int_0^{T^*} | f_{\hat{X}}(s) - (\varphi_s^n, \Gamma_1(s,\cdot)) |^2 \, ds \longrightarrow 0, \tag{11.8.6}$$

$$\mathbb{E}\int_0^{T^*}\int_U | g_{\hat{X}}(s,y) - (\varphi_s^n, \Gamma_2(s,\cdot,y)) |^2 \, ds\nu(dy) \longrightarrow 0. \tag{11.8.7}$$

Proof Since the discounted wealth process of an approximating sequence (x_n, φ^n) equals

$$\hat{X}^n(t) = x_n + \int_0^t \langle(\varphi_s^n, \Gamma_1(s,\cdot)), dW(s)\rangle + \int_0^t\int_U (\varphi_s^n, \Gamma_2(s,\cdot,y))\tilde{\pi}(ds,dy), \quad t \in [0,T^*], \tag{11.8.8}$$

the hedging error equals

$$\begin{aligned}\mathbb{E}[(\hat{X}^n(T^*) - \hat{X})^2] = \mathbb{E}\Big[\Big(x_n - \mathbb{E}(\hat{X}) &+ \int_0^{T^*} \big\langle(\varphi_s^n, \Gamma_1(s,\cdot)) - f_{\hat{X}}(s), dW(s)\big\rangle \\ &+ \int_0^{T^*}\int_U \big((\varphi_s^n, \Gamma_2(s,\cdot,y)) - g_{\hat{X}}(s,y)\big)\,\tilde{\pi}(ds,dy)\Big)^2\Big].\end{aligned}$$

Using the independence and the zero mean property of the stochastic integrals in (11.8.8) we obtain

$$\mathbb{E}[(\hat{X}^n(T^*) - \hat{X})^2] = (x_n - E(\hat{X}))^2 + \mathbb{E}\Big[\Big(\int_0^{T^*} \langle(\varphi_s^n, \Gamma_1(s,\cdot)) - f_{\hat{X}}(s), dW(s)\rangle\Big)^2\Big]$$

$$+ \mathbb{E}\Big[\Big(\int_0^{T^*}\int_U \big((\varphi_s^n, \Gamma_2(s,\cdot,y)) - g_{\hat{X}}(s,y)\big)\,\tilde{\pi}(ds,dy)\Big)^2\Big]$$

$$= (x_n - \mathbb{E}(\hat{X}))^2 + \mathbb{E}\int_0^{T^*} \mid f_{\hat{X}}(s) - (\varphi_s^n, \Gamma_1(s,\cdot)) \mid^2 ds$$

$$+ \mathbb{E}\int_0^{T^*}\int_U \mid g_{\hat{X}}(s,y) - (\varphi_s^n, \Gamma_2(s,\cdot,y)) \mid^2 ds\nu(dy),$$

so the assertion follows. □

Our method of examining approximate completeness in the sequel involves decompositions (11.8.4). Since

$$\mathbb{E}[\hat{X}^2] = \mathbb{E}\int_0^{T^*} \mid f_{\hat{X}}(t) \mid^2 dt + \mathbb{E}\int_0^{T^*}\int_U \mid g_{\hat{X}}(t,y) \mid^2 dt\nu(dy).$$

Let us introduce the product Hilbert space $\mathcal{H}$ of predictable processes $(f(t), g(t,y), t \in [0,T^*], y \in U)$ such that

$$\mathbb{E}\int_0^{T^*} f^2(t)dt + \mathbb{E}\int_0^{T^*}\int_U g^2(t,y)dt\nu(dy) < +\infty,$$

with scalar product

$$((f_1,g_1),(f_2,g_2))_{\mathcal{H}} = \mathbb{E}\int_0^{T^*} f_1(t)f_2(t)dt + \mathbb{E}\int_0^{T^*}\int_U g_1(t,y)g_2(t,y)dt\nu(dy).$$

The following result will be our basic tool for examining approximate completeness in the sequel.

Proposition 11.8.2 *Let $\tilde{\mathcal{A}} \subseteq \mathcal{A}$ be a Banach space such that the operator*

$$\mathcal{K}\varphi(t,y) = ((\varphi_t, \Gamma_1(t,\cdot)); (\varphi_t, \Gamma_2(t,\cdot,y))), \quad \varphi \in \tilde{\mathcal{A}}, \quad t \in [0,T^*], y \in U, \quad (11.8.9)$$

is a linear bounded operator from $\tilde{\mathcal{A}}$ into $\mathcal{H}$. Then the market with discounted prices (11.8.1) *is approximately complete in the class $\tilde{\mathcal{A}}$ if and only if*

$$Ker\ \mathcal{K}^* = \{0\},$$

where $\mathcal{K}^*$ stands for the adjoint operator of $\mathcal{K}$.

Proof In view of Proposition 11.8.1, approximate completeness is equivalent to the statement that the image of $\mathcal{K}$ is dense in $\mathcal{H}$. However, for $(f, g) \in \mathcal{H}$:

$$(\mathcal{K}\varphi, (f, g))_{\mathcal{H}} = (\varphi, \mathcal{K}^*(f, g))_{\tilde{\mathcal{A}}}, \quad \varphi \in \mathcal{H},$$

where the right side denotes the value of the functional $\mathcal{K}^*(f, g)$ on the element φ. Therefore $Im\mathcal{K}$ is dense in $\mathcal{H}$ if and only if the implication

$$\mathcal{K}^*(f, g) = 0 \quad \Longrightarrow \quad (f, g) = 0$$

holds. □

11.8.1 HJM Model

Now we examine approximate completeness in the HJM model

$$df(t, T) = \alpha(t, T)dt + \langle \sigma(t, T), dZ(t) \rangle, \tag{11.8.10}$$

with a d-dimensional Lévy process Z. We begin with a result that approximate completeness may appear if the support of the Lévy measure of Z is infinite. In the opposite case the concepts of approximate completeness and completeness coincide.

Theorem 11.8.3 *(a) If the market* (11.8.10) *is approximately complete then, for each* $t \in [0, T^*]$, $\mathbb{P}$*-almost surely, the vectors*

$$\Sigma^1(t, T), \Sigma^2(t, T), \ldots, \Sigma^d(t, T)$$

are linearly independent as functions of $T \in (0, +\infty)$.

(b) Let the jumps of Z take values in a finite set. Then the market (11.8.10) *is approximately complete if and only if it is complete.*

Proof (*a*) If the vectors $\Sigma^1(t, T), \Sigma^2(t, T), \ldots, \Sigma^d(t, T)$ are linearly dependent with positive $d\mathbb{P} \times dt$ measure then one can find $f_{\hat{X}}$ such that for some $\varepsilon > 0$ and any $\varphi \in \mathcal{A}$,

$$\mathbb{E} \int_0^{T^*} \mid f_{\hat{X}}(s) - (\varphi_s, \Gamma_1(s, \cdot)) \mid^2 ds > \varepsilon.$$

A precise construction of $f_{\hat{X}}$ can be deduced from the proof of Theorem 11.5.2 (a).

(*b*) In view of Theorem 11.5.2 the market is complete if and only if the vectors

$$\Sigma(t, T) := (\Sigma^1(t, T), \ldots, \Sigma^d(t, T)), \; e^{-\langle \Sigma(t,T), y_1 \rangle} - 1, \; \ldots \; , e^{-\langle \Sigma(t,T), y_n \rangle} - 1$$

are linearly independent. Above $y_1, \ldots, y_n$ denote all possible values of jumps of Z. If the linear independence of the preceding set fails, then one can construct $f_{\hat{X}}$ and $g_{\hat{X}}$ such that for some $\varepsilon > 0$ and any $\varphi \in \mathcal{A}$,

$$\mathbb{E}\int_0^{T^*} | f_{\hat{X}}(s) - (\varphi_s, \Gamma_1(s), \cdot) |^2 ds + \mathbb{E}\int_0^{T^*}\int_U | g_{\hat{X}}(s,y) - (\varphi_s, \Gamma_2(s,\cdot,y)) |^2 ds\nu(dy) > \varepsilon.$$

In the construction one can follow the proof of Theorem 11.5.2 (b). □

We assume in the sequel that Z does not have the Wiener part. We will also restrict admissible strategies to measure valued strategies requiring that measures have densities with respect to some finite measure μ on $[0,+\infty)$ with infinite support, i.e.

$$\varphi(t, dT) = \varphi(t,T)\mu(dT), \quad t \in [T^*].$$

The case in which the support of μ equals $[a,b]$, with some $0 < a < b < +\infty$, corresponds to trading with bonds with maturities from $[a,b]$ only. In this framework the operator $\mathcal{K}\colon \mathcal{A} \longrightarrow \mathcal{H}$ in (11.8.9) takes the reduced form

$$(\mathcal{K}\varphi)(t,y) := \int_0^{+\infty} \Gamma_2(t,T,y)\varphi(t,T)\mu(dT), \tag{11.8.11}$$

and the space $\mathcal{H}$ is given by

$$\mathcal{H} := \{g = g(t,y) : \mathbb{E}\int_0^{T^*}\int_U g^2(t,y)dt\nu(dy) < +\infty\}.$$

More formally, $\mathcal{H} = L^2(\Omega \times [0,T^*] \times U, \mathcal{P} \times \mathcal{B}(U), \mathbb{P} \times dt \times \nu)$.

Now we formulate and prove the main result of this section.

Theorem 11.8.4 *Let in the HJM model* (11.8.10) *forward rates be positive,*

$$supp\{\nu\} \subseteq \mathbb{R}^d_+, \quad \Sigma(t,T) \in \mathbb{R}^d_+, \quad t \in [0,T^*], \quad T > 0 \tag{11.8.12}$$

and

$$\operatorname*{ess\ sup}_{s\in[0,T^*],\ T\geq s,\ \omega\in\Omega} | \Sigma(s,T) | < +\infty \tag{11.8.13}$$

hold. Let $\tilde{\mathcal{A}}$ be the space of $\mathcal{M}$-valued strategies with densities, i.e.

$$\varphi_t(dT) = \varphi(t,T)\mu(dT),$$

equipped with the norm

$$\tilde{\mathcal{A}} := \left\{\varphi = \varphi(t,T) : | \varphi |^2_{\tilde{\mathcal{A}}} := \mathbb{E}\int_0^{T^*}\int_0^{+\infty} \varphi^2(t,T)dt\,\mu(dT) < +\infty\right\}. \tag{11.8.14}$$

Then the following statements are true.

(a) *The market is approximately complete in the class $\tilde{\mathcal{A}}$ if and only if the following implication holds: for almost all (ω, t)*

$$\int_U (e^{-\langle \Sigma(t,T),y\rangle} - 1)h(t,y)\nu(dy) = 0, \quad \textit{for } \mu\textit{-almost all } T \in [t, +\infty)$$

$$\Longrightarrow \quad h(t,y) = 0 \quad \nu - a.s.. \qquad (11.8.15)$$

(b) *If $d = 1$ and for almost all (ω, t), there exists an interval $[a(\omega), b(\omega)]$, contained in the support of μ and such that*

$$\{\Sigma(t, T, \omega) : T > t\} \supseteq [a(\omega), b(\omega)], \quad \mathbb{P} - a.s.,$$

then the market is approximately complete in the class $\tilde{\mathcal{A}}$.

Proof The proof is divided into two steps. In Step 1 we show that the operator $\mathcal{K}$ is bounded. In Step 2 we determine $\mathcal{K}^*$ and prove (a) and (b).

Step 1: To prove that $\mathcal{K}$ is bounded note that for $\varphi \in \tilde{\mathcal{A}}$ we have

$$\begin{aligned}
| \mathcal{K}\varphi |_{\mathcal{H}}^2 &= \mathbb{E} \int_0^{T^*} \int_U (\varphi_s, \Gamma_2(s, \cdot, y))^2 ds\nu(dy) \\
&= \mathbb{E} \int_0^{T^*} \int_U \Big(\int_0^{+\infty} \varphi(s,T)\Gamma_2(s,T,y)\mu(dT) \Big)^2 ds\nu(dy) \\
&\le \mathbb{E} \int_0^{T^*} \int_U \Big(\int_0^{+\infty} \varphi^2(s,T)\mu(dT) \Big)\Big(\int_0^{+\infty} \Gamma_2^2(s,T,y)\mu(dT) \Big) ds\nu(dy).
\end{aligned}$$

Since $\Gamma_2(t, T, y) = \hat{P}(t-, T)[e^{-\langle \Sigma(t,T),y\rangle} - 1]$ (see Proposition 11.2.1) and forward rates are positive, we continue the estimation as follows

$$| \mathcal{K}\varphi |_{\mathcal{H}}^2 \le \mathbb{E} \int_0^{T^*} \Big(\int_0^{+\infty} \varphi^2(s,T)\mu(dT) \Big)\Big(\int_0^{+\infty} \hat{P}^2(s-,T)\psi(\Sigma(s,T))\mu(dT) \Big) ds,$$

where ψ is the following function

$$\psi(x) := \int_U | e^{-\langle x,y\rangle} - 1 |^2 \nu(dy), \quad x \in \mathbb{R}_+^d.$$

Since $1 - e^{-z} \le z; z \ge 0$ we obtain

$$\begin{aligned}
\int_{|y|\le 1} | e^{-\langle x,y\rangle} - 1 |^2 \nu(dy) &= \int_{|y|\le 1} \left(\frac{e^{-\langle x,y\rangle} - 1}{\langle x,y\rangle} \right)^2 | \langle x,y\rangle |^2 \nu(dy) \\
&\le | x |^2 \int_{|y|\le 1} | y |^2 \nu(dy), \quad x \in \mathbb{R}_+^d.
\end{aligned}$$

Clearly,

$$\int_{|y|>1} | e^{-\langle x,y\rangle} - 1 |^2 \nu(dy) \le \nu(\{| y | > 1\}) < +\infty, \quad x \in \mathbb{R}_+^d,$$

so

$$\psi(x) \le A + B \mid x \mid^2, \quad x \in \mathbb{R}_+^d, \tag{11.8.16}$$

where $A := \nu(\{\mid y \mid > 1\})$ and $B := \int_{|y|\le 1} \mid y \mid^2 \nu(dy)$. Therefore we obtain

$$\begin{aligned}
\mid \mathcal{K}\varphi \mid_{\mathcal{H}}^2 &\le \mathbb{E}\int_0^{T^*} \Big(\int_0^{+\infty} \varphi^2(s,T)\mu(dT)\Big)\Big(\int_0^{+\infty} (A + B \mid \Sigma(s,T) \mid^2)\mu(dT)\Big)ds \\
&\le \mu([0,+\infty))(A + B \operatorname*{ess\,sup}_{\omega\in\Omega, 0\le s\le T^*, s\le T} \mid \Sigma(s,T) \mid^2) \mid \varphi \mid_{\tilde{\mathcal{A}}}^2,
\end{aligned}$$

and the assertion follows.

Step 2: We determine $\mathcal{K}^*$. For $\varphi \in \tilde{\mathcal{A}}$ and $g \in \mathcal{H}$, we have

$$\begin{aligned}
(\mathcal{K}(\varphi), g)_{\mathcal{H}} &= \mathbb{E}\left(\int_0^{T^*}\int_U \mathcal{K}\varphi(t,y)\cdot g(t,y)dt\nu(dy)\right) \\
&= \mathbb{E}\left(\int_0^{T^*}\int_U \left(\int_0^{+\infty} \Gamma_2(t,T,y)\varphi(t,T)\mu(dT)\right)\cdot g(t,y)dt\nu(dy)\right) \\
&= \mathbb{E}\left(\int_0^{T^*}\int_0^{+\infty} \left(\int_U \Gamma_2(t,T,y)\cdot g(t,y)\nu(dy)\right)\varphi(t,T)dt\mu(dT)\right) \\
&= (\varphi, \mathcal{K}^*(g))_{\tilde{\mathcal{A}}},
\end{aligned}$$

so it follows that $\mathcal{K}^*g$ can be identified with the random field

$$(\mathcal{K}^*g)(t,T) = \int_U \Gamma_2(t,T,y)\cdot g(t,y)\nu(dy).$$

To prove (a), note that the condition

$$\mathcal{K}^*g = 0$$

means that $dt \times \mu(dT)$ – almost surely,

$$\int_U \hat{P}(t-,T)\left[e^{-\langle\Sigma(t,T),y\rangle} - 1\right]g(t,y)\nu(dy) = 0,$$

or, equivalently, that

$$\int_U \left[e^{-\langle\Sigma(t,T),y\rangle} - 1\right]g(t,y)\nu(dy) = 0.$$

The result follows. Part (b) follows from Proposition 11.8.5. □

Proposition 11.8.5 *Let h be a function from $L^2((0,+\infty),\nu)$, where ν satisfies $\int_0^{+\infty}(y^2 \wedge 1)\nu(dy) < +\infty$. If there exist $0 < \lambda_1 < \lambda_2 < +\infty$ such that*

$$H(\lambda) := \int_0^{+\infty} (e^{-\lambda y} - 1)h(y)\nu(dy) = 0,$$

on a dense subset of (λ_1,λ_2), then $h = 0$, as an element of $L^2((0,+\infty),\nu)$.

We establish first the following technical lemma.

Lemma 11.8.6 *Let us assume that* $h \in L^2((0, +\infty), \nu)$, *with* ν *satisfying* $\int_0^{+\infty}(y^2 \wedge 1)\nu(dy) < +\infty$. *Then the function*

$$H(\lambda) := \int_0^{+\infty} (e^{-\lambda y} - 1)h(y)\nu(dy), \quad \lambda > 0$$

is differentiable and

$$H'(\lambda) = -\int_0^{+\infty} e^{-\lambda y}\, y\, h(y)\nu(dy), \quad \lambda > 0. \tag{11.8.17}$$

Proof We split H into two parts

$$H(\lambda) = H_0(\lambda) + H_\infty(\lambda),$$

where

$$H_0(\lambda) := \int_0^1 (e^{-\lambda y} - 1)h(y)\nu(dy), \quad \lambda > 0,$$

$$H_\infty(\lambda) := \int_1^{+\infty} (e^{-\lambda y} - 1)h(y)\nu(dy), \quad \lambda > 0.$$

We show the result for H_0. The argument for H_∞ is similar and easier. Let us denote

$$g(\lambda, y) := (e^{-\lambda y} - 1)h(y).$$

Then

$$\frac{\partial g(\lambda, y)}{\partial \lambda} = -ye^{-\lambda y}h(y)$$

and

$$\left|\frac{\partial g(\lambda, y)}{\partial \lambda}\right| \le y \mid h(y) \mid,$$

so

$$\begin{aligned}\int_0^1 \left|\frac{\partial g(\lambda, y)}{\partial \lambda}\right| \nu(dy) &\le \int_0^1 y \mid h(y) \mid \nu(dy)\\ &\le \left(\int_0^1 y^2\nu(dy)\right)^{\frac{1}{2}} \left(\int_0^1 h^2(y)\nu(dy)\right)^{\frac{1}{2}} < +\infty.\end{aligned}$$

By the classical result on differentiation of integrals with respect to a parameter the formula (11.8.17) follows. □

Proof of Proposition 11.8.5 It follows from Lemma 11.8.6 that

$$H'(\lambda) = -\int_0^{+\infty} e^{-\lambda y}\, y\, h(y)\nu(dy), \quad \lambda \in (\lambda_1, \lambda_2).$$

Thus the Laplace transform of $yh(y)\nu(dy)$ vanishes on a dense subset of the interval (λ_1, λ_2), so it vanishes for all $\lambda > 0$ and thus $yh(y) = 0$, ν-almost surely. □

Remark 11.8.7 The severe assumptions required in Theorem 11.8.4 were used to show that $\mathcal{K}$ is a bounded operator.

11.8.2 Factor Model

Theorem 11.8.4 allows us to solve the approximate completeness problem for factor models

$$f(t,T) = G(T-t, X(t)), \tag{11.8.18}$$

where

$$G(t,x) = G_1(t)x + G_2(t), \tag{11.8.19}$$

$$dX(t) = a(t)dt + b(t)dZ(t), \tag{11.8.20}$$

and Z is a real-valued Lévy process with no Wiener part.

Proposition 11.8.8 *Let $G(t,x)$ have the decomposition* (11.8.19) *satisfying*

$$G_1(t) \geq 0, \quad G_2(t) \geq 0, \quad t \geq 0, \quad \int_0^{+\infty} |\, G_1(t) \,|\, dt < +\infty.$$

Let the factor (11.8.20) *be a positive process, $b(t) > 0, t \geq 0$, and the Lévy measure of Z has a nonzero concentration point and supported by $(0, +\infty)$. Then the model* (11.8.18) *is not complete and is approximately complete in the class of admissible strategies $\tilde{\mathcal{A}}$ defined by* (11.8.14).

Specifically, the result is true for the short-rate factor

$$dR^x(t) = (a + bR^x(t))dt + dZ(t), \quad R^x(0) = x, \quad t \geq 0, \tag{11.8.21}$$

where $b < 0$.

Proof The fact that the model is not complete follows from Theorem 11.7.1 (*b*). Passing to the HJM parametrization yields $\sigma(t,T) = G_1(T-t)b(t)$, so

$$\Sigma(t,T) = \int_t^T G_1(u-t)b(t)du = b(t)\int_0^{T-t} G_1(s)ds, \quad t < T.$$

It follows that for each $t \in [0, T^*]$ there exists an interval contained in the set

$$\{\Sigma(t,T); T > t\},$$

so, by Theorem 11.8.4 (*b*), the model is approximately complete.

For the process satisfying (11.8.21) we have shown in Proposition 11.7.3 that $G_1(t) = e^{bt}$. So, G_1 with negative b is integrable and the reasoning above applies. □

11.8.3 Affine Model

Let us consider the affine model

$$P(t,T) = e^{-C(T-t)-D(T-t)R(t)}, \quad t \in [0,T^*],\ T \geq t,$$

with short rate of the form

$$dR(t) = (aR(t)+b)dt + R(t-)^{\frac{1}{\alpha}} dZ^{\alpha}(t), \quad R(0) = x \geq 0, \quad a \in \mathbb{R},\ b \geq 0, \tag{11.8.22}$$

where Z^{α} is a stable martingale with index $\alpha \in (1,2)$. Let us recall that in this case

$$\hat{P}(s,T) = e^{-C(T-s)-D(T-s)R(s)-\int_0^s R(u)du}, \tag{11.8.23}$$

and that

$$\Gamma_2(t,T,y) = \hat{P}(t-,T)(e^{-D(T-t)R(t-)^{\frac{1}{\alpha}}y} - 1), \quad t \in [0,T^*], T \geq 0, y \in (0,+\infty) \tag{11.8.24}$$

(see Proposition 11.2.1). In Section 11.6 we have shown that this model is not complete (see Theorem 11.6.1).

Theorem 11.8.9 *Let μ be a finite measure on $[0,+\infty)$ such that its support is dense in the set*

$$\{D(v), v \geq 0\}.$$

Let $\tilde{\mathcal{A}}$ be the space of $\mathcal{M}$-valued strategies $\varphi_t(\cdot)$ such that

$$\varphi_t(dT) = \varphi(t,T)\mu(dT),$$

equipped with the norm

$$\mid \varphi \mid^2_{\tilde{\mathcal{A}}} := \operatorname*{ess\,sup}_{\omega\in\Omega, t\in[0,T^*]} \int_0^{+\infty} \varphi^2(t,T)\mu(dT) < +\infty. \tag{11.8.25}$$

Then the affine model with short rate (11.8.22) is approximately complete in the class $\tilde{\mathcal{A}}$.

Proof We show first that the operator $\mathcal{K}$ given by

$$(\mathcal{K}\varphi)(t,y) := \int_0^{+\infty} \hat{P}(t-,T)(e^{-D(T-t)R(t-)^{\frac{1}{\alpha}}y} - 1)\varphi(t,T)\mu(dT),$$
$$t \in [0,T^*],\ T \geq 0$$

is a bounded operator from $\tilde{\mathcal{A}}$ into $\mathcal{H}$.

Using (11.8.23) and (11.8.24) we obtain for $\varphi \in \tilde{\mathcal{A}}$,

$$
\begin{aligned}
\mid \mathcal{K}\varphi \mid^2_{\mathcal{H}} &= \mathbb{E}\int_0^{T^*}\int_U (\varphi_s, \Gamma_2(s,\cdot,y))^2 ds\nu(dy) \\
&\leq \mathbb{E}\int_0^{T^*}\int_U \Big(\int_0^{+\infty} \varphi^2(s,T)\mu(dT)\Big)\Big(\int_0^{+\infty}\Gamma_2^2(s,T,y)\mu(dT)\Big) ds\nu(dy) \\
&\leq \mathbb{E}\int_0^{T^*}\Big(\int_0^{+\infty} \varphi^2(s,T)\mu(dT)\Big)\Big(\int_0^{+\infty} e^{-2\int_0^s R(u)du}\psi(R^{\frac{1}{\alpha}}(s-)D(T-s))\mu(dT)\Big)ds,
\end{aligned}
$$

where

$$
\psi(x) := \int_U \mid e^{-xy} - 1 \mid^2 \nu(dy), \quad x \geq 0.
$$

However, for arbitrary $x > 0$,

$$
\psi(x) = \int_0^{+\infty} (1 - e^{-xy})^2 \frac{1}{y^{1+\alpha}} dy = x^\alpha c,
$$

where

$$
c := \int_0^{+\infty} (1 - e^{-u})^2 \frac{1}{u^{1+\alpha}} du < +\infty.
$$

Recall that in Theorem 10.2.7 we showed that $D(\cdot)$ solves

$$
D'(v) = -c_\alpha D^\alpha(v) + aD(v) + 1, \quad D(0) = 0,
$$

and $C'(v) = bD(v), C(0) = 0$, where $c_\alpha = \frac{1}{\alpha(\alpha-1)}\Gamma(2-\alpha) > 0$. Note that $D(v) \uparrow x_\alpha$ as $v \uparrow +\infty$, where x_α is a solution of the equation $K(x) = 0, x \geq 0$, where

$$
K(x) := -c_\alpha x^\alpha + ax + 1.
$$

Therefore

$$
\begin{aligned}
|\mathcal{K}\varphi|^2_{\mathcal{H}} &\leq c\mathbb{E}\int_0^{T^*}\Big(\int_0^{+\infty} \varphi^2(s,T)\mu(dT)\Big)\Big(\int_0^{+\infty} e^{-2\int_0^s R(u)du} D^\alpha(T-s)R(s-)\mu(dT)\Big)ds \\
&\leq c\, x_\alpha\ \mu([0,+\infty))\ \mathbb{E}\int_0^{T^*}\Big(\int_0^{+\infty}\varphi^2(s,T)\mu(dT)\Big)R(s-)e^{-2\int_0^s R(u)du} ds \\
&\leq c\, x_\alpha\ \mu([0,+\infty))\ \mid \varphi \mid^2_{\tilde{\mathcal{A}}}\ \mathbb{E}\Big(\int_0^{T^*} R(s-)e^{-2\int_0^s R(u)du} ds\Big).
\end{aligned}
$$

However,

$$
\begin{aligned}
\mathbb{E}\Big(\int_0^{T^*} R(s-)e^{-2\int_0^s R(u)du} ds\Big) &= \mathbb{E}\Big(\int_0^{T^*} R(s)e^{-2\int_0^s R(u)du} ds\Big) \\
&= \mathbb{E}\int_0^{T^*} \frac{d}{ds}\Big[-\frac{1}{2}e^{-2\int_0^s R(u)du}\Big]ds = \frac{1}{2}\mathbb{E}\Big(1 - e^{-2\int_0^{T^*} R(u)du}\Big) \leq \frac{1}{2}.
\end{aligned}
$$

Finally,

$$| \mathcal{K}\varphi |^2_{\mathcal{H}} \le \frac{1}{2} c\, x_\alpha\, \mu([0, +\infty))\ | \varphi |^2_{\tilde{\mathcal{A}}}$$

and the boundedness of $\mathcal{K}$ follows.

In the present situation

$$\mathcal{K}^* g(t, T) = \int_0^{+\infty} \hat{P}(t-, T)(e^{-D(T-t)R(t-)^{\frac{1}{\alpha}} y} - 1) g(t, y) \nu(dy),$$

so the condition $\mathcal{K}^* g = 0$ means that $\mathbb{P} - a.s.$ for almost all $t \in [0, T^*]$ and for μ-almost all $T \ge t$

$$\int_0^{+\infty} (e^{-D(T-t)R(t-)^{\frac{1}{\alpha}} y} - 1) g(t, y) \nu(dy) = 0.$$

However, for $t > 0$, $R(t-) > 0$ and on the interval $(t, +\infty)$ the function $D(T - t)$ of the T argument is increasing. Since the support of μ is dense in $\{D(v), v \ge 0\}$, by Proposition 11.8.5 we obtain that $g = 0$. □

Part IV

Stochastic Equations in the Bond Market

12

Stochastic Equations for Forward Rates

Four types of stochastic equations describing the movement of forward rates are introduced: the Heath–Jarrow–Morton equation, Morton's equation, the Heath–Jarrow–Morton–Musiela equation and the Morton–Musiela equation.

12.1 Heath–Jarrow–Morton Equation

In the Heath–Jarrow–Morton approach to the bond market the forward rates are represented in the form

$$
\begin{aligned}
f(t,T) &= f(0,T) + \int_0^t \alpha(s,T)ds + \int_0^t \langle \sigma(s,T), dZ(s)\rangle, \qquad (12.1.1)\\
&= f(0,T) + \int_0^t \alpha(s,T)ds + \sum_{j=1}^{d} \int_0^t \sigma^j(s,T)dZ^j(s), \quad t \in [0,T^*],\ T > 0,
\end{aligned}
$$

with $0 < T^* < +\infty$ or, equivalently

$$
df(t,T) = \alpha(t,T)dt + \langle \sigma(t,T), dZ(t)\rangle, \quad t \in [0,T^*],\ T > 0.
$$

Here $Z(s) = (Z^1(s), \ldots, Z^d(s))$ is a d-dimensional Lévy process, $\sigma(s,T) = (\sigma^1(s,T), \ldots, \sigma^d(s,T))$ is the d-dimensional volatility field and $\alpha(s,T)$ is the drift term. The representation does require the specification of α, σ and Z. To ensure that the forward rates have such basic properties like mean reversion and positivity, it is convenient to represent f as a solution of stochastic equations, rather than as a sum of two integrals. It is natural to assume that the volatility is a function of forward rates, say

$$
\sigma(t,T) = g(t,T,f(t-,T)), \qquad t \in [0,T^*],\ T > 0,
$$

where g is a fixed deterministic function $g(t,T,z)$, $0 \le t \le T < +\infty$, $z \ge 0$. Recall that the HJM model given by (12.1.1) satisfies (MP) if and only if

$$\alpha(t,T) = \langle DJ\Big(\int_t^T \sigma(t,s)ds\Big), \sigma(t,T)\rangle, \tag{12.1.2}$$

where $J\colon \mathbb{R}^d \to \mathbb{R}$ stands for the Laplace transform of Z and DJ for its derivative. This leads to the equation

$$\begin{aligned} df(t,T) = &\Big\langle DJ\Big(\int_t^T g(t,s,f(t-,s))ds\Big), g(t,T,f(t-,T))\Big\rangle dt \\ &+ \langle g(t,T,f(t-,T)), dZ(t)\rangle, \qquad t \in [0,T^*],\ T > 0, \end{aligned} \tag{12.1.3}$$

which will be called the *Heath–Jarrow–Morton equation* (HJM equation).

12.2 Morton's Equation

Morton's equation is a particular case of the HJM equation (12.1.3) with one-dimensional Lévy process and linear function G, i.e. volatility has the form

$$\sigma(t,T) = \lambda(t,T)f(t-,T), \qquad t \in [0,T^*],\ T \ge 0, \tag{12.2.1}$$

where λ is a deterministic, positive and continuous function. Although (12.2.1) describes a simple form of dependence of volatility on the forward rate, the drift becomes nonlinear and nonlocal, i.e.

$$\alpha(t,T) = J'\left(\int_t^T \lambda(t,u)f(t-,u)du\right)\lambda(t,T)f(t-,T), \qquad t \in [0,T^*],\ T \ge 0. \tag{12.2.2}$$

The resulting equation,

$$df(t,T) = J'\left(\int_t^T \lambda(t,u)f(t-,u)du\right)\lambda(t,T)f(t-,T)dt + \lambda(t,T)f(t-,T)dZ(t), \tag{12.2.3}$$

will be called Morton's equation. The problem of solvability of (12.2.3), with $\lambda(t,T) \equiv 1$, has been first stated by Morton in [95] in the case when Z was a Wiener process and solved with negative answer: linear volatility implies that there is no solution to (12.2.3) in that case (see [95] Section 4.7 or Filipović [52], Section 7.4). This fact was one of the main reasons why the well known Brace–Gątarek–Musiela (BGM) model was formulated in terms of Libor rates and not in terms of forward rates (see Brace, Gątarek and Musiela [15]).

12.3 The Equations in the Musiela Parametrization

It is sometimes convenient to change the bond market notation by passing from the (t, T) parametrization used so far, which we call the *natural frame*, to the new one (t, x) introduced by Musiela and which will be called the *moving frame*. For a current time point t and maturity T we use *time to maturity* $x = T - t$. The forward rate in the moving frame is given by

$$r(t,x) := f(t,t+x), \qquad t \in [0,T^*],\ x \geq 0.$$

We will concentrate on the case in which

$$\sigma(s,T) = (\sigma^j(s,T), j=1,\dots,d) = (g^j(T-s, f(s-,T)), j=1,\dots,d).$$

Assume that the function valued process $r(t,\cdot)$, $t \in [0,T^*]$ takes values in a Hilbert space H of some functions defined on $[0,+\infty)$, such as

$$L^{2,\gamma} := \left\{ h\colon [0,+\infty) \to \mathbb{R} : \int_0^{+\infty} \mid h(x) \mid^2 e^{\gamma x} dx < +\infty \right\}, \tag{12.3.1}$$

or

$$H^{1,\gamma} \left\{ h\colon [0,+\infty) \to \mathbb{R} : \int_0^{+\infty} (\mid h(x) \mid^2 + \mid h'(x) \mid^2) e^{\gamma x} dx < +\infty \right\}. \tag{12.3.2}$$

We will show that the solution of the HJM equation regarded in the Musiela parametrization satisfies the so-called *Heath–Jarrow–Morton–Musiela* (HJMM) stochastic evolution equation

$$\begin{aligned} dr(t,x) &= \Big(\frac{\partial}{\partial x} r(t,x) + F(r(t))(x)\Big) dt + \sum_{j=1}^{d} G^j(r(s-))(x) dZ^j(s), \\ r(0,\cdot) &= r_0. \end{aligned} \tag{12.3.3}$$

The transformations F, G^j, $j = 1,\dots,d$ are of the form

$$G^j(h)(x) = g^j(x, h(x)), j = 1,\dots d, \quad F(h)(x) = \langle DJ(\int_0^x G(h(v))dv), Gh(x)\rangle$$

for $h \in H$ and $x \geq 0$. Moreover, $\frac{\partial}{\partial x}$ denotes the generator A of the shift semigroup $S(t)$:

$$S(t)h(x) = h(t+x), \ \ t \geq 0, \ \ x \geq 0.$$

For basic definitions related to stochastic evolution equations, we refer to Appendix C where the concepts of *strong, weak* and *mild solutions* are defined. As weak and mild solutions usually coincide we often speak about solutions meaning weak and/or mild solutions. Strong solutions exist very rarely.

We prove now that r is a mild solution of the equation (12.3.3), that is,

$$r(t) = S(t)r_0 + \int_0^t S(t-s)F(r(s))ds + \sum_{j=1}^d \int_0^t S(t-s)G^j(r(s-))dZ^j(s). \quad (12.3.4)$$

Assume that $r(t,x) = f(t,x)$, $r(t-,x) = f(t-,t+x)$, $t \geq 0, x \geq 0$ and that the integral version of (12.1.1),

$$f(t,T) = f(0,t) + \int_0^t \alpha(s,T)ds + \sum_{j=1}^d \int_0^t \sigma^j(s,T)dZ^j(s), \qquad (12.3.5)$$

where

$$\sigma^j(s,T) = g^j(T-s, f(s-,T)), \quad \alpha(s,T) = \langle DJ\Big(\int_s^T \sigma(s,v)dv\Big), \sigma(s,T)\rangle, \ s \leq T, \qquad (12.3.6)$$

holds with full probability for all $0 \leq s \leq T < +\infty$.

Thus inserting $T = t + x$ into (12.3.5) yields

$$r(t,x) = f(t,t+x) = f(0,t+x) + \int_0^t \alpha(s,t+x)ds + \sum_{j=1}^d \int_0^t \sigma^j(s,t+x)dZ^j(s).$$

Moreover,

$$\sigma^j(s,t+x) = g^j(t+x-s, f(s-,t+x)) = g^j(t+x-s, r(s-,t+x-s)).$$

Consequently,

$$\begin{aligned}
\sum_{j=1}^d \int_0^t \sigma(s,t+x)dZ^j(s) &= \sum_{j=1}^d \int_0^t g^j(t+x-s, f(s-,t+x))dZ^j(s) \\
&= \sum_{j=1}^d \int_0^t g^j(t+x-s, r(s-,t+x-s))dZ^j(s) \\
&= \sum_{j=1}^d \int_0^t S(t-s)G^j(r(s-))(x)dZ^j(s).
\end{aligned}$$

Similarly,

$$\begin{aligned}
&\alpha(s,t+x) \\
&= \left\langle DJ\left(\int_s^{t+x} g(u-t, f(t-,u)du), g(t+x-s, f(s-,t+x)\right)\right\rangle \\
&= \left\langle DJ\left(\int_0^{t-s+x} g(v+s-t, r(t-,v+s-t))dv), g(t-s+x, r(s-,t-s+x)\right)\right\rangle \\
&= S(t-s)F(r(s))(x),
\end{aligned}$$

and

$$\int_0^t \alpha(s,t+x)ds = \int_0^t S(t-s)F(r(s))(x)ds.$$

Addition of the obtained formulae yields the required identity.

In a similar way Morton's equation in the Musiela parameterization becomes:

$$dr(t,x) = \left(\frac{\partial}{\partial x}r(t,x) + J'\Big(\int_0^x r(s,u)\lambda(u)du\Big)\lambda(x)r(t,x)\right)dt$$
$$+ \lambda(x)r(t-,x)dZ(t), \quad t \in [0,T^*], \quad x \geq 0,$$

and will be called the *Morton–Musiela* equation. Here λ is a fixed function and the corresponding one-dimensional volatility $\sigma(t,T,f(t-,T)) = \lambda(T-t)f(t-,T)$. This corresponds to the function $g(x,z) = \lambda(x)z$. Its mild version is of the form

$$r(t,x) = r(0,x) + \int_0^t S_{t-s}\left(J'\left(\int_0^x r(s,u)\lambda(u)du\right)\lambda(x)r(s,x)\right)ds$$
$$+ \int_0^t S_{t-s}\Big(\lambda(x)r(s-,x)\Big)dZ(s), \qquad t \in [0,T^*], \quad x \geq 0. \quad (12.3.7)$$

The main advantage of passing to the Musiela parametrization is that the forward rate process becomes Markovian and evolves in a fixed state space. General results on Markov processes and stochastic evolution equations can be used.

In the present part of the book the existence and uniqueness questions for the equations introduced are settled. They should provide a starting point to investigate asymptotic behaviour, time-reversion of forward curves and more special questions such as consistency of the models.

The equation (12.3.3) was intensively studied in the case in which Z is a Wiener process in $\mathbb{R}^d$ (see e.g. Filipović [54], Peszat and Zabczyk [100] and references therein). Then the function DJ is linear, i.e. $DJ(z) = Qz, z \in \mathbb{R}^d$, where Q is the covariance matrix of Z. There are also several results for the case of general, also infinite dimensional, Lévy process Z (see e.g. Peszat and Zabczyk [100], Filipović and Tappe [56], Rusinek [109], Rusinek [111], Rusinek [112], Marinelli [90], Filipović, Tappe and Teichmann [57], Peszat and Zabczyk [101]). In particular, in [90] the local solvability of (12.3.3) was studied for a Lévy process Z having exponential moments. An extensive study of the HJM equations in Banach spaces can be found in Brzeźniak and Tayfun [23].

13

Analysis of the HJMM Equation

In this chapter we study the solvability of the HJMM equation through the general theory of stochastic evolution equations. We establish the existence of local and global solutions. In particular, we deal with local solutions of the Morton–Musiela equation.

13.1 Existence of Solutions to the HJMM Equation

In this section we examine the existence of weak solutions (see Appendix C) of the HJMM equation

$$dr(t,x) = \Big(\frac{\partial}{\partial x} r(t,x) + F(r(t))(x)\Big)dt + G(r(t-))(x)dZ(t), \quad r(0,\cdot) = r_0, \quad t \in [0,T^*], \tag{13.1.1}$$

with a one-dimensional Lévy process Z and coefficients of the form

$$G(r)(x) = g(x, r(x)), \quad x \geq 0, \ r \in H, \tag{13.1.2}$$

$$F(r)(x) = J'\left(\int_0^x g(v, r(v))dv\right) g(x, r(x)), \quad x \geq 0, \ r \in H. \tag{13.1.3}$$

In the preceding $g\colon \mathbb{R} \times \mathbb{R} \longrightarrow \mathbb{R}$ is some function and J' denotes the derivative of the Laplace exponent of the process Z. Recall that for a Lévy process Z with characteristic triplet (a, q, ν), J is given by

$$J(z) = -az + \frac{1}{2}qz^2 + \int_{\mathbb{R}} (e^{-zy} - 1 + zy\mathbf{1}_{(-1,1)}(y))\, \nu(dy), \qquad z \in \mathbb{R},$$

and

$$J'(z) = -a + qz + \int_{\mathbb{R}} y(\mathbf{1}_{(-1,1)}(y) - e^{-zy})\, \nu(dy), \qquad z \in \mathbb{R},$$

providing that the preceding integrals exist. We focus on local and global solutions of equation (13.1.1) living in the spaces $H = L^{2,\gamma}_+([0,+\infty))$ and $H^{1,\gamma}_+([0,+\infty))$, (see (12.3.1) and (12.3.2)). In view of Theorem C.1.4, in Appendix C, we need conditions for local Lipschitz property and linear growth of the transformations

$$F\colon H \longrightarrow H, \quad G\colon H \longrightarrow H.$$

They will be formulated in terms of the function g and the characteristic triplet of the process Z. Our analysis uses the following decomposition of the Laplace exponent J

$$J(z) = -az + \frac{1}{2}qz^2 + J_1(z) + J_2(z) + J_3(z) + J_4(z), \quad z \in \mathbb{R}, \tag{13.1.4}$$

where

$$J_1(z) := \int_{-\infty}^{-1} (e^{-zy} - 1)\nu(dy), \qquad J_2(z) := \int_{-1}^{0} (e^{-zy} - 1 + zy)\nu(dy),$$

$$J_3(z) := \int_{0}^{1} (e^{-zy} - 1 + zy)\nu(dy), \qquad J_4(z) := \int_{1}^{+\infty} (e^{-zy} - 1)\nu(dy).$$

It is natural to require that forward rates are positive processes. Therefore, before passing to the existence results, we establish a theorem providing necessary and sufficient conditions on the volatility function g implying the positivity of the solutions of the HJMM equation.

Theorem 13.1.1 *Assume that the coefficients G and F of the equation* (13.1.1) *are locally Lipschitz in H, where $H = L^{2,\gamma}$ or $H = H^{1,\gamma}$. Then* (13.1.1) *is positivity preserving if and only if*

$$r + g(x,r)u \geq 0 \qquad \text{for all } r \geq 0,\ x \geq 0,\ u \in \operatorname{supp}\nu, \tag{13.1.5}$$

$$g(x,0) = 0 \qquad \text{for all } x \geq 0. \tag{13.1.6}$$

Proof We will use Theorem C.1.5 in a similar way to Peszat and Zabczyk [100]. Let us consider the Lévy–Itô decomposition of Z

$$Z(t) = at + qW(t) + Z_0(t) + Z_1(t), \text{ where}$$
$$Z_0(t) := \int_0^t \int_{|y|\leq 1} y\tilde{\pi}(ds,dy),\ Z_1(t) := \int_0^t \int_{|y|>1} y\pi(ds,dy),$$

and a sequence of its approximations of the form

$$Z^n(t) = at + qW(t) - tm_n + (Z^n_0(t) + Z_1(t)),$$

with $Z^n_0(t) := \int_0^t \int_{\{\frac{1}{n}<|y|\leq 1\}} y\pi(dy)$ and $m_n := \mathbb{E}[Z^n_0(1)]$.

The equation (13.1.1) preserves positivity if and only if for each n does the equation

$$dr_n(t,x) = \Big(\frac{d}{dx}r_n(t,x) + \Big(J'\Big(\int_0^x g(y, r_n(t,y))dy\Big) + a - m_n\Big)g(x, r_n(t,x))\Big)dt$$
$$+ g(x, r_n(t-,x))(dZ_0^n(t) + dZ_1(t) + qdW(t)). \qquad (13.1.7)$$

Since the sum $Z_0^n(t) + Z_1(t)$ is a compound Poisson process with jumps greater than $\frac{1}{n}$, the driving noise in (13.1.7) between the jumps is a Wiener process only. Thus we may use Theorem C.1.5. The conditions

$$\int_0^{+\infty} \Big(J'\Big(\int_0^x g(v,\varphi(v))dv\Big) + a - m_n\Big)g(x,\varphi(x))f(x)e^{\gamma x}dx \geq 0,$$

$$\int_0^{+\infty} g(x,\varphi(x))f(x)e^{\gamma x}dx = 0$$

are satisfied for any $\varphi, f \in L^{2,\gamma}$ such that $< \varphi, f >= 0$ if and only if $g(x,0) = 0$. The solution remains positive in the moment of jump of Z^n if and only if

$$r + g(x,r)u \geq 0, \qquad r \geq 0, \quad u \in \operatorname{supp}\{\nu\} \cup \Big[\frac{1}{n}, +\infty\Big).$$

Passing to the limit $n \to +\infty$ we obtain (13.1.5). □

13.1.1 Local Solutions

For the solvability of (13.1.1) in $L^{2,\gamma}_+$ we will need the following conditions on g.

$$(G1) \begin{cases} (i) \quad \text{The function g is continuous on } \mathbb{R}^2_+ \text{ and} \\ \qquad g(x,0) = 0,\ g(x,y) \geq 0, \quad x, y \geq 0. \\ (ii) \quad \text{For all } x, y \geq 0 \text{ and } u \in \operatorname{supp} \nu, \\ \qquad y + g(x,y)u \geq 0. \\ (iii) \quad \text{There exists a constant } C > 0 \text{ such that} \\ \qquad | g(x,u) - g(x,v) | \leq C \mid u - v \mid, \qquad x, u, v \geq 0. \end{cases}$$

Conditions $(G1(i))$ and $(G1(ii))$ ensure that the solution of (13.1.1) is positive (see Theorem 13.1.1) while $(G1(iii))$ is needed for the Lipschitz property of F and G.

Theorem 13.1.2 *Assume that* $(G1)$ *holds and either* Z *is a Wiener process or for some* $z_0 > 0$,

$$(L1) \qquad \int_{-\infty}^{-1} | y |^2 e^{z_0|y|}\nu(dy) < +\infty, \quad \text{and} \quad \int_1^{+\infty} y^2\nu(dy) < +\infty.$$

Then for arbitrary initial condition $r_0 \in L^{2,\gamma}_+$ *there exists a unique local solution of* (13.1.1) *in* $L^{2,\gamma}_+$.

Condition $(L1)$ in the theorem implies that

$$\int_{|y|>1} | y |^2 \nu(dy) < +\infty,$$

so Z satisfies

$$\mathbb{E}[Z^2(t)] < +\infty, \quad t \in [0, T^*].$$

Consequently, Z is a square integrable martingale with drift.

For the existence of local solutions in $H^{1,\gamma}_+$ we will need more stringent conditions on the function g.

$$(G2) \quad \begin{cases} (i) & \text{The functions } g'_x, g'_y \text{ are continuous on } \mathbb{R}^2_+ \text{ and} \\ & g'_x(x,0) = 0, \qquad x \geq 0. \\ (ii) & \sup_{x,y\geq 0} | g'_y(x,y) | < +\infty. \\ (iii) & \text{There exists a constant } C > 0 \text{ such that} \\ & | g'_x(x,u) - g'_x(x,v) | + | g'_y(x,u) - g'_y(x,v) | \leq C | u - v |, \quad x,u,v \geq 0. \end{cases}$$

Theorem 13.1.3 *Assume that* $(G1)$ *and* $(G2)$ *hold and for some* $z_0 > 0$,

$$(L2) \qquad \int_{-\infty}^{-1} | y |^3 e^{z_0|y|} \nu(dy) < +\infty, \quad \textit{and} \quad \int_1^{+\infty} y^3 \nu(dy) < +\infty.$$

Then for arbitrary initial condition $r_0 \in H^{1,\gamma}_+$ *there exists a unique local solution of* (13.1.1) *in* $H^{1,\gamma}_+$.

Since $(L2)$ implies that

$$\int_{|y|>1} | y |^3 \nu(dy) < +\infty,$$

we see that Z satisfies

$$\mathbb{E}[| Z(t) |^3] < +\infty, \quad t \geq 0.$$

We pass to the proofs of the theorems.

Proof of Theorem 13.1.2 We use the fact that locally Lipschitz coefficients of a general evolution equation imply the existence of its local solutions (see Peszat and Zabczyk [100]). We show thus that F and G given by (13.1.2) are locally Lipschitz in $L^{2,\gamma}$.

Step 1: We show that condition $(L1)$ is satisfied if and only if the function J' is locally Lipschitz.

Differentiation of (13.1.4) yields

$$J'(z) = -a + qz + J_1'(z) + J_2'(z) + J_3'(z) + J_4'(z), \qquad z \in \mathbb{R},$$

with

$$J_1'(z) := -\int_{-\infty}^{-1} ye^{-zy}\nu(dy), \qquad J_2'(z) := \int_{-1}^{0} y(1-e^{-zy})\nu(dy),$$

$$J_3'(z) := \int_0^1 y(1-e^{-zy})\nu(dy), \qquad J_4'(z) := -\int_1^{+\infty} ye^{-zy}\nu(dy),$$

providing that the preceding integrals exist (see for instance Lemma 8.1 and 8.2 in Rusinek [109]). Similarly, $J''(z) = q + J_1''(z) + J_2''(z) + J_3''(z) + J_4''(z)$, $z \in \mathbb{R}$, where

$$J_1''(z) := \int_{-\infty}^{-1} y^2 e^{-zy}\nu(dy), \qquad J_2''(z) := \int_{-1}^{0} y^2 e^{-zy}\nu(dy),$$

$$J_3''(z) := \int_0^1 y^2 e^{-zy}\nu(dy), \qquad J_4''(z) := \int_1^{+\infty} y^2 e^{-zy}\nu(dy).$$

The function J' is Lipschitz on $[0, z_0)$ for some $z_0 > 0$ if and only if J'' is bounded on $[0, z_0)$. Since J_2'' and J_3'' are bounded on $[0, z_0)$ one needs conditions that yield that J_1'' and J_4'' are bounded. This leads to $L(1)$.

Step 2: By $(G1(iii))$ it is clear that G is Lipschitz in $L^{2,\gamma}$. We prove that if J' is locally Lipschitz and $(G1)$ holds then F is locally Lipschitz as well.

For any $r, \bar{r} \in L^{2,\gamma}$ we have

$$\begin{aligned} | F(r) - F(\bar{r}) |_{L^{2,\gamma}}^2 &= \int_0^{+\infty} \Bigg[J'\left(\int_0^x g(y, r(y))dy\right) g(x, r(x)) \\ &\qquad - J'\left(\int_0^x g(y, \bar{r}(y))dy\right) g(x, \bar{r}(x))\Bigg]^2 e^{\gamma x} dx \\ &\le 2I_1 + 2I_2, \end{aligned}$$

where

$$I_1 := \int_0^{+\infty} \left[J'\left(\int_0^x g(y, r(y))dy\right) - J'\left(\int_0^x g(y, \bar{r}(y))dy\right)\right]^2 g^2(x, r(x)) e^{\gamma x} dx,$$

$$I_2 := \int_0^{+\infty} \left[J'\left(\int_0^x g(y, \bar{r}(y))dy\right)\right]^2 \Big(g(x, \bar{r}(x)) - g(y, r(x))dx\Big)^2 e^{\gamma x} dx.$$

By $(G1(iii))$ we have

$$\begin{aligned} \int_0^x g(y, r(y))dy &= \int_0^x \Big(g(y, r(y)) - g(y, 0)\Big)dy \le C\int_0^x r(y)dy \\ &\le C\left(\int_0^{+\infty} e^{-\gamma y} dy\right)^{1/2} \left(\int_0^{+\infty} | r(y) |^2 e^{\gamma y} dy\right)^{1/2} \le \frac{C}{\sqrt{\gamma}} | r |_{L^{2,\gamma}}. \end{aligned}$$

Denoting by $D = D(| r |_{L^{2,\gamma}}, | \bar{r} |_{L^{2,\gamma}})$ the local Lipschitz constant of J', we obtain

$$\begin{aligned} I_1 &\le D \int_0^{+\infty} \left[\int_0^x \Big(g(y, r(y)) - g(y, \bar{r}(y)) \Big) dy \right]^2 g^2(x, r(x)) e^{\gamma x} dx \\ &\le D \mid g(\cdot, r(\cdot)) - g(\cdot, \bar{r}(\cdot)) \mid_{L^{2,\gamma}}^2 \int_0^{+\infty} g^2(x, r(x)) e^{\gamma x} dx \\ &\le D \mid g(\cdot, r(\cdot)) - g(\cdot, \bar{r}(\cdot)) \mid_{L^{2,\gamma}}^2 \cdot \int_0^{+\infty} \Big(g(x, r(x)) - g(x, 0) \Big)^2 e^{\gamma x} dx \\ &\le DC^2 \int_0^{+\infty} (r(x) - \bar{r}(x))^2 e^{\gamma x} dx \cdot C^2 \int_0^{+\infty} r^2(x) e^{\gamma x} dx \\ &\le DC^4 \mid r - \bar{r} \mid_{L^{2,\gamma}}^2 \mid r \mid_{L^{2,\gamma}}^2 . \end{aligned}$$

Similarly, for a local boundary B of J' we get

$$I_2 \le BC^2 \mid r - \bar{r} \mid_{L^{2,\gamma}}^2 ,$$

and thus the local Lipschitz property of F follows. □

Proof of Theorem 13.1.3: We prove that F and G are locally Lipschitz in $H^{1,\gamma}$.

Step 1: We show that $(L2)$ is equivalent to the fact that J'' is locally Lipschitz. Using (13.1.4) we obtain $J'''(z) = J_1'''(z) + J_2'''(z) + J_3'''(z) + J_4'''(z)$, where

$$J_1'''(z) := -\int_{-\infty}^{-1} y^3 e^{-zy} \nu(dy), \qquad J_2'''(z) := -\int_{-1}^{0} y^3 e^{-zy} \nu(dy),$$

$$J_3'''(z) := -\int_0^1 y^3 e^{-zy} \nu(dy), \qquad J_4'''(z) := -\int_1^{+\infty} y^3 e^{-zy} \nu(dy).$$

For fixed $z_0 > 0$ the functions J_2''', J_3''' are bounded on $[0, z_0)$. It follows that J''' is bounded on $[0, z_0)$ if and only if J_1''' and J_4''' are bounded on $[0, z_0)$, which leads to $(L2)$. Let us notice that $(L2)$ implies $(L1)$ and, consequently, that J' is also locally Lipschitz.

Step 2: We show that if J' and J'' are locally Lipschitz and $(G1)$, $(G2)$ hold then F and G are locally Lipschitz in $H_+^{1,\gamma}$.

To get the required estimation for G we need to estimate

$$I_0 := \int_0^{+\infty} \left[g_y'(x, r(x)) r'(x) - g_y'(x, \bar{r}(x)) \bar{r}'(x) \right]^2 e^{\gamma x} dx.$$

Using Lemma 15.2.3 we obtain the following inequalities

$$\begin{aligned}
I_0 &\leq 2\int_0^{+\infty} \left|g_y'(x, r(x))\right|^2 \left|r'(x) - \bar{r}'(x))\right|^2 e^{\gamma x}dx \\
&+ \int_0^{+\infty} \left|g_y'(x, r(x)) - g_y'(x, \bar{r}(x))\right|^2 \left|\bar{r}'(x))\right|^2 e^{\gamma x}dx \\
&\leq 2 \sup_{x,r\geq 0} \mid g_y'(x,r) \mid^2 \int_0^{+\infty} \mid r'(x) - \bar{r}'(x) \mid^2 e^{\gamma x}dx \\
&+ 2C^2 \int_0^{+\infty} \mid r(x) - \bar{r}(x) \mid^2 \mid \bar{r}'(x) \mid^2 e^{\gamma x}dx \\
&\leq 2 \sup_{x,r\geq 0} \mid g_y'(x,r) \mid^2 \cdot \mid r - \bar{r} \mid^2_{H^{1,\gamma}} + 2C^2 \frac{4}{\gamma} \mid r - \bar{r} \mid^2_{H^{1,\gamma}} \cdot \mid \bar{r} \mid^2_{H^{1,\gamma}},
\end{aligned}$$

and thus the local Lipschitz property of G follows.

To show the same for F it is sufficient to show the Lipschitz estimation for the formula

$$\begin{aligned}
I := \int_0^{+\infty} \Bigg[\frac{d}{dx}\Bigg\{\Bigg(J'\left(\int_0^x g(y, r(y))dy\right) g(x, r(x)) \\
-J'\left(\int_0^x g(y, \bar{r}(y))dy\right) g(x, \bar{r}(x))\Bigg)\Bigg\}\Bigg]^2 e^{\gamma x}dx.
\end{aligned}$$

By explicit calculations we obtain

$$I \leq 3I_1 + 3I_2 + 3I_3,$$

where

$$I_1 := \int_0^{+\infty} \left[J''\left(\int_0^x g(y, r(y))dy\right) g^2(x, r(x)) - J''\left(\int_0^x g(y, \bar{r}(y))dy\right) g^2(x, \bar{r}(x))\right]^2 e^{\gamma x}dx,$$

$$I_2 := \int_0^{+\infty} \left[J'\left(\int_0^x g(y, r(y))dy\right) g_x'(x, r(x)) - J'\left(\int_0^x g(y, \bar{r}(y))dy\right) g_x'(x, \bar{r}(x))\right]^2 e^{\gamma x}dx,$$

$$I_3 := \int_0^{+\infty} \left[J'\left(\int_0^x g(y, r(y))dy\right) g_y'(x, r(x)) \cdot r'(x) - J'\left(\int_0^x g(y, \bar{r}(y))dy\right) g_y'(x, \bar{r}(x)) \cdot \bar{r}'(x)\right]^2 e^{\gamma x}dx.$$

With the use of $(G1)$ and $(G2)$ we can estimate I_1 and I_2 in a similar way as we did in Step 2 of the proof of Theorem 13.1.2. In fact, by the estimation of I_1 a new fourth-power term of $\bar{r}$ appears, which can be estimated as follows

$$\int_0^{+\infty} \mid \bar{r}(x) \mid^4 e^{\gamma x}dx \leq \sup_{x\geq 0} \mid \bar{r}(x) \mid^2 \int_0^{+\infty} \mid \bar{r}(x) \mid^2 e^{\gamma x}dx \geq C \mid \bar{r} \mid^4_{H^{1,\gamma}}$$

(see Lemma 15.2.3). To estimate I_3, we need additional inequalities for

$$\int_0^{+\infty} \left[g_y'(x, r(x))r'(x) - g_y'(x, \bar{r}(x))\bar{r}'(x)\right]^2 e^{\gamma x} dx,$$

which is exactly I_0 and is estimated in the preceding, and for

$$I_4 := \int_0^{+\infty} \left[g_y'(x, \bar{r}(x))\bar{r}'(x)\right]^2 e^{\gamma x} dx.$$

Estimation for I_4 follows from the bound on g_y'. □

13.1.2 Global Solutions

We pass now to the existence of global solutions of (13.1.1) in the spaces $L_+^{2,\gamma}$ and in $H_+^{1,\gamma}$. As one suspects, we need stronger conditions for the function g and the Lévy process Z than those formulated for local solutions in the previous section.

Theorem 13.1.4 *Assume that* $(G1)$ *holds and that*

$$q = 0, \quad supp\{\nu\} \subseteq [0, +\infty), \quad \int_0^{+\infty} \max\{y, y^2\}\nu(dy) < +\infty. \tag{13.1.8}$$

Then for arbitrary $r_0 \in L_+^{2,\gamma}$ *the equation (13.1.1) has a unique global solution in* $L_+^{2,\gamma}$.

The requirements for the function g are the same as in Theorem 13.1.2 but the process Z is a square integrable subordinator with drift.

For global existence in $H_+^{1,\gamma}$ we need additional conditions for g.

$$(G3) \quad \begin{cases} (i) \quad \text{Partial derivatives } g_y', g_{xy}'', g_{yy}'' \text{ are bounded on } \mathbb{R}_+^2. \\ (ii) \quad 0 \le g(x,y) \le c\sqrt{y}, \qquad x, y \ge 0. \\ (iii) \quad | g_x'(x,y) | \le h(x), \qquad x, y \ge 0, \text{ for some } h \in L_+^{2,\gamma}. \end{cases}$$

Theorem 13.1.5 *Assume that conditions* $(G1)$, $(G2)$ *and* $(G3)$ *are satisfied and*

$$q = 0, \quad supp\{\nu\} \subseteq [0, +\infty), \quad \int_0^{+\infty} \max\{y, y^3\}\nu(dy) < +\infty. \tag{13.1.9}$$

Then for arbitrary $r_0 \in H_+^{1,\gamma}$ *there exists a unique global solution of* (13.1.1) *in* $H_+^{1,\gamma}$.

The class of Lévy processes in Theorem 13.1.5 is much narrower than that required for the existence of local solutions in Theorem 13.1.3. Now Z must not have the Wiener part and has positive jumps only. It is of finite variation and

$$\mathbb{E}[| Z(t) |^3] < +\infty, \quad t \ge 0.$$

Thus Z is a subordinator with drift that is integrable in the third power.

We pass to the proofs of the theorems.

Proof of Theorem 13.1.4 As local Lipschitz conditions for F and G in $L^{2,\gamma}$ were shown in the proof of Theorem 13.1.2, our goal now is to show the linear growth condition for the transformation F. The analogous result for G is trivial.

Step 1: We show that, under (13.1.8), J' is bounded on $[0,+\infty)$.

It follows from (13.1.4) that $J'(0)$ is finite if and only if both $J_1'(0)$ and $J_4'(0)$ are finite, which is equivalent to

$$\int_{|y|>1} |\, y \,|\, \nu(dy) < +\infty. \tag{13.1.10}$$

Moreover, it follows that

$$\lim_{z\to+\infty} |\, J_1'(z)\,| = +\infty, \qquad \lim_{z\to+\infty} |\, J_2'(z)\,| = +\infty,$$

and that

$$|\, J_3' \,| \text{ is bounded} \iff \int_0^1 y\nu(dy) < +\infty,$$

$$|\, J_4'(z) \,| \text{ is bounded} \iff \int_1^{+\infty} y\nu(dy) < +\infty.$$

Consequently, J' is bounded on $[0,+\infty)$ if and only if Z has neither Wiener part nor negative jumps and $\int_0^{+\infty} y\nu(dy) < +\infty$.

Step 2: We show that if J' is bounded on $[0,+\infty)$ then F has linear growth in $L^{2,\gamma}_+$.

Let $C_1 := \sup_{z\geq 0} J'(z) < +\infty$. The assertion follows from the estimation

$$\begin{aligned}
&|\, F(r)\,|_{L^{2,\gamma}} \\
&= \int_0^{+\infty}\left[J'\left(\int_0^x g(y,r(y))dy\right)g(x,r(x))\right]^2 e^{\gamma x}dx \leq C_1\int_0^{+\infty}[g(x,r(x))]^2\, e^{\gamma x}dx \\
&\leq C_1\int_0^{+\infty}[g(x,r(x))-g(x,0)]^2 e^{\gamma x}dx \leq C_1C^2\int_0^{+\infty} r^2(x)e^{\gamma x}dx \leq C_1C^2\,|\, r\,|^2_{L^{2,\gamma}}.
\end{aligned}$$

□

Proof of Theorem 13.1.5 Since the local Lipschitz property of F and G in $H^{1,\gamma}$ were shown in the proof of Theorem 13.1.3, we prove now the linear growth of F and G.

Step 1: We show that (13.1.9) implies that J'' is bounded on $[0,+\infty)$.

By similar analysis as in the proof of Theorem 13.1.3, one obtains that

$$\lim_{z\to+\infty} |\, J_1''(z)\,| = +\infty, \qquad \lim_{z\to+\infty} |\, J_2''(z)\,| = +\infty,$$

and

$$| J_3'' | \text{ is bounded}, \quad | J_4''(z) | \text{ is bounded} \iff \int_1^{+\infty} y^2 \nu(dy) < +\infty.$$

So, J'' is bounded on $[0, +\infty)$ if and only if the Wiener part of Z disappears, there are no negative jumps and $\int_1^{+\infty} y^2 \nu(dy) < +\infty$. From Step 1 of the proof of Theorem 13.1.4 we see that, under (13.1.9), also J' is bounded on $[0, +\infty)$.

Step 2: We prove that if J' and J'' are bounded on $[0, +\infty)$ and $(G1), (G2), (G3)$ hold then G and F have linear growth.

The linear growth of G follows from the estimation

$$\begin{aligned}
&\int_0^{+\infty} \left| \frac{d}{dx} g(x, r(x)) \right|^2 e^{\gamma x} dx \\
&= \int_0^{+\infty} \left[g_x'(x, r(x)) + g_y'(x, r(x)) r'(x) \right]^2 e^{\gamma x} dx \\
&\leq 2 \int_0^{+\infty} [h(x)]^2 e^{\gamma x} dx + 2 \sup_{x,r \geq 0} | g_y'(x, r) |^2 \int_0^{+\infty} | r'(x) |^2 e^{\gamma x} dx \\
&\leq 2 | h |_{L^{2,\gamma}}^2 + 2 \sup_{x,r \geq 0} | g_y'(x, r) |^2 \cdot | r |_{H^{1,\gamma}}^2 .
\end{aligned} \tag{13.1.11}$$

To show the linear growth of F let us start with the inequality

$$\begin{aligned}
&\int_0^{+\infty} \left| \frac{d}{dx} \left(J'\left(\int_0^x g(v, r(v)) dv \right) g(x, r(x)) \right) \right|^2 e^{\gamma x} dx \\
&\leq 2 \int_0^{+\infty} \left| J''\left(\int_0^x g(v, r(v)) dv \right) g^2(x, r(x)) \right|^2 e^{\gamma x} dx \\
&\quad + \int_0^{+\infty} \left| J'\left(\int_0^{+\infty} g(v, r(v)) dv \right) [g_x'(x, r(x)) + g_y'(x, r(x)) r'(x)] \right|^2 e^{\gamma x} dx.
\end{aligned}$$

The second integral can be estimated in the same way as (13.1.11). The linear growth of the first integral follows from the inequality

$$\int_0^{+\infty} | g(x, r(x)) |^4 e^{\gamma x} dx \leq \sup_{x,r \geq 0} \left| \frac{g(x, r)}{\sqrt{r}} \right|^4 \int_0^{+\infty} | r(x) |^2 e^{\gamma x} dx. \qquad \square$$

13.1.3 Applications to the Morton–Musiela Equation

In this section we use general results from the previous sections to examine the existence of solutions of the HJMM equation in the specific case in which

$$g(x, y) = \lambda(x) y, \qquad x, y \geq 0,$$

where $\lambda(\cdot)$ is a continuous nonnegative bounded function on $[0,+\infty)$. As we already know we arrive then at Morton–Musiela equation

$$dr(t,x) = \Big(\frac{\partial}{\partial x} r(t,x) + F(r(t))(x)\Big)dt + G(r(t-))(x)dZ(t), \quad r(0,\cdot) = r_0, \quad t \in [0,T^*], \tag{13.1.12}$$

with

$$G(r)(x) = \lambda(x)r(x), \qquad F(r)(x) = J'\Big(\int_0^x \lambda(v)r(v)\Big)\lambda(x)r(x), \qquad x \geq 0. \tag{13.1.13}$$

As before, the state spaces are $L^{2,\gamma}_+$ and $H^{1,\gamma}_+$.

It turns out that the applicability of general results on the existence of global solutions is of limited use because F hardly ever satisfies the linear growth condition. This is why we have a special chapter on Morton–Musiela equations in which we develop the methodology introduced by Morton leading to better results.

Proposition 13.1.6 *If the drift transformation F given by* (13.1.13) *is of linear growth in $L^{2,\gamma}$, then J' is bounded on* $[0,+\infty)$. *In particular, the characteristic triplet of Z satisfies*

$$q = 0, \quad supp\{\nu\} \subseteq [0,+\infty) \quad and \quad \int_0^{+\infty} y\nu(dy) < +\infty. \tag{13.1.14}$$

It follows that only subordinators with drift make F of linear growth in $L^{2,\gamma}$. However, in Section 15.1 we show, with the use of alternative methods, that solutions of (13.1.12)–(13.1.13) exist also for more general Lévy processes.

Proof Assume, to the contrary, that J' is unbounded and define

$$r_n(x) = n\mathbf{1}_{[1,3]}(x), \qquad n = 1, 2, \ldots.$$

Since for sufficiently large $z \geq 0$ the function $(J'(z))^2$ is increasing, we have, for large n

$$\frac{\mid F(r_n)\mid^2}{\mid r_n \mid^2} = \frac{\int_1^3 \Big(J'\Big(\int_1^x \lambda(y)ndy\Big)\Big)^2 \lambda^2(x)n^2 e^{\gamma x}dx}{n^2\int_1^3 e^{\gamma x}dx} \geq \frac{\underline{\lambda}^2 \int_1^3 \Big(J'\Big(\underline{\lambda}n(x-1)\Big)\Big)^2 e^{\gamma x}dx}{\int_1^3 e^{\gamma x}dx},$$

where $\underline{\lambda} := \inf_{x\geq 0}\lambda(x)$. Since

$$\int_1^3 \Big(J'\Big(\underline{\lambda}n(x-1)\Big)\Big)^2 e^{\gamma x}dx \geq \int_2^3 \Big(J'\Big(\underline{\lambda}n(x-1)\Big)\Big)^2 e^{\gamma x}dx$$
$$\geq \Big(J'(\underline{\lambda}n)\Big)^2 \int_2^3 e^{\gamma x}dx \underset{n}{\longrightarrow} +\infty,$$

the main claim holds. Splitting $J'(z)$ as in the proof of Theorem 13.1.2 one can show that J' is bounded on $[0, +\infty)$ if and only if (13.1.14) holds. □

The existence of local solutions can be deduced from local Lipschitz properties studied for a general function g in Section 13.1.1. This requires assumptions on the function λ as well as on the jumps of the Lévy process. The following theorem is a direct consequence of Theorem 13.1.2.

Theorem 13.1.7 *Assume that:*

($\Lambda 0$) λ *is continuous and* $\inf_{x\geq 0}\lambda(x) = \underline{\lambda} > 0,\ \ \sup_{x\geq 0}\lambda(x) = \bar{\lambda} < +\infty,$

($\Lambda 1$) $\operatorname{supp} \nu \subseteq [-\frac{1}{\bar{\lambda}}, +\infty),$

($L1$) $\int_1^{+\infty} y^2 \nu(dy) < +\infty,$

hold. Then there exists a unique local weak solution to the equation (13.1.12)–(13.1.13) *taking values in the space* $L^{2,\gamma}_{+}$.

The positivity assumptions $(G1)(i)$, $(G1)(ii)$ required in Theorem 13.1.2 follow from $(\Lambda 0), (\Lambda 1)$ while $(G1)(iii)$ follows from $(\Lambda 0)$. In the formulation of the theorem a simplified, but under $(\Lambda 0), (\Lambda 1)$, equivalent version of the condition (L1) from Theorem 13.1.2 is used. Notice that the process Z now is required to be a square integrable martingale with drift but without negative large jumps.

Similarly, as a consequence of Theorem 13.1.3 we obtain the following local existence result in $H^{1,\gamma}_{+}$.

Theorem 13.1.8 *Assume that conditions* $(\Lambda 0), (\Lambda 1)$,

($\Lambda 2$) λ, λ' *are bounded and continuous on* $\mathbb{R}_{+}$,

and

($L2$) $\int_1^{+\infty} y^3 \nu(dy) < +\infty,$

are satisfied. Then there exists a unique local weak solution to the equation (13.1.12) taking values in the space $H^{1,\gamma}_{+}$.

Comparing with Theorem 13.1.7 we additionally require for the process Z that

$$\mathbb{E}[|\, Z(t)\, |^3] < +\infty, \quad t > 0.$$

14

Analysis of Morton's Equation

This chapter is concerned with the solvability of the Morton equation. Necessary conditions and sufficient conditions on the Lévy process and the volatility function are specified for the equation to have a solution. An important role in the proofs is played by an operator version of the equation.

14.1 Results

Here we study the solvability of the Morton equation,

$$df(t,T) = J'\left(\int_t^T \lambda(t,u)f(t-,u)du\right)\lambda(t,T)f(t-,T)dt + \lambda(t,T)f(t-,T)dZ(t), \tag{14.1.1}$$

on the bounded domain

$$(t,T) \in \mathcal{T} := \left\{(t,T) \in \mathbb{R}^2 : 0 \leq t \leq T \leq T^*\right\}, \tag{14.1.2}$$

with an initial condition

$$f(0,T) = f_0(T), \quad T \in [0,T^*]. \tag{14.1.3}$$

Like Morton [95] we treat (14.1.1) as a family of stochastic equations indexed by T. This approach arises from the classical formulation of the HJM model. In Chapter 15 the solution will be treated as a function-valued process.

Solutions of (14.1.1)–(14.1.3) will be searched in the class of *random fields* $f(t,T)$, $(t,T) \in \mathcal{T}$ such that

$$f(\cdot,T) \text{ is adapted and càdlàg on } [0,T] \text{ for all } T \in [0,T^*], \tag{14.1.4}$$

$$f(t,\cdot) \text{ is continuous on } [t,T^*] \text{ for all } t \in [0,T^*], \tag{14.1.5}$$

$$\mathbb{P}\left(\sup_{(t,T)\in\mathcal{T}} f(t,T) < +\infty\right) = 1. \tag{14.1.6}$$

Random fields satisfying (14.1.4)–(14.1.6) will be called *regular*.

Results on the existence and nonexistence of solutions will be established under the following standing assumptions.

(A1) The initial curve f_0 is positive and continuous on $[0, T^*]$.

(A2) The support of the Lévy measure is contained in the interval $(-1/\bar{\lambda}, +\infty)$ and

$$\int_1^\infty y\nu(dy) < +\infty,$$

where

$$\underline{\lambda} := \inf_{(t,T)\in\mathcal{T}} \lambda(t,T) > 0, \qquad \bar{\lambda} := \sup_{(t,T)\in\mathcal{T}} \lambda(t,T) < +\infty. \tag{14.1.7}$$

$$\text{(A3)} \begin{cases} \text{For each } 0 < t \le T^* \text{ the process } \int_0^t \lambda(s,T)dZ(s);\ T \in [t, T^*] \text{ is continuous.} \\ \text{The field } \mid \int_0^t \lambda(s,T)dZ(s) \mid \text{ is bounded on } \mathcal{T}. \end{cases}$$

Under $(A1)$–$(A3)$ the solvability of (14.1.1) depends on the growth of the function J'.

Theorem 14.1.1 *Assume that for some $a > 0$, $b \in \mathbb{R}$,*

$$(J1) \qquad J'(z) \ge a\ln^3 z + b, \qquad \forall z > 0.$$

For arbitrary $\kappa \in (0,1)$, there exists a positive constant K such that if

$$f_0(T) > K, \quad \forall T \in [0, T^*], \tag{14.1.8}$$

then, with probability greater or equal to κ, there are no regular solutions $f: \mathcal{T} \longrightarrow \mathbb{R}_+$ of equation (14.1.1).

Theorem 14.1.2 *Assume that*

$$(J2) \qquad \limsup_{z\to\infty} \left(\ln z - \bar{\lambda}T^*J'(z)\right) = +\infty.$$

(i) *Then there exists a regular field $f: \mathcal{T} \longrightarrow \mathbb{R}_+$ which solves* (14.1.1).

(ii) *If, in addition, $\int_1^\infty y^2\nu(dy) < +\infty$ holds, then the solution f is unique in the class of regular fields.*

The key step in the investigation of equation (14.1.1) is to reformulate it into an alternative fixed point problem:

$$f(t,T) = \mathcal{A}f(t,T), \quad (t,T) \in \mathcal{T}, \tag{14.1.9}$$

where

$$\mathcal{A}f(t,T) := a(t,T)e^{\int_0^t J'\left(\int_s^T \lambda(s,u)f(s,u)du\right)\lambda(s,T)ds}, \qquad (t,T) \in \mathcal{T},$$

and

$$a(t,T) := f_0(T)e^{\int_0^t \lambda(s,T)dZ(s) - \frac{q^2}{2}\int_0^t \lambda^2(s,T)ds + \int_0^t \int_{-1/\bar{\lambda}}^{+\infty} \left(\ln(1+\lambda(s,T)y) - \lambda(s,T)y\right)\pi(ds,dy)}. \tag{14.1.10}$$

The operator $\mathcal{A}$ will be called Morton's operator and the equation (14.1.9) *Morton's operator equation*. It was originally introduced by Morton in [95] for the case when Z was a Wiener process. Solutions of (14.1.9) will be searched also in the class of fields satisfying (14.1.4)–(14.1.6).

The rest of the chapter is organized as follows. In the next subsection we comment on the assumptions (A1)–(A3). Then, in Section 14.2, applications are presented and the theorems are reformulated in terms of the characteristics (a, q, ν) of the noise process Z. The proofs of the theorems will be presented in the final two sections. In particular, in Section 14.3 we prove the equivalence of the equations (14.1.1) and (14.1.9), the result that is also used in Section 14.4.

14.1.1 Comments on Assumptions $(A1)$–$(A3)$

Let us notice that in the equation (14.1.1) or equivalently in (14.1.9) intervene the values of $J'(z)$ for all nonnegative z. To make the equations well posed one needs therefore to know that $J'(z)$ exists for $z \geq 0$. We have the following result.

Proposition 14.1.3 *Under* (A2) *the function* J' *is well defined on* $[0, +\infty)$.

Proof By (A2) the function J can be written in the form

$$J(z) = -az + \frac{1}{2}qz^2 + J_1(z) + J_2(z) + J_3(z),$$

where

$$J_1(z) := \int_{-1/\bar{\lambda}}^{0} (e^{-zy} - 1 + zy)\, \nu(dy), \quad J_2(z) := \int_0^1 (e^{-zy} - 1 + zy)\, \nu(dy),$$

$$J_3(z) := \int_1^{\infty} (e^{-zy} - 1)\, \nu(dy).$$

The fact that ν is a Lévy measure implies that $J(z)$ is well defined for $z \geq 0$. Moreover, it also implies that for $z > 0$ the functions J_1, J_2, J_3 have derivatives of any order (see Lemma 8.1 and 8.2 in Rusinek [109]) and

$$J_1'(z) = \int_{-1/\bar{\lambda}}^{0} y(1 - e^{-zy})\nu(dy),\ J_2'(z) = \int_0^1 y(1 - e^{-zy})\nu(dy),\ J_3'(z)$$
$$= -\int_1^{\infty} ye^{-zy}\nu(dy). \tag{14.1.11}$$

It is clear that the condition

$$\int_1^\infty y\nu(dy) < +\infty \tag{14.1.12}$$

implies that $J'(0)$ exists and then

$$J'(0) = -a + J_3'(0) = -a - \int_1^\infty y\nu(dy). \qquad \square$$

Taking into account Morton's operator equation and writing (14.1.10) in the form

$$a(t,T) = f_0(T)\; e^{\int_0^t \lambda(s,T)dZ(s) - \frac{q^2}{2}\int_0^t \lambda^2(s,T)ds} \cdot \prod_{s\le t}(1+\lambda(s,T)\Delta Z(s))e^{-\lambda(s,T)\Delta Z(s)}, \quad (t,T)\in\mathcal{T},$$

we see that (A1) and (A2) are necessary for (14.1.9) to have positive solutions.

Condition (A3) together with (A1) and (A2) is needed to show the regularity of the field a in (14.1.10) (see Proposition 14.3.7 for details). Condition (A3) is satisfied if, for instance, $\lambda(\cdot,\cdot)$ is constant or, more generally, if it is of the form

$$\lambda(t,T) = \sum_{n=1}^N a_n(t)b_n(T),$$

where $\{a_n(\cdot)\}, \{b_n(\cdot)\}$ are continuous functions. The assumption that $\lambda(\cdot,\cdot)$ is continuous does not imply, in general, (A3) (see Brzeźniak, Peszat and Zabczyk [22], Kwapień, Marcus and Rosiński [86] for counterexamples).

14.2 Applications of the Main Theorems

Here we present Theorem 14.1.1 and Theorem 14.1.2 in terms of the characteristic triplet (a,q,ν) of the driving Lévy process Z. In the formulation of all results we implicitly assume that (A1), (A2) and (A3) are satisfied. As in the previous sections, the Laplace exponent J of Z is examined starting from the decomposition

$$J(z) = -az + \frac{1}{2}qz^2 + J_1(z) + J_2(z) + J_3(z), \quad z \ge 0,$$

where

$$J_1(z) := \int_{-1/\bar\lambda}^0 (e^{-zy} - 1 + zy)\,\nu(dy), \quad J_2(z) := \int_0^1 (e^{-zy} - 1 + zy)\,\nu(dy),$$

$$J_3(z) := \int_1^\infty (e^{-zy} - 1)\,\nu(dy).$$

The first theorem, which generalizes the result of Morton (see [95]) shows that for the existence of regular solutions to (14.1.9) the Gaussian part of Z must disappear and, what is rather surprising, Z must not have negative jumps.

Theorem 14.2.1 *If the Lévy exponent J of Z is such that $q > 0$ or $\nu\{(-\frac{1}{\bar{\lambda}}, 0)\} > 0$ then (J1) holds. Consequently, there are no regular solutions to* (14.1.1).

Proof If $q > 0$ then J' satisfies

$$J'(z) \geq -a + qz + J_3'(0), \qquad z \geq 0.$$

If $\nu\{(-\frac{1}{\bar{\lambda}}, 0)\} > 0$ then

$$J'(z) \geq -a + J_1'(z) + J_3'(0), \qquad z \geq 0,$$

and since

$$J_1'''(z) = -\int_{-1/\bar{\lambda}}^{0} y^3 e^{-zy}\, \nu(dy) \geq 0,$$

J_1' is convex and as such it satisfies the inequality

$$J_1'(z) \geq J_1''(0)z + J_1'(0).$$

In both cases $(J1)$ holds and thus the assertion follows from Theorem 14.1.1. □

Subordinators with drift provide a class of Lévy processes in which (14.1.1) has a solution.

Proposition 14.2.2 *If the process Z is a sum of a subordinator and a linear function then (J2) holds and* (14.1.1) *has a regular solution. In particular, this is true if Z is a compound Poisson process with drift and positive jumps only.*

In this case condition (J2) is trivially satisfied because J' is bounded. This follows from the calculation

$$J'(z) = -a + \int_0^1 y(1 - e^{-zy})\nu(dy) - \int_1^\infty y e^{-zy} \nu(dy) \leq -a + \int_0^1 y\nu(dy).$$

The converse implication is not true. If (14.1.1) has a regular solution, then Z does not have to be a subordinator with drift. In the following we present an example.

Example 14.2.3 Let Z be a purely jump process with the Lévy measure

$$\nu(dy) = \frac{1}{y^2 \mid \ln y \mid^{\gamma}} \mathbf{1}_{(0,\frac{1}{2})}(y)dy,$$

where $\gamma > 0$. We show the following.

(a) There exists a regular solution of (14.1.9) and (J2) holds for any $\gamma > 0$.
(b) Z is a subordinator $\Longleftrightarrow \gamma > 1$.

To see this, we find J_2 explicitly. Calculations yield

$$J_2(z) = \int_0^{\frac{1}{z}} y(1-e^{-zy})\frac{1}{y^2 \mid \ln y \mid^\gamma} dy = \int_0^{\frac{z}{2}} \frac{1-e^{-u}}{u} \cdot \frac{1}{\mid \ln \frac{u}{z} \mid^\gamma} du.$$

For large z, we have

$$J_2(z) \le c \int_0^{\frac{1}{2}} \frac{1}{\mid \ln \frac{z}{u} \mid^\gamma} du + \int_{\frac{1}{2}}^{\frac{z}{2}} \frac{1}{u} \cdot \frac{1}{\mid \ln \frac{z}{u} \mid^\gamma} du.$$

The first integral tends to 0 with $z \to +\infty$. The second can be written in the form

$$\int_{\frac{1}{2}}^{\frac{z}{2}} \frac{1}{u} \cdot \frac{1}{\mid \ln \frac{z}{u} \mid^\gamma} du = \int_4^{2z} \frac{1}{v \mid \ln v \mid^\gamma} dv.$$

As a consequence,

$$\lim_{z\to+\infty} \frac{J_2'(z)}{\ln z} = 0$$

and thus

$$\lim_{z\to+\infty} \left(\ln z - aJ_2'(z)\right) = +\infty$$

for any $a > 0$. Thus (J2) holds and a solution exists.

To prove (b) we have to check that $\int_0^1 y\nu(dy) < +\infty$, which is straightforward.

In the sequel we consider Morton's equation with Lévy processes without Gaussian part and without negative jumps. Criteria for the existence of solutions of (14.1.1) may will be formulated in terms of the Lévy measure of Z. In this case, the problem is related to the behaviour of the distribution function

$$U_\nu(x) := \int_0^x y^2 \nu(dy), \qquad x \ge 0$$

of the modified Lévy measure $y^2\nu(dy)$ near the origin and will be examined using a Tauberian theorem.

For the formulation we need the concept of slowly varying functions. A positive function M *varies slowly at* 0 if for any fixed $x > 0$,

$$\frac{M(tx)}{M(t)} \longrightarrow 1, \qquad \text{as } t \longrightarrow 0. \tag{14.2.1}$$

Typical examples are constants or, for arbitrary γ and small positive t, functions

$$M(t) = \left(\ln \frac{1}{t}\right)^\gamma.$$

If M varies slowly at zero, then for any $\varepsilon > 0$ the following estimation holds (see Lemma 2 p. 277 in Feller [51])

$$t^{\varepsilon} < M(t) < t^{-\varepsilon} \tag{14.2.2}$$

for all positive t sufficiently small. If

$$\frac{f(x)}{g(x)} \longrightarrow 1, \qquad \text{as } x \longrightarrow 0,$$

then we write $f(x) \sim g(x)$.

Theorem 14.2.4 *Assume that for some $\rho \in (0, +\infty)$,*

$$U_{\nu}(x) \sim x^{\rho} \cdot M(x), \qquad \text{as } x \to 0, \tag{14.2.3}$$

where M is a slowly varying function at 0.

(i) *If $\rho > 1$ then (J2) holds and there exists a regular solution of* (14.1.1).
(ii) *If $\rho < 1$ then (J1) holds and there is no regular solution of* (14.1.1).
(iii) *If $\rho = 1$, the measure ν has a density and*

$$M(x) \longrightarrow 0 \quad \text{as } x \to 0, \quad \text{and} \quad \int_0^1 \frac{M(x)}{x}\,dx = +\infty, \tag{14.2.4}$$

then (J2) holds and there exists a regular solution of (14.1.1).

For the proof of Theorem 14.2.4 we need an auxiliary result.

Proposition 14.2.5 *The following conditions are equivalent*

(i) $\int_0^1 y\nu(dy) = +\infty$,
(ii) $\lim_{z\to\infty} J_2'(z) = +\infty$.
Moreover, if the measure ν has a density and

$$U_{\nu}(x) \sim x \cdot M(x), \quad \text{as } x \to 0, \tag{14.2.5}$$

where M is such that $M(x) \to c > 0$ as $x \to 0$, then each of the preceding conditions is equivalent to

(iii) $\int_0^1 \frac{M(x)}{x} dx = +\infty$.

Proof The equivalence of (i) and (ii) follows directly from the dominated convergence theorem. We show the equivalence of (i) and (iii). In virtue of (14.2.5) we have

$$c\int_0^1 \frac{U_{\nu}(x)}{x^2}dx \le \int_0^1 \frac{M(x)}{x}dx \le C\int_0^1 \frac{U_{\nu}(x)}{x^2}dx$$

for some constants $0 < c < C$. However, (i) holds if and only if the last integral diverges. In fact, integration by parts yields

$$\int_0^1 \frac{U_\nu(x)}{x^2}dx = \int_0^1 \left(\int_0^x y^2 g(y)dy\right)\cdot\frac{1}{x^2}dx = \left(-\frac{1}{x}\int_0^x y^2 g(y)dy\right)\Big|_0^1 + \int_0^1 yg(y)dy$$

$$= \lim_{x\to 0} M(x) - \int_0^1 y^2 g(y)dy + \int_0^1 yg(y)dy = c + \int_0^1 y^2\nu(dy) + \int_0^1 y\nu(dy),$$

where g is a density of ν. □

Proof of Theorem 14.2.4 Fix $\rho \in (0, +\infty)$. Let us notice that

$$J_2''(z) = \int_0^1 y^2 e^{-zy}\nu(dy) = \int_0^1 e^{-zy}\mu(dy)$$

is a Laplace transform of the measure $\mu(dy) := y^2\nu(dy)$. Thus it follows from Tauberian theorem (see Theorem 2, p. 445 in Feller [51]) that the condition (14.2.3) is equivalent to

$$\lim_{z\to\infty} \frac{J_2''(z)}{\Gamma(\rho+1)z^{-\rho}\cdot M(\frac{1}{z})} = 1, \tag{14.2.6}$$

where Γ stands for the gamma function.

(i) Assume that $\rho > 1$ and fix $\varepsilon > 0$ such that $\rho - \varepsilon > 1$. Using(14.2.2) we can find $z_0 > 0$ such that $M(\frac{1}{z}) < z^\varepsilon$ for all $z > z_0$. In virtue of (14.2.6) for any $z > z_0$ we have the following estimation

$$J_2'(z) \le J_2'(z_0) + 2\int_{z_0}^z v^{-\rho} M\left(\frac{1}{v}\right)dv \le J_2'(z_0) + 2\int_{z_0}^{+\infty} v^{\varepsilon-\rho}dv < +\infty.$$

Thus condition (J2) holds because

$$J'(z) = -a + J_2'(z) + J_3'(z), \qquad z \ge 0,$$

and J_2' is a bounded function. The assertion follows from Theorem 14.1.2.

To prove (ii) and (iii) notice first that in view of Proposition 14.2.5:

$$\lim_{z\to\infty} J_2'(z) = +\infty.$$

As a consequence of (14.2.2) when $\rho \in (0, 1)$ and the assumption (14.2.4) when $\rho = 1$, we have

$$\lim_{z\to\infty}\int_a^z v^{-\rho}\cdot M\left(\frac{1}{v}\right)dv = +\infty, \qquad \rho \in (0, 1]$$

for any $a > 0$. By the d'Hospital formula, for any $a > 0$,

$$\lim_{z\to\infty} \frac{J_2'(z)}{\Gamma(\rho+1)\int_a^z v^{-\rho}\cdot M(\frac{1}{v})dv} = \lim_{z\to\infty}\frac{J_2''(z)}{\Gamma(\rho+1)\, z^{-\rho}\cdot M(\frac{1}{z})} = 1, \quad \rho\in(0,1].$$

(ii) $\rho \in (0,1)$; Fix any $\varepsilon > 0$ such that $\rho + \varepsilon < 1$. By (14.2.2) we can find a constant $a > 0$ such that for any $v > a$ we have $M(\frac{1}{v}) > v^{-\varepsilon}$. Then for z sufficiently large the following estimation holds

$$J_2'(z) \geq (1-\varepsilon)\Gamma(\rho+1)\int_a^z v^{-\rho} M(\frac{1}{v})dv \geq (1-\varepsilon)\Gamma(\rho+1)\int_a^z v^{-\rho}v^{-\varepsilon}dv$$

$$= \frac{(1-\varepsilon)\Gamma(\rho+1)}{1-(\rho+\varepsilon)}\left(z^{1-(\rho+\varepsilon)} - a^{1-(\rho+\varepsilon)}\right).$$

Consequently, J' satisfies $(J1)$ and the assertion follows from Theorem 14.1.1.
(iii) $\rho = 1$; Let $c > 0$ be any positive constant. Using (14.2.4) we can fix a constant $a > 0$ such that

$$0 < 1 - 2c \max_{v\in[a,\infty]} M\left(\frac{1}{v}\right) < 1.$$

For large z satisfying

$$\frac{J_2'(z)}{\int_a^z \frac{1}{v}\cdot M(\frac{1}{v})dv} \leq 2,$$

we have the estimates

$$\ln z - cJ_2'(z) = \ln z - c\frac{J_2'(z)}{\int_a^z \frac{1}{v}\cdot M\left(\frac{1}{v}\right)dv}\int_a^z \frac{1}{v}\cdot M\left(\frac{1}{v}\right)dv$$

$$\geq \ln z - 2c\int_a^z \frac{1}{v}\cdot M\left(\frac{1}{v}\right)dv$$

$$\geq \ln z - 2c[\ln z - \ln a]\max_{v\in[a,z]} M\left(\frac{1}{v}\right)$$

$$\geq \left(1 - 2c\max_{v\in[a,\infty]} M\left(\frac{1}{v}\right)\right)\ln z + 2c\ln a\cdot\max_{v\in[a,\infty]} M\left(\frac{1}{v}\right) \underset{z\to\infty}{\longrightarrow} \infty.$$

Thus condition (J2) holds because

$$J'(z) = -a + J_2'(z) + J_3'(z), \qquad z \geq 0$$

and J_2' is a bounded function. The assertion follows from Theorem 14.1.2. □

We present now two examples for which the conditions

$$M(x) \longrightarrow 0 \qquad \text{as } x \to 0, \tag{14.2.7}$$

$$\int_0^1 \frac{M(x)}{x}dx = +\infty \tag{14.2.8}$$

are not simultaneously satisfied but the existence problem can be solved with the use of Theorem 14.1.2.

Example 14.2.6 Let ν be a measure with the density

$$\nu(dx) = \frac{1}{x^2} \cdot \frac{(\ln \frac{1}{x})^{\gamma} + \gamma(\ln \frac{1}{x})^{\gamma-1}}{(\ln \frac{1}{x})^{2\gamma}} \cdot \mathbf{1}_{(0,1)}(x), \qquad \gamma > 1.$$

It can be checked that the function U_ν is given by

$$U_\nu(x) = \int_0^x y^2 \cdot g(y)dy = x \cdot \frac{1}{(\ln \frac{1}{x})^{\gamma}}.$$

It is clear that the function

$$M(x) := \frac{1}{(\ln \frac{1}{x})^{\gamma}}, \quad \gamma > 1$$

varies slowly at zero and that (14.2.7) holds. However, condition (14.2.8) is not satisfied and thus Theorem 14.2.4 does not cover this case. We can explicitly show that J_2' is bounded and use Theorem 14.1.2. We have

$$J_2'(z) = \int_0^1 y(1 - e^{-zy})g(y)dy = \int_1^{+\infty} \frac{1 - e^{-\frac{z}{x}}}{x} \cdot \frac{(\ln x)^{\gamma} + \gamma(\ln x)^{\gamma-1}}{(\ln x)^{2\gamma}} dx$$

$$\leq \int_1^{+\infty} \frac{1}{x} \cdot \frac{(\ln x)^{\gamma} + \gamma(\ln x)^{\gamma-1}}{(\ln x)^{2\gamma}} dx < +\infty.$$

Example 14.2.7 Let ν be given by

$$\nu(dy) = \frac{1}{y^{1+\rho}} \mathbf{1}_{(0,1)}(y)\, dy, \qquad \rho \in (0, 2).$$

Then

(a) if $\rho \in (1, 2)$ then equation (14.1.1) has no regular solutions,
(b) if $\rho \in (0, 1)$, or
(c) $\rho = 1$ and $\bar{\lambda}T^* < 1$ then equation (14.1.1) has a regular solution.

Indeed, for $\rho \in (0, 2)$ we have

$$U_\nu(x) = \frac{1}{2 - \rho} x^{2-\rho}, \qquad x \in (0, 1),$$

and thus (a) and (b) follow from Theorem 14.2.4. If $\rho = 1$ than (c) cannot be deduced from Theorem 14.2.4 because the function $M(x) \equiv 1$ does not tend to zero. However, we have

$$J_2'(z) = \int_0^1 y(1 - e^{-zy}) \frac{1}{y^2}\, dy = \int_0^z \frac{1 - e^{-v}}{\frac{v}{z}} \frac{1}{z}\, dv = \int_0^z \frac{1 - e^{-v}}{v}\, dv,$$

and consequently

$$\lim_{z\to\infty} \frac{\ln z}{\bar{\lambda}T^* J_2'(z)} \overset{d'H}{=} \lim_{z\to\infty} \frac{\frac{1}{z}}{\frac{1-e^{-z}}{z}\cdot \bar{\lambda}T^*} = \lim_{z\to\infty} \frac{1}{\bar{\lambda}T^*(1-e^{-z})} = \frac{1}{\bar{\lambda}T^*} > 1.$$

This condition clearly implies $(J2)$ and (c) follows from Theorem 14.1.2.

14.3 Proof of Theorem 14.1.1

14.3.1 Outline of the Proof

The main idea of the proof of Theorem 14.1.1 on nonexistence is based on a *majorizing function* method for the equation (14.1.9), which will be shown in Section 14.3.2 to be equivalent to (14.1.1). This method was introduced by Morton in [95] to show that (14.1.1) with a Wiener noise does not have regular solutions.

Since the proof of the theorem is rather involved, we outline first its general idea when $\lambda(t, T) \equiv 1$.

Assume that we want to prove that a solution f_1 of the equation

$$f_1(t,T) = e^{\int_0^t J'(\int_s^T f_1(s,u)du)ds} g_1(t,T), \qquad (t,T) \in \mathcal{T}, \tag{14.3.1}$$

with a bounded deterministic nonnegative function g_1, explodes in some point (x, y) of the domain $\mathcal{T}$, that is, $\lim_{(t,T)\to(x,y)} f_1(t,T) = +\infty$. Let us consider an auxiliary function f_2 such that

$$\lim_{(t,T)\uparrow(x,y)} f_2(t,T) = +\infty, \tag{14.3.2}$$

and for any $0 < \delta < y$ the function f_2 is a unique solution of the equation

$$f_2(t,T) = e^{\int_0^t R(\int_s^T f_2(s,u)du)ds} g_2(t,T), \qquad (t,T) \in \mathcal{T}_{x,y-\delta}, \tag{14.3.3}$$

where $\mathcal{T}_{x,y} := \{(t,T) \in \mathcal{T} : 0 \le t \le x, 0 \le T \le y\}$, $R\colon \mathbb{R}_+ \to \mathbb{R}_+$, $g_2\colon \mathcal{T}_{x,y} \to \mathbb{R}_+$ is bounded. The function R should be related to J' such that the following estimation holds

$$f_1(t,T) \ge e^{\int_0^t R(\int_s^T f_1(s,u)du)ds} \hat{g}_1(t,T), \qquad (t,T) \in \mathcal{T}_{x,y-\delta}, \tag{14.3.4}$$

where $\hat{g}_1$ is a bounded, nonnegative function on $\mathcal{T}_{x,y}$. The function R for which the preceding estimation holds can be constructed because $(J1)$ holds. The key step of the method is the relation between (14.3.3) and (14.3.4). The following implication is true

$$\hat{g}_1(t,T) \ge g_2(t,T) \quad (t,T) \in \mathcal{T}_{x,y-\delta} \quad \Rightarrow \quad f_1(t,T) \ge f_2(t,T) \quad (t,T) \in \mathcal{T}_{x,y-\delta}, \tag{14.3.5}$$

which, in view of (14.3.2), yields the explosion of f_1 at (x, y). The function f_2 is called a majorizing function for f_1. Although our original equation (14.1.9) is not deterministic we will find a deterministic majorizing function that is valid for each $\omega \in \Omega$. The property that $\hat{g}_1$ majorizes g_2 will be obtained by imposing additional constraints on the initial condition f_0.

To apply the method described in the preceding we need to find a point $(x, y) \in \mathcal{T}$ and the functions R and g_2 such that the corresponding majorizing function explodes in (x, y). We examine functions $R = R_{\alpha,\gamma} : \mathbb{R}_+ \longrightarrow \mathbb{R}_+$ given by

$$R(z) = R_{\alpha,\gamma}(z) := \alpha \ln^3 \left(\gamma(z + e^2)\right), \qquad z \geq 0,\ \alpha > 0, \gamma \geq 1 \tag{14.3.6}$$

and

$$g_2(t,T) = g(t,T) := \begin{cases} e^{-\int_0^t R_{\alpha,\gamma}\left(\int_s^T h(s,u)du\right)ds} \cdot h(t,T) & \text{for } (t,T) \neq (x,y), \\ 0 & \text{for } (t,T) = (x,y), \end{cases} \tag{14.3.7}$$

with $h \colon \mathcal{T}_{x,y} \longrightarrow \bar{\mathbb{R}}_+ := \mathbb{R}_+ \cup \{+\infty\}$ given by

$$h(t,T) := e^{\varphi(t,T)}, \quad \text{where} \quad \varphi(t,T) := \begin{cases} \frac{1}{x-t+y-T} & \text{for } (t,T) \neq (x,y), \\ +\infty & \text{for } (t,T) = (x,y), \end{cases} \tag{14.3.8}$$

where $0 < x < y \leq T^*$, and show that they satisfy all the required properties for certain parameters α and γ.

14.3.2 Equivalence of Equations (14.1.1) and (14.1.9)

Proposition 14.3.1 *Assume that f is a regular field and conditions (A1) and (A2) are satisfied. Then f is a solution of* (14.1.1) *if and only if it solves* (14.1.9).

Proof Let us notice, that for each T the solution $f(t,T)$, $t \in [0,T]$ of (14.1.1) is a stochastic exponential and by the Doléans-Dade exponential formula (see Theorem 4.4.6) it can be written in the form

$$\begin{aligned} f(t,T) = f_0(T)\ & e^{\int_0^t J'\left(\int_s^T \lambda(s,u) f(s-,u)du\right)\lambda(s,T)ds + \int_0^t \lambda(s,T)dZ(s) - \frac{q^2}{2}\int_0^t \lambda^2(s,T)ds} \\ & \cdot \prod_{s \leq t} (1 + \lambda(s,T)\Delta Z(s)) e^{-\lambda(s,T)\Delta Z(s)}, \qquad (t,T) \in \mathcal{T}. \end{aligned} \tag{14.3.9}$$

Under assumptions (A1) and (A2) we can write equation (14.3.9) in the form

$$\begin{aligned} f(t,T) = f_0(T)\ & e^{\int_0^t J'\left(\int_s^T \lambda(s,u) f(s-,u)du\right)\lambda(s,T)ds + \int_0^t \lambda(s,T)dZ(s) - \frac{q^2}{2}\int_0^t \lambda^2(s,T)ds} \\ & \cdot e^{\int_0^t \int_{-1/\bar{\lambda}}^{+\infty} \left(\ln(1+\lambda(s,T)y) - \lambda(s,T)y\right)\pi(ds,dy)}, \qquad (t,T) \in \mathcal{T}, \end{aligned} \tag{14.3.10}$$

or equivalently as

$$f(t,T) = a(t,T)e^{\int_0^t J'\left(\int_s^T \lambda(s,u)f(s-,u)du\right)\lambda(s,T)ds}, \qquad (t,T) \in \mathcal{T}. \tag{14.3.11}$$

We show now that we can replace $f(s-,u)$ in (14.3.11) by $f(s,u)$. To do this we prove that for each $(t,T) \in \mathcal{T}$,

$$\int_0^t J'\left(\int_s^T \lambda(s,u)f(s,u)du\right)\lambda(s,T)ds = \int_0^t J'\left(\int_s^T \lambda(s,u)f(s-,u)du\right)\lambda(s,T)ds.$$

Let us start with the observation that for $T \in [0,T^*]$, moments of jumps of the process $f(\cdot,T)$ are the same as for $a(\cdot,T)$. Moreover, it follows from (14.1.10) that the set of jumps of $a(\cdot,T)$ is independent of T and is contained in the set

$$\mathcal{Z} := \{t \in [0,T^*] : \triangle Z(t) \neq 0\}.$$

Thus if $s \notin \mathcal{Z}$ then

$$J'\left(\int_s^T \lambda(s,u)f(s,u)du\right)\lambda(s,T) = J'\left(\int_s^T \lambda(s,u)f(s-,u)du\right)\lambda(s,T).$$

By Theorem 2.8 in Applebaum [2] the set $\mathcal{Z}$ is at most countable, so the assertion follows. □

14.3.3 Auxiliary Results

The following properties of the function J' will be needed in the sequel.

Proposition 14.3.2 *(i) If (A2) holds then J_1', J_2', J_3', and thus J' as well, are increasing, real-valued functions on the interval $[0,+\infty)$.*
(ii) J' is a Lipschitz function on $[0,+\infty)$ if and only if

$$\int_1^\infty y^2\nu(dy) < +\infty. \tag{14.3.12}$$

They follow directly from the formulae (14.1.11) and

$$J_1''(z) = \int_{-1/\bar{\lambda}}^0 y^2e^{-zy}\nu(dy), \quad J_2''(z) = \int_0^1 y^2e^{-zy}\nu(dy), \quad J_3''(z) = \int_1^\infty y^2e^{-zy}\nu(dy). \tag{14.3.13}$$

Proposition 14.3.3 *Let $\alpha > 0$, $\gamma \geq 1$ be fixed constants such that $\alpha\gamma > 2$ and $\gamma T^* > 1$. Choose $(x,y) \in \mathcal{T}$ such that $0 < x < y < \frac{\alpha}{2} \wedge T^*$ and $\gamma(y-x) > 1$. Let the functions $h, R_{\alpha,\gamma}$ be given by (14.3.8) and (14.3.6), respectively. Then the function $g\colon \mathcal{T}_{x,y} \longrightarrow \mathbb{R}_+$ defined by the formula (14.3.7) is continuous.*

In the proof of Proposition 14.3.3 we will use the following inequality

$$\ln^3\left(\frac{1}{b-a}\int_a^b p(x)dx + e^2\right) \geq \frac{1}{b-a}\int_a^b \ln^3\left(p(x)+e^2\right)dx, \tag{14.3.14}$$

valid for any positive integrable function p on the interval (a,b), $a < b$. It can be proven by an application of the Jensen's inequality to the concave function $\ln^3(z+e^2)$.

Proof of Proposition 14.3.3 We need to show the continuity of g only in the point (x,y). Thus consider any point $(t,T) \in \mathcal{T}_{x,y}$ that is close to (x,y), i.e. s.t. $(t,T) \neq (x,y)$ and $\gamma(T-t) > 1$. Using the monotonicity of $R_{\alpha,\gamma}$ and (14.3.14) we obtain the following estimation

$$e^{-\int_0^t R_{\alpha,\gamma}\left(\int_s^T h(s,u)du\right)ds} \cdot h(t,T) = e^{-\alpha\int_0^t \ln^3\left(\gamma(\int_s^T h(s,u)du+e^2)\right)ds} \cdot h(t,T)$$

$$\leq e^{-\alpha\int_0^t \ln^3\left(\gamma\int_s^T h(s,u)du+e^2\right)ds} \cdot h(t,T) \leq e^{-\alpha\int_0^t \ln^3\left(\frac{\gamma(T-t)}{T-s}\int_s^T h(s,u)du+e^2\right)ds} \cdot h(t,T)$$

$$\leq e^{-\alpha\int_0^t \ln^3\left(\frac{1}{T-s}\int_s^T h(s,u)du+e^2\right)ds} \cdot h(t,T) \leq e^{-\alpha\int_0^t \frac{1}{T-s}\int_s^T \ln^3(h(s,u)+e^2)duds} \cdot h(t,T)$$

$$\leq e^{-\frac{\alpha}{T}\int_0^t\int_s^T \ln^3 h(s,u)duds} \cdot h(t,T) = e^{-\frac{\alpha}{T}\int_0^t\int_s^T \varphi^3(s,u)duds+\varphi(t,T)}.$$

One can check that

$$\int_0^t\int_s^T \frac{1}{(x-s+y-u)^3}\,du\,ds = \frac{t}{2}\cdot\frac{-T^2-Tt-ty+2Ty+2Tx-tx}{(x-t+y-T)(x+y-2t)(x+y-T)(x+y)},$$

and thus

$$-\frac{\alpha}{T}\int_0^t\int_s^T \varphi^3(s,u)duds + \varphi(t,T)$$
$$= \left(1 - \frac{\alpha t\left(-T^2-Tt-ty+2Ty+2Tx-tx\right)}{2T(x+y-2t)(x+y-T)(x+y)}\right)\varphi(t,T).$$

Passing to the limit we obtain

$$\lim_{t\to x,T\to y}\left(-T^2-Tt-ty+2Ty+2Tx-tx\right) = y^2-x^2,$$

$$\lim_{t\to x,T\to y}(x+y-2t) = y-x, \qquad \lim_{t\to x,T\to y}(x+y-T) = x.$$

Hence

$$\lim_{t\to x,T\to y}\left(1 - \frac{\alpha t\left(-T^2-Tt-ty+2Ty+2Tx-tx\right)}{2T(x-t+y-T)(x+y-2t)(x+y-T)(x+y)}\right) = 1-\frac{\alpha}{2y} < 0,$$

and consequently $\lim_{t\to x,T\to y} g(t,T) = 0$. □

Proposition 14.3.4 *Fix $\alpha > 0$, $\gamma \geq 1$ s.t. $\alpha\gamma > 2$ and $\gamma T^* > 1$. Let $0 < x < y < \frac{\alpha}{2} \wedge T^*$, $\gamma(y - x) > 1$, $0 < \delta < y$ and $g\colon \mathcal{T}_{x,y-\delta} \longrightarrow \mathbb{R}_+$ be a bounded function. Assume that there exists a bounded function $h\colon \mathcal{T}_{x,y-\delta} \longrightarrow \mathbb{R}_+$, which solves the equation*

$$h(t,T) = e^{\int_0^t R_{\alpha,\gamma}\left(\int_s^T h(s,u)du\right)ds} \cdot g(t,T), \qquad \forall (t,T) \in \mathcal{T}_{x,y-\delta}, \tag{14.3.15}$$

where $R_{\alpha,\gamma}$ is given by (14.3.6). Then h is uniquely determined in the class of bounded functions on $\mathcal{T}_{x,y-\delta}$.

For the proof of Proposition 14.3.4 we use the following auxiliary result.

Lemma 14.3.5 *Let $0 < t_0 \leq T_0 < +\infty$ and define a set*

$$A := \left\{(t,T) : t \leq T,\ 0 \leq t \leq t_0,\ t \leq T \leq T_0\right\}.$$

If $d\colon A \longrightarrow \mathbb{R}_+$ is a bounded function satisfying

$$d(t,T) \leq K \int_0^t \int_s^T d(s,u)duds \qquad \forall (t,T) \in A, \tag{14.3.16}$$

where $0 < K < \infty$ then $d(t,T) \equiv 0$ on A.

Proof Assume that d is bounded by a constant $M > 0$ on A. We show inductively that

$$d(t,T) \leq MK^n \frac{(tT)^n}{(n!)^2}, \qquad \forall (t,T) \in A. \tag{14.3.17}$$

The formula (14.3.17) is valid for $n = 0$. Assume that it is true for some n and show that it is true for $n + 1$. We have the following estimation

$$d(t,T) \leq K \int_0^t \int_s^T MK^n \frac{(su)^n}{(n!)^2} duds = MK^{n+1} \frac{1}{(n!)^2} \int_0^t s^n \Big(\int_s^T u^n du\Big) ds$$

$$= MK^{n+1} \frac{1}{(n!)^2} \int_0^t s^n \left(\frac{T^{n+1} - s^{n+1}}{n+1}\right) ds \leq MK^{n+1} \frac{1}{(n!)^2} \int_0^t s^n \frac{T^{n+1}}{n+1} ds$$

$$= MK^{n+1} \frac{1}{(n!)^2} \frac{t^{n+1}}{(n+1)} \frac{T^{n+1}}{(n+1)} = MK^{n+1} \frac{(tT)^{n+1}}{((n+1)!)^2}.$$

Letting $n \longrightarrow \infty$ in (14.3.17) we see that $d(t,T) = 0$. □

It can be verified that

$$d := R'_{\alpha,\gamma}(0) = \frac{3\alpha(2 + \ln\gamma)^2}{e^2}$$

and that $R_{\alpha,\gamma}$ is concave. Thus

$$|R_{\alpha,\gamma}(z_1) - R_{\alpha,\gamma}(z_2)| \le d|z_1 - z_2|, \qquad z_1, z_2 \ge 0. \tag{14.3.18}$$

Proof of Proposition 14.3.4 Assume that $h_1, h_2 \colon \mathcal{T}_{x,y-\delta} \longrightarrow \mathbb{R}_+$ are bounded solutions of (14.3.15). Then the function $\mid h_1 - h_2 \mid$ is bounded and satisfies

$$\begin{aligned} \mid h_1(t,T) - h_2(t,T) \mid &\le \mid g \mid \cdot \mid e^{\int_0^t R_{\alpha,\gamma}\left(\int_s^T h_1(s,u)du\right)ds} \\ &- e^{\int_0^t R_{\alpha,\gamma}\left(\int_s^T h_2(s,u)du\right)ds} \mid, \quad \forall (t,T) \in \mathcal{T}_{x,y-\delta}, \end{aligned}$$

where

$$\mid g \mid = \sup_{(t,T)\in\mathcal{T}_{x,y-\delta}} \mid g(t,T) \mid .$$

As a consequence of the inequality $\mid e^x - e^y \mid \le \max\{e^x, e^y\} \mid x - y \mid$ for $x, y \in \mathbb{R}$ we have

$$\begin{aligned} \mid h_1(t,T) - h_2(t,T) \mid \le K \int_0^t \Bigg| R_{\alpha,\gamma}\left(\int_s^T h_1(s,u)du\right) \\ - R_{\alpha,\gamma}\left(\int_s^T h_2(s,u)du\right)\Bigg| ds, \quad \forall (t,T) \in \mathcal{T}_{x,y-\delta}, \end{aligned}$$

where

$$K := \mid g \mid \sup_{(t,T)\in\mathcal{T}_{x,y-\delta}} \max_{i=1,2} \left\{ e^{\int_0^t R_{\alpha,\gamma}\left(\int_s^T h_i(s,u)du\right)ds} \right\} < \infty.$$

In virtue of (14.3.18) we have

$$\mid h_1(t,T) - h_2(t,T) \mid \le dK \int_0^t \int_s^T \mid h_1(s,u) - h_2(s,u) \mid duds, \qquad \forall (t,T) \in \mathcal{T}_{x,y-\delta}.$$

In view of Lemma 14.3.5, with $t_0 = \min\{x, y-\delta\}$, $T_0 = y - \delta$, we have $h_1(t,T) = h_2(t,T)$ for all $(t,T) \in \mathcal{T}_{x,y-\delta}$. □

Proposition 14.3.6 *Fix $\alpha > 0$, $\gamma \ge 1$ s.t. $\alpha\gamma > 2$, $\gamma T^* > 1$ and the function $R_{\alpha,\gamma}$ given by* (14.3.6). *Choose (x,y) s.t. $0 < x < y < \frac{\alpha}{2} \wedge T^*$, $\gamma(y-x) > 1$ and δ s.t. $0 < \delta < y$. Let $f_1 \colon \mathcal{T}_{x,y-\delta} \longrightarrow \mathbb{R}_+$, where be a bounded function satisfying inequality*

$$f_1(t,T) \ge e^{\int_0^t R_{\alpha,\gamma}\left(\int_s^T f_1(s,u)du\right)ds} \cdot g_1(t,T), \qquad \forall (t,T) \in \mathcal{T}_{x,y-\delta}, \tag{14.3.19}$$

where $g_1 \colon \mathcal{T}_{x,y-\delta} \longrightarrow \mathbb{R}_+$. Let $f_2 \colon \mathcal{T}_{x,y-\delta} \longrightarrow \mathbb{R}_+$ be a bounded function solving equation

$$f_2(t,T) = e^{\int_0^t R_{\alpha,\gamma}\left(\int_s^T f_2(s,u)du\right)ds} \cdot g_2(t,T), \qquad \forall (t,T) \in \mathcal{T}_{x,y-\delta}, \tag{14.3.20}$$

where $g_2 \colon \mathcal{T}_{x,y-\delta} \longrightarrow \mathbb{R}_+$ *is a bounded function. Moreover, assume that*

$$g_1(t,T) \geq g_2(t,T) \geq 0, \qquad \forall (t,T) \in \mathcal{T}_{x,y-\delta}. \tag{14.3.21}$$

Then $f_1(t,T) \geq f_2(t,T)$ *for all* $(t,T) \in \mathcal{T}_{x,y-\delta}$.

Proof Let us define the operator $\mathcal{K}$ acting on bounded functions on $\mathcal{T}_{x,y-\delta}$ by

$$\mathcal{K}k(t,T) := e^{\int_0^t R_{\alpha,\gamma}\left(\int_s^T k(s,u)du\right)ds} \cdot g_2(t,T), \qquad (t,T) \in \mathcal{T}_{x,y-\delta}. \tag{14.3.22}$$

Let us notice that in view of (14.3.19), (14.3.21) and (14.3.22) we have

$$\mathcal{K}f_1(t,T) \leq e^{\int_0^t R_{\alpha,\gamma}\left(\int_s^T f_1(s,u)du\right)ds} \cdot g_1(t,T) \leq f_1(t,T), \qquad \forall (t,T) \in \mathcal{T}_{x,y-\delta}. \tag{14.3.23}$$

It is clear that the operator $\mathcal{K}$ is order preserving, i.e.

$$k_1(t,T) \leq k_2(t,T) \;\; \forall (t,T) \in \mathcal{T}_{x,y-\delta} \;\; \Longrightarrow \;\; \mathcal{K}k_1(t,T) \leq \mathcal{K}k_2(t,T) \;\; \forall (t,T) \in \mathcal{T}_{x,y-\delta}. \tag{14.3.24}$$

Let us consider the sequence of functions: $f_1, \mathcal{K}f_1, \mathcal{K}^2 f_1, \ldots$. In virtue of (14.3.23) and (14.3.24) we see that $f_1 \geq \mathcal{K}f_1 \geq \mathcal{K}^2 f_1 \geq \ldots$. Thus this sequence is pointwise convergent to some function $\bar{f}$ and it is bounded by f_1, so applying the dominated convergence theorem in the formula

$$\mathcal{K}^{n+1} f_1(t,T) = e^{\int_0^t R_{\alpha,\gamma}\left(\int_s^T \mathcal{K}^n f_1(s,u)du\right)ds} \cdot g_2(t,T), \qquad \forall (t,T) \in \mathcal{T}_{x,y-\delta}$$

we obtain

$$\bar{f}(t,T) = e^{\int_0^t R_{\alpha,\gamma}\left(\int_s^T \bar{f}(s,u)du\right)ds} \cdot g_2(t,T), \qquad \forall (t,T) \in \mathcal{T}_{x,y-\delta}.$$

Moreover, $\bar{f}$ is bounded and thus, in view of Proposition 14.3.4, we have $\bar{f} = f_2$. As a consequence $f_1 \geq f_2$ on $\mathcal{T}_{x,y-\delta}$. □

As a consequence of $(A1)$–$(A3)$ we have the following regularity of the field a in (14.1.10):

Proposition 14.3.7 *Assume that the conditions (A1), (A2), (A3) are satisfied. Then the field* $\{a(t,T);\ (t,T) \in \mathcal{T}\}$ *given by* (14.1.10) *is bounded from below and above by strictly positive random constants. Moreover,* $a(\cdot,T)$ *is adapted and càdlàg on* $[0,T]$ *for all* $T \in [0,T^*]$ *and* $a(t,\cdot)$ *is continuous on* $[t,T^*]$ *for all* $t \in [0,T^*]$.

A rather technical proof can be found in Barski and Zabczyk [7], proposition 2.3.

14.3.4 Conclusion of the Proof

Let us notice that for $0 < \tilde{\alpha} < a$ and any $\tilde{\gamma} \geq 1$ we have

$$\lim_{z \to +\infty} \frac{\tilde{\alpha} \ln^3 \left(\tilde{\gamma}(z + e^2)\right)}{a \ln^3 z} = \frac{\tilde{\alpha}}{a} < 1.$$

Thus $(J1)$ implies that for $0 < \tilde{\alpha} < a$ and any $\tilde{\gamma} \geq 1$ there exists $\tilde{\beta} \in \mathbb{R}$ such that

$$J'(z) \geq \tilde{\alpha} \ln^3 \left(\tilde{\gamma}(z + e^2)\right) + \tilde{\beta} = R_{\tilde{\alpha},\tilde{\gamma}}(z) + \tilde{\beta}, \qquad z \geq 0. \tag{14.3.25}$$

Now fix parameters $\tilde{\alpha}, \tilde{\beta}, \tilde{\gamma}$ such that (14.3.25) holds and

$$0 < \tilde{\alpha} < a, \quad \tilde{\gamma} \geq 1, \quad \tilde{\gamma}(\underline{\lambda} \wedge 1) \geq 1, \quad \underline{\lambda}\tilde{\alpha}\tilde{\gamma}(\underline{\lambda} \wedge 1) > 2, \quad \tilde{\gamma}(\underline{\lambda} \wedge 1)T^* > 1.$$

It can be checked that for a constant $c > 0$ s.t. $\gamma(c \wedge 1) \geq 1$ we have

$$R_{\alpha,\gamma}(cz) \geq R_{\alpha,\gamma(c\wedge 1)}(z), \qquad z \geq 0. \tag{14.3.26}$$

Let us assume that there exists a regular solution of (14.1.9). Using (14.3.25), (14.1.7) and (14.3.26) the forward rate f satisfies the following inequality

$$f(t,T) = e^{\int_0^t J'\left(\int_s^T \lambda(s,u) f(s,u) du\right)\lambda(s,T) ds} a(t,T) \geq e^{\int_0^t R_{\tilde{\alpha},\tilde{\gamma}}\left(\int_s^T \lambda(s,u) f(s,u) du\right)\lambda(s,T) ds + \tilde{\beta} t} a(t,T)$$

$$\geq e^{\underline{\lambda} \int_0^t R_{\tilde{\alpha},\tilde{\gamma}}\left(\underline{\lambda} \int_s^T f(s,u) du\right) ds} e^{\tilde{\beta} t} a(t,T) \geq e^{\int_0^t R_{\underline{\lambda}\tilde{\alpha},\tilde{\gamma}(\underline{\lambda}\wedge 1)}\left(\int_s^T f(s,u) du\right) ds} e^{\tilde{\beta} t} a(t,T)$$

$$= e^{\int_0^t R_{\alpha,\gamma}\left(\int_s^T f(s,u) du\right) ds} e^{\tilde{\beta} t} a(t,T). \tag{14.3.27}$$

The preceding constants $\alpha := \underline{\lambda}\tilde{\alpha}$, $\gamma := \tilde{\gamma}(\underline{\lambda} \wedge 1)$ satisfy $\alpha > 0$, $\gamma \geq 1$, $\alpha\gamma > 2$, $\gamma T^* > 1$. Choose $(x, y) \in \mathcal{T}$ such that $0 < x < y < \frac{\alpha}{2} \wedge T^*$, $\gamma(y - x) > 1$ and fix three deterministic functions $h \colon \mathcal{T}_{x,y} \longrightarrow \bar{\mathbb{R}}_+$, $R_{\alpha,\gamma} \colon \mathbb{R}_+ \longrightarrow \mathbb{R}_+$, $g \colon \mathcal{T}_{x,y} \longrightarrow \mathbb{R}_+$ given by (14.3.8), (14.3.6) and (14.3.7), respectively. It is clear that they satisfy the equation

$$h(t,T) = e^{\int_0^t R_{\alpha,\gamma}\left(\int_s^T h(s,u) du\right) ds} \cdot g(t,T), \qquad \forall (t,T) \in \mathcal{T}_{x,y}. \tag{14.3.28}$$

In virtue of Proposition 14.3.3 the function g is continuous on $\mathcal{T}_{x,y}$ and thus bounded. It follows from Proposition 14.3.7 that if the constant K is sufficiently large, then with probability arbitrarily close to 1,

$$e^{\tilde{\beta} t} a(t,T) \geq g(t,T), \qquad \forall (t,T) \in \mathcal{T}_{x,y}. \tag{14.3.29}$$

Let us fix $0 < \delta < y$ and consider inequality (14.3.27) and equality (14.3.28) on the set $\mathcal{T}_{x,y-\delta}$. Then the function h is continuous. In virtue of Proposition 14.3.6 we have

$$f(t,T) \geq h(t,T) = e^{\frac{1}{(x-t+y-T)}}, \qquad \forall (t,T) \in \mathcal{T}_{x,y-\delta}.$$

For any sequence $(t_n, T_n) \in \mathcal{T}_{x,y}$ satisfying $t_n \uparrow x$, $T_n \uparrow y$, define a sequence $\delta_n := \frac{y-T_n}{2}$. Then

$$f(t,T) \geq e^{\frac{1}{(x-t+y-T)}}, \qquad \forall (t,T) \in \mathcal{T}_{x,y-\delta_n},$$

and consequently $\lim_{n\to\infty} f(t_n, T_n) = +\infty$, which contradicts the assumption that f is regular. □

14.4 Proof of Theorem 14.1.2

We use again the equivalent equation (14.1.9)

$$f = \mathcal{A}f,$$

where, for field f,

$$\mathcal{A}f(t,T) := a(t,T) \cdot e^{\int_0^t J'\left(\int_s^T \lambda(s,u)f(s,u)du\right)\lambda(s,T)ds}, \qquad (t,T) \in \mathcal{T} \tag{14.4.1}$$

and

$$a(t,T) := f_0(T)e^{\int_0^t \lambda(s,T)dZ(s) - \frac{q^2}{2}\int_0^t \lambda^2(s,T)ds + \int_0^t \int_{-1/\bar{\lambda}}^{+\infty}\left(\ln(1+\lambda(s,T)y) - \lambda(s,T)y\right)\pi(ds,dy)}.$$

The proof will be divided into two steps. We fix $\omega \in \Omega$ and treat (14.4.1) as a deterministic transformation with a positive and bounded function a. In Step 1, in Proposition 14.4.1, one shows that under the assumption (J2) the operator $\mathcal{A}$ preserves the boundedness of random fields on $\mathcal{T}$. In Step 2, one shows that the limit of an increasing sequence of the iterations of the operator $\mathcal{A}$ on the initial function 0 is a solution of the equation.

Step 1:

Proposition 14.4.1 *Assume that the function J' satisfies (J2). Then there exists a positive constant c such that if*

$$h(t,T) \leq c, \quad \forall (t,T) \in \mathcal{T}$$

for a nonnegative function h, then

$$\mathcal{A}h(t,T) \leq c, \quad \forall (t,T) \in \mathcal{T}. \tag{14.4.2}$$

Proof Let us assume that $h(t,T) \leq c$ for all $(t,T) \in \mathcal{T}$ for some positive c. Using the fact that J' is increasing and λ positive, we have

$$\mathcal{A}h(t,T) \leq a(t,T) \cdot e^{J'(\bar{\lambda}cT^*)\int_0^t \lambda(s,T)ds}.$$

By Proposition 14.3.7, $a(\cdot,\cdot)$ is bounded by a positive constant $K = K(\omega)$ and we arrive at the inequality

$$\mathcal{A}h(t,T) \leq Ke^{J'(\bar{\lambda}cT^*)\int_0^t \lambda(s,T)ds}, \quad (t,T) \in \mathcal{T}.$$

It is therefore enough to find a positive constant c such that

$$\ln K + J'(\bar{\lambda} c T^*) \cdot \int_0^t \lambda(s,T)ds \le \ln c, \quad (t,T) \in \mathcal{T}.$$

If the function J' is negative on $[0,+\infty)$ then it is enough to take $c = K$. If J' takes positive values then it is enough to find a positive arbitrarily large constant c such that

$$\ln K + \bar{\lambda} T^* \cdot J'(\bar{\lambda} c T^*) \le \ln c, \quad (t,T) \in \mathcal{T}.$$

Existence of such c is an immediate consequence of the assumption (14.1.2). □

Step 2: Part (i): The operator $\mathcal{A}$ is order-preserving, i.e.

$$h_1 \le h_2 \quad \Longrightarrow \quad \mathcal{A}h_1 \le \mathcal{A}h_2.$$

The sequence $h_0 \equiv 0$, $h_{n+1} := \mathcal{A}h_n$ is thus monotonically increasing to $\bar{h}$ and by the monotone convergence theorem we have

$$\bar{h}(t,T) = \mathcal{A}\bar{h}(t,T), \qquad \forall (t,T) \in \mathcal{T}.$$

Moreover, since $h_0 \le c$, where $c = c(\omega)$ is given by Proposition 14.4.1, $\bar{h}$ is bounded and thus (14.1.6) is satisfied. Moreover, by the boundedness of $\bar{h}$ it follows that the process

$$\int_0^t J'\left(\int_s^T \lambda(s,u)\bar{h}(s,u)du\right)\lambda(s,T)ds \tag{14.4.3}$$

is continuous wrt. $(t,T) \in \mathcal{T}$ for fixed ω. It is also adapted wrt. t. If we replace $\bar{h}(s,u)$ in the preceding formula by any bounded field $k(s,u)$ that is adapted wrt. s then (14.4.3) becomes adapted wrt. t. As a consequence, $\bar{h}(\cdot,T)$ is adapted as a limit of the adapted sequence $\{h_n(\cdot,T)\}$. In virtue of Proposition 14.3.7, the field $\bar{h}$ satisfies (14.1.4) and (14.1.5).

Part (ii): The function J' is Lipschitz on $[0,+\infty)$ and therefore we can repeat all arguments from the proof of Proposition 14.3.4 and the result follows. □

15

Analysis of the Morton–Musiela Equation

Existence and nonexistence results for the Morton-Musiela equation are established.

15.1 Formulation and Comments on the Results

We are concerned here with the Morton–Musiela equation

$$dr(t,x) = \left(\frac{\partial}{\partial x}r(t,x) + J'\Big(\int_0^x r(s,u)\lambda(u)du\Big)\lambda(x)r(t,x)\right)dt$$

$$+\,\lambda(x)r(t-,x)dZ(t), \quad t \in [0,T^*], \quad x \geq 0, \qquad (15.1.1)$$

already introduced in Section 12.3. The preceding $\lambda(\cdot)$ is a deterministic bounded function on $[0,+\infty)$. Our aim now is to provide conditions for the existence of solutions

$$r(t) = r(t,x), \quad x \geq 0, \quad t \in [0,T^*]$$

of (15.1.1) in spaces $L^{2,\gamma}_+$ or $H^{1,\gamma}_+$. As for the Morton equation in the previous chapter the central role here will be played by two logarithmic growth conditions on the function J' introduced already in Chapter 14 (see Section 14.1), i.e.

$$(J1) \quad \text{For some } a>0,\ b\in\mathbb{R},\ \ J'(z) \geq a(\ln z)^3 + b, \qquad \text{for all } z>0,$$

$$(J2) \qquad \limsup_{z\to\infty}\left(\ln z - \bar{\lambda}T^*J'(z)\right) = +\infty, \qquad 0 < T^* < +\infty,$$

where

$$\bar{\lambda} := \sup_{x\geq 0}\lambda(x) < +\infty.$$

We will show, roughly speaking, that the solutions of (15.1.1) explode if $(J1)$ holds and solutions exist if $(J2)$ is satisfied. For proving this we use the fact that equation (15.1.1) is equivalent to the *Morton–Musiela operator equation*, which will be introduced in the proofs. This equivalence allows one to use methods developed in the previous chapter.

Our first result on solutions in the state space $H_+^{1,\gamma}$ is of negative type.

Theorem 15.1.1 *Assume that the conditions*

$(\Lambda 0)$ λ *is continuous and* $\inf_{x\geq 0}\lambda(x) = \underline{\lambda} > 0, \quad \sup_{x\geq 0}\lambda(x) = \bar{\lambda} < +\infty,$

$(\Lambda 1)$ $\operatorname{supp}\nu \subseteq [-\frac{1}{\bar{\lambda}}, +\infty),$

$(\Lambda 3)$ $\lambda, \lambda', \lambda''$, *are bounded and continuous on* $\mathbb{R}_+$,

$(L0)$ $\int_1^{+\infty} y\nu(dy) < +\infty,$

$(J1)$ $J'(z) \geq a(\ln z)^3 + b,$ for some $a > 0,\ b \in \mathbb{R},$ and all $z > 0$

are satisfied.
Then, for some $k > 0$ and all $r_0(\cdot) \in H_+^{1,\gamma}$ such that $r_0(x) \geq k$, $\forall x \in [0, T^*]$, the equation (15.1.1) *does not have solutions in $H_+^{1,\gamma}$ on the interval* $[0, T^*]$.

As we proved in Theorem 13.1.8, equation (15.1.1) has local solutions in $H_+^{1,\gamma}$. It follows thus from Theorem 15.1.1 that under $(\Lambda 0)$, $(\Lambda 1)$, $(\Lambda 2)$, $(\Lambda 3)$, (L0), (J1) and (L2), each such solution explodes.

First, we formulate the main existence results in the spaces $L_+^{2,\gamma}$ and $H_+^{1,\gamma}$ and outline the methods of proving them. The proofs are given in the next subsection.

Theorem 15.1.2 *Assume that* $(\Lambda 0)$, $(\Lambda 1)$ *and the following conditions hold*

$(\Lambda 2)$ λ, λ'*are bounded and continuous on* $\mathbb{R}_+$,

$(L0)$ $\int_1^{+\infty} y\nu(dy) < +\infty,$

$(J2)$ $\limsup_{z\to\infty} \left(\ln z - \bar{\lambda}T^* J'(z)\right) = +\infty, \qquad 0 < T^* < +\infty.$

(a) If $r_0 \in L_+^{2,\gamma}$ then there exists a solution to (15.1.1) *taking values in the space* $L_+^{2,\gamma}$.

(b) Assume, in addition, that

$(\Lambda 3)$ $\lambda, \lambda', \lambda''$, *are bounded and continuous on* $\mathbb{R}_+$,

Z has positive jumps only, i.e. $\operatorname{supp}\{\nu\} \subseteq [0, +\infty)$ *and*

$(L1)$ $\int_1^{\infty} y^2\nu(dy) < \infty.$

If $r_0 \in H_+^{1,\gamma}$ then there exists a unique solution to (15.1.1) *taking values in the space* $H_+^{1,\gamma}$.

15.1.1 Comments on the Results

Let us recall important classes of Lévy processes for which $(J1)$ or $(J2)$ is satisfied. Condition $(J1)$ is satisfied if the Wiener part of Z is not degenerated, i.e. $q \neq 0$, or Z admits negative jumps, i.e. $\nu((-\infty, 0)) > 0$ (see Theorem 14.2.1). Hence $(J2)$ may be satisfied only if Z does not have the Wiener part nor negative jumps. In fact, if this is the case, then

$$J'(z) = -a + \int_0^1 y(1 - e^{-zy})\nu(dy) + \int_1^{+\infty} ye^{-zy}\nu(dy),$$

and (J2) depends only on the behaviour of ν close to zero. To see this, note that

$$\sup_{z\geq 0}\int_1^{+\infty} ye^{-zy}\nu(dy) < +\infty, \quad z \geq 0,$$

so only small jumps of Z are essential for $(J2)$. Let us assume that for some $\rho \in (0,+\infty)$,

$$\int_0^x y^2\nu(dy) \sim x^\rho \cdot M(x), \qquad as\ x \to 0 \tag{15.1.2}$$

is satisfied, where M is a slowly varying function at 0 (see (14.2.1)). Then

(a) if $\rho > 1$, then (J2) holds,
(b) if $\rho < 1$, then (J1) holds,
(c) if $\rho = 1$, the measure ν has a density and

$$M(x) \longrightarrow 0 \quad \text{as } x \to 0, \quad and \quad \int_0^1 \frac{M(x)}{x}\,dx = +\infty, \tag{15.1.3}$$

then (J2) holds (see Theorem 14.2.4).

Recall that (J2) is satisfied in the case in which Z is a subordinator with drift (see Proposition 14.2.2). As explained in Example 14.2.3, the converse implication is, however, not true.

15.2 Proofs of Theorems 15.1.1 and 15.1.2

15.2.1 Equivalence Results

In the proofs of the theorems we use Theorem 15.2.1 and Theorem 15.2.2 which state that equations (15.1.1) and (15.2.1) are equivalent under rather mild conditions. As their proofs are technical we refer the reader to Barski and Zabczyk [6].

A random field $r(t,x)$, $t \in [0,T^*], x \geq 0$, is said to be a solution, in $L^{2,\gamma}$, respectively in $H^{1,\gamma}$, to the *Morton–Musiela operator equation*:

$$r(t,x) = a(t,x)e^{\int_0^t J'\left(\int_0^{t-s+x}\lambda(v)r(s,v)dv\right)\lambda(t-s+x)ds}, \quad x \geq 0, \quad t \in [0,T^*], \tag{15.2.1}$$

where

$$a(t,x) := r_0(t+x)e^{\int_0^t \lambda(t-s+x)dZ(s) - \frac{q^2}{2}\int_0^t \lambda^2(t-s+x)ds}$$

$$\cdot \prod_{0\leq s\leq t}(1+\lambda(t-s+x)(Z(s)-Z(s-)))\,e^{-\lambda(t-s+x)(Z(s)-Z(s-))},$$

$$x \geq 0, \ t \in (0,T^*], \tag{15.2.2}$$

if $r(t,\cdot)$ $,t \in [0,T^*]$, is $L^{2,\gamma}-$, respectively $H^{1,\gamma}$-valued, bounded and adapted process such that, for each $t \in [0,T^*]$, equation (15.2.1) holds for almost all $x > 0$, in the case of $L^{2,\gamma}$, and for all $x \geq 0$, in the case of $H^{1,\gamma}$. We refer to Section 14.1 where a similar concept is introduced for Morton's equation.

Theorem 15.2.1 *Let r be a solution of* (15.1.1) *in the state space $H^{1,\gamma}_+$. Then $r(\cdot,\cdot)$ is a solution of* (15.2.1) *in $H^{1,\gamma}_+$.*

Theorem 15.2.2 *Assume that conditions* $(\Lambda 0)$, $(\Lambda 1)$ *and*

(*L*0) $\int_1^{+\infty} y\nu(dy) < +\infty,$

are satisfied.

(a) If

$(\Lambda 2)$ *λ, λ' are bounded and continuous on $\mathbb{R}_+$,*

and $r(\cdot)$ is a bounded solution in $L^{2,\gamma}_+$ of (15.2.1)*, then $r(\cdot)$ is a càdlàg process in $L^{2,\gamma}_+$ and solves* (15.1.1).

(b) If

$(\Lambda 3)$ *$\lambda, \lambda', \lambda''$, are bounded and continuous on $\mathbb{R}_+$,*

Z has positive jumps only, i.e. supp$\{\nu\} \subseteq [0,+\infty)$ and

(*L*1) $\int_1^{\infty} y^2\nu(dy) < \infty,$

and $r(\cdot)$ is a bounded solution in $H^{1,\gamma}_+$ of (15.2.1)*, then $r(\cdot)$ is càdlàg in $H^{1,\gamma}_+$ and solves* (15.1.1).

As a consequence, equations (15.1.1) and (15.2.1) are equivalent in $H^{1,\gamma}_+$, while each solution of (15.2.1) in $L^{2,\gamma}_+$ solves also (15.1.1).

15.2.2 Proof of Theorem 15.1.1

Assume to the contrary that r is a solution of (15.1.1) on $[0, T^*]$ in the space $H^{1,\gamma}_+$. Then the solution in the natural frame $f(t,T) = r(t, T-t), 0 \leq t \leq T \leq T^*$ satisfies Morton's equation:

$$f(t,T) = f_0(T) + \int_0^t J'\left(\int_s^T \lambda(v-s)f(s,v)dv\right)\lambda(T-s)f(s,T)ds$$
$$+ \int_0^t \lambda(T-s)f(s-,T)dZ(s), \qquad (15.2.3)$$

a particular form of equation 14.1.1 studied in Section 14.1 as here $\lambda(\cdot)$ depends on one variable only. Assumptions $(\Lambda 0)$, $(\Lambda 1)$, $(\Lambda 3)$, $(L0)$ imply the conditions $(A1)$–$(A3)$ required in Section 14.1. We check the regularity conditions (14.1.4)–(14.1.6). Since r is adapted and càdlàg in $H^{1,\gamma}_+$, it follows that

(a) $f(\cdot, T)$ is adapted and càdlàg for each $T \in [0, T^*]$,
(b) $f(t, \cdot)$ is continuous.

Using Lemma 15.2.3 and the fact that r is bounded on $[0, T^*]$, as a càdlàg process in $H_+^{1,\gamma}$, we obtain

$$\sup_{t\in[0,T^*],x\geq 0} \mid r(t,x) \mid = \sup_{t\in[0,T^*]} \sup_{x\geq 0} \mid r(t,x) \mid \leq 2\left(\frac{1}{\gamma}\right)^{\frac{1}{2}} \sup_{t\in[0,T^*]} \mid r \mid_{H_+^{1,\gamma}} < +\infty,$$

which clearly implies that

$$(c) \qquad \sup_{0\leq t\leq T\leq T^*} f(t,T) < +\infty.$$

It follows, however, from Theorem 14.1.1 that, under (J1), for sufficiently large $k > 0$ there is no solution of (15.2.3) in the class of regular random fields satisfying (a)–(c), a contradiction. □

Lemma 15.2.3 *If $r \in H^{1,\gamma}$ then*

$$\sup_{x\geq 0} |r(x)| \leq 2\Big(\frac{1}{\gamma}\Big)^{1/2} \mid r \mid_{H^{1,\gamma}} .$$

Proof Integrating by parts yields

$$\int_0^x y \frac{dr(y)}{dy} \, dy = yr(y)|_0^x - \int_0^x r(y) \, dy,$$

and thus

$$\begin{aligned}
|xr(x)| &\leq \Big|\int_0^x y \frac{dr(y)}{dy} \, dy\Big| + \Big|\int_0^x r(y) \, dy\Big| \\
&\leq \Big(\int_0^{+\infty} e^{-\gamma y} y^2 \, dy\Big)^{1/2} \Big(\int_0^{+\infty} e^{\gamma y} \Big(\frac{dr(y)}{dy}\Big)^2 dy\Big)^{1/2} \\
&\quad + \Big(\int_0^{+\infty} e^{-\gamma y} \, dy\Big)^{1/2} \Big(\int_0^{+\infty} e^{\gamma y} r^2(y) \, dy\Big)^{1/2} \\
&\leq \Big(\frac{2}{\gamma^3}\Big)^{1/2} \mid r \mid_{H^{1,\gamma}} + \Big(\frac{1}{\gamma}\Big)^{1/2} \mid r \mid_{L^{2,\gamma}} .
\end{aligned}$$

In particular,

$$\lim_{x\to+\infty} r(x) = 0.$$

Moreover,

$$\begin{aligned}
|r(x) - r(0)| &= \Big|\int_0^x \frac{dr}{dy}(y) \, dy\Big| \leq \int_0^x e^{-\gamma y/2} e^{\gamma y/2} \Big|\frac{dr}{dy}(y)\Big| \, dy \\
&\leq \Big(\int_0^{+\infty} e^{-\gamma y} \, dy\Big)^{1/2} \Big(\int_0^{+\infty} e^{\gamma y} \Big(\frac{dr}{dy}(y)\Big)^2 dy\Big)^{1/2} \leq \Big(\frac{1}{\gamma}\Big)^{1/2} \mid r \mid_{H^{1,\gamma}} .
\end{aligned}$$

Consequently,

$$| r(0) | \leq \left(\frac{1}{\gamma}\right)^{1/2} | r |_{H^{1,\gamma}},$$

and therefore

$$\sup_{x\geq 0} |r(x)| \leq 2\left(\frac{1}{\gamma}\right)^{1/2} | r |_{H^{1,\gamma}} .$$

□

15.2.3 Proof of Theorem 15.1.2

Proof of the existence part

Define the operator $\mathcal{K}$, acting on functions of two variables, by

$$\mathcal{K}(h)(t,x) = a(t,x)e^{\int_0^t J'\left(\int_0^{t-s+x} \lambda(v)h(s,v)dv\right)\lambda(s,t-s+x)ds}, \quad x \geq 0, \quad t \in [0,T^*], \tag{15.2.4}$$

where $a(t,x)$ is given by (15.2.2). Then the equation (15.2.1) can be compactly written in the form

$$r(t,x) = \mathcal{K}(r)(t,x), \qquad t \in [0,T^*], \quad x \geq 0.$$

Let us consider the sequence of random fields

$$h_0 \equiv 0, \qquad h_{n+1} := \mathcal{K}h_n, \qquad n = 1,2,\ldots. \tag{15.2.5}$$

Let us write a in the form

$$a(t,x) = r_0(t+x)I_1(t,x)I_2(t,x),$$

where

$$I_1(t,x) := \int_0^t \lambda(t-s+x)dZ(s), \quad t \in [0,T^*],\ x \geq 0, \tag{15.2.6}$$

$$I_2(t,x) := \int_0^t \int_{-\frac{1}{\lambda}}^{+\infty} \left[\ln(1+\lambda(t-s+x)y) - \lambda(t-s+x)y\right]\pi(ds,dy),$$
$$t \in [0,T^*], x \geq 0. \tag{15.2.7}$$

One can show that under $(\Lambda 2)$ the field b, defined by $b(t,x) := I_1(t,x)I_2(t,x)$, is bounded, i.e.

$$\sup_{t\in[0,T^*],x\geq 0} b(t,x) < \bar{b}, \tag{15.2.8}$$

where $\bar{b} = \bar{b}(\omega) > 0$ (for details see Barski and Zabczyk [6, p. 2681–2682]).

It can be shown by induction that if $r_0 \in L^{2,\gamma}_+$ then $h_n(t)$ is a bounded process in $L^{2,\gamma}_+$ for each n. Indeed assume this for h_n and show for h_{n+1}. In view of (15.2.8) we have

$$h_{n+1}(t,x) \leq r_0(t+x)\,\bar{b}\,e^{\bar{\lambda}\int_0^t |J'(\int_0^{t-s+x}\lambda(v)h_n(s,v)dv)|ds}$$

$$\leq r_0(t+x)\,\bar{b}\,e^{\bar{\lambda}T^*\left|J'(\frac{\bar{\lambda}}{\sqrt{\gamma}}\sup_t|h_n(t)|_{L^{2,\gamma}})\right|},$$

and thus $h_{n+1}(t)$ is bounded in $L^{2,\gamma}_+$. It follows from the assumption $\underline{\lambda} > 0$ and the fact that J' is increasing that the sequence $\{h_n\}$ is monotonically increasing and thus there exists $\bar{h} : [0,T^*] \times [0,+\infty) \longrightarrow \mathbb{R}_+$ such that

$$\lim_{n\to+\infty} h_n(t,x) = \bar{h}(t,x), \qquad 0 \leq t \leq T^*, x \geq 0. \tag{15.2.9}$$

Passing to the limit in (15.2.5), by the monotone convergence, we obtain

$$\bar{h}(t,x) = \mathcal{K}\bar{h}(t,x), \qquad 0 \leq t \leq T^*, x \geq 0.$$

It turns out that properties of the field $\bar{h}$ strictly depend on the growth of the function J'. In the following text we show that $\bar{h}(t)$ is a bounded process in $L^{2,\gamma}_+$, i.e. $\bar{h}(t), t \in [0,T^*]$ is a non-exploding solution of (15.2.1) in $L^{2,\gamma}_+$. Additional assumptions guarantee that $\bar{h}(t)$ is bounded in $H^{1,\gamma}_+$ and that the solution is unique.

We need an auxiliary result.

Proposition 15.2.4 *Assume that J' satisfies (J2). If $r_0 \in L^{2,\gamma}_+$ then there exists a positive constant c_1 such that if*

$$\sup_{t\in[0,T^*]} |\, h(t)\,|_{L^{2,\gamma}_+} \leq c_1$$

then

$$\sup_{t\in[0,T^*]} |\, \mathcal{K}h(t)\,|_{L^{2,\gamma}_+} \leq c_1.$$

Proof By elementary arguments using (15.2.8), for any $t \in [0,T^*]$, we have

$$|\, \mathcal{K}h(t,\cdot)\,|^2_{L^{2,\gamma}_+} = \int_0^{+\infty} |r_0(t+x)b(t,x)|^2 e^{2\int_0^t J'(\int_0^{t-s+x}\lambda(v)h(s,v)dv)\lambda(t-s+x)ds} e^{\gamma x}dx$$

$$\leq \bar{b}^2 \int_0^{+\infty} |r_0(t+x)|^2 e^{2J'\left(\frac{\bar{\lambda}}{\sqrt{\gamma}}\cdot\sup_t|h(t)|_{L^{2,\gamma}_+}\right)\int_0^t\lambda(t-s+x)ds} e^{\gamma x}dx$$

$$\leq \bar{b}^2 \cdot |\, r_0\,|^2_{L^{2,\gamma}_+} \cdot \sup_{s\in[0,t],x\geq 0} e^{2J'\left(\frac{\bar{\lambda}}{\sqrt{\gamma}}\cdot\sup_t|h(t)|_{L^{2,\gamma}_+}\right)\int_0^t\lambda(t-s+x)ds}.$$

This implies

$$\sup_t \mid \mathcal{K}h(t) \mid_{L^{2,\gamma}_+} \leq \bar{b} \cdot \mid r_0 \mid_{L^{2,\gamma}_+} \cdot \sup_{t\in[0,T^*],s\in[0,t],x\geq 0} e^{J'\left(\frac{\bar{\lambda}}{\sqrt{\gamma}} \cdot \sup_t |h(t)|_{L^{2,\gamma}_+}\right) \int_0^t \lambda(t-s+x)ds},$$

and thus it is enough to find a constant c_1 such that

$$\ln\left(\bar{b} \cdot \mid r_0 \mid_{L^{2,\gamma}_+}\right) + \sup_{t\in[0,T^*],s\in[0,t],x\geq 0} J'\left(\frac{\bar{\lambda}c_1}{\sqrt{\gamma}}\right) \int_0^t \lambda(t-s+x)ds \leq \ln c_1. \quad (15.2.10)$$

If $J'(z) \leq 0$ for each $z \geq 0$ then we put $c_1 = \bar{b} \cdot \mid r_0 \mid_{L^{2,\gamma}_+}$. If J' takes positive values then it is enough to find large c_1 such that

$$\ln\left(\bar{b} \cdot \mid r_0 \mid_{L^{2,\gamma}_+}\right) \leq \ln c_1 - \bar{\lambda} T^* J'\left(\frac{\bar{\lambda}c_1}{\sqrt{\gamma}}\right).$$

Existence of such c_1 is a consequence of (J2). □

Continuation of the proof of Theorem 15.1.2:
Since $\bar{h}(\cdot, x)$ is adapted for each $x \geq 0$ as a pointwise limit, we only need to show that $\bar{h}(t)$ is a bounded process in $L^{2,\gamma}_+$, resp. $H^{1,\gamma}_+$. Then $\bar{h}$ solves (15.1.1) in virtue of Theorem 15.2.2.

(*a*) Let c_1 be a constant given by Proposition 15.2.4. By the Fatou lemma we have

$$\sup_{t\in[0,T^*]} \int_0^{+\infty} \mid \bar{h}(t,x) \mid^2 e^{\gamma x} dx \leq \sup_{t\in[0,T^*]} \liminf_{n\to+\infty} \int_0^{+\infty} \mid h_n(t,x) \mid^2 e^{\gamma x} dx \leq c_1^2,$$

and hence $\bar{h}(t)$ is bounded in $L^{2,\gamma}_+$.

(*b*) In view of (*a*) we need to show that $h'_x(t)$ is bounded in $L^{2,\gamma}$. Differentiating the equation $\bar{h} = \mathcal{K}\bar{h}$ yields

$$\begin{aligned}\bar{h}'(t,x) &= r_0'(t+x)b(t,x)F_1(t,x) + r_0(t+x)b'_x(t,x)F_1(t,x)\\ &\quad + r_0(t+x)b(t,x)F_1(t,x)F_2(t,x),\end{aligned}$$

where

$$F_1(t,x) := e^{\int_0^t J'(\int_0^{t-s+x} \lambda(v)\bar{h}(s,v)dv)\lambda(t-s+x)ds},$$

$$F_2(t,x) := \int_0^t J''\left(\int_0^{t-s+x} \lambda(v)\bar{h}(s,v)dv\right) \lambda^2(t-s+x)\bar{h}(s,t-s+x)ds$$

$$+ \int_0^t J'\left(\int_0^{t-s+x} \lambda(v)h(s,v)dv\right) \lambda'_x(t-s+x)ds.$$

Assumption (Λ3) implies that $b(\cdot,\cdot)$ and $b'_x(\cdot,\cdot)$ are bounded on $(t,x) \in [0,T^*] \times [0,+\infty)$. Since $r_0 \in H^{1\gamma}_+$, it is enough to show that

$$\sup_{t\in[0,T^*],x\geq 0} F_1(t,x) < +\infty, \qquad \sup_{t\in[0,T^*],x\geq 0} F_2(t,x) < +\infty.$$

We have

$$\sup_{t\in[0,T^*],x\geq 0} F_1(t,x) \leq e^{\left|J'\left(\frac{\bar\lambda}{\sqrt{\gamma}}\sup_t|\bar h(t)|_{L^{2,\gamma}_+}\right)\right|\bar\lambda T^*} < +\infty.$$

It follows from Proposition 14.2.2 that (J2) excludes the Wiener part of the noise as well as negative jumps. Thus J'' reduces to the form $J''(z) = \int_0^{+\infty} y^2 e^{-zy}\nu(dy)$ and $0 \leq J''(0) < +\infty$ due to the assumption (L1). Since J'' is decreasing, the following estimation holds

$$\sup_{t\in[0,T^*],x\geq 0} F_2(t,x) \leq J''(0)T^*\bar\lambda^2 \sup_{t\in[0,T^*],x\geq 0}\int_0^t \bar h(s,t-s+x)ds$$

$$+ T^*\left|J'\left(\frac{\bar\lambda}{\sqrt{\gamma}}\sup_t \mid \bar h(t)\mid_{L^{2,\gamma}_+}\right)\right| \cdot \sup_{x\geq 0}\lambda'(x),$$

and it is enough to show that $\bar h$ is bounded on $\{(t,x), t\in[0,T^*], x\geq 0\}$. In view of the fact that $\bar h = \mathcal{K}\bar h$ we obtain

$$\sup_{t\in[0,T^*],x\geq 0} \bar h(t,x) \leq \sup_{x\geq 0} r_0(x)\cdot \sup_{t\in[0,T^*],x\geq 0} b(t,x)\cdot e^{\left|J'\left(\frac{1}{\sqrt{\gamma}}\sup_t|\bar h(t)|_{L^{2,\gamma}_+}\right)\right|\bar\lambda T^*} < +\infty.$$

Proof of the uniqueness part

We need the following auxiliary result which is a simple modification of Lemma 14.3.5.

Lemma 15.2.5 *Let $d\colon [0,T^*]\times[0,+\infty)\to\mathbb{R}_+$ be a bounded function satisfying*

$$d(t,x) \leq C\int_0^t\int_0^{t-s+x} d(s,v)dvds, \tag{15.2.11}$$

where $C>0$ is a fixed constant. Then $d(t,x)=0$ for all $(t,x)\in[0,T^*]\times[0,+\infty)$.

To start the proof of the theorem assume that r_1, r_2 are two solutions of the equation (15.1.1) in $H^{1,\gamma}_+$. Then they are bounded processes in $H^{1,\gamma}$ and, in view of Theorem 15.2.1, satisfy (15.2.1). Define

$$d(t,x) :=\mid r_1(t,x) - r_2(t,x)\mid, \qquad 0\leq t\leq T^*, x\geq 0.$$

Denote $B := \sup_{t\in[0,T^*],x\geq 0} b(t,x)$. The following estimation holds

$$d(t,x) \leq r_0(t+x)b(t,x)\Big[e^{\int_0^t J'\left(\int_0^{t-s+x}\lambda(s,v)r_1(s,v)dv\right)\lambda(s,t-s+x)ds}$$

$$+ e^{\int_0^t J'\left(\int_0^{t-s+x}\lambda(s,v)r_2(s,v)dv\right)\lambda(s,t-s+x)ds}\Big]$$

$$\leq \sup_{x\geq 0} r_0(x)\cdot B\cdot\left[e^{\bar\lambda T^*\left|J'\left(\frac{\bar\lambda}{\sqrt{\gamma}}\sup_t|r_1(t)|_{L^{2,\gamma}_+}\right)\right|} + e^{\bar\lambda T^*\left|J'\left(\frac{\bar\lambda}{\sqrt{\gamma}}\sup_t|r_2(t)|_{L^{2,\gamma}_+}\right)\right|}\right] < +\infty,$$

and thus d is bounded on $[0, T^*] \times [0, +\infty)$. In view of the inequality $| e^x - e^y | \le e^{x \vee y} | x - y |; \ x, y \ge 0$ and the fact that J'' is decreasing with $0 \le J''(0) < +\infty$ due to assumption (L1), we have

$$d(t,x) \le \sup_{x \ge 0} r_0(x) \cdot B e^{\max\left\{\int_0^t J'\left(\int_0^{t-s+x} \lambda(s,v) r_1(s,v) dv\right) \lambda(s,t-s+x) ds; \int_0^t J'\left(\int_0^{t-s+x} \lambda(s,v) r_2(s,v) dv\right) \lambda(s,t-s+x) ds\right\}}$$

$$\cdot \left| \int_0^t J'\left(\int_0^{t-s+x} \lambda(s,v) r_1(s,v) dv\right) \lambda(s,t-s+x) ds - \int_0^t J'\left(\int_0^{t-s+x} \lambda(s,v) r_2(s,v) dv\right) \lambda(s,t-s+x) ds \right|$$

$$\le \sup_{x \ge 0} r_0(x) \cdot B e^{\bar{\lambda} T^* \max\left\{\left|J'\left(\frac{\bar{\lambda}}{\sqrt{\gamma}} \sup_t |r_1(t)|_{L^{2,\gamma}_+}\right)\right|; \left|J'\left(\frac{\bar{\lambda}}{\sqrt{\gamma}} \sup_t |r_2(t)|_{L^{2,\gamma}_+}\right)\right|\right\}}.$$

$$\cdot J''(0) \bar{\lambda}^2 \int_0^t \int_0^{t-s+x} | r_1(s,v) - r_2(s,v) | dv ds$$

$$= C \int_0^t \int_0^{t-s+x} d(s,v) dv ds, \qquad (t,x) \in [0, T^*] \times [0, +\infty).$$

It follows from Lemma 15.2.5 that $r_1 = r_2$ on $[0, T^*] \times [0, +\infty)$. □

Appendix A

Following Itô [73] and Kunita [84], we present and prove the general martingale representation theorem when the filtration is generated by a general Lévy process without the Wiener part.

A.1 Martingale Representation for Jump Lévy Processes

We assume that Z is of pure jump type,

$$Z(t) = at + \int_t \int_{|u| \leq 1} y\tilde{\pi}(ds, du) + \int_t \int_{|u|>1} y\pi(ds, du), \quad t \geq 0,$$

and $\mathcal{F}_t = \sigma\{Z(s), s \leq t\}$ is its natural filtration. Let ν be the intensity measure and $\tilde{\pi}$ the compensated random measure of Z. Denote by $\Psi_{1,2}$ the class of predictable processes $g(s, y), s \in [0, T^*], y \in U$, satisfying

$$\int_0^{T^*} \int_U (| g(s, y) |^2 \wedge | g(s, y) |) ds\nu(dy) < +\infty, \quad \mathbb{P} - a.s.$$

For $g \in \Psi_{1,2}$, one defines the integral

$$\int_0^t \int_U g(s, y)\tilde{\pi}(ds, dy) := \int_0^t \int_U g\mathbf{1}_{\{|g|\leq 1\}}\tilde{\pi}(ds, dy) + \int_0^t \int_U g\mathbf{1}_{\{|g|>1\}}\tilde{\pi}(ds, dy),$$

which is a local martingale. In this setting Theorem 6.1.1 has the following form.

Theorem A.1.1 *Let M be a real-valued $\mathbb{P}$-local martingale on $[0, T^*]$ adapted to the filtration $\mathcal{F}_t = \sigma\{Z(s), s \leq t\}$. Then there exists a process $\psi \in \Psi_{1,2}$ such that*

$$M_t = M_0 + \int_0^t \int_U \psi(s, y)\tilde{\pi}(ds, dy), \qquad t \in [0, T^*]. \tag{A.1.1}$$

Moreover, the process ψ is unique, i.e. if $\psi' \in \Psi_{1,2}$ and (A.1.1) holds with ψ', then

$$\psi = \psi', \quad d\mathbb{P} \times dt \times d\nu - \ a.s.$$

If M is a square integrable martingale then (A.1.1) *holds and*

$$\mathbb{E}M_t^2 = \mathbb{E}M_0^2 + \mathbb{E}\left(\int_0^t \int_U \mid \psi(s,y) \mid^2 ds\nu(dy)\right) < +\infty.$$

The proof will be divided into several parts. First, one introduces chaos processes and establishes the representation theorem for them in the form of the so-called multiple Itô–Wiener chaoses. Then one treats the case of square integrable martingales. If M is a square integrable martingale then

$$M_t = \mathbb{E}(M_{T^*}|\mathcal{F}_t), \quad t \in [0, T^*]$$

and M_{T^*} is a square integrable random variable and thus an element of the space $L^2(\Omega, \mathcal{F}_{T^*}, \mathbb{P})$. One first represents X as the orthogonal series

$$M_{T^*} = \sum_{k=0} X_k$$

of the k-multiple Itô–Wiener chaoses. The sum of the representations is the required representation in the square integrable case. The final part is devoted to the general case when M is a general local martingale. It is technically rather complicated and starts from the case when M is positive (see Kunita [84]).

A.1.1 Multiple Chaos Processes

Denote by $\bar{\nu}$ the product measure $dt\nu(du)$. It follows from the basic properties of the compensated random measure that if $E_1, E_2, \ldots, E_k$, are disjoint subsets of $[0, T^*] \times U$ such that

$$\iint_{E_j} dt\, \nu(du) = \bar{\nu}(E_j) < +\infty, \quad j = 1, 2, \ldots, k,$$

then the random variables

$$\int_0^{T^*} \int_U \mathbf{1}_{E_j}(s,u)\tilde{\pi}(ds,du) = \tilde{\pi}(E_j),\ j = 1, \ldots, k,$$

are independent with zero expectations. The product

$$\prod_{j=1}^{k} \int_0^{T^*} \int_U \mathbf{1}_{E_j}(s,u)\tilde{\pi}(ds,du)$$

is the simplest example of the k-multiple Itô–Wiener integral, also called k-multiple chaos,

$$I_k(f) = \iint_{[0,T^*)\times U} \cdots \iint_{[0,T^*)\times U} f\big((t_1,u_1), \ldots, (t_k,u_k)\big)\tilde{\pi}(dt_1,du_1)\ldots\tilde{\pi}(dt_k,du_k), \tag{A.1.2}$$

when

$$f\big((t_1,u_1),\ldots,(t_k,u_k)\big) = \mathbf{1}_{E_{i_1}}(t_1,u_1)\cdot\ldots\cdot\mathbf{1}_{E_{i_k}}(t_k,u_k).$$

The concept can be extended to all functions $f\big((t_1,u_1),\ldots,(t_k,u_k)\big)$ belonging to the Hilbert space L^2_k equipped with the norm,

$$|f|_{L^2_k} = \Bigg(\iint\limits_{[0,T^*)\times U}\cdots\iint\limits_{[0,T^*)\times U} f^2\big((t_1,u_1),\ldots,(t_k,u_k)\big)\bar{\nu}(dt_1,du_1)\ldots\bar{\nu}(dt_k,du_k)\Bigg)^{1/2}. \tag{A.1.3}$$

The extension is defined in an obvious way for the so-called *special elementary functions of order k* that are of the form

$$f\big((t_1,u_1),\ldots,(t_k,u_k)\big) := \sum_{(i_1,\ldots,i_k)\in I} a_{i_1,\ldots,i_k}\mathbf{1}_{E_{i_1}}(t_1,u_1)\cdot\ldots\cdot\mathbf{1}_{E_{i_k}}(t_k,u_k), \tag{A.1.4}$$

where $E_1,\ldots,E_m$, $m \geq k$, is an arbitrary sequence of disjoint subsets of $[0,T^*]\times U$, I stands for the set of all sequences $(i_1,\ldots,i_k)$ with pairwise different elements from $\{1,2,\ldots,m\}$ and $a_{i_1,\ldots,i_k}\in\mathbb{R}$. Namely, for such f one defines the *k-multiple integral of order k*

$$I_k(f) = \iint\limits_{[0,T^*]\times U}\cdots\iint\limits_{[0,T^*]\times U} f\big((t_1,u_1),\ldots,(t_k,u_k)\big)\tilde{\pi}(dt_1,du_1)\ldots\tilde{\pi}(dt_k,du_k),$$

by the formula

$$I_k(f) := \sum_{(i_1,\ldots,i_k)\in I} a_{i_1,\ldots,i_k}\tilde{\pi}(E_{i_1})\ldots\tilde{\pi}(E_{i_k}).$$

One checks directly that if f is a special elementary function of order k, then

$$\mathbb{E}|I_k(f)|^2 = |f|_{L^2_k}. \tag{A.1.5}$$

To perform extension to arbitrary $f\in L^2_k$, note that special elementary functions are dense in L^2_k and therefore for arbitrary $f\in L^2_k$ there exists a sequence (f_n) of special elementary function converging, in the sense of L^2_k, to f. It follows from (A.1.5) that also random variables $I_k(f_n)$ converge, in the sense of $L^2(\Omega,\mathcal{F}_{T^*})$, to a random variable. The limiting random variable is independent of the choice of the approximating sequence and is identified as $I_k(f)$.

We establish now the main properties of the multiple integrals needed in the proof of the chaos expansion theorem.

A function $g\colon \big([0,T^*\times U\big)^k \to \mathbb{R}$, is called *symmetric* if it is invariant under permutations of its arguments, i.e.

$$g\big((t_1,u_1),\ldots,(t_k,u_k)\big) = g\big((t_{p(1)},u_{p(1)}),\ldots,(t_{p(k)},u_{p(k)})\big)$$

for any permutation p of the set $\{1,2,\ldots,k\}$. It is clear that $I_k(f)$ is invariant after any permutation of its arguments. The following *symmetrization* $\hat{g}$ of g is clearly a symmetric function

$$\hat{g}\big((t_1,u_1),\ldots,(t_k,u_k)\big) := \frac{1}{k!}\sum_p g\big((t_{p(1)},u_{p(1)}),\ldots,(t_{p(k)},u_{p(k)})\big).$$

The preceding sum is taken over all permutations p of the set $\{1,2,\ldots,k\}$. The following proposition is crucial for the extension.

Proposition A.1.2 *i) For arbitrary functions f,g belonging to L_k^2,*

$$I_k(f) = I_k(\hat{f}), \tag{A.1.6}$$

$$\mathbb{E}\big[I_k(f)I_k(g)\big] = k! \iint_{[0,T^*]\times U} \cdots \iint_{[0,T^*]\times U} \hat{f}\big((t_1,u_1),\ldots,(t_k,u_k)\big) \cdot \hat{g}\big((t_1,u_1),\ldots,(t_k,u_k)\big)\, \bar{\nu}(dt_1,du_1)\ldots\bar{\nu}(dt_k,du_k) \tag{A.1.7}$$

$$= k! \quad (\hat{f},\hat{g})_{L_k^2}. \tag{A.1.8}$$

ii) If $f\in L_k^2,\ g\in L_l^2,\ k\neq l,$ *then* $\mathbb{E}\big[I_k(f)I_l(g)\big] = 0. \qquad$ (A.1.9)

Proof We can assume that f,g are special elementary functions. First, we prove (A.1.6). If f is of the form (A.1.4) then

$$\hat{f}\big((t_1,u_1),\ldots,(t_k,u_k)\big) = \frac{1}{k!}\sum_{(i_1,\ldots,i_k)\in I} a_{i_1,\ldots,i_k}\sum_p \mathbf{1}_{E_{i_1}}(t_{p(1)},u_{p(1)})\ldots\mathbf{1}_{E_{i_k}}(t_{p(k)},u_{p(k)})$$

and therefore

$$I_k(\hat{f}) = \frac{1}{k!}\sum_{(i_1,\ldots,i_k)\in I} a_{i_1,\ldots,i_k}\sum_p \big[\tilde{\pi}(E_{i_1})\ldots\tilde{\pi}(E_{i_k})\big]$$
$$= I_k(f).$$

Now we prove (A.1.7). Let g be of the form

$$g\big((t_1,u_1),\ldots,(t_k,u_k)\big) = \sum_{(j_1,\ldots,j_k)\in I} b_{j_1,\ldots,j_k}\mathbf{1}_{E_{j_1}}(t_1,u_1)\cdot\ldots\cdot\mathbf{1}_{E_{j_k}}(t_k,u_k).$$

Then

$$\mathbb{E}\big[I_k(f)I_k(g)\big] = \mathbb{E}\Big(\sum_{(i_1,\ldots,i_k)\in I} a_{i_1,\ldots,i_k}\tilde{\pi}(E_{i_1})\ldots\tilde{\pi}(E_{i_k})\cdot\sum_{(j_1,\ldots,j_k)\in I} b_{j_1,\ldots,j_k}\tilde{\pi}(E_{j_1})\ldots\tilde{\pi}(E_{j_k})\Big).$$

One writes $(j_1,\dots,j_k) \simeq (i_1,\dots,i_k)$ if $(j_1,\dots,j_k)$ can be obtained from $(i_1,\dots,i_k)$ by permutation of its elements. Taking into account that $\tilde{\pi}(E_i), \tilde{\pi}(E_j)$ are independent for disjoint sets E_i, E_j and

$$\mathbb{E}\big[\tilde{\pi}(E_i)\big] = 0, \quad \mathbb{E}\big[(\tilde{\pi}(E_i))^2\big] = \bar{\nu}(E_i), \qquad i = 1,2,\dots,m,$$

we have

$$\begin{aligned}&\mathbb{E}\big[\tilde{\pi}(E_{i_1})\dots\tilde{\pi}(E_{i_k})\cdot\tilde{\pi}(E_{j_1})\dots\tilde{\pi}(E_{j_k})\big]\\ &\quad = \mathbb{E}\big[(\tilde{\pi}(E_{i_1})\dots\tilde{\pi}(E_{i_k}))^2\big] \quad \text{if} \quad (j_1,\dots,j_k) \simeq (i_1,\dots,i_k),\end{aligned}$$

and 0 otherwise. Consequently,

$$\begin{aligned}\mathbb{E}\big[I_k(f)I_k(g)\big] &= \mathbb{E}\Big(\sum_{(i_1,\dots,i_k)\in I} a_{i_1,\dots,i_k}\cdot\sum_{(j_1,\dots,j_k)\simeq(i_1,\dots,i_k)} b_{j_1,\dots,j_k}\cdot\big(\tilde{\pi}(E_{i_1})\dots\tilde{\pi}(E_{i_k})\big)^2\Big)\\ &= \sum_{(i_1,\dots,i_k)\in I} a_{i_1,\dots,i_k}\cdot\sum_{(j_1,\dots,j_k)\simeq(i_1,\dots,i_k)} b_{j_1,\dots,j_k}\cdot\bar{\nu}(E_{i_1})\dots\bar{\nu}(E_{i_k}).\end{aligned} \tag{A.1.10}$$

However,

$$\begin{aligned}&\iint_{E_1}\dots\iint_{E_k}\hat{f}\big((t_1,u_1),\dots,(t_k,u_k)\big)\cdot\hat{g}\big((t_1,u_1),\dots,(t_k,u_k)\big)\bar{\nu}(dt_1,du_1)\dots\bar{\nu}(dt_k,du_k)\\ &= \frac{1}{(k!)^2}\iint_{E_1}\dots\iint_{E_k}\sum_p\sum_{(i_1,\dots,i_k)\in I} a_{i_1\dots i_k}\mathbf{1}_{E_{i_1}}(t_{p(1)},u_{p(1)})\dots\mathbf{1}_{E_{i_k}}(t_{p(k)},u_{p(k)})\\ &\quad\cdot\sum_q\sum_{(j_1,\dots,j_k)\in I} b_{j_1\dots j_k}\mathbf{1}_{E_{j_1}}(t_{q(1)},u_{q(1)})\dots\mathbf{1}_{E_{j_k}}(t_{q(k)},u_{q(k)})\,\bar{\nu}(dt_1,du_1)\dots\bar{\nu}(dt_k,du_k).\end{aligned}$$

For the product

$$\mathbf{1}_{E_{i_1}}(t_{p(1)},u_{p(1)})\dots\mathbf{1}_{E_{i_k}}(t_{p(k)},u_{p(k)})\cdot\mathbf{1}_{E_{j_1}}(t_{q(1)},u_{q(1)})\dots\mathbf{1}_{E_{j_k}}(t_{q(k)},u_{q(k)})$$

to be different from zero, one must have that $(j_1,\dots,j_k) \simeq (i_1,\dots,i_k)$. In addition, the permutation q should be such that the product is equal to

$$\mathbf{1}_{E_{i_1}}(t_{p(1)},u_{p(1)})\dots\mathbf{1}_{E_{i_k}}(t_{p(k)},u_{p(k)}).$$

Therefore

$$\begin{aligned}&\iint_{[0,T^*]\times U}\dots\iint_{[0,T^*]\times U}\hat{f}\big((t_1,u_1),\dots,(t_k,u_k)\big)\\ &\quad\cdot\hat{g}\big((t_1,u_1),\dots,(t_k,u_k)\big)\,\bar{\nu}(dt_1,du_1)\dots\bar{\nu}(dt_k,du_k)\end{aligned}$$

$$= \frac{1}{(k!)^2} \iint\limits_{[0,T^*]\times U} \cdots \iint\limits_{[0,T^*]\times U} \sum_{p} \sum_{(i_1,\dots,i_k)\in I} a_{i_1\dots i_k} \mathbf{1}_{E_{i_1}}(t_{p(1)}, u_{p(1)}) \dots \mathbf{1}_{E_{i_k}}(t_{p(k)}, u_{p(k)})$$

$$\cdot \sum_{(j_1,\dots,j_k)\simeq(i_1,\dots,i_k)} b_{j_1\dots j_k}\, \bar{\nu}(dt_1, du_1)\dots\bar{\nu}(dt_k, du_k)$$

$$= \frac{1}{(k!)^2} \sum_{(i_1,\dots,i_k)\in I} a_{i_1\dots i_k}\big[k!\ \bar{\nu}(E_{i_1})\dots\bar{\nu}(E_{i_k})\big] \sum_{(j_1,\dots,j_k)\simeq(i_1,\dots,i_k)} b_{j_1\dots j_k}.$$

Taking into account (A.1.10), the proof of (A.1.7) is complete.

To prove (A.1.9) assume that $k < l$. Then

$$\mathbb{E}\Big[\tilde{\pi}(E_{i_1})\dots\tilde{\pi}(E_{i_k})\cdot\tilde{\pi}(E_{j_1})\dots\tilde{\pi}(E_{j_l})\Big] = 0,$$

because in the sequence $E_{j_1},\dots,E_{j_l}$ there exists a set, say $E_{\bar{j}}$, disjoint from all the others and the expectation of the product is a multiple of $\mathbb{E}[\tilde{\pi}(E_{\bar{j}})] = 0$. □

A.1.2 Representation of Chaoses

Let $f \in L^2_k$. The following integral

$$I^t_k(f) := \iint\limits_{[0,t]\times U} \iint\limits_{[0,t]\times U} \cdots \iint\limits_{[0,t]\times U} f\big((t_1,u_1),(t_2,u_2),\dots,(t_k,u_k)\big)\tilde{\pi}(dt_1,du_1)\dots\tilde{\pi}(dt_k,du_k),$$

is well defined for any $t \in [0,T^*]$ because it can be represented as the multiple integral

$$I^t_k(f) = I_k(F_t),$$

with

$$F_t\big((t_1,u_1),\dots,(t_k,u_k)\big) = f\big((t_1,u_1),\dots,(t_k,u_k)\big)\mathbf{1}_{[0,t]}(t_1)\mathbf{1}_{[0,t]}(t_2)\dots\mathbf{1}_{[0,t]}(t_k)$$

and $F_t \in L^2_k$. Now we show that the process $t \to I^t_k(f), t \in [0,T^*]$ is a square integrable martingale and can be represented as a stochastic integral.

Theorem A.1.3 *For arbitrary $f \in L^2_k$ there exists a predictable process $R^f(t,u)$, $t \in [0,T^*], u \in U$ such that*

$$I^t_k(f) = \int_0^t \int_U R^f(s,u)\tilde{\pi}(ds,du), \quad t \in [0,T^*], \tag{A.1.11}$$

and

$$\mathbb{E}\big[(I^t_k(f))^2\big] = \mathbb{E}\Big(\int_0^t \int_U (R^f(s,u))^2\, ds\, \nu(du)\Big) = |\, f\,|^2_k \quad t \in [0,T^*]. \tag{A.1.12}$$

The arguments in the proof will involve *iterated Itô–Wiener integrals* of the function f. More precisely, we prove that

$$I_k^t(f) = \int_0^t \int_U R^f(t_k, u_k)\tilde{\pi}(dt_k, du_k), \quad t \in [0, T^*],$$

where $R_n, n = 1, 2, \ldots, k$ is a sequence of processes iteratively defined by

$$R_1\big((t_1, u_1), \ldots, (t_k, u_k)\big) := \bar{f}\big((t_1, u_1), \ldots, (t_k, u_k)\big), \tag{A.1.13}$$

$$R_{n+1}\big((t_{n+1}, u_{n+1}), \ldots, (t_k, u_k)\big) := \int_0^{t_{n+1}} \int_U R_n\big((t_n, u_n), \ldots, (t_k, u_k)\big)\tilde{\pi}(dt_n, du_n), \tag{A.1.14}$$

where $\bar{f}_k$ is the *truncated symmetrization* of f defined by

$$\bar{f}_k\big((t_1, u_1), \ldots, (t_k, u_k)\big) := k!\ \mathbf{1}_{\triangle_k}(t_1, \ldots, t_k)\hat{f}\big((t_1, u_1), \ldots, (t_k, u_k)\big).$$

The preceding $\triangle_k$ is a simplex given by

$$\triangle_k := \{(t_1, t_2, \ldots, t_k) : 0 < t_1 < t_2 < \cdots < t_k \le T\}.$$

Moreover,

$$\mathbb{E}\Big[\int_0^t \int_U \big(R_k(t_k, u_k)\big)^2 dt_k\ \nu(du_k)\Big] < +\infty.$$

We need also the following auxiliary result.

Proposition A.1.4 *For* $f \in L_k^2$,

$$I_k^t(f) = I_k^t(\hat{f}) = I_k^t(\bar{f}), \quad t \in [0, T^*].$$

Proof To simplify notation we assume that $t = T^*$. In view of Proposition A.1.2 only the latter identity requires a proof. It is sufficient to show that $\hat{\bar{f}} = \hat{f}$. Note that

$$\hat{\bar{f}}\big((t_1, u_1), \ldots, (t_k, u_k)\big) = \frac{1}{k!}\sum_p k!\,\mathbf{1}_{\triangle_k}(t_{p(1)}, \ldots, t_{p(k)})\hat{f}_k\big((t_{p(1)}, u_{p(1)}), \ldots, (t_{p(k)}, u_{p(k)})\big).$$

However, for each permutation p,

$$\hat{f}\big((t_{p(1)}, u_{p(1)}), \ldots, (t_{p(k)}, u_{p(k)})\big) = \hat{f}\big((t_1, u_1), \ldots, (t_k, u_k)\big),$$

and therefore

$$\begin{aligned}\hat{\bar{f}}\big((t_1, u_1), \ldots, (t_k, u_k)\big) &= \hat{f}\big((t_1, u_1), \ldots, (t_k, u_k)\big)\sum_p \mathbf{1}_{\triangle_k}(t_{p(1)}, \ldots, t_{p(k)}) \\ &= \hat{f}\big((t_1, u_1), \ldots, (t_k, u_k)\big),\end{aligned}$$

because for arbitrary vector $(s_1, \ldots, s_k)$ with different coordinates $(s_{p(1)}, \ldots, s_{p(k)}) \in \triangle_k$ for exactly one permutation. □

Proof of Theorem A.1.3 We prove the result for special elementary functions of order k, with supports in $\triangle_k$ (see (A.1.4)) where

$$E_j = [a_j, b_j) \times U_j, \qquad b_j \le a_{j+1},\ j = 1, 2, \dots, k-1,\ U_j \subseteq U, \tag{A.1.15}$$

which are dense in L^2_k. We restrict to functions of the form

$$f\big((t_1, u_1), \dots, (t_k, u_k)\big) = \prod_{j=1}^{k} \mathbf{1}_{E_j}(t_j, u_j). \tag{A.1.16}$$

Notice that for $f \in L^2_k$ satisfying (A.1.16), (A.1.15) the truncated symmetrization is identical with the original function, i.e.

$$f = \bar{f} = k! \hat{f} \mathbf{1}_{\triangle_k}.$$

Starting from $R^f_1\big((t_1, u_1), \dots, (t_k, u_k)\big) = f\big((t_1, u_1), \dots, (t_k, u_k)\big)$ we obtain

$$\begin{aligned} R^f_2\big((t_2, u_2), \dots, (t_k, u_k)\big) &= \int_0^{t_2} \int_U \mathbf{1}_{E_1}(t_1, u_1) \cdot \dots \cdot \mathbf{1}_{E_k}(t_k, u_k)\, \tilde{\pi}(dt_1, du_1) \\ &= \tilde{\pi}(E_1) \cdot \mathbf{1}_{E_2}(t_2, u_2) \cdot \dots \cdot \mathbf{1}_{E_k}(t_k, u_k). \end{aligned}$$

Further integration provides a general formula for any $n = 1, 2, \dots, k$

$$R^f_n\big((t_n, u_n), \dots, (t_k, u_k)\big) = \tilde{\pi}(E_1) \cdot \dots \cdot \tilde{\pi}(E_{n-1}) \cdot \mathbf{1}_{E_n}(t_n, u_n) \cdot \dots \cdot \mathbf{1}_{E_k}(t_k, u_k). \tag{A.1.17}$$

As a consequence we obtain

$$\begin{aligned} \int_0^t \int_U R^f_k(t_k, u_k)\, \tilde{\pi}(dt_k, du_k) &= \tilde{\pi}(E_1) \cdot \dots \cdot \tilde{\pi}(E_{k-1}) \int_0^t \int_U \mathbf{1}_{E_k}(t_k, u_k)\, \tilde{\pi}(dt_k, du_k) \\ &= \tilde{\pi}(E_1) \cdot \dots \cdot \tilde{\pi}(E_{k-1}) \int_0^t \int_U \mathbf{1}_{[a_k, b_k] \times U_k}(t_k, u_k) \tilde{\pi}(dt_k, du_k) \\ &= \tilde{\pi}(E_1) \cdot \dots \cdot \tilde{\pi}(E_{k-1}) \tilde{\pi}([t \wedge a_k, t \wedge b_k] \times U_k). \end{aligned}$$

However, we can calculate directly

$$\begin{aligned} I^t_k(f) &= \iint\limits_{[0,t]\times U} \iint\limits_{[0,T]\times U} \cdots \iint\limits_{[0,T]\times U} \mathbf{1}_{E_1}(t_1, u_1) \cdot \dots \cdot \mathbf{1}_{E_k}(t_k, u_k) \tilde{\pi}(dt_1, du_1) \dots \tilde{\pi}(dt_k, du_k) \\ &= \tilde{\pi}(E_1) \cdot \dots \cdot \tilde{\pi}(E_{k-1}) \tilde{\pi}([t \wedge a_k, t \wedge b_k] \times U_k), \end{aligned} \tag{A.1.18}$$

which yields (A.1.11). To prove (A.1.12) we proceed similarly as in Proposition A.1.2. We use the independence of random variables $\tilde{\pi}(E_1), \dots, \tilde{\pi}(E_k)$ which, in view of (A.1.17), yields

$$\mathbb{E}\Big[\big(R_k^f(t_k,u_k)\big)^2\Big] = \mathbb{E}\Big[\big(\tilde{\pi}(E_1)\cdot\ldots\cdot\tilde{\pi}(E_{k-1})\big)^2\Big]\cdot \mathbf{1}_{E_k}(t_k,u_k)$$
$$= \bar{\nu}(E_1)\cdot\ldots\cdot\bar{\nu}(E_{k-1})\cdot\mathbf{1}_{E_k}(t_k,u_k),$$

and

$$\int_0^t\int_U \mathbb{E}\Big[\big(R_k^f(t_k,u_k)\big)^2\Big]dt_k\nu(du_k) = \bar{\nu}(E_1)\cdot\ldots\cdot\bar{\nu}(E_{k-1})\int_0^t\int_U \mathbf{1}_{E_k}(t_k,u_k)dt_k\ \nu(du_k)$$
$$= \bar{\nu}(E_1)\cdot\ldots\cdot\bar{\nu}(E_{k-1})\cdot\nu([t\wedge a_k, t\wedge b_k]\times U_k). \tag{A.1.19}$$

Finally, by (A.1.18) and (A.1.19), we obtain

$$\mathbb{E}\Big[(I_k^t(f))^2\Big] = \bar{\nu}(E_1)\cdot\ldots\cdot\bar{\nu}(E_{k-1})\cdot\nu([t\wedge a_k, t\wedge b_k]\times U_k)$$
$$= \mathbb{E}\Big(\int_0^t\int_U \big(R_k^f(t_k,u_k)\big)^2\ dt_k\ \nu(du_k)\Big).$$

In fact, let f be an arbitrary element of L_k^2 and f_n a sequence of functions considered in the preceding converging to f. Then $R_k^{f_n}$ is a sequence of predictable fields such that the sequence

$$\mathbb{E}\int_0^t\int_U (R_k^{f_n}(s,u) - R_k^{f_m}(s,u)^2 ds\nu(du) = |f_n - f_m|_k^2 \tag{A.1.20}$$

tends to 0 as $m,n\to+\infty$. Therefore a subsequence $R_k^{f_{n_j}}$ converges almost surely, on the product space, $\Omega\times[0,T^*]\times U$ to a predictable field g with the property

$$\mathbb{E}\int_0^t\int_U (R_k^{f_n}(s,u) - g(s,u))^2 ds\nu(du) = |f_n - g|_k^2. \tag{A.1.21}$$

Thus it is enough to define $R_k^f = g$. □

A.1.3 Chaos Expansion Theorem

We prove now the chaos expansion theorem, from which the representation theorem for square integrable martingales will be derived.

Let $\hat{L}_k^2$ denote the Hilbert space of symmetric functions f of $2k$-variables $(t_1,u_1),\ldots,(t_k,u_k)$ with $(t_i,u_i)\in[0,T^*]\times U =: \bar{U}$, equipped with the norm

$$|\,f\,|_k^2 = k!\int\limits_{\bar{U}^k}\cdots\int f^2\big((t_1,u_1),\ldots,(t_k,u_k)\big)\bar{\nu}(dt_1,du_1)\ldots\bar{\nu}(dt_k,du_k).$$

Since $\hat{L}_k^2$ is a closed subspace of L_k^2, it is also a Hilbert space. The transformation

$$I_k\colon \hat{L}_k^2 \longrightarrow H := L^2(\Omega,\mathcal{F}_T^*,P)$$

is an isometry and this is why in the sequel we will use the analysis $\hat{L}^2_k$ instead of L^2_k. By H_k we denote the subspace of H consisting of all integrals $I_k(f), f \in \hat{L}^2_k$. It follows that H_k is a closed subspace of H. Let H_0 be the real line with usual metric.

Theorem A.1.5 *The Hilbert space H is a direct sum of the subspaces H_k, $k = 0, 1, \ldots$, i.e.*

$$H = \sum_{k \geq 0} \oplus H_k. \tag{A.1.22}$$

That is, arbitrary $X \in L^2(\Omega, \mathcal{F}_{T^*}, P)$ can be represented as the sum

$$X = \mathbb{E}(X) + \sum_{k=1}^{+\infty} I_k(f_k), \tag{A.1.23}$$

where (f_k) is a sequence of symmetric functions from the spaces $L^2_k, k = 1, 2, \ldots$. The series in (A.1.23) *consists of orthogonal random variables and converges in mean square.*

Proof It follows from Proposition A.1.2, formula (A.1.9), that the subspaces $H_k, k = 0, 1, \ldots$, are mutually orthogonal and the identity (A.1.22) means, in addition,

$$\mathbb{E}[X^2] = (\mathbb{E}X)^2 + \sum_{k=1}^{-\infty} k! \int \cdots \int_{\bar{U}^k} f_k^2\big((t_1, u_1), \ldots, (t_k, u_k)\big) \bar{\nu}(dt_1, du_1) \ldots \bar{\nu}(dt_k, du_k).$$

It is clear that $\sum_{k \geq 0} \oplus H_k$ is a closed subspace of H and therefore it is sufficient to show that it is dense in H. Taking into account that the σ-field $\mathcal{F}_T^*$ is generated by the Lévy process $Z(t), t \in [0, T^*]$, it is not difficult to see that random variables of the form

$$f(\pi(E_1), \ldots, \pi(E_m)),$$

where f is a bounded continuous function and sets $E_j, j = 1, 2, \ldots, m$ are pairwise disjoint, consitute a dense set in the space H. Indeed, a function of Z can be approximated by a function of its increments. The same is true if we replace functions f by linear combinations of polynomials

$$Y = \pi(E_1)^{p_1} \pi(E_2)^{p_2} \ldots \pi(E_k)^{p_k}.$$

Taking into account that elements of H_m can be approximated by linear combinations of

$$Z = \pi(F_1)\pi(F_2) \ldots \pi(F_m),$$

with pairwise disjoint sets $F_1, \ldots, F_m$, it is enough to show that random variables Y can be approximated in H by linear combinations of random variables of the form Z.

To do so, set $\kappa := \bar{\nu}(\bigcup E_j)$. We can assume that $\kappa < +\infty$. Choose $\varepsilon \in (0, \kappa)$ and let $F_1, F_2, \ldots, F_m$ be a subdivision of $E_1, \ldots, E_k$ so fine that

$$\bar{\nu}(F_j) \leq \frac{\varepsilon}{\kappa}, \quad j = 1, 2, \ldots, m. \tag{A.1.24}$$

Note that sets $E_1, \ldots, E_k$ are covered by disjoint families $(F_{j_1(1)}, \ldots, F_{j_{m_1}(1)}), \ldots,$ $(F_{j_1(k)}, \ldots, F_{j_{m_k}(k)})$, and therefore Y is a sum of elements of the form

$$\pi(F_{l(1)})^{k_1} \ldots \pi(F_{l(r)})^{k_r}$$

with

$$l(1) < l(2) < \ldots < l(r) \leq m. \tag{A.1.25}$$

Therefore

$$Y = \sum_R \pi(F_{l(1)})^{k_1} \ldots \pi(F_{l(r)})^{k_r},$$

where the sum is over a finite family R of sequences of the form $\big((l(1), \ldots, l(r)),$ $(k_1, \ldots, k_r)\big)$ satisfying (A.1.25). Since the random variables $\pi(G)$ take values $0, 1, 2, \ldots$ we have that

$$Y \geq \sum_R \pi(F_{l(1)}) \ldots \pi(F_{l(r)}) =: V.$$

However, Y is strictly greater than V if $\pi(F_l) \geq 2$ for some l. Consequently,

$$\begin{aligned}
\mathbb{P}(Y > V) &= \mathbb{P}(\pi(F_l) \geq 2 \text{ for some } l) \\
&\leq \sum_{l=1}^{m} \mathbb{P}(\pi(F_l) \geq 2) \leq \sum_{l=1}^{m} \big(\bar{\nu}(F_l)\big)^2 \\
&\leq \frac{\varepsilon}{\kappa} \sum_{l=1}^{m} \big(\bar{\nu}(F_l)\big) \leq \varepsilon.
\end{aligned}$$

Thus Y can be approximated in probability by V and thus, for appropriate subsequence, also almost surely and, since $Y \geq V$, by Lebesgue's dominated convergence theorem also in H. □

A.1.4 Representation of Square Integrable Martingales

We prove here Theorem A.1.1 for the case in which M is a square integrable martingale and Z is a Lévy process without Gaussian part.

Proof Let $X := M_{T^*}$. By Theorem A.1.5, for some sequence $f_k \in \hat{L}^2_k$,

$$X = \mathbb{E}(X) + \sum_{k=1}^{+\infty} I_k^{T^*}(f_k),$$

and, consequently,

$$M_t = \mathbb{E}[X \mid \mathcal{F}_t]$$
$$= \mathbb{E}(X) + \sum_{k=1}^{+\infty} \mathbb{E}[I_k^{T^*}(f_k) \mid \mathcal{F}_t].$$

In view of Theorem A.1.3

$$I_k^{T^*}(f_k) = \int_0^{T^*} \int_U R_k^f(s,u)\tilde{\pi}(ds,du).$$

Consequently,

$$\mathbb{E}[I_k^{T^*}(f_k) \mid \mathcal{F}_t] = \int_0^t \int_U R_k^f(s,u)\tilde{\pi}(ds,du),$$

and

$$M_t = \mathbb{E}(M_0) + \sum_{k=1}^{+\infty} \int_0^t \int_U R_k^f(s,u)\tilde{\pi}(ds,du),$$

with the series on the right side converging in mean square because $\sum I_k^{T^*}(f_k)$ converges in H. But

$$\mathbb{E}[(X - \mathbb{E}(X))^2] = \sum_{k=1}^{+\infty} \mathbb{E}\big((I_k^{T^*}(f_k))^2\big) < +\infty,$$

and the integrals $I_k^{T^*}(f_k), k = 1, 2, \ldots$ are orthogonal, i.e.

$$\mathbb{E}\big(I_k^{T^*}(f_k) I_l^{T^*}(f_l)\big) = 0, \quad \text{for } k \neq l.$$

Therefore, for $k \neq l$

$$\mathbb{E}\Big(\int_0^{T^*} \int_U R_k^f(s,u)\tilde{\pi}(ds,du) \cdot \int_0^{T^*} \int_U R_l^f(s,u)\tilde{\pi}(ds,du)\Big)$$
$$= \mathbb{E}\Big(\int_0^{T^*} \int_U R_k^f(s,u) R_l^f(s,u)\, ds\, \nu(du)\Big) = 0.$$

Thus

$$\sum_{k=1}^{+\infty} \mathbb{E}(\mid I_k^{T^*}(f_k) \mid^2) = \mathbb{E}\Big(\int_0^{T^*} \int_U \sum_{k=1}^{+\infty} (R_k^f)^2(s,u)\, ds\, \nu(du)\Big) < +\infty,$$

and the sum

$$\sum_{k=1}^{+\infty} R_k^f(s,u)$$

of predictable integrands R_k^f has a subseries converging $d\mathbb{P} \times ds \times du$ – almost surely to a predictable g. Consequently,

$$\sum_{k=1}^{+\infty} \int_0^t \int_U R_k^f(s,u)\tilde{\pi}(ds,du) = \int_0^t \int_U g(s,u)\tilde{\pi}(ds,du). \qquad \square$$

A.1.5 Representations of Local Martingales

To prove the general version of Theorem A.1.1 we start with auxiliary results dealing with strictly positive local martingales.

Proposition A.1.6 *Let g, h be predictable processes such that $h \in \Psi_2$ is bounded and $g(t,y)h(t,y) = 0, t \in [0,T^*], y \in U$. Assume that $A(t), t \in [0,T^*]$ is a continuous process of finite variation. Then the process α given by*

$$\alpha_t := e^{X_t},\ X_t := \int_0^t \int_U g(s,y)\pi(ds,dy) + \int_0^t \int_U h(s,y)\tilde{\pi}(ds,dy) - A_t, \quad t \in [0,T^*]$$

is a local martingale if and only if the following two conditions are satisfied

$$e^g - 1 \in \Psi_1, \tag{A.1.26}$$

$$A_t = \int_0^t \int_U (e^{g(s,y)} - 1)ds\nu(dy) + \int_0^t \int_U (e^{h(s,y)} - 1 - h(s,y))ds\nu(dy), \quad t \in [0,T^*]. \tag{A.1.27}$$

Proof Application of the Itô formula provides the integral representation of α. We have

$$\begin{aligned}
\alpha_t &= 1 + \int_0^t \alpha_{s-}dX_s + \frac{1}{2}\int_0^t \alpha_{s-}d\langle X^c, X^c\rangle_s + \sum_{s\in[0,t]} (\alpha_s - \alpha_{s-} - \alpha_{s-}\Delta X_s)\\
&= 1 + \int_0^t \int_U \alpha_{s-}g(s,y)\pi(ds,dy) + \int_0^t \int_U \alpha_{s-}h(s,y)\tilde{\pi}(ds,dy) - \int_0^t \alpha_{s-}dA_s\\
&\quad + \int_0^t \int_U \alpha_{s-}(e^{g(s,y)+h(s,y)} - 1 - g(s,y) - h(s,y))\pi(ds,dy),\ t \in [0,T^*].
\end{aligned}$$

Since $g(s,y)h(s,y) = 0$ we can split the last integral into the form

$$\int_0^t \int_U \alpha_{s-}(e^{g(s,y)} - 1 - g(s,y))\pi(ds,dy) + \int_0^t \int_U \alpha_{s-}(e^{h(s,y)} - 1 - h(s,y))\pi(ds,dy).$$

Then, by compensation, we finally obtain

$$\alpha_t = 1 + \int_0^t \int_U \alpha_{s-}(e^{h(s,y)} - 1)\tilde{\pi}(ds, dy) - \int_0^t \alpha_{s-} dA_s$$
$$+ \int_0^t \int_U \alpha_{s-}(e^{g(s,y)} - 1)\pi(ds, dy)$$
$$+ \int_0^t \int_U \alpha_{s-}(e^{h(s,y)} - 1 - h(s, y))ds\nu(dy), \quad t \in [0, T^*]. \tag{A.1.28}$$

Now we show the sufficiency of (A.1.26) and (A.1.27). In view of (A.1.26) we can rearrange (A.1.28) to the form

$$\alpha_t = 1 + \int_0^t \int_U \alpha_{s-}(e^{h(s,y)} - 1)\tilde{\pi}(ds, dy) + \int_0^t \int_U \alpha_{s-}(e^{g(s,y)} - 1)\tilde{\pi}(ds, dy)$$
$$- \int_0^t \alpha_{s-} dA_s + \int_0^t \int_U \alpha_{s-}(e^{g(s,y)} - 1)\nu(dy)ds$$
$$+ \int_0^t \int_U \alpha_{s-}(e^{h(s,y)} - 1 - h(s, y))ds\nu(dy), \quad t \in [0, T^*].$$

Further, (A.1.27) implies that the last two lines in the preceding disappear and, consequently, α is a local martingale.

To show the necessity of (A.1.26) and (A.1.27) let us write (A.1.28) in the form

$$\alpha_t - 1 - \int_0^t \int_U \alpha_{s-}(e^{h(s,y)} - 1)\tilde{\pi}(ds, dy)$$
$$= - \int_0^t \alpha_{s-} dA_s + \int_0^t \int_U \alpha_{s-}(e^{g(s,y)} - 1)\pi(ds, dy)$$
$$+ \int_0^t \int_U \alpha_{s-}(e^{h(s,y)} - 1 - h(s, y))ds\nu(dy), \quad t \in [0, T^*]. \tag{A.1.29}$$

Since the left side is a local martingale and the right side is a finite variation process it follows from Proposition 4.2.9 that the right side belongs to $\mathcal{A}_{loc}$. It follows from Proposition 4.2.8 that

$$- \int_0^t \alpha_{s-} dA_s + \int_0^t \int_U \alpha_{s-}(e^{h(s,y)} - 1 - h(s, y))ds\nu(dy) \in \mathcal{A}_{loc},$$

which implies that

$$\int_0^t \int_U \alpha_{s-}(e^{g(s,y)} - 1)\pi(ds, dy) \in \mathcal{A}_{loc}.$$

In view of Theorem 4.2.13 there exists a compensator of the preceding process and it has the form $\int_0^t \int_U \alpha_{s-}(e^{g(s,y)} - 1)ds\nu(dy)$ which means that (A.1.26) holds. Compensation of the right side of (A.1.29) gives that the process

$$-\int_0^t \alpha_{s-}dA_s + \int_0^t \int_U \alpha_{s-}(e^{g(s,y)} - 1)ds\nu(dy)$$
$$+ \int_0^t \int_U \alpha_{s-}(e^{h(s,y)} - 1 - h(s,y))ds\nu(dy), \quad t \in [0, T^*]$$

is a local martingale. Since it is also a continuous process of finite variation it follows from Proposition 4.2.10 that it disappears, which means that (A.1.27) is satisfied. □

For the second auxiliary result we need the following result on continuous compensators of jump processes.

Proposition A.1.7 *Let X be a process adapted to the filtration generated by a Lévy process Z with càdlàg paths satisfying*

$$\triangle X(t) \neq 0 \quad \Longrightarrow \quad \triangle Z(t) \neq 0. \tag{A.1.30}$$

If the process

$$Y(t) := \sum_{s\in[0,t];\triangle X(s)\neq 0} h(s, \triangle X(s)),$$

where $h\colon \Omega \times \mathbb{R}_+ \times U \to \mathbb{R}$, is adapted and of locally integrable variation then its compensator is continuous.

Proof Let us define the function f by

$$\triangle X(s) = f(s, \triangle Z(s)), \qquad s \geq 0.$$

By (A.1.30) we have $f(s, 0) = 0$ and hence we can represent Y in the form

$$\begin{aligned} Y(t) &= \sum_{s\in[0,t];f(s,\triangle Z(s))\neq 0} h(s,f(s, \triangle Z(s))) \\ &= \sum_{s\in[0,t];\triangle Z(s)\neq 0;f(s,\triangle Z(s))\neq 0} h(s,f(s, \triangle Z(s))) \\ &= \sum_{s\in[0,t];\triangle Z(s)\neq 0} h(s,f(s, \triangle Z(s)))\mathbf{1}_{\{f(s,\triangle Z(s))\neq 0\}} \\ &= \int_0^t \int_{U\setminus\{0\}} h(s,f(s,y))\mathbf{1}_{\{f(s,y)\neq 0\}}\pi(ds,dy), \quad t \geq 0, \end{aligned}$$

where $\pi(ds, dy)$ stands for the jump measure of Z. It follows that the compensator of Y is given by

$$\int_0^t \int_{U\setminus\{0\}} h(s,f(s,y))\mathbf{1}_{\{f(s,y)\neq 0\}}ds\nu(dy), \quad t \geq 0,$$

which is clearly continuous. □

Lemma A.1.8 *Let α be a local martingale such that $\alpha_t > 0, t \in [0, T^*]$ and $\alpha_0 = 1$. Then there exists a unique process ψ, where $e^{\psi} - 1 \in \Psi_{1,2}$, such that*

$$\alpha_t = e^{X_t}, \quad t \in [0, T^*], \tag{A.1.31}$$

where

$$\begin{aligned} X_t := & \int_0^t \int_U \psi_1(s, y)\pi(ds, dy) - \int_0^t \int_U (e^{\psi_1(s,y)} - 1)ds\nu(dy) \\ & + \int_0^t \int_U \psi_2(s, y)\tilde{\pi}(ds, dy) \\ & - \int_0^t \int_U (e^{\psi_2(s,y)} - 1 - \psi_2(s, y))ds\nu(dy), \quad t \in [0, T^*] \end{aligned}$$

and

$$\psi_1(s, y) = \psi(s, y)\mathbf{1}_{\{|\psi(s,y)|>1\}}, \quad \psi_2(s, y) = \psi(s, y)\mathbf{1}_{\{|\psi(s,y)|\leq 1\}}, \quad s \in [0, T^*], y \in U.$$

However, if $e^{\psi} - 1 \in \Psi_{1,2}$ then the process α given by (A.1.31) *is a strictly positive local martingale.*

Before we present the proof let us comment on the formulation of the assertion.

Remark A.1.9 The condition $e^{\psi} - 1 \in \Psi_{1,2}$ is equivalent to the pair of conditions

$$e^{\psi_1} - 1 \in \Psi_1, \quad \psi_2 \in \Psi_2,$$

and thus the integrals in the definition of the process X make sense.

Remark A.1.10 Application of the Itô formula shows that (A.1.31) is equivalent to the fact that α solves the Doléans-Dade equation (see Theorem 4.4.6)

$$\alpha_t = 1 + \int_0^t \int_U \alpha_{s-}(e^{\psi(s,y)} - 1)\tilde{\pi}(ds, dy), \quad t \in [0, T^*].$$

Proof of Lemma A.1.8 The process $X_t := \ln \alpha_t, t \in [0, T^*]$ is a semimartingale with jump times equal to these of α. Since, for each $n = 1, 2, \ldots$ the process

$$P_t^n := \sum_{s \in [0,t]} \Delta X_s \mathbf{1}_{\{1<|\Delta X_s|\leq n\}}, \quad t \in [0, T^*],$$

has bounded jumps it follows that its variation is locally bounded, that is, $P^n \in \mathcal{A}_{loc}$. By Theorem 4.2.13 there exists a compensator C^n of P^n, that is, the process

$$M_t^n := P_t^n - C_t^n, \quad t \in [0, T^*]$$

is a local martingale. In view of Theorem 4.3.3 the jumps of M^n are also bounded and hence it is a local square integrable martingale. It follows from Theorem A.1.1 for local square integrable martingales that

$$M_t^n = \int_0^t \int_U g_n(s,y)\tilde{\pi}(ds,dy), \quad t \in [0,T^*],$$

for some $g_n \in \Psi_2$. By Proposition A.1.7 C^n is continuous, so we have $\triangle P_t^n = g_n(t, \triangle Z_t)$, where Z stands for the Lévy process with jump measure π, and thus

$$P_t^n = \int_0^t \int_U g_n(s,y)\pi(ds,dy), \quad t \in [0,T^*].$$

Since $g_m = g_n \mathbf{1}_{\{|g_n| \leq m\}}$ for $m < n$ we can define a process g by $g(s,y)\mathbf{1}_{\{|g(s,y)| \leq n\}} = g_n(s,y), t \in [0,T^*], y \in U$. Then we have

$$\sum_{s \in [0,t]} \triangle X_s \mathbf{1}_{\{|\triangle X_s| > 1\}} = \int_0^t \int_U g(s,y)\pi(ds,dy), \quad t \in [0,T^*].$$

Now let us cut off from X the jumps exceeding 1 and define

$$X_t^0 = X_t - \sum_{s \in [0,t]} \triangle X_s \mathbf{1}_{\{|\triangle X_s| > 1\}} = X_t - \int_0^t \int_U g(s,y)\pi(ds,dy), \quad t \in [0,T^*].$$

Since $| \triangle X_t^0 | \leq 1$ it follows from Theorem 4.3.3 that it is a special semimartingale and the martingale part M in its canonical decomposition $X_t^0 = M_t + A_t, _{t \in [0,T^*]}$ is a local square integrable martingale. Again application of Theorem A.1.1 for local square integrable martingales yields

$$M_t = \int_0^t \int_U h(s,y)\tilde{\pi}(ds,dy), \quad t \in [0,T^*]$$

for some $h \in \Psi_2$. Notice also that h is bounded because the jumps of M are bounded. Moreover, by Proposition A.1.7, the process A is continuous. Finally we obtain

$$X_t = \int_0^t \int_U h(s,y)\tilde{\pi}(ds,dy) + \int_0^t \int_U g(s,y)\pi(ds,dy) + A_t, \quad t \in [0,T^*], \tag{A.1.32}$$

where $g(s,y)h(s,y) = 0$ because X^0 and $\int_0^t \int_U g(s,y)\pi(ds,dy)$ do not jump together. Writing

$$\alpha_t = e^{X_t}, \quad t \in [0,T^*], \tag{A.1.33}$$

we are in the position to apply Proposition A.1.6, which provides that $e^g - 1 \in \Psi_1$ and

$$-A_t = \int_0^t \int_U (e^{g(s,y)} - 1)ds\nu(dy) + \int_0^t \int_U (e^{h(s,y)} - 1 - h(s,y))ds\nu(dy), \quad t \in [0,T^*].$$

Plugging this into (A.1.32) then coming back to (A.1.33) and defining $\psi(s,y) := h(s,y) + g(s,y)$ we obtain the result.

The converse implication follows from Remark A.1.10. □

Proof of Theorem A.1.1 Let M be a martingale. For $\varepsilon > 0$ define two strictly positive martingales $M^{1,\varepsilon}(t) := E[M^+_{T^*} \mid \mathcal{F}_t] + \varepsilon$, $M^{2,\varepsilon}(t) := E[M^-_{T^*} \mid \mathcal{F}_t] + \varepsilon$, $t \in [0, T^*]$. Application of Lemma A.1.8 and Remark A.1.10 to the normalized martingales $\frac{M_t^{1,\varepsilon}}{M_0^{1,\varepsilon}}$, $\frac{M_t^{2,\varepsilon}}{M_0^{2,\varepsilon}}$ provides the existence of processes ψ^1, ψ^2 where $e^{\psi^1} - 1, e^{\psi^2} - 1 \in \Psi_{1,2}$, such that

$$M_t^{1,\varepsilon} = M_0^{1,\varepsilon} + \int_0^t \int_U M_{s-}^{1,\varepsilon}(e^{\psi^1(s,y)} - 1)\tilde{\pi}(ds, dy),$$

$$M_t^{2,\varepsilon} = M_0^{2,\varepsilon} + \int_0^t \int_U M_{s-}^{2,\varepsilon}(e^{\psi^2(s,y)} - 1)\tilde{\pi}(ds, dy), \quad t \in [0, T^*].$$

It follows that

$$M_t = M_t^{1,\varepsilon} - M_t^{1,\varepsilon} = M_0 + \int_0^t \int_U \phi(s, y)\tilde{\pi} ds, dy, \quad t \in [0, T^*], \tag{A.1.34}$$

with $\psi(s, y) := M_{s-}^{1,\varepsilon}(e^{\psi^1(s,y)} - 1) - M_{s-}^{2,\varepsilon}(e^{\psi^2(s,y)} - 1)$ and it is clear that $\psi \in \Psi_{1,2}$.

If M is a local martingale with localizing sequence $\{\tau_n\}$ then, in view of the first part of the proof, for each $M_t^n := M(t \wedge \tau_n)$ we have

$$M_t^n = M_0 + \int_0^t \int_U \psi^n(s, y)\tilde{\pi}(ds, dy), \qquad t \in [0, T^*], \tag{A.1.35}$$

for $\psi^n \in \Psi_{1,2}$. From uniqueness we conclude that there exists a process ψ such that

$$\psi(s, y)\mathbf{1}_{\{[0,\tau_n]\}}(s) = \psi^n(s, y), \quad s \in [0, T^*], \ n = 1, 2, \ldots.$$

Since for each n hold

$$\begin{aligned}
&\int_0^{\tau_n \wedge T^*} \int_U \Big(\mid \psi(s, y) \mid^2 \wedge \mid \psi(s, y) \mid \Big) ds\nu(dy) \\
&= \int_0^{T^*} \int_U \Big(\mid \psi^n(s, y) \mid^2 \wedge \mid \psi^n(s, y) \mid \Big) ds\nu(dy) < +\infty,
\end{aligned}$$

we see that $\psi \in \Psi_{1,2}$. Letting $n \to +\infty$ in (A.1.35) we obtain the assertion. □

Appendix B

The concept of generators of semigroups of linear transformations is recalled and their forms for the semigroups corresponding to stochastic equations are derived.

B.1 Semigroups and Generators

A family $S(t)$, $t \geq 0$, of linear, bounded transformations from a Banach space E into E satisfying

$$S(0) = I, \; S(t+s)x = S(s)(S(t)x), \qquad s,t \geq 0, \; x \in E, \tag{B.1.1}$$

$$\lim_{t \downarrow 0} S(t)x = x, \;\; x \in E \tag{B.1.2}$$

is called a C_0, or strongly continuous, semigroup. The *infinitesimal generator* of the C_0-semigroup $S(t), t \geq 0$, is defined by

$$Ax := \lim_{h \downarrow 0} \frac{S(h)x - x}{h}, \qquad x \in D(A),$$

where

$$D(A) := \left\{ x \in H : \exists \lim_{h \downarrow 0} \frac{S(h)x - x}{h} \right\}.$$

A *core* of a linear operator $\mathcal{A}$ with domain $D(\mathcal{A})$ is any linear subset Λ of $D(\mathcal{A})$ such that the closure of the set $\{(f, \mathcal{A}(f)) : f \in \Lambda\}$ is exactly $\{(f, \mathcal{A}(f)) : f \in D(\mathcal{A})\}$. Generators are uniquely determined by their values on a core.

Let A be a linear operator on a Hilbert space H with dense domain $D(A)$. Then the domain $D(A^*)$ of its adjoint operator A^* consists of all $y \in H$ such that there exists $z \in H$ that for all $x \in D(A)$, $\langle Ax, y \rangle = \langle x, z \rangle$. If this is the case then $A^*y = z$.

If $(S(t))$ is a C_0-semigroup on a Hilbert space H, then its adjoint semigroup $(S^*(t))$ is determined by the formula

$$\langle S^*(t)y, x\rangle = \langle y, S(t)x\rangle, \quad x, y \in H, \ t \geq 0.$$

Its generator is the adjoint operator A^*.

Transition semigroups $(\mathcal{P}_t)$ of Markovian processes evolving in closed subsets K of $\mathbb{R}^n$ are usually studied on various spaces of functions defined on K, like $C_b(K)$, the space of all bounded continuous functions on K. When $K = [0, +\infty)$ the transition semigroup will be considered here on the space $C_0([0, +\infty))$ of all bounded continuous functions vanishing at $+\infty$.

B.1.1 Generators for Equations with Lévy Noise

Consider solutions $X^x(t)$ on $\mathbb{R}^n$ of the following stochastic equation:

$$dX(t) = F(X(t))dt + G(X(t-))dZ(t), \quad X(0) = x \in \mathbb{R}^n, \quad t > 0, \tag{B.1.3}$$

where $F\colon \mathbb{R}^n \longrightarrow \mathbb{R}^n$, $G(x)$, for each $x : \mathbb{R}^n$ is a linear transformation from $\mathbb{R}^d$ into $\mathbb{R}^n$ and Z is a Lévy process in $\mathbb{R}^d$ with characteristic triplet (a, Q, ν). According to the Lévy–Itô decomposition (see (5.2.3)),

$$Z(t) = at + W(t) + \int_0^t \int_{|y|\leq 1} y\,\tilde{\pi}(ds, dy) + \int_0^t \int_{|y|>1} y\,\pi(ds, dy), \quad t > 0. \tag{B.1.4}$$

Let $\mathcal{P}_t$ be the transition semigroup of the solution X^x

$$\mathcal{P}_t f(x) := \mathbb{E}(f(X^x(t))), \quad t \geq 0, x \in \mathbb{R}^n, \ f \in C_b(\mathbb{R}^n).$$

Assume that $\mathcal{P}_t$ is a strongly continuous semigroup on E, a Banach subspace of $C_b(\mathbb{R}^n)$. If there exists the limit

$$\frac{1}{h}(\mathcal{P}_h f(x)) - f(x)) \tag{B.1.5}$$

as $h \to 0$, uniformly in x and, as a function of x, it does belong to E then one says that f is in the domain $D(\mathcal{A})$ of the generator of the transition semigroup and the limit is, by definition, the value of the generator $\mathcal{A}$ on f and is denoted by $\mathcal{A}f(x)$. If, instead of uniform convergence one requires existence of the limit for each x then the limit is called the weak generator of the transition semigroup or, with some abuse of language, of the equation.

Here is a typical result on generators and its proof can be easily adapted to equations acting on some subsets K of $\mathbb{R}^n$. The proof is based on the Itô formula

$$f(X(t)) - f(X(0)) = \sum_{j=1}^{n} \int_0^t \frac{\partial f}{\partial x_j}(X(s-))dX^j(s)$$

$$+ \frac{1}{2} \sum_{1 \le j,l \le n} \int_0^t \frac{\partial^2 f}{\partial x_j \partial x_l}(X(s-))d[X^j, X^l]^c(s) \tag{B.1.6}$$

$$+ \sum_{0<s\le t} \Big\{ f(X(s)) - f(X(s-)) - \sum_{j=1}^{n} \frac{\partial f}{\partial x_j}(X(s-))\Delta X^j(s) \Big\}, \tag{B.1.7}$$

where $f\colon \mathbb{R}^n \longrightarrow \mathbb{R}$ (see (4.4.17)). In the following Proposition we assume that the coefficients are regular enough so that the stochastic equation has a unique solution, for instance that F and G are functions satisfying Lipschitz conditions. By $C_b^2(\mathbb{R}^n)$ one denotes the space of functions bounded and continuous together with their first and second derivatives

Proposition B.1.1 *Let E be a Banach subspace of $C_b(\mathbb{R}^n)$ such that the transition semigroup $(\mathcal{P}_t)$ corresponding to the solution of the equation* (B.1.3), *is strongly continuous in E. Then its generator is given by the formula:*

$$\mathcal{A}f(x) = \langle Df(x), F(x) + G(x)a\rangle + \frac{1}{2} Trace\Big\{D^2 f(x) G(x) Q G^*(x)\Big\}$$

$$+ \int_U \Big\{ f(x + G(x)y) - f(x) - \mathbf{1}_{\{|y|\le 1\}}(y)\langle Df(x), G(x)y\rangle \Big\} \nu(dy) \tag{B.1.8}$$

for functions f such that all three terms in the formula (B.1.8) *are in E and, in addition, $G^*(x)Df(x)$, $x \in \mathbb{R}^n$ is bounded.*

Proof Using (B.1.4) we write (B.1.6) in the form

$$f(X(t)) - f(x) = I_1 + I_2 + I_3 + I_4 + I_5 + I_6 + I_7,$$

where

$$I_1 := \int_0^t \langle Df(X(s-)), F(X(s))\rangle ds, \quad I_2 := \int_0^t \langle Df(X(s-)), G(X(s))a\rangle ds,$$

$$I_3 := \int_0^t \langle Df(X(s-)), G(X(s-))dW(s)\rangle,$$

$$I_4 := \int_0^t \int_{|y|\le 1} \langle Df(X(s-)), G(X(s-))y\rangle \tilde{\pi}(ds, dy),$$

$$I_5 := \int_0^t \int_{|y|>1} \langle Df(X(s-)), G(X(s-))y\rangle \pi(ds, dy),$$

$$I_6 := \frac{1}{2} \sum_{1\leq j,l\leq n} \int_0^t \frac{\partial^2 f}{\partial x_j \partial x_l}(X(s-))d[X^j, X^l]^c(s),$$

$$I_7 := \sum_{0<s\leq t} \Big\{ f(X(s)) - f(X(s-)) - \sum_{j=1}^{n} \frac{\partial f}{\partial x_j}(X(s-))\triangle X^j(s) \Big\}.$$

The processes I_3, I_4, are local martingales and boundedness of the function $G^*(x)Df(x)$, $x \in \mathbb{R}^n$ is sufficient condition for them to be square integrable martingales and thus of expectation 0.

First we determine I_6. Writing (B.1.3) componentwise

$$X^j(t) = x^j + \int_0^t F^j(X(s))ds + \sum_{k=1}^{d} \int_0^t G^{j,k}(X(s-))dZ^k(s), \quad j = 1, 2, \ldots, n,$$

we see that quadratic covariation satisfies

$$\begin{aligned}
[X^j(t), X^l(t)] &= \left[\sum_{k=1}^{d} \int_0^t G^{j,k}(X(s-))dZ^k(s), \sum_{k'=1}^{d} \int_0^t G^{j,k'}(X(s-))dZ^{k'}(s) \right] \\
&= \sum_{k,k'=1}^{d} \left[\int_0^t G^{j,k}(X(s-))dZ^k(s), \int_0^t G^{j,k'}(X(s-))dZ^{k'}(s) \right] \\
&= \sum_{k,k'=1}^{d} \int_0^t G^{j,k}(X(s-))G^{j,k'}(X(s-))d[Z^k, Z^{k'}](s).
\end{aligned}$$

But

$$[Z, Z](t) = \big([Z^k, Z^{k'}](t) \big)_{k,k'} = tQ + \Big(\sum_{s\leq t} \triangle Z^k(s)\triangle Z^{k'}(s) \Big)_{k,k'},$$

so

$$\begin{aligned}
[X^j(t), X^l(t)]^c &= \sum_{k,k'=1}^{d} q^{k,k'} \int_0^t G^{j,k}(X(s))G^{l,k'}(X(s))ds \\
&= \int_0^t \Big(\sum_{k,k'=1}^{d} G^{j,k}(X(s))G^{l,k'}(X(s))q^{k,k'} \Big) ds,
\end{aligned}$$

where $q^{k,k'}$ stands for the corresponding element of Q. Consequently, in the multidimensional notation, we obtain

$$[X,X]^c(t) = \int_0^t G(X(s))QG^*(X(s))ds.$$

Hence

$$\begin{aligned} I_6 &= \frac{1}{2} \sum_{1\le j,l\le n} \int_0^t \frac{\partial^2 f}{\partial x_j \partial x_l}(X(s)) \sum_{k,k'=1}^{d} G^{j,k}(X(s))G^{l,k'}(X(s))q^{k,k'} ds \\ &= \frac{1}{2} \int_0^t Tr\Big\{D^2 f(X(s))G(X(s))QG^*(X(s))\Big\} ds. \end{aligned} \tag{B.1.9}$$

Since $\triangle X(t) = G(X(t-))\triangle Z(t)$,

$$\begin{aligned} I_7 &= \sum_{0<s\le t} \Big\{f(X(s)) - f(X(s-)) - \langle Df(X(s-)), \triangle X(s)\rangle\Big\} \\ &= \int_0^t \int_U \Big\{f(X(s-) + G(X(s-))y) - f(X(s-)) \\ &\quad - \langle Df(X(s-)), G(X(s-))y\rangle\Big\} \pi(ds,dy) \\ &= \int_0^t \int_U \Big\{f(X(s-) + G(X(s-))y) - f(X(s-)) \\ &\quad - \mathbf{1}_{\{|y|\le 1\}} \langle Df(X(s-)), G(X(s-))y\rangle\Big\} \pi(ds,dy) \\ &\quad - \int_0^t \int_U \mathbf{1}_{\{|y|>1\}} \langle Df(X(s-)), G(X(s-))y\rangle\Big\} \pi(ds,dy), \end{aligned}$$

therefore,

$$\begin{aligned} I_5 + I_7 = \int_0^t \int_U &\Big\{f(X(s-) + G(X(s-))y) - f(X(s-)) \\ &- \mathbf{1}_{\{|y|\le 1\}} \langle Df(X(s-)), G(X(s-))y\rangle\Big\} \pi(ds,dy). \end{aligned}$$

Denoting the martingale $I_3 + I_4$ by $M(t)$ one arrives at the following identity:

$$\begin{aligned} f(X(t)) - f(x) = I_1 + I_2 + M(t) &+ \frac{1}{2}\int_0^t Tr\Big\{D^2 f(X(s))G(X(s))QG^*(X(s))\Big\} ds \\ &+ \int_0^t \int_U \Big\{f(X(s-) + G(X(s-))y) - f(X(s-)) \\ &- \mathbf{1}_{\{|y|\le 1\}} \langle Df(X(s-)), G(X(s-))y\rangle\Big\} \pi(ds,dy). \end{aligned}$$

Taking expectations one gets:

$$\begin{aligned}\mathcal{P}_t f(x) - f(x) &= \mathbb{E}\int_0^t \langle Df(X(s-)), F(X(s-)) + G(X(s-))a\rangle ds \\ &+ \frac{1}{2}\mathbb{E}\int_0^t Tr\Big\{D^2 f(X(s))G(X(s))QG^*(X(s))\Big\}ds \\ &+ \mathbb{E}\int_0^t \Big[\int_U \Big\{f(x + G((X(s-))y) - f(X(s-)) \\ &- \mathbf{1}_{\{|y|\le 1\}}(y)\langle Df(X(s-)), G(X(s-))y\rangle\Big\}\nu(dy)\Big]ds.\end{aligned}$$

Denote

$$\begin{aligned}\mathcal{A}_1 f(x) &= \langle Df(x), F(x) + G(x)a\rangle, \\ \mathcal{A}_2 f(x) &= \frac{1}{2}Tr\Big\{D^2 f(x)G(x)QG^*(x)\Big\}, \\ \mathcal{A}_3 f(x) &= \int_U \Big\{f(x + G(x)y) - f(x) - \mathbf{1}_{\{|y|\le 1\}}(y)\langle Df(x), G(x)y\rangle\Big\}\nu(dy).\end{aligned}$$

Then

$$\begin{aligned}\frac{1}{t}(\mathcal{P}_t f(x) - f(x)) - \mathcal{A}f(x) &= \frac{1}{t}\mathbb{E}\int_0^t [\mathcal{A}_1 f(X(s-)) - \mathcal{A}_1 f(x)]ds \\ &+ \frac{1}{t}\mathbb{E}\int_0^t [\mathcal{A}_2 f(X(s-)) - \mathcal{A}_2 f(x)]ds \\ &+ \frac{1}{t}\mathbb{E}\int_0^t [\mathcal{A}_3 f(X(s-)) - \mathcal{A}_3 f(x)]ds.\end{aligned}$$

Equivalently

$$\begin{aligned}\frac{1}{t}(\mathcal{P}_t f(x) - f(x)) - \mathcal{A}f(x) &= \frac{1}{t}\int_0^t [\mathcal{P}_s\mathcal{A}_1 f(x) - \mathcal{A}_1 f(x)]ds \\ &+ \frac{1}{t}\int_0^t [\mathcal{P}_s\mathcal{A}_2 f(x)) - \mathcal{A}_2 f(x)]ds \\ &+ \frac{1}{t}\int_0^t [\mathcal{P}_s\mathcal{A}_3 f(x)) - \mathcal{A}_3 f(x)]ds, \quad x \in \mathbb{R}^n.\end{aligned}$$

Since, by assumptions, $\mathcal{A}_i f,\ i = 1,2,3$ are in E and the semigroup $(\mathcal{P}_t)$ is C_0 there, therefore the preceding expression converges uniformly to 0 as required. □

In Chapter 12, on affine term structures, the following equation:

$$dX(t) = F(X(t))dt + G(X(t-))dZ(t), \quad X(0) = x \ge 0, \quad t > 0 \tag{B.1.10}$$

on $[0,+\infty)$, with the one-dimensional Lévy process having only nonnegative jumps and with nonnegative G, was regarded. In the next proposition, $C_0([0,+\infty))$ consists of all bounded continuous function on $[0,+\infty)$ vanishing at $+\infty$, equipped with the supremum norm. By $C_b^2([0,+\infty))$ and $C_c^2([0,+\infty))$ we denote the spaces of all twice continuously differentiable functions having two first derivatives bounded or, respectively, with bounded supports. We have the following result on the generator to (B.1.10) with the proof as of Proposition B.1.1.

Proposition B.1.2 *Assume that the transition semigroup corresponding to the equation* (B.1.10)*, is strongly continuous on* $C_0([0,+\infty))$. *Then its generator is given by the formula:*

$$\mathcal{A}f(x) = f'(x)(F(x) + G(x)a) + \frac{1}{2}qf''(x)G^2(x)$$
$$+ \int_{[0,+\infty)} \left\{ f(x + G(x)y) - f(x) - \mathbf{1}_{\{|y|\le 1\}}(y)f'(x)G(x)y \right\} \nu(dy) \quad \text{(B.1.11)}$$

for all $f \in C_b^2[0,+\infty)$ *for which all three terms in* (B.1.11) *are in* $C_0([0,+\infty))$. *In particular, the formula is true for* $f \in C_c^2([0,+\infty))$.

Appendix C

Existence results for stochastic evolution equations and conditions for their positivity are formulated.

C.1 General Evolution Equations

Here we recall basic concepts and existence results for the *evolution equation*

$$dX(t) = (AX(t) + F(X(t)))dt + G(X(t-))dZ(t), \qquad X(0) = x_0 \in H, \quad t \in [0, T^*] \tag{C.1.1}$$

in a separable Hilbert space H. In the preceding Z is a U-valued Lévy process and $X(t-)$ stands for the left limit of X at t. We will mainly deal with the case in which the process Z is finite dimensional, say $Z = (Z^1, \ldots, Z^d)$ and $U = \mathbb{R}^d$, with the scalar product $\langle \cdot, \cdot \rangle_d$. Moreover, $F\colon H \longrightarrow H$ is usually a nonlinear operator acting in the space H and $G\colon H \longrightarrow L(U, H)$ is a transformation from the space H into the space of linear operators acting from U into H. The transformation A is an H-valued, linear, usually unbounded operator defined on some dense linear subspace $D(A)$ of H. If Z is a finite dimensional process then the equation (C.1.1) becomes

$$dX(t) = (AX(t) + F(X(t)))dt + \sum_{j=1}^{d} G^j(X(t-))dZ^j(t), \qquad X(0) = x_0 \in H,$$
$$t \in [0, T^*]. \tag{C.1.2}$$

An important class of operators A is formed by infinitesimal generators of semigroups of linear operators (see Appendix B, Section B.1). Let A generate a C_0-semigroup $S(t), t \geq 0$, in H. A process X is a *strong solution* of (C.1.1) if $X(t) \in D(A)$ for $t \in [0, T^*]$ and

$$X(t) = x_0 + \int_0^t (AX(s) + F(X(s)))ds + \int_0^t G(X(s-))dZ(s), \tag{C.1.3}$$

providing that the preceding integrals exist. Since strong solutions exist very rarely, weaker forms of solutions are considered. A process X is a *weak solution* of (C.1.1) if for all $y \in D(A^*)$ and $t \in [0, T^*]$,

$$\langle X(t), y\rangle = \langle x_0, y\rangle + \int_0^t \langle X(s), A^* y\rangle ds + \int_0^t \langle F(X(s)), y\rangle ds + \int_0^t \langle G^*(X(s-))y, dZ(s)\rangle, \tag{C.1.4}$$

where $*$ indicates adjoint operators. Let us recall that $y \in D(A^*)$ if there exists $z \in H$ such that for all $x \in D(A)$, $< Ax, y >=< x, z >$ and then one sets $A^*(y) = z$ (see Appendix B, Section B.1). Finally, X is a *mild solution* of (C.1.1) if

$$X_t = S_t x_0 + \int_0^t S(t-s)F(X(s))ds + \int_0^t S(t-s)G(X(s-))dZ(s), \quad t \in [0, T^*]. \tag{C.1.5}$$

The preceding integrals must be well defined, which requires some conditions for the transformations F and G as well as for the solution X. In particular, one requires that solutions should be càdlàg, implying that the integrands in the stochastic integrals are predictable processes – an important property required in the definition of the stochastic integral. The relation between the types of the introduced solutions will be clarified in the sequel.

Example C.1.1 First order SPDE's on $[0, +\infty)$.
In the book we study equations either in the Hilbert space $H = L^{2,\gamma}$, of measurable, square integrable functions h on $[0, +\infty)$ with the norm

$$\mid h \mid_{L^{2,\gamma}} := \left(\int_0^{+\infty} \mid h(x) \mid^2 e^{\gamma x} dx \right)^{\frac{1}{2}} < +\infty, \tag{C.1.6}$$

or in the Hilbert space $H = H^{1,\gamma}$, of absolutely continuous functions h on $[0, +\infty)$ such that

$$\mid h \mid_{H^{1,\gamma}} := \left(\int_0^{+\infty} \left(\mid h(x) \mid^2 + \mid h'(x) \mid^2 \right) e^{\gamma x} dx \right)^{\frac{1}{2}} < +\infty, \tag{C.1.7}$$

with $\gamma > 0$. Both spaces can be extended by including constant functions. When we would like to prove existence of nonnegative solutions we study existence of solutions in the subsets of the spaces consisting of nonnegative functions only denoted, respectively, by $L^{2,\gamma}_+$ and $H^{1,\gamma}_+$.

The operator A in (C.1.1) will be the infinitesimal generator of the so-called shift semigroup on $H = L^{2,\gamma}$ or on $H = H^{1,\gamma}$:

$$S(t)h(x) = h(t + x), \quad t \geq 0, x \geq 0, \; h \in H. \tag{C.1.8}$$

It turns out that $D(A)$ consists of absolutely continuous functions h such that their first derivative $\frac{d}{dx}h$ belongs to the space H (see e.g. Zabczyk [123] p. 194). In addition,

$$Ah(x) = \frac{d}{dx}h(x), \ x \geq 0, \ \ h \in D(A).$$

The domain of its adjoint operator A^* consists of absolutely continuous functions h such that $h(0) = 0$, $h \in H$ and

$$A^*h(x) = -\frac{d}{dx}h(x), \ x \geq 0, \ \ h \in D(A).$$

Equation (C.1.1) can be then formally written as:

$$dX(t,x) = \Big(\frac{\partial}{\partial x}X(t,x)dt + F(X(t))(x)\Big) + G(X(t-))dZ(t), \qquad t \in [0,T^*], x \geq 0,$$
$$X(0,\cdot) = x(\cdot) \in H.$$

Solutions of (C.1.5) are searched in the class of càdlàg processes satisfying some integrability conditions. To formulate a typical existence result (see e.g. Theorem 9.29 of Peszat and Zabczyk [100]) we have to introduce some conditions on F, G, namely *linear growth condition*

$$\exists c > 0, \ \mid F(x) \mid \leq c(1+ \mid x \mid), \quad \mid G(x) \mid_{L(U,H)} \leq c(1+ \mid x \mid), \quad x \in H \qquad \text{(C.1.9)}$$

and *Lipschitz condition*

$$\exists c > 0, \ \mid F(x) - F(y) \mid \leq c \mid x - y \mid, \quad \mid G(x) - G(y) \mid_{L(U,H)} \leq c \mid x - y \mid, \quad x, y \in H. \qquad \text{(C.1.10)}$$

Theorem C.1.2 *If F and G satisfy linear growth condition and Lipschitz condition, the semigroup S is of contraction type, then there exists a unique càdlàg mild solution of* (C.1.1) *on any interval* $[0, T^*]$ *and it does coincide with the weak solution of the equation. If, in addition, Z is square integrable, then there exists a constant C such that*

$$\sup_{t \in [0,T^*]} \mathbb{E} \mid X(t) \mid^2 \leq C(1 + |x_0|^2). \qquad \text{(C.1.11)}$$

Remark C.1.3 The proof uses, in a standard way, a fixed point argument when the process Z is square integrable. If Z is a general Lévy process one uses the decomposition

$$\begin{aligned} Z(t) &= a_N t + W(t) + \int_0^t \int_{|y| \leq N} y\tilde{\pi}(ds, dy) + \int_0^t \int_{|y|>N} y\pi(ds, dy) \\ &= Z_N(t) + \int_0^t \int_{|y|>N} y\pi(ds, dy), \quad t \geq 0, \ N = 1, 2, \ldots. \end{aligned}$$

The processes Z_N are square integrable Lévy processes and if N is less than M then

$$Z_N(t) = Z_M(t), \quad t < \tau_N,$$

where $\tau_N := \inf\{s > 0 : |Z(s) - Z(s-)| > N\}$. If X_N is a solution of the equation (C.1.1) with Z replaced by Z_N then $X_N(t) = X_M(t)$ for $t < \tau_N$. Thus the solution X of (C.1.1) in the general case is given by

$$X(t) = \lim_{N\to+\infty} X_N(t), \quad t \in [0, T^*].$$

It turns out that in Theorem C.1.2 one can replace the Lipschitz condition by the *local Lipschitz condition*, which says that for each $R > 0$ there exists $c_R > 0$ such that

$$| F(x) - F(y) | + | G(x) - G(y) |_{L(U,H)} \leq c_R|x - y|, \qquad |x| \leq R,\ |y| \leq R. \tag{C.1.12}$$

The resulting stronger version of the theorem will be successfully applied in the sequel.

Theorem C.1.4 *Assume that* (C.1.9) *and* (C.1.12) *are satisfied. Then there exists a unique càdlàg weak solution to the equation* (C.1.1).

Proof We can assume that Z is square integrable (see Remark C.1.3). Let F_n, G_n, $n = 1, 2, \ldots$ be such that

(i) $F_n(x) = F(x)$ and $G_n(x) = G(x)$ if $|x| \leq n$,
(ii) for all $x \in H$,

$$| F_n(x) | + | G_n(x) | \leq c(1 + |x|),$$

(iii) there is a constant c_n such that for all $x, y \in H$,

$$| F_n(x) - F_n(y) | + | G_n(x) - G_n(y) | \leq c_n|x - y|.$$

By Theorem C.1.2 the equation obtained from (C.1.1) by replacing F and G by F_n and G_n, has a unique càdlàg solution X_n starting from any $x_0 \in H$ and satisfies the estimation

$$\sup_{t\leq T^*} \mathbb{E} | X_n(t) |^2 \leq C\left(1 + | x_0 |^2\right) \tag{C.1.13}$$

for some $C > 0$. Let

$$\tau_n := \inf\{t \leq T^* : | X_n(t) | > n\}.$$

On the time interval $[0, \tau_n)$, the trajectories of X_n are contained in the ball $B(0, n)$ in H with center at 0 and radius n, and therefore X_n satisfies (C.1.1). For $m > n$ the solutions X_m and X_n coincide on $[0, \tau_n)$. Define $X(t) := X_n(t)$ if $t < \tau_n$. Note that X is well defined. To finish the proof it is enough to show that

$$\lim_{n\to\infty} \mathbb{P}\big(\sup_{t\le T^*} \mid X_n(t) \mid > n\big) = 0.$$

Let n be such that $\mid x_0 \mid \le n/3$ for $t \le T^*$. Then

$$\mathbb{P}\left(\sup_{t\le T^*} \mid X_n(t) \mid > n \right) \le \mathbb{P}\left(\sup_{t\le T^*} \left| \int_0^t S(t-s)F_n(X_n(s))ds \right| > \frac{n}{3}\right)$$
$$+ \mathbb{P}\left(\sup_{t\le T^*} \left| \sum_{i=1}^d \int_0^t S(t-s)G_n^i(X_n(s-))dZ^i(s) \right| > \frac{n}{3}\right) := I_1 + I_2.$$

However, for a constant $\hat{c}$ independent of n,

$$\sup_{t\le T^*} \left| \int_0^t S(t-s)F_n(X_n(s))ds \right| \le \hat{c}\left(1 + \int_0^{T^*} \mid X_n(s) \mid_H ds\right),$$

and hence, by Chebyshev's inequality and (C.1.13), there is a constant $\hat{\hat{c}}$ such that

$$\begin{aligned} I_1 &\le \frac{3\hat{c}}{n}\left(1 + \int_0^{T^*} \mathbb{E} \mid X_n(s) \mid ds\right) \\ &\le \frac{3\hat{c}}{n}\left(1 + \int_0^{T^*} \left(\mathbb{E} \mid X_n(s) \mid^2\right)^{1/2} ds\right) \\ &\le \frac{3\hat{\hat{c}}}{n}\left(1 + \mid x_0 \mid^2\right)^{1/2}. \end{aligned}$$

Hence $I_1 \to 0$ as $n \to \infty$. By Kotelenez's inequality (see e.g. Peszat and Zabczyk [100]) and (C.1.13) there is a constant $\tilde{c}$ such that

$$\begin{aligned} I_2 &\le \left(\frac{3}{n}\right)^2 \tilde{c}\, \mathbb{E} \int_0^{T^*} \mid G_n(X_n(s)) \mid^2 ds \\ &\le 2c\left(\frac{3}{n}\right)^2 \tilde{c} \int_0^{T^*} \left(1 + \mathbb{E} \mid X_n(s) \mid^2\right) ds \\ &\le \tilde{\tilde{c}}\left(\frac{3}{n}\right)^2 \left(1 + \mid x_0 \mid^2\right). \end{aligned}$$

Since $I_2 \to 0$ with $n \to \infty$, the assertion follows. □

To establish the positivity of solutions of (C.1.1) we will use the following direct extension, to the locally Lipschitz case, of the result of Milian (see [91]).

Theorem C.1.5 *In addition to the assumptions of Theorem C.1.4, assume that $H = L^2(E, \mu)$, with μ being a σ-finite measure on E, that the semigroup $S(t)$, $t \ge 0$*

preserves positivity and that Z *is a Wiener process. If for each nonnegative* $g \in H \cap C_c^\infty(E)$ *and each nonnegative* $h \in H \cap C(E)$ *such that* $\langle g, h \rangle_H = 0$, *the following hold*

$$\langle F(g), h \rangle_H \geq 0, \tag{C.1.14}$$

$$\langle G(g), h \rangle_H = 0, \tag{C.1.15}$$

then $X \geq 0$. *Conversely, if all solutions to* (C.1.1), *starting from nonnegative initial conditions, stay nonnegative, then* (C.1.14) *and* (C.1.15) *hold.*

References

[1] Aihara, S. I., A. Bagchi: Stochastic hyperbolic dynamics for infinite-dimensional forward rates and option pricing (2005), *Mathematical Finance* 15, 27–47.

[2] Applebaum, D.: Lévy Processes and Stochastic Calculus (2004), Cambridge University Press.

[3] Barski, M. Incompleteness of the bond market with Lévy noise under the physical measure (2015), *Advances in Mathematics of Finance, Banach Center Publications* 104, 61–84.

[4] Barski, M., Jakubowski, J., Zabczyk, J.: On incompleteness of bond markets with infinite number of random sources (2011), *Mathematical Finance* 21, 3, 541–556.

[5] Barski, M., Zabczyk, J.: Completeness of bond market driven by Lévy processes (2010), *International Journal of Theoretical and Applied Finance* 13, 635–656.

[6] Barski, M., Zabczyk, J.: Heat–Jarrow–Morton–Musiela equation with Lévy perturbation (2012), *Journal of Differential Equations* 253, 9, 2657–2697.

[7] Barski, M., Zabczyk, J.: Forward rate models with linear volatilities (2012), *Finance and Stochastics* 16, 3, 537–560.

[8] Barski, M., Zabczyk, J.: On generalized CIR equations (2019), arXiv: 1902.08976.

[9] Barski, M., Zabczyk, J.: On CIR equations with general factors (forthcoming), SIAM Journal on Financial Mathematics.

[10] Bayraktar, E., Chen, L., Poor, H. V.: Consistency problem for jump diffusion models, arXiv:cs/0501055v1 [cs.IT] January 23, 2005.

[11] Berestycki, J., Döring, L., Mytnik, L., Zambotti, L.: Hitting properties and non-uniqueness for SDE driven by stable processes (2015), *Stochastic Processes and Their Applications* 150, 918–949.

[12] Bertoin, J.: Lévy Processes (1996), Cambridge University Press.

[13] Bichteler, K.: Stochastic Integration with Jumps (2011), Cambridge University Press.

[14] Bichteler, K., Lin, S. J.: On the stochastic Fubini theorem (1995), *Stochastics and Stochastic Reports* 54, 271–279.

[15] Brace, A., Gątarek, D., Musiela, M.: The market model of interest rate dynamics (1997), *Mathematical Finance* 7, 127–147.

[16] Björk, T.: Arbitrage Theory in Continuous Time (2009), Oxford University Press.

[17] Björk, T.: Interest rate theory (1996), Lecture Notes in Mathematics 1656, 53–122.

[18] Björk, T.: On the term structure of discontinuous interest rates (1995), *Obozrenie Prikladnoi i Promyshlennoi Matematiki* 2, 626–657 (in Russian).

[19] Björk, T., Di Masi, G., Kabanov, Y., Runggaldier, W.: Towards a general theory of bond markets (1997), *Finance and Stochastics* 1, 141–174.

[20] Björk, T., Kabanov, Y., Runggaldier, W.: Bond market structure in the presence of marked point process (1997), *Mathematical Finance* 7, 211–239.

[21] Borwein, P. B., Erdélyi, T.: Generalizations of Müntz's theorem via a Remez-type inequality for Müntz spaces (1997), *Journal of the American Mathematical Society* 10, 2, 327–349.

[22] Brzeźniak, Z., Peszat, S., Zabczyk, J.: Continuity of stochastic convolutions (2001), *Czechoslovak Mathematical Journal* 51, 679–684.

[23] Brzeźniak, Z., Kok, T.: Stochastic evolution equations in Banach spaces and applications to the Heath–Jarrow–Morton–Musiela equations (2018), *Finance and Stochastics* 22, 4, 959–1006.

[24] Carmona, R. A., Tehranchi, M. R.: A characterization of hedging portfolios for interest rate contingent claims (2004), *Annals of Applied Probability* 14, 1267–1294.

[25] Carmona, R. A., Tehranchi, M. R.: Interest Rate Models: An Infinite Dimensional Stochastic Analysis Perspective (2006), Springer.

[26] Chatelain, M., Stricker, C.: On componentwise and vector stochastic integration (1994), *Mathematical Finance* 4, 57–65.

[27] Cherny A. S. Vector stochastic integrals in the fundamental theorem of asset pricing (1998) Proceedings of the Workshop on Mathematical Finance, INRIA, 149–163.

[28] Cont, R.: Modeling term structure dynamics: An infinite dimensional approach (2005), *International Journal of Theoretical and Applied Finance* 8, 3, 357–380.

[29] Cont, R., Tankov, P.: Financial Modelling with Jump Processes (2004), Chapman & Hall.

[30] Cox, J., Ingersoll, J., Ross, S.: A theory of the term structure of interest rates (1985), *Econometrica* 53, 385–408.

[31] Dawson, D. A., Zenghu, L.: "Skew convolution semigroups and affine Markov processes (2006), *Annals of Probability*, 34, 1103–1142.

[32] Da Prato, G. An Introduction to Infinite-Dimensional Analysis (2006), Springer.

[33] Da Prato, G., Zabczyk, J. Stochastic Equations in Infinite Dimensions (1992), Cambridge University Press.

[34] Davis, M. H. A.: The representation of martingales of a jump process (1976), *SIAM Journal on Control and Optimization* 14, 623–638.

[35] De Acosta, A.: Exponential moments of vector valued random series and triangular arrays (1980), *Annals of Probability* 8, 381–389.

[36] Delbaen, F., Schachermayer, W.: A general version of the fundamental theorem of asset pricing (1994), *Mathematische Annalen* 300, 463–520.

[37] Delbaen, F., Schachermayer, W.: The fundamental theorem of asset pricing for unbounded stochastic processes (1998), *Mathematische Annalen* 312, 215–250.

[38] Dellacherie, C.: Intégrales stochastiques par rapport aux processus de Wiener ou de Poisson (1974), (Séminaire de Probabilités, 494. VIII, University of Strasbourg, 1972–1973), Lecture Notes in Mathematics 381, 25–26.

[39] Dellacherie, C.: Corrections to Intégrales stochastiques par rapport aux processus de Wiener ou de Poisson (Séminaire de Probabilités, VIII, (University of Strasbourg,

Année Universitaire 1972–1973, Lecture Notes in Mathematics, 1974, 381, 25–26). Séminaire de Probabilités, IX (Seconde Partie, University of Strasbourg, Strasbourg, Annés Universitaires 1973–1974 et 1974–1975) (1975), Lecture Notes in Mathematics 465–494.

[40] De Donno, M., Pratelli, M.: On the use of measured-valued strategies in bond markets (2004), *Finance and Stochastics*, 8, 87–109.

[41] De Donno, M.: A note on completeness in large financial markets (2004), *Mathematical Finance*, 14, 2, 295–315.

[42] Duffie, D., Filipović, D. and Schachermeyer, W.: Affine processes and applications in finance (2003), *Annals of Applied Probability*, 13, 3, 984–1053.

[43] Dunford, N., Schwartz, J. T.: Linear Operators, Part I (1964), Interscience.

[44] Dunford, N., Schwartz, J. T.: Linear Operators, Part II (1963), Interscience.

[45] Dybvig, P., Ingersoll, J., Ross, S.: Long forward and zero coupon rates can never fall (1996), *Journal of Business*, 69, 1–25.

[46] Dynkin, E. B.: Markov Processes (1965), Springer.

[47] Eberlein, E., Jacod, J., Raible, S.: Lévy term structure models: No-arbitrage and completeness (2005), *Finance and Stochastics*, 9, 67–88.

[48] Eberlein, E., Raible, S.: Term structure models driven by general Lévy processes (1999), *Mathematical Finance*, 9, 31–53.

[49] Eberlein, E., Özkan, F.: The Lévy LIBOR model (2005), *Finance and Stochastics*, 9, 327–348.

[50] Ekeland, I., Taflin, E.: A theory of bond portfolios (2005), *Annals of Applied Probability*, 15, 2, 1260–1305.

[51] Feller, W.: An Introduction to Probability Theory and Its Applications (1970), John Willey and Sons.

[52] Filipović, D.: Term-Structure Models: A Graduate Course (2009), Springer-Verlag.

[53] Filipović, D.: A general characterization of one factor affine term structure models (2001), *Finance and Stochastics*, 5, 3, 389–412.

[54] Filipović, D.: Consistency problems for Heath–Jarrow–Morton interest rate models (2001), *Lecture Notes in Mathematics, Vol. 1760.*

[55] Filpović, D.: Separable term structures and the maximal degree problem (2002), *Mathematical Finance*, 12, 4, 341–349.

[56] Filipović, D., Tappe, S.: Existence of Lévy term structure models (2008), *Finance and Stochastics*, 12, 83–115.

[57] Filipović, D., Tappe, S., Teichmann, J.: Term structure models driven by Wiener process and Poisson measures: Existence and positivity (2010), *SIAM Journal on Financial Mathematics*, 1, 523–554.

[58] Filipović, D., Zabczyk, J.: Markovian term structure models in discrete time (1999), *Preprint, Institute of Matematics, Polish Academy of Sciences*, 601, 1–12.

[59] Filipović, D., Zabczyk, J.: Markovian term structure models in discrete time (2002), *The Annals of Applied Probability*, 12, 710–729.

[60] Fu, Z., Li, Z.: Stochastic equations of non-negative processes with jumps (2010), *Stochastic Processes and Their Applications*, 120, 306–330.

[61] Gantmacher, F. , Krein, M.: Oscillation Matrices and Kernels and Small Vibrations of Mechanical Systems (2002), rev. ed., AMS Chelsea Publishing.

[62] Gątarek, D.: Some remarks on the market model of interest rates (1996), *Control and Cybernetics*, 25, 6, 1233–1244.

[63] Gikhman, I., Skorohod, A.V.: The theory of stochastic processes (1974), Vols. I–II, Springer.

[64] Glasserman, P., Kou, S. G.: The term structure of simple forward rates with jump risk (2003), *Mathematical Finance*, 13, 383–410.

[65] Grzywacz, T., Leżaj, Ł., Trajan, B.: Transition densities of subordinators (2018), arXiv: 1812.06793 [math.Pr].

[66] Harrison, J. M., Pliska, S. R.: Martingales and stochastic integrals in the theory of continuous trading (1981), *Stochastic Processes and Their Applications*, 11, 3, 215–260.

[67] Heath, D., Jarrow, R., Morton, A.: Bond pricing and the term structure of interest rates: A new methodology for contingent claim valuation (1992), *Econometrica*, 60, 77–105.

[68] Hubalek, F.: A counterexample involving exponential-affine Laplace transforms (2001), unpublished, TU Vienna.

[69] Hubalek, F., Klein, I., Teichmann, J.: A general proof of the Dybvig–Ingersoll–Ross theorem: Long forward rates can never fall (2002), *Mathematical Finance*, 12, 447–451.

[70] Hunt, P. J., Kennedy, J. E.: Financial Derivatives in Theory and Practice (2005), John Wiley & Sons, Ltd.

[71] Ikeda, N., Watanabe, S.: Stochastic Differential Equations and Diffusion Processes (1981), Kodansha, Ltd.

[72] Ikeda, N., Watanabe, S.: On the uniqueness and non-uniqueness of a class of non-linear equations and explosion problem for branching processes (1970), Journal of the Faculty of Science University Tokyo Sect. I, 17, 187–214.

[73] Itô, K.: Spectral type of the shift transformation of differential processes with stationary increments (1956), *Transactions of the American Mathematical Society*, 81, 253–63.

[74] Jacod, J.: Calcul Stochastique et Problém des Martingales (1979), Lecture Notes in Mathematics 714, Springer.

[75] Jacod, J., Shiryaev, A. N.: Limit Theorems for Stochastic Processes (2002), Springer.

[76] Jakubowski, J., Zabczyk, J.: Exponential moments for HJM models with jumps (2007), *Finance and Stochastics*, 11, 429–445.

[77] Jarrow, R. A., Madan, D. B.: A characterization of complete security markets on a Brownian filtration (1991), *Mathematical Finance*, 1, 31–43.

[78] Jeanblanc, K., Yor, M., Chesney, M.: Mathematical Methods for Financial Markeets (2009), Springer.

[79] Kallenberg, O.: Foundations of Modern Probability (2001), 2nd ed., Springer-Verlag.

[80] Karatzas, I., Shreve, S. E.: Methods of Mathematical Finance (1998), Springer.

[81] Kawazu, K., Watanabe, S.: Branching processes with immigration and related topics (1971), *Theory of Probability and Its Applications* 16, 36–54.

[82] Kinney, J. H.: "Continuity properties of sample functions of Markov processes (1953), *Transactions of the American Mathematical Society*, 74, 280–302.

[83] Knapp, A.W.: Basic algebra (2006), Birkhäuser.

[84] Kunita, H.: Representation of martingales with jumps and applications to mathematical finance (2004), Advanced Studies in Pure Mathematics 41, *Stochastic Analysis and Related Topics*, 209–232.

[85] Kunita, H., Watanabe, S.: On square-integrable martingales (1967), *Nagoya Mathematical Journal* 30, 209–245.

[86] Kwapień, S., Marcus, M. B., Rosiński, J.: Two results on continuity and boundness of stochastic convolutions (2006), *Annales de l'Institut Henri Poincare* 42, 553–566.

[87] Kyprianou, A. E., Pardo, J. C.: Continuous-state branching processes and self-similarity (2008), *Journal of Applied Probability*, 45, 1140–1160.

[88] Li, Z., Mytnik, L.: Strong solutions for stochastic differential equations with jumps (2011), *Annales de l'Institut Henri Poincaré Probability et Statistics*, 47, 1055–1087.

[89] Liptser, R. S., Shiryaev, A. N.: Statistics of Random Processes, I: General Theory, vol. 2 (2001), 2nd ed., Springer-Verlag.

[90] Marinelli, C.: Local well-posedness of Musiela's SPDE with Lévy noise (2010), *Mathematical Finance* 20, 341–363.

[91] Milian, A.: Comparison theorems for stochastic evolution equation (2002), *Stochastics and Stochastics Reports*, 72, 79–108.

[92] Métivier, M.: Semimartingales: A Course on Stochastic Processes (1982), Walter de Gruyter.

[93] Mikulevicius, R., Rozovskii, B. L.: Normalized stochastic integrals in topological vector spaces (1998), *Séminaire de probabilités XXXII (Lecture Notes in Mathematics)*, Springer, 137–165.

[94] Mikulevicius, R., Rozovskii, B. L.: Martingales problems for stochastic PDEs (1999), Stochastic Partial Differential Equations: Six Perspectives, ed. R. Carmona, B. Rozovski (Mathematical Surveys and Monographs, 64), American Mathematical Society, 243–326.

[95] Morton, A.: Arbitrage and martingales (1989), PhD dissertation, Cornell University.

[96] Müntz, C. H.: Über den Approximationssatz von Weierstrass (1914). In H. A. Schwarz's Festschrift, 303–312.

[97] Musiela, M.: Stochastic PDEs and term structure models (1993), *Journées International de Finance, IGR-AFFI, La Baule.*

[98] Musiela, M., Gołdys, B.: Infinite dimensional diffusions, Kolmogorov equations and interest rate models (2001), *Option Pricing, Interest Rates and Risk Management, Handbook of Mathematical Finance*, ed. E. Jouini, J. Cvitanic, M. Musiela, Cambridge University Press, 314–335.

[99] Özkan, F.: Lévy processes in credit risk and market models (2002), Ph thesis, Freiburg University.

[100] Peszat, S., Zabczyk, J.: Stochastic Partial Differential Equations with Lévy Noise (2007), Cambridge University Press.

[101] Peszat, S., Zabczyk, J.: Heath–Jarrow–Morton–Musiela equation of bond market (2007), www.impan.pl/Preprints/p677.pdf.

[102] Protter, P.: Stochastic Integration and Differential Equations (2005), Springer.

[103] Pham, H.: A predictable decomposition in infinite asset model with jumps: Application to hedging and optimal investment (2003), *Stochastics and Stochastic Reports*, 5, 343–368.

[104] Pontriagin, L. S.: Ordinary Differential Equations (1962), Pergamon.

[105] Raible, S.: Lévy processes in finance: Theory, numerics and empirical facts (2000), PhD thesis, Freiburg University.

[106] Rogers, L. C. G., Williams, D.: Diffusions, Markov Processes and Martingales (2000), Cambridge University Press.

[107] Ringer, N., Tehranchi, M.: Optimal portfolio choice in the bond market (2006), *Finance and Stochastics*, 10, 4, 553–573.

[108] Rosiński, J.: Remarks on strong exponential integrability of vector-valued random series and triangular arrays (1995), *Annals of Probability*, 23, 464–473.

[109] Rusinek, A.: Properties of solutions to the Musiela equation with Lévy noise (2011), PhD dissertation, Institute of Mathematics, Polish Academy of Sciences.

[110] Rusinek, A.: A note on HJMM models on a space of square integrable functions (2012), manuscript.

[111] Rusinek, A.: Mean reversion for HJMM forward rate models (2010), *Advances in Applied Probability*, 42, 371–391.

[112] Rusinek, A.: Invariant measures for forward rate HJM model with Lèvy noise (2006), Preprint IMPAN 669, www.impan.pl/Preprints/p669.pdf.

[113] Rusinek, A.: Forward rate models on the space of square integrable functions (2016), *Rocznik Naukowy WZ Ciechanów*, Vol. X, 1–4.

[114] Sato, K. I.: Lévy Processes and Infinite Divisible Distributions (1999), Cambridge University Press.

[115] Schachermayer, W.: A Hilbert space proof of the fundamental theorem of asset pricing in finite discrete time (1992), *Insurance: Mathematics and Economics*, 11, 4, 249–257.

[116] Shiryaev, A. N.: Essentials of Stochastic Finance, Facts, Models, Theory (1999), vol. 3, World Scientific Publishing,

[117] Shreve, S.: Stochastic Calculus for Finance II (2004), Springer.

[118] Słomiński, L. Private communication.

[119] Taflin, E.: Bond market completeness and attainable contingent claims (2005), *Finance and Stochastics*, 9, 429–452.

[120] Taflin, E.: Generalized integrands and bond portfolios: Pitfalls and counterexamples (2011), *Annals of Applied Probability*, 21, 266–282.

[121] Vasiček, O.: An equilibrium characterization of the term structure (1977), *Journal of Financial Economics*, 5, 177–188.

[122] Yosida, K.: Functional Analysis (1980), Springer.

[123] Zabczyk, J.: Mathematical Control Theory: An Introduction (1992), Birkhäuser (2nd ed. forthcoming in 2020).

[124] Zanzotto, P. A.: On solutions of one-dimensional stochastic differential equations driven by stable Lévy motion, *Stochastic Processes and Their Applications* (1997), 68, 209–228.

[125] Zongfei, F., Zenghu, L.: "Stochastic equations of non-negative processes with jumps (2010), *Stochastic Processes and Their Applications*, 120, 306–330.

Index

Bold numbers refer to PART I of the book.